Lecture Notes in Mathematics

Edited by A. Dold, B. Eckmann and F. Takens

1455

J.-P. Françoise   R. Roussarie (Eds.)

Bifurcations of Planar Vector Fields

Proceedings of a Meeting held in Luminy, France, Sept. 18–22, 1989

Springer-Verlag
Berlin Heidelberg New York London
Paris Tokyo Hong Kong Barcelona

Editors

Jean-Pierre Françoise
Université de Paris VI
U.F.R. 920, Mathématiques
45–46, 5ème étage
4 Place Jussieu
75252 Paris, France

Robert Roussarie
Université de Bourgogne
Laboratoire de Topologie
U.F.R. Sciences et Techniques
Bât. Mirande, BP 138
21004 Dijon, France

Mathematics Subject Classification (1980): 34CXX, 58F14, 32C05

ISBN 3-540-53509-8 Springer-Verlag Berlin Heidelberg New York
ISBN 0-387-53509-8 Springer-Verlag New York Berlin Heidelberg

Printing and binding: Druckhaus Beltz, Hemsbach/Bergstr.
2146/3140-543210 – Printed on acid-free paper

PREFACE

The meeting held in Luminy in September 18-22, 1988 brought together most of the world's specialists in bifurcations of vector fields of the plane. The main subjects of the theory were discussed, including:

- Finiteness of the number of limit cycles of ordinary differential equations in the plane. The problem of Dulac is that of determining whether polynomial vector fields have a finite number of limit cycles. One solution is presented in this volume in the framework of new and much farther-reaching methods for the study of differential equations, such as accelero-summation.
- Multiplicity of polycycles. Their definition seems to be a first step towards the solution of Hilbert's 16th problem produced to prove the existence of a uniform bound, dependent only on degree, for the number of limit cycles.
- Zeroes of abelian integrals. This is a topic which links up directly to real algebraic geometry. It intervenes in an infinitesimal version of Hilbert's 16th problem, and also in the question of enumeration of critical points of the period for which is important in the study of bifurcations.
- Numerical simulation and symbolic computation on computer in the study of differential equations.
- The work (in particular of Chinese groups of researchers) on quadratic equations, that pick up again classical methods of bifurcation theory such as the method of rotations.
- Modelling of predator-prey ecological systems. The subject is in widespread use in biomathematics to describe biological cycles.
- The use of methods of non-standard analysis in the study of bifurcation with delay.

The articles in this volume will initiate the reader quickly to the most recent result in this field at the interface of fundamental mathematics and of its applications, currently in full development.

We enjoyed the support of the Centre National de la Recherche Scientifique, of the Ministère des Affaires Etrangères (Direction Générale des relations culturelles, scientifiques et techniques), of the Société Mathématique de France through the intermediary of the Centre International de Rencontre Mathématiques at Luminy, of the Union des Assurances de Paris and of the Université de Bourgogne.

We are grateful to Miss Courtial and Mrs. Gadenne for their assistance in the preparation of the manuscripts.

We thank Springer-Verlag for the care and competence shown in the publication of these proceedings.

LISTE DES ARTICLES

Retard à la bifurcation : du local au global

Bernard Candelpergher
Laboratoire de Mathématiques
Parc Valrose
06034 Nice Cedex

Francine Diener
Laboratoire de Mathématiques
Parc Valrose
06034 Nice Cedex

Marc Diener
U.F.R. de Mathématiques
Université Paris 7
75251 Paris Cedex 05

C'est Claude Lobry qui, lors d'un colloque en 1985 [8] avait attiré notre attention sur le phénomène de "retard à la bifurcation" par des expérimentations numériques où il devinait une intervention des canards. Nous abordons ici cette question sur une équation modèle dûe à G. Wallet, où le lien avec la question de sommation de séries divergentes apparait de manière remarquablement élémentaire, et à travers laquelles se profile la confirmation d'une très ancienne conjecture de J–P. Ramis sur le caractère Gevrey des séries intervenant dans diverses études de canards. Nous remercions vivement F. Pham et J. Ecalle pour nous avoir patiemment aidés à découvrir et appliquer les méthodes de sommation.

1 Etude numérique du retard à la bifurcation

Considérons l'exemple de G. Wallet :

$$\left\{\begin{array}{lcl} \mu' & = & \varepsilon \\ u_1' & = & \mu u_1 - u_2 + \varepsilon \\ u_2' & = & u_1 + \mu u_2 \end{array}\right. \qquad (1)$$

pour $\varepsilon = 0.05$. Le terme de "retard à la bifurcation" est dû à l'écart entre le comportement des trajectoires d'un système tel que (1) et le raisonnement (heuristique et inapproprié) suivant :

Comme ε est petit, on étudie le système (1) pour $\varepsilon = 0$. Dans ce cas $\mu = Cste$ est un simple paramètre, et le système satisfait par $u = (u_1, u_2)$ présente un unique point stationnaire : $(0,0)$. Ce point stationnaire est un foyer stable pour $\mu < 0$ et instable pour $\mu > 0$. On peut penser que, si l'on choisit $\varepsilon = 0.05$, et une condition "initiale" (μ_-, u_-) telle $\mu_- = -1.3$, et $\|u_-\| = 1$ (Fig. 1), μ va croître lentement avec la variable indépendante, disons t ($'$ $= d/dt$), u va s'enrouler rapidement vers le

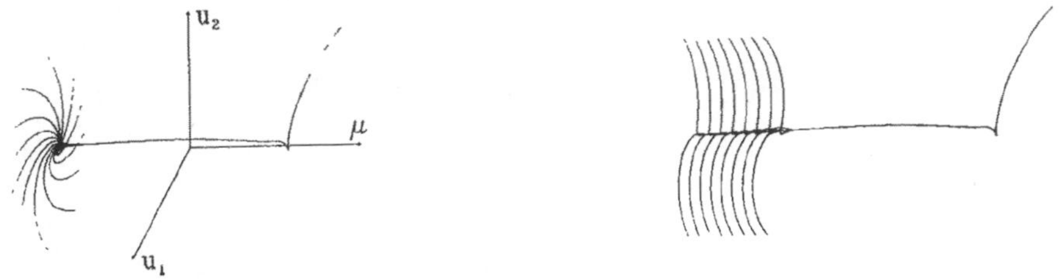

Figure 1: Trajectoires issues de points tels que $\|u\| = 1$, pour $\mu = -1.3$ et divers angles (à gauche), ou divers $\mu \leq -1.3$ (à droite). On observera la sortie en $\mu \approx +1$ (c'est le retard), commune à toutes ces trajectoire (c'est la butée).

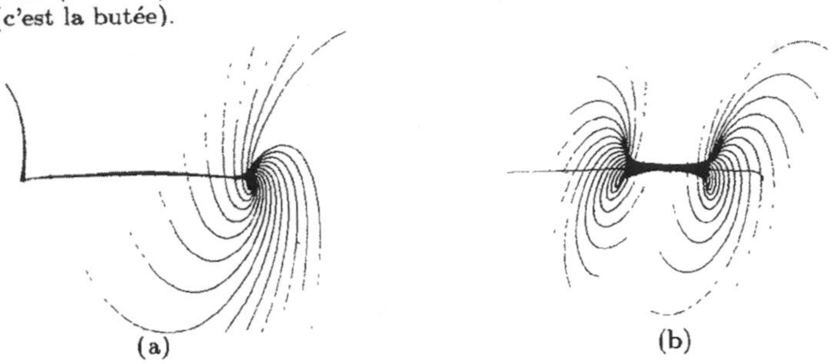

Figure 2: (a) Superposition de trajectoires d'un même point pour diverses valeurs petites de $\varepsilon > 0$: les comportements diffèrent appréciablement lors de la déstabilisation, qui a toutefois toujours lieu pour $\mu \approx +1$. (b) La relation entrée-sortie $\mu_+ = -\mu_-$ pour $|\mu_\pm| \ll 1$.

"point stationnaire" $\left(-\frac{\mu\varepsilon}{1+\mu^2}, \frac{\varepsilon}{1+\mu^2}\right) \approx (0,0)$, tant que $\mu < 0$, puis se dérouler *dès que μ franchira la valeur de bifurcation* $\mu = 0$.

C'est cette dernière affirmation, en italiques, qui est démentie par l'expérience (et la théorie). L'erreur du raisonnement heuristique réside dans le fait que pour obtenir une portion de trajectoire correspondant à une variation appréciable de μ, il convient de faire parcourir à la variable t un intervalle de l'ordre de $1/\varepsilon$: pour le problème envisagé, le système (1) est une fausse perturbation régulière du cas $\varepsilon = 0$: on constate en effet un *retard dans la destabilisation de la solution*, qui ne quitte l'équilibre instable que pour μ voisin de $+1$ (Fig. 1).

Outre ce retard à la bifurcation, nous observons un second phénomène, nouveau à notre connaissance : toutes les trajectoires calculées se déstabilisent ensemble : bien qu'issues de points bien distincts , elles quittent l'équilibre pour une même valeur de μ et sous un même angle. C'est ce retard maximal, ou *butée*, que nous souhaitons également expliquer, d'autant que la valeur $\mu \approx +1$ est insensible au choix de $\varepsilon > 0$ petit. Observons toutefois que l'"angle de sortie" est, lui, sensible au choix de $\varepsilon > 0$ petit (Fig. 2.a).

Nous n'avons envisagé jusqu'ici que le cas de trajectoires telles que $\mu_- \ll -1$. Pour $-1 \ll \mu_- \ll 0$, on observe également un retard, mais celui-ci est modulé par la valeur de μ_- : il y a destabilisation pour $\mu_+ \approx -\mu_-$ (Fig. 2.b), et donc existence d'une fonction entrée-sortie : ceci sera facile à analyser en termes de canards dans le prochain

Figure 3: Trajectoires des deux cercles $\mu_{\pm} = \pm 1.3, \|u_{\pm}\| = 1$; à gauche le tracé "exact", à droite un schéma topologique.

paragraphe. Sur la figure 3 ont été représentées des trajectoires de points (μ_+, u_+) avec $\mu_+ = +1.3$ et toujours $\|u_+\| = 1$: on trouve une symétrie prévisible. On a également représenté schématiquement les "cylindres", réunion des trajectoires des deux cercles $\mu_{\pm} = \pm 1.3$, $\|u_{\pm})\| = 1$: ils constituent deux "trompettes" enlacées ; ce schéma permet de comprendre qu'il n'y a aucune obstruction $\mathcal{C}^\infty$ pour que la torsade centrale soit de longueur infiniment petite, ce qui suggère le caractère analytique du phénomène de retard, que nous allons étudier maintenant sur un plan plus théorique.

2 Relation avec les canards

Posons dorénavant $x := \mu$ et supposons $\varepsilon > 0$ infinitésimal. Réécrivons le système (1), après changement de temps $t \mapsto \tau = \varepsilon t$, en

$$\left\{ \begin{array}{rcl} \dot{x} & = & 1 \\ \varepsilon \dot{u_1} & = & x u_1 - u_2 + \varepsilon \\ \varepsilon \dot{u_2} & = & u_1 + x u_2 \end{array} \right. \tag{2}$$

où $\dot{} = d/d\tau$. C'est un champ lent-rapide de $\mathbf{R} \times \mathbf{R}^2$ ou $\mathbf{R} \times \mathbf{C}$ [5] du type

$$\left\{ \begin{array}{rcl} \dot{x} & = & 1 \\ \varepsilon \dot{u} & = & f(x,u) \end{array} \right. \tag{3}$$

dont les images des solutions sont des graphes de solutions de l'équation

$$\varepsilon \frac{du}{dx} = f(x,u) \tag{4}$$

où $u = u_1 + i u_1$, et où f est supposée $\mathcal{C}^1$ (donc analytique en u), et $\mathcal{S}^1$ en u (i.e. f'_u est S-continue). La courbe $\mathcal{L}$ de $\mathbf{R}_x \times \mathbf{C}_u$ d'équation ${}^o f(x,u) = 0$ est appelée la *courbe lente* de (3) ou (4) . Dans l'exemple (2) il s'agit de la droite $u = 0$. Un point standard (x_0, u_0) de $\mathcal{L}$ est dit *attractif* si la partie réelle de ${}^o f'_u(x_0, u_0)$ est strictement négative, et *répulsif* si cette même partie réelle est strictement positive. Dans l'exemple (2) les points de la courbe lente tels que $x < 0$ sont attractifs, et ceux tels que $x > 0$ sont répulsifs. On dit qu'une solution $x \mapsto \bar{u}(x)$ de (4) est *lente* sur l'intervalle $I =]x_-, x_+[$ si, pour tout $x \in I$, $f(x, \bar{u}(x)) \simeq 0$, c'est-à-dire si elle reste infiniment voisine de $\mathcal{L}$ sur cet intervalle. Enfin on dit que $\bar{u}$ est un *canard* sur I [2,14], si $\bar{u}$ est lente sur I, et s'il existe un x_0 standard, $x_- \ll x_0 \ll x_+$ tel que ${}^o\bar{u}(x)$ est attractif pour tout $x \ll x_0$ et répulsive pour tout $x \gg x_0$ ($x \in I$). Si $u_0 = {}^o\bar{u}(x_0)$, on dit encore que $\bar{u}$ est un canard au point (x_0, u_0).

Le phénomène de retard à la bifurcation correspond donc précisément à l'existence de canards. Or cette existence pour des systèmes lents-rapides tels que (3) n'est pas élémentaire. Etablie sur un exemple par Shishkova [12], elle a été prouvée par Neishtadt [9,10], comme nous l'indiquons ci-dessous. Nous y reviendrons au prochain paragraphe. Notons que même pour le cas du système (2), où, en posant toujours $u = u_1 + iu_2$, les (images des) trajectoires satisfont à l'équation

$$\varepsilon \frac{du}{dx} = (x+i)u + \varepsilon \tag{5}$$

et ont donc pour expression

$$u(x) = e^{\frac{(x+i)^2}{2\varepsilon}} \left(K + \int_0^x e^{-\frac{(t+i)^2}{2\varepsilon}} dt \right) \quad , \quad K \in \mathbb{C}$$

il n'est guère plus simple de déduire l'existence de canards de cette quadrature que d'utiliser la sommation des séries divergentes que nous proposons ci-dessous.

Un point des théories existentes des canards s'adapte toutefois ici sans difficultés : c'est le calcul de la *fonction entrée-sortie* [1]. Comme pour toute solution lente, on peut [5] associer à chaque canard de (4) un *point* (standard) *d'entrée dans le halo* de la courbe lente, et un *point de sortie du halo*, caractérisés, dans le cas du système (2), par leur abscisse x_- et x_+ (éventuellement égales à $\pm\infty$). Voici comment sont réliées x_- et x_+ (pour le système (2) on aura $x_- = -x_+$) :

Proposition 2.1 (Fonction entrée-sortie)

On suppose que le système (2) possède un canard dont l'image est le graphe d'une fonction $x \mapsto \bar{u}(x)$ définie pour tout $x \in]a,b[$, a et b standard. Soient $x_-, x_+ \in]a,b[$ standard tels que

$$\int_{x_-}^{x_+} {}^o f'_u(x, {}^o\bar{u})dx = 0.$$

Alors toute solution de (4) qui atteint le halo de $\mathcal{L}$ au point d'abscisse x_- en ressort au point d'abscisse x_+.

Preuve : Pour $r = |u|$, considérer la *loupe de Benoit* $R = \varepsilon \log r$. □

Application à l'exemple (1) : Ici ${}^o f'_u(x, {}^o\bar{u}) = (x+i)$: la fonction entrée-sortie est une simple symétrie autour de l'origine des abscisses.

Notation : Soit $\varepsilon \neq 0$ infinitésimal. On dira que $\mu \in \mathbb{C}$ est un *ε-exponentiellement petit*, s'il existe $h \in \mathbb{C}$ de partie réelle positive non infinitésimale tel que $\mu = e^{-\frac{|h|}{|\varepsilon|}}$ ou si $\mu = 0$. On notera $\mathcal{M}_\varepsilon$ ou simplement $\mathcal{M}$ l'ensemble (externe) des ε-exponentiellement petits.

Proposition 2.2 (Stabilité des canards par microperturbation)

Soient f C^1 et S^1 en u, et l deux fonctions localement lipschitziennes. Supposons que l'équation $\varepsilon \frac{du}{dx} = f(x,u)$ admette un canard $x \mapsto \bar{u}(x)$ au point (x_0, u_0) et que pour tout (x,u) infiniment proche du canard on ait $|l(x,u)| \in \mathcal{M}_\varepsilon$. Alors l'équation suivante admet un canard au point (x_0, u_0).

$$\varepsilon \frac{du}{dx} = f(x,u) + l(x,u).$$

Preuve : Considérons, pour (x,u) tel que $0 < |u - \bar{u}(x)| < 1$, le revêtement défini par

$$(x, U) = (x, \varepsilon \log(u - \bar{u}(x)))$$

à valeurs sur Re$U < 0$. L'équation $\varepsilon du/dx = f(x,u)$ devient

$$\frac{dU}{dx} = \frac{f(x,u) - f(x,\bar{u}(x))}{u - \bar{u}(x)} + \frac{l(x,u)}{u - \bar{u}(x)}$$

Or, pour Re$U < 0$ appréciable, on a $u = \bar{u}(x) + e^{U/\varepsilon} \simeq \bar{u}(x)$, et donc, comme $f'_u(x,u) \not\simeq 0$, $f(x,u) - f(x,\bar{u}(x)) = f'_u(x,u(x))[u - \bar{u}(x)](1 + \phi)$, avec $\phi \simeq 0$.

D'autre part, par le principe de Fehrele [6], il existe k réel positif non infiniment petit tel que $|l(x,u)| < e^{-\frac{k}{\varepsilon}}$ pour tout (x,u) infiniment voisin du graphe du canard. Donc, toujours pour Re$U < 0$,

$$\left|\frac{l(x,u)}{u - \bar{u}(x)}\right| = \left|\frac{l(x,u)}{e^{U/\varepsilon}}\right| \leq e^{-\frac{\mathrm{Re}U + k}{\varepsilon}}.$$

Donc, pour $-k \ll \mathrm{Re}U \ll 0$, $\frac{l(x,u)}{u - \bar{u}(x)} \simeq 0$ et finalement

$$\frac{dU}{dx} \simeq f'_u(x, \bar{u}(x))$$

d'où $U(x) \simeq U_0 + \int_{x_0}^{x} f'_u(\xi, \bar{u}(\xi))d\xi$ tant que cette expression garde sa partie réelle comprise entre $-k$ et 0, et non infiniment voisine de ces valeurs, ce qui est assuré sur tout un voisinage standard $[x_-, x_+]$ de x_0 pourvu que l'on choisisse U_0 satisfaisant à ces conditions. En revenant à l'échelle (x,u) initiale, on voit que la solution issue de $u(x_0) := \bar{u}(x_0) + e^{\frac{U_0}{\varepsilon}}$ est, sur $[x_-, x_+]$, le canard cherché. □

Remarque : On notera que l'équation $\varepsilon \frac{du}{dx} = (x+i)u + \varepsilon$ n'est pas une microperturbation (au sens de la proposition (2.2)) de l'équation linéaire $\varepsilon \frac{du}{dx} = (x+i)u$ qui, elle, possède bien un canard évident. La proposition (2.2) ne peut donc pas lui être appliquée. Cependant, le changement d'inconnue $\varepsilon u = u_1 - \frac{\varepsilon}{x+i}$ transforme cette équation en

$$\varepsilon \frac{du_1}{dx} = (x+i)u_1 - \frac{\varepsilon^2}{(x+i)^2}$$

dont le terme perturbatif $-\frac{\varepsilon^2}{(x+i)^2}$ n'est plus d'ordre ε, mais ε^2. Un nouveau changement d'inconnue $\varepsilon u_1 = u_2 + \varepsilon^2/(x+i)^3$ permettrait de diminuer encore l'ordre de grandeur du terme perturbatif. On comprend qu'il est ainsi facile de construire une suite (u_n) de changements de variables de telle sorte que l'équation obtenue après N changements d'inconnue soit une microperturbation de l'équation à canard évident $\varepsilon\frac{du}{dx} = (x+i)u$. Bien entendu, N sera infiniment grand : $N = \frac{1}{\varepsilon}$ par exemple convient. C'est à partir de cette remarque facile à comprendre que Neishtadt est parvenu à montrer l'existence de canards et expliquer le retard à la bifurcation.

J. Martinet a attiré notre attention sur le fait qu'un point de vue Gevrey sur la question permet d'obtenir cette existence des canards de manière très naturelle. C'est l'objet du prochain paragraphe.

3 Le point de vue Gevrey

La difficulté dans la preuve du théorème de Neishtadt réside dans le fait que, très généralement, la suite des changements d'inconnue mentionnée conduit à une série divergente, et sa preuve a consisté à maîtriser par des majorations la somme des $\frac{1}{\varepsilon}$ premiers termes de cette série. Le fait que cela soit possible peut se comprendre, dans le cas analytique, par les propriétés du développement des solutions lentes au voisinage d'un point (x_0, u_0) où $\Re(f'_u)(x_0, u_0)$ est inversible : un tel développement est en effet Gevrey (d'ordre 1). Il est donc naturel d'appliquer des méthodes de sommation qui sont bien adaptées à une telle situation et d'obtenir une preuve par aveu spontané ...

Theorème 3.1 *Soit $F : \mathbf{R}\times\mathbf{C}\times\mathbf{R} \longrightarrow \mathbf{C}$, $(x,u,\varepsilon)\mapsto F(x,u,\varepsilon)$, une fonction standard, analytique au voisinage d'un point standard $(x_0,u_0,0)$ tel que $F(x_0,u_0,0) = 0$. On considère la situation où la partie réelle de $F'_u(x,u_0,0)$ est du signe de $x - x_0$, et on suppose que la partie imaginaire $\omega(x_0)$ de $F'_u(x_0,u_0,0)$ est non nulle. Soit $\varepsilon > 0$ infinitésimal, et posons $f(x,u) := F(x,u,\varepsilon)$. Il existe $R \gg 0$ et une solution $\bar{u}$ de*

$$\varepsilon\frac{du}{dx} = f(x,u) \tag{6}$$

définie pour tout $|x - x_0| < R$, telle que $\bar{u}(x_0) \simeq u_0$ et $f(x,\bar{u}(x)) \simeq 0$ pour tout $|x - x_0| < R$.

Rappelons qu'un développement $\hat{u} = \sum_{p\geq 0}\hat{u}_p\varepsilon^{p+1}$, où les $\hat{u}_p$ appartiennent à une algèbre normée (ici les fonctions analytiques définies sur un voisinage de x_0 munies de la norme uniforme), est dit *Gevrey d'ordre k* (ici $k = 1$) si et seulement s'il existe des réels C et M tels que pour tout $p \geq 0$ on ait $|\hat{u}_p| \leq CM^p(p\,!)^k$ [11]. Une fonction $\tilde{u}(x,\varepsilon)$ définie pour $x \in I =]x_-, x_+[$ pour ε dans un secteur de sommet 0 (et, ici, contenant le demi-axe réel positif) est appelée une *solution-Gevrey* de $\varepsilon\frac{du}{dx} = F(x,u,\varepsilon)$ si $\tilde{\mu}(x,\varepsilon) := \varepsilon\frac{d\tilde{u}(x,\varepsilon)}{dx} - F(x,\tilde{u}(x,\varepsilon),\varepsilon)$ est *exponentiellement petite*, c'est-à-dire que pour elle et chacune de ses dérivées $\tilde{\mu}^{(n)}$ par rapport à x, il existe des constantes $h > 0$ et K telles que $\tilde{\mu}^{(n)}(x,\varepsilon) \leq Ke^{-h/\varepsilon}$. Supposons F standard et soit $\varepsilon > 0$ infiniment petit. Posons $\tilde{u}(x) = \tilde{u}(x,\varepsilon)$. On dira que la fonction $\tilde{u}$ est un *canard-Gevrey* au point standard

(x_0, u_0), $x_- \ll x_0 \ll x_+$ et $\tilde{u}(x_0) \simeq u_0$, si, de plus, la partie réelle de $F'_u(x, u_0, 0)$ est du signe de $x - x_0$ et $f(x, \tilde{u}(x)) \simeq 0$.

Preuve : Quitte à procéder au changement d'inconnue $v = u - \hat{u}_{-1}$ où $\hat{u}_{-1}$ est la fonction implicite locale définie par $F(x, \hat{u}_{-1}(x), 0) = 0$ telle que $\hat{u}_{-1}(x_0) = u_0$, on peut supposer que $F(x, u_0, 0) = 0$ pour tout x dans un voisinage de x_0, puisque $F'_u(x_0, u_0, 0)$ est inversible ($\omega(x_0) \neq 0$). Cette même hypothèse de régularité de F permet d'appliquer un théorème de Y. Sibuya [13] assurant que l'équation (6) $\varepsilon\frac{du}{dx} = F(x, u, \varepsilon)$ admet une solution formelle $\hat{u} = \sum_{p\geq 1} \hat{u}\varepsilon^p$, où les $\hat{u}_p$ sont des fonctions analytiques définies sur un même voisinage compact de x_0, et que cette solution formelle est Gevrey d'ordre 1.

On en déduit dès lors une solution–Gevrey $\tilde{u}(x)$ admettant $\hat{u}$ pour développement asymptotique, au moyen d'une sommation de Borel–Laplace-tronquée. Rappelons brièvement comment :

La transformation de Borel $\mathcal{B}$ convertit la série formelle $\hat{u}$ en la série

$$\hat{\mathbf{u}} = \mathcal{B}\left(\sum_{p\geq 0} \hat{u}_p \varepsilon^{p+1}\right) := \sum_{p\geq 0} \hat{u}_p \frac{\lambda^p}{p\,!},$$

qui, du fait que $\hat{u}$ est Gevrey d'ordre 1, est une série convergente et définit donc une fonction $\mathbf{u}$ sur un voisinage de x_0 indépendant du choix de ε voisin de 0. Il est dès lors possible d'appliquer à $\mathbf{u}$ une transformation de Laplace-*tronquée*, c'est-à-dire étendue à un chemin d'intégration Γ (indépendant du choix de ε) joignant $\lambda = 0$ à un point non nul λ_0 dans le domaine de $\mathbf{u}$.

La transformation de Borel est un homomorphisme de l'algèbre des séries Gevrey d'ordre 1 (munie du produit formel des séries) dans l'algèbre des séries convergentes (munie du produit de convolution), et cet homomorphisme commute avec la dérivation par rapport au "paramètre" x : $\mathcal{B}\left(\frac{d\hat{u}}{dx}\right) = \frac{d}{dx}\mathcal{B}(\hat{u})$ (puisque, par les formules de Cauchy, la dérivée en x d'une série Gevrey (en puissances de ε à coefficients analytiques) est encore Gevrey). La transformation de Laplace $\mathcal{L}$ ayant les mêmes propriétés (du produit de convolution vers le produit ordinaire cette fois), on est assuré que $\bar{u} := \mathcal{L} \circ \mathcal{B}(\hat{u})$ satisfait aux mêmes équations algébro–différentielles que $\hat{u}$, et, par densité, les mêmes équations analytiquo–différentielles. Toutefois, faute de pouvoir choisir $\lambda_0 = \infty$, la transformation de Laplace-tronquée n'est un homomorphisme qu'à exponentiellement petit près, et donc, si $\varepsilon\frac{d\hat{u}}{dx} - F(x, \hat{u}, \varepsilon) = 0$, on n'aura pour $\tilde{u} := \mathcal{L}_\Gamma \circ \mathcal{B}(\hat{u})$ que $\varepsilon\frac{d\tilde{u}}{dx} - F(x, \tilde{u}(x), \varepsilon) =: \tilde{\mu}(x, \varepsilon)$ où $\tilde{\mu}$ est une fonction exponentiellement petite. Par transfert, on peut supposer que $\tilde{u}$ et $\tilde{\mu}$ sont standard et donc, si $\varepsilon > 0$ est infinitésimal, $\tilde{\mu}(x, \varepsilon) \in \mathcal{M}_\varepsilon$. La fin de la preuve est alors la conséquence du corollaire suivant de la proposition (2.2).

Corollaire 3.1 *Soit $\varepsilon > 0$ infinitésimal ; posons $f(x, u) := F(x, u, \varepsilon)$. Si l'équation $\varepsilon\frac{du}{dx} = f(x, u)$ admet un canard-Gevrey au point (x_0, u_0), alors elle admet également un canard en ce point.*

Preuve : Soit $\bar{u}$ le canard-Gevrey défini sur un voisinage standard $]x_-, x_+[$ de x_0 ; posons $(x, v) = (x, u - \bar{u}(x))$. Par hypothèse $\varepsilon\frac{du}{dx} = f(x, \bar{u}(x)) + \mu(x)$, où $\mu(x)$ est une fonction exponentiellement petite sur $]x_-, x_+[$, et il existe donc des constantes $k > 0$ et $k' > 0$ telles que $\mu(x) \leq e^{-\frac{k}{\varepsilon}}$ et $\mu'(x) \leq e^{-\frac{k'}{\varepsilon}}$. On a donc

$$\varepsilon \frac{dv}{dx} = f(x, \bar{u}(x) + v) - f(x, \bar{u}(x)) - \mu(x) = vg(x, v) - \mu(x) \tag{7}$$

où $g(x, v) \simeq f'_u(x, \bar{u}(x)$ pour tout $v \simeq 0$. L'équation (7) est une microperturbation de l'équation $\varepsilon \frac{dv}{dx} = vg(x, v) =: h(x, v)$ qui admet la solution lente évidente $v(x) = 0$; c'est un canard de cette équation, puisque $({}^o h)'_v(x, 0) \simeq h'_v(x, 0) = g(x, 0) \simeq f'_u(x, \bar{u}(x)) \simeq ({}^o f)'_u(x, ({}^o \bar{u})(x))$, qui est par hypothèse du signe de $x - x_0$ sur $]x_-, x_+[$. D'après la proposition 2.2, l'équation (7) admet donc elle aussi un canard au point (x_0, u_0). □

4 Local ou global ?

Jusqu'à présent nous avons parlé du problème de l'existence locale des canards, qui correspond, comme nous l'avons dit, à la présence d'un retard à la bifurcation. Abordons à présent le problème de leur taille maximale, c'est-à-dire le problème de déterminer la longueur du retard, pour une condition initiale donnée.

Une fois encore, nous obtiendrons ce renseignement en sommant la série formelle $\sum_{p \geq 0} u_p(x) \varepsilon^{p+1}$, généralement divergente, qui représente la solution canard considérée. Mais cette fois il ne suffit plus d'une *sommation tronquée* qui fournit une solution-Gevrey, il convient de calculer par sommation les *solutions exactes*, lorsque cela est possible.

Dans ce paragraphe, nous verrons comment mener ce programme à bien dans le cas simple d'une équation linéaire de la forme

$$\varepsilon u' = (x + i)u + \varepsilon g(x) \tag{8}$$

où g est une fonction entière. Le cas non linéaire sera abordé au paragraphe suivant mais par une méthode différente : on utilisera alors, non plus une sommation par rapport à ε, comme nous allons le faire à présent, mais une sommation par rapport à une variable ad hoc remplaçant x.

La première étape consiste à déterminer la série formelle $\sum_{p \geq 0} u_p(x) \varepsilon^{p+1}$ qui vérifie l'équation, en portant cette série dans l'équation et en déterminant les coefficients $u_0, u_1, \ldots$ de proche en proche. On trouve en particulier $u_0(x) = -g(x)/(x + i)$. On calcule ensuite la transformée de Borel $\mathcal{B}$ de cette série, par rapport à la variable ε : $\frac{\lambda^p}{p\,!} = \mathcal{B}(\varepsilon^{p+1})$. On obtient ainsi la série

$$\hat{\mathbf{u}}(x, \lambda) = \sum_{p \geq 0} u_p(x) \frac{\lambda^p}{p\,!}$$

qui est convergente au voisinage de $\lambda = 0$, puisque comme nous l'avons dit la série initiale est Gevrey. Désignons par $\mathbf{u}(x, \lambda)$ le prolongement analytique de la fonction ainsi

$$\begin{array}{ccc} \sum_{p\geq 0} \frac{(-1)^{p-1}}{(x+i)^{2p+1}} 3\cdot 5 \cdots (2p-1)\varepsilon^{p+1} & \overset{\mathcal{B}}{\longrightarrow} & \hat{\mathbf{u}}(x,\lambda) = \sum_{p\geq 0} \frac{(-1)^{p-1}}{(x+i)^{2p+1}} \frac{3\cdot 5\cdots(2p-1)}{p\,!}\lambda^p \\ \uparrow & & | \\ \text{développement} & & \text{prolongement} \\ \text{asymptotique} & & \text{analytique} \\ | & & \downarrow \\ u(x,\varepsilon) = -\int_d e^{-\frac{\lambda}{\varepsilon}} \frac{1}{\sqrt{2\lambda+(x+i)^2}} d\lambda & \underset{\mathcal{L}}{\longleftarrow} & \mathbf{u}(x,\lambda) = \frac{-1}{\sqrt{2\lambda+(x+i)^2}} \end{array}$$

Figure 4: La sommation dans le cas de l'équation (5).

définie. Il s'agit d'étudier les singularités de cette fonctions de λ (ici x est à considérer comme un paramètre) et sa croissance à l'infini. Ces renseignements obtenus, on calcule une solution de l'équation en posant

$$u(x,\varepsilon) = \int_d e^{-\frac{\lambda}{\varepsilon}} \mathbf{u}(x,\lambda) d\lambda$$

où d est une demi droite ne contenant aucune singularité de $\mathbf{u}$ et sur laquelle $\mathbf{u}$ est à croissance exponentielle en λ, c'est-à-dire sur laquelle il existe $K \in \mathbf{R}$ et $C \in \mathbf{R}$ tels que $|\mathbf{u}(x,\lambda)| \leq Ce^{K|\lambda|}$. Cette méthode de sommation [7,4,3] est schématisée sur la figure 4 dans le cas de l'équation (5).

En pratique, la partie délicate de ce procédé de sommation est le prolongement analytique $\hat{\mathbf{u}} \longrightarrow \mathbf{u}$. A l'exception de quelques cas particuliers (tels que l'équation (5)), il est impossible de le calculer explicitement. On adopte alors une autre stratégie : on détermine la transformée de Borel non plus de la série initiale mais de l'équation elle-même. On obtient :

$$\partial_x \mathbf{u}(x,\lambda) = (x+i)[\partial_\lambda \mathbf{u}(x,\lambda) + \mathbf{u}(x,0)\delta] + g(x)\delta \tag{9}$$

où δ est la mesure de Dirac en $\lambda = 0$, image par la transformation de Borel de la "série" constante 1. Comme

$$\mathbf{u}(x,0) = u_0(x) = -g(x)/(x+i),$$

on a simplement :

$$\left\{ \begin{array}{rcl} \partial_x \mathbf{u} & = & (x+i)\partial_\lambda \mathbf{u} \\ \mathbf{u}(x,0) & = & -g(x)/(x+i). \end{array} \right. \tag{10}$$

La fonction $\mathbf{u}$ que nous cherchons à déterminer est donc simplement la solution de cette équation, c'est-à-dire la fonction

$$\mathbf{u}(x,\lambda) = \frac{g(\sqrt{2\lambda+(x+i)^2} - i)}{\sqrt{2\lambda+(x+i)^2}}. \tag{11}$$

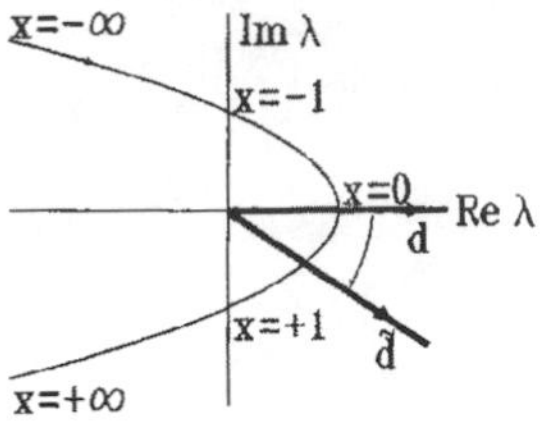

Figure 5: La singularité de **u** se déplace sur une parabole dans le plan de λ, lorsque x varie.

Les singularités de cette fonction proviennent d'une part du pôle ramifié $\lambda = -(x+i)^2/2$ et d'autre part des singularités de g.

Proposition 4.1 (Existence de la butée)
Si g est une fonction entière à croissance exponentielle et si $g(-i) \not\simeq 0$, l'équation (8) possède une solution $\bar{u}(x,\varepsilon)$ infinitésimale pour tout $x \ll 1$. Cette solution $\bar{u}$ devient infiniment grande dès que $x \gg 1$. Enfin si u_1 est une autre solution infinitésimale en un point $x_- \ll 1$, elle reste infiniment voisine de $\bar{u}$ tant que $x \ll x_+ = -x_-$.

Cette proposition explique le phénomène de butée observée lors de l'étude numérique du premier paragraphe. Elle indique que, pour une solution de cette famille d'équations, le plus long retard possible est 1 puisque toutes les solutions devenues lentes avant -1 restent lentes jusqu'en 1 puis cessent de l'être, en quittant l'équilibre pour des valeurs de x proches de 1, tout en restant infiniment voisines les unes des autres tant qu'elles restent limités.

On retrouve bien entendu ce même comportement symétriquement pour les trajectoires lentes au-delà de $x = 1$.

Preuve : Comme la fonction g est entière, la fonction **u** donnée par la formule (11) possède, pour chaque valeurs du paramètre x, une unique singularité au point $\lambda = -(x+i)^2/2$. Lorsque x varie, cette singularité se déplace sur une parabole, comme indiqué sur la figure 5.

On voit que pour toute valeur de $x < 0$, la demi-droite réelle positive d ne rencontre aucune singularité de **u** (et **u** est à croissance exponentielle par hypothèse) ; la fonction

$$\bar{u}(x,\varepsilon) = \int_d e^{-\frac{\lambda}{\varepsilon}} \mathbf{u}(x,\lambda) d\lambda$$

est donc une solution de l'équation, définie pour tout $x < 0$. Lorsque x prend des valeurs positives, on peut encore définir cette solution par une formule analogue, à condition de faire pivoter la droite d en $\tilde{d}$, comme indiqué sur la figure 5. Plus précisément l'intégrale

$$\int_{\tilde{d}} e^{-\frac{\lambda}{\varepsilon}} \mathbf{u}(x,\lambda) d\lambda$$

représente la même fonction $\bar{u}(x,\varepsilon)$ aussi longtemps que la singularité, située sur la parabole, reste au dessus de la demi droite $\tilde{d}$. Donc, dès que $x > 0$, on a :

$$\bar{u}(x,\varepsilon) = \int_d e^{-\frac{\lambda}{\varepsilon}} \mathbf{u}(x,\lambda)d\lambda + \int_{\tilde{d}-d} e^{-\frac{\lambda}{\varepsilon}} \mathbf{u}(x,\lambda)d\lambda.$$

Estimons le second terme de cette somme que nous appelons ***correction exponentielle.***

Lemme 4.1 *Pour tout $\varepsilon > 0$ infinitésimal, on a*

$$\int_{\tilde{d}-d} e^{-\frac{\lambda}{\varepsilon}} \mathbf{u}(x,\lambda)d\lambda \simeq -2\sqrt{2\varepsilon\pi}g(-i)e^{\frac{(x+i)^2}{2\varepsilon}}.$$

Preuve : (du lemme)
Notons $I = \int_{\tilde{d}-d} e^{-\frac{\lambda}{\varepsilon}} \frac{g(\sqrt{2\lambda+(x+i)^2}-i)}{\sqrt{2\lambda+(x+i)^2}} d\lambda$ et posons $s = 2\lambda + (x+i)^2$. On a :

$$I = -2e^{\frac{(x+i)^2}{2\varepsilon}} \int_0^\infty e^{\frac{s}{2\varepsilon}} \frac{g(\sqrt{s}-i)}{\sqrt{s}} ds$$

d'où, en posant $S = \frac{s}{2\varepsilon}$,

$$I = -2\sqrt{2\varepsilon}e^{\frac{(x+i)^2}{2\varepsilon}} \int_0^\infty e^{-S} S^{-\frac{1}{2}} g(\sqrt{2\varepsilon S} - i)dS.$$

Comme $g(\sqrt{2\varepsilon S} - i) \simeq g(-i)$ pour tout S limité et $\varepsilon \simeq 0$, on peut évaluer cette intégrale par le théorème d'approximation dominée [6] et la formule d'Euler de $\Gamma(1/2)$; on obtient :

$$I \simeq -2\sqrt{2\varepsilon}e^{\frac{(x+i)^2}{2\varepsilon}} g(-i)\sqrt{\pi}$$

ce qui achève la preuve du lemme. □

On déduit du lemme que la correction exponentielle est une quantité infinitésimale (c'est même un exponentiellement petit) aussi longtemps que $\mathrm{Re}((x+i)^2/2) < 0$.Elle devient par contre infiniment grande dès que cette partie réelle est positive. On voit donc que son influence est négligeable aussi longtemps que $x \ll 1$ et devient, au contraire, prépondérante lorsque $x \gg 1$. La solution $\bar{u}(x,\varepsilon)$ a donc le comportement annoncé.

Il reste à montrer que si u_1 est une autre solution infinitésimale en un point $x_- \ll -1$, elle reste infiniment voisine de $\bar{u}$ tant que $x \ll x_+ = -x_-$. Compte tenu de la forme très particulière de l'équation (8) considérée, la différence $u_1 - \bar{u}$ est de la forme

$$u_1(x,\varepsilon) - \bar{u}(x,\varepsilon) = Ce^{\frac{(x+i)^2}{2\varepsilon}},$$

c'est donc une fonction paire de x. □

Remarques :

- On notera que si la fonction $g(x)$ possède une singularité en un point $x = x_0$, la fonction $\mathbf{u}(x, \lambda)$ donnée par la formule (11) a, pour chaque valeur de x, deux singularités, l'une située comme précédemment sur la parabole $\lambda = -\frac{(x+i)^2}{2}$ et l'autre sur la parabole $\lambda = -\frac{(x+i)^2}{2} + \frac{(x_0+i)^2}{2}$, image de la précédente par une translation. Si les demi-droites d et $\tilde{d}$ ne rencontrent pas cette seconde parabole, les résultats de la proposition s'étendent sans difficulté. Sinon la présence de singularité de g modifie en général la position de la butée.

- On peut genéraliser les calculs précédents à des équations du type :

$$\varepsilon u' = (x+i)u + \sum_{n\geq 0} g_n(x)\varepsilon^{n+1}.$$

L'équation associée en $\mathbf{u}$ est alors :

$$\left\{\begin{array}{rcl} \partial_x \mathbf{u} & = & (x+i)\partial_\lambda \mathbf{u} + \sum_{n\geq 1} g_n(t)\frac{\lambda^n}{n\,!} \\ \mathbf{u}(x,0) & = & -g_0(x)/(x+i). \end{array}\right. \tag{12}$$

A noter cependant que la position de la butée peut-être repoussée à $+\infty$ par une simple perturbation de l'équation d'ordre ε^p. Par exemple pour n'importe quel $p \geq 2$, l'équation

$$\varepsilon u' = (x+i)u + \varepsilon + \frac{(-1)^{p+1}}{(x+i)^{2p}} 3 \cdot 5 \cdot \cdots \cdot (2p-1)\varepsilon^{p+1}$$

possède une solution infinitésimale définie pour tout $x \in \mathbb{R}$ (dont le développemennt asymptotique est un polynôme de ε de degré p).

- Lorsqu'on applique cette méthode de sommation à une équation non linéaire, par exemple

$$\varepsilon u' = (x+i)u + \varepsilon + u^2$$

on obtient une équation aux dérivés partielles pour la fonction $\mathbf{u}$ qui est non linéaire également et dont on ne connaît plus l'expression analytique des solutions :

$$\left\{\begin{array}{rcl} \partial_x \mathbf{u} & = & (x+i)\partial_\lambda \mathbf{u} + \mathbf{u} * \mathbf{u} \\ \mathbf{u}(x,0) & = & -1/(x+i). \end{array}\right. \tag{13}$$

Ce cas est étudié au prochain paragraphe par une autre méthode.

5 Perturbations non linéaires

Dans ce paragraphe, nous allons étudier l'influence des termes non linéaires de l'équation sur la longueur du retard. L'idée est qu'en général c'est le terme linéaire, dans nos exemples $(x+i)u$, qui détermine cette longueur, ici égale à 1. Nous allons détailler la méthode sur un cas particulier simple, celui de l'équation

$$\varepsilon \frac{du}{dx} = (x+i)u + f_0(\varepsilon) + f_2(\varepsilon)u^2 \tag{14}$$

où l'on suppose que f_0 et f_2 sont des fonctions de ε qui possèdent les développements

$$\begin{aligned} f_0(\varepsilon) &= c_1\varepsilon + c_2\varepsilon^2 + \cdots \\ f_2(\varepsilon) &= d_0 + d_1\varepsilon + d_2\varepsilon^2 + \cdots \ . \end{aligned} \tag{15}$$

Pour traiter ces perturbations non linéaires, nous n'utilisons pas le même type de sommation qu'aux paragraphes précédents. Il s'agit d'une sommation par rapport à la variable de l'équation, une fois celle-ci mise sous *forme préparée* [7]. On procède de la façon suivante :

L'équation différentielle $\varepsilon \frac{du}{dx} = (x+i)u$ est conjuguée à l'équation $\frac{du}{dz} = u$ par le changement de variable

$$z = \frac{(x+i)^2}{2\varepsilon}. \tag{16}$$

Celui-ci transforme donc l'équation (14) en l'équation préparée :

$$\frac{du}{dz} = u + g_0(\varepsilon)z^{-\frac{1}{2}} + g_2(\varepsilon)z^{-\frac{1}{2}}u^2 \tag{17}$$

avec $g_p(\varepsilon) = (2\varepsilon)^{-\frac{1}{2}} f_p(\varepsilon)$, avec $p = 0$ ou 2. Cette équation possède une solution formelle

$$u(z,\varepsilon) = \sum_{n\geq 1} d_n(\varepsilon) z^{-\frac{n}{2}}$$

(qui, après réarrangement des termes, n'est autre que la solution formelle en puissances de ε de l'équation (14)). On va sommer cette série par transformation de Borel en la variable z : $\mathcal{B}(z^{-p}) = \zeta^{p-1}/\Gamma(p)$.

Theorème 5.1 *L'équation (14) possède une solution infinitésimale définie pour tout* $x \ll 1$.

Preuve : Par la tranformation de Borel en z, l'équation (17) devient

$$(\zeta+1)\mathbf{u} = a_0(\varepsilon)\zeta^{-\frac{1}{2}} + a_2(\varepsilon)\zeta^{-\frac{1}{2}} * \mathbf{u} * \mathbf{u} \tag{18}$$

avec $a_p(\varepsilon) = -(2\varepsilon\pi)^{-\frac{1}{2}} f_p(\varepsilon)$. On va montrer qu'on peut construire une solution $\mathbf{u}$ de l'équation (18) qui est analytique dans $\mathbb{C} \setminus \mathbb{R}_-$ et à croissance exponentielle à l'infini dans tout secteur ouvert S (Fig. 6). Plus précisément, nous allons montrer le lemme suivant :

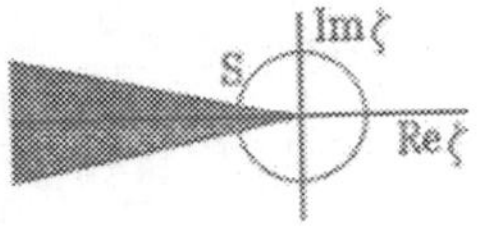

Figure 6: Un secteur ouvert S de sommet $\zeta = 0$ et d'adhérence contenue dans $\mathbf{C} \setminus \mathbf{R}_-$.

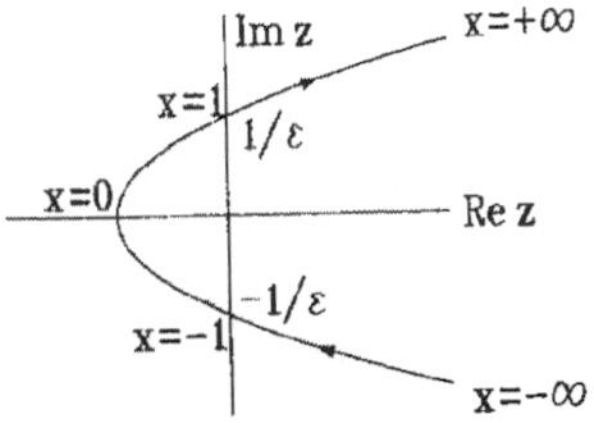

Figure 7: La parabole $z = \frac{(x+i)^2}{2\varepsilon}$.

Lemme 5.1 *L'équation (18) possède une solution analytique dans* $\mathbf{C} \setminus \mathbf{R}_-$ *telle que pour tout secteur ouvert* S *de sommet* $\zeta = 0$ *et d'adhérence contenue dans* $\mathbf{C} \setminus \mathbf{R}_-^*$, *il existe des constantes positives* M *et* L *(indépendantes de* ε *et* ζ*) telles que :*

$$|\mathbf{u}(\zeta)| \leq L\varepsilon^{\frac{1}{2}} |\zeta|^{-\frac{1}{2}} e^{M|\zeta|}.$$

Supposons ce lemme démontré (nous l'établissons ci-dessous). L'intégrale de Laplace

$$u(z) = \int_{\mathbf{R}_+} e^{-z\zeta} \mathbf{u}(\zeta) d\zeta$$

définit alors une fonction analytique dans le demi-plan $\mathrm{Re}(z) > M$. Examinons la solution $u(x, \varepsilon)$ de l'équation (14) que l'on obtient ainsi. Lorsque x varie de $-\infty$ à $+\infty$, la variable z décrit la parabole $z = \frac{(x+i)^2}{2\varepsilon}$ (Fig. 7). Le nombre ε étant infiniment petit, l'inégalité $\mathrm{Re}(\frac{(x+i)^2}{2\varepsilon}) > M$ reste vraie tant que $x \ll -1$. On peut donc poser, pour tout $x \ll -1$:

$$u(x, \varepsilon) = \int_d e^{-\frac{(x+i)^2}{2\varepsilon}\zeta} \mathbf{u}(\zeta) d\zeta \quad \text{où} \quad d =]0, \infty[. \tag{19}$$

Pour les valeurs de x infiniment voisines de -1 ou supérieures à -1, on prolonge analytiquement cette fonction en faisant tourner la demi-droite d'intégration d en $\tilde{d}$ comme

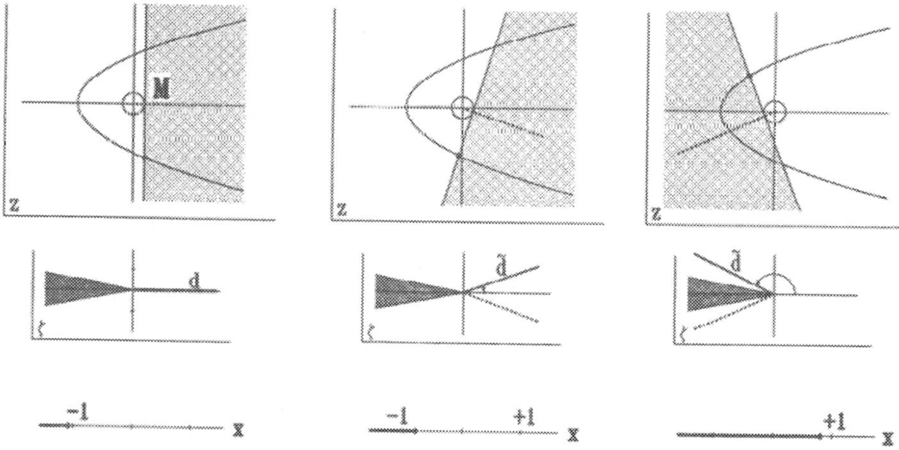

Figure 8: On prolonge analytiquement la fonction $u(x,\varepsilon)$ en faisant tourner la droite d en $\tilde{d}$: l'intégrale (19) est définie pourvu que $\mathrm{Re}\frac{(x+i)^2}{2\varepsilon} > M$, c'est-à-dire pourvu que $\frac{(x+i)^2}{2\varepsilon}$ soit dans le demi-plan hachuré, qui tourne avec d (mais en sens inverse).

indiqué sur la figure 8. La demi-droite $\tilde{d}$ peut tourner tant qu'elle ne rencontre pas $\mathbf{R}_-$ car $\mathbf{u}$ est analytique dans $\mathbb{C}\setminus\mathbf{R}_-$ en vertu du lemme 5.1, c'est-à-dire aussi longtemps que $x<1$. On obtient ainsi une solution $u(x,\varepsilon)$ de (14), analytique pour tout $x<1$. Il reste à voir que cette solution est infiniment petite aussi longtemps que x n'est pas infiniment proche de 1. En vertu du lemme 5.1, si la droite $\tilde{d}$ est standard (proche de $\mathbf{R}_-$, mais non infiniment proche), les constantes L et M peuvent être choisies standard par transfert et dans ce cas $\mathbf{u}$ est majoré par un infiniment petit. La solution $\mathbf{u}$ reste donc infinitésimale tant que $x \ll 1$. □

Montrons à présent le lemme 5.1.

Preuve : La preuve consiste à résoudre l'équation (18) par perturbation en introduisant un paramètre α : on cherche une solution de l'équation

$$(\zeta+1)\mathbf{u} = \alpha \;\; a_0(\varepsilon)\zeta^{-\frac{1}{2}} + a_2(\varepsilon)\zeta^{-\frac{1}{2}} * \mathbf{u} * \mathbf{u} \tag{20}$$

sous la forme

$$\mathbf{u} = \alpha\mathbf{u}_1 + \alpha^2\mathbf{u}_2 + \alpha^3\mathbf{u}_3 + \ldots + \alpha^n\mathbf{u}_n + \ldots \tag{21}$$

En effet si cette série est convergente, on obtient pour $\alpha = 1$ une solution de (18).

En identifiant les puissances de α dans l'équation (20), celle-ci se décompose en une suite d'équations qui permettent de calculer les $\mathbf{u}_n$ par récurrence :

$$\begin{array}{rcl} (\zeta+1)\mathbf{u}_1 & = & a_0(\varepsilon)\zeta^{-\frac{1}{2}} \\ (\zeta+1)\mathbf{u}_2 & = & a_2(\varepsilon)\zeta^{-\frac{1}{2}} * \mathbf{u}_1 * \mathbf{u}_1 \\ . & = & . \\ (\zeta+1)\mathbf{u}_n & = & a_2(\varepsilon)\zeta^{-\frac{1}{2}} * \sum_{i+j=n} \mathbf{u}_i * \mathbf{u}_j. \end{array} \tag{22}$$

Remarquons que $\mathbf{u}_1$ a une ramification en 0 et un pôle simple en -1, les $\mathbf{u}_n$ se déduisant les uns des autres au moyen de convolutions et de la multiplication par $\frac{1}{(\zeta+1)}$, vont donc avoir toutes leurs singularités sur $\{0, -1, -2, -3, ., -n, .\}$.

Lemme 5.2 *Pour tout secteur S, il existe une constante positive K_S, indépendante de ε, telle que :*

$$|u(\zeta)| \leq K_S^n |a_0|^n |a_2|^{n-1} \frac{|\zeta|^{\frac{n}{2}-1}}{n\Gamma(n/2)} \tag{23}$$

pour tout $n \geq 1$ et tout ζ dans S.

Preuve : On va montrer par récurrence sur n, que l'on a une majoration du type suivant :

$$|u_n(\zeta)| \leq C^n \frac{|\zeta|^{\frac{n}{2}-1}}{\Gamma(n/2)} g(n) \tag{24}$$

où $g(n)$ est une suite donnée par une récurrence simple et C une constante, toutes deux indépendantes de ζ dans S (mais fonctions de ε que l'on va déterminer).

• Pour $n = 1$ on a explicitement : $\mathbf{u}_1 = a_0(\varepsilon)\frac{\zeta^{-\frac{1}{2}}}{\zeta+1}$.

La fonction $|\frac{\zeta^{-\frac{1}{2}}}{\zeta+1}|$ est majorée dans S par $K'_S|\zeta|^{-\frac{1}{2}}$, il suffit donc de poser :

$$C = \Gamma(\tfrac{1}{2})|a_0(\varepsilon)|K'_S \quad \text{et} \quad g(1) = 1. \tag{25}$$

• Supposons la majoration (24) vérifiée pour $i < n$ et montrons la au rang n ; ceci va déterminer la récurrence qui définit $g(n)$. On a

$$\mathbf{u}_n = \frac{a_2(\varepsilon)}{\zeta+1}\zeta^{-\frac{1}{2}} * \sum_{i+j=n} \mathbf{u}_i * \mathbf{u}_j.$$

Pour tous i et j inférieurs à n, on majore le produit de convolution

$$\mathbf{u}_i * \mathbf{u}_j(\zeta) = \int_0^1 \mathbf{u}_i(\zeta(1-t))\mathbf{u}_j(\zeta t)\zeta dt$$

à l'aide de (24) et de l'identité :

$$\int_0^1 (1-t)^{\alpha-1} t^{\beta-1} dt = \frac{\Gamma(\alpha)\Gamma(\beta)}{\Gamma(\alpha+\beta)}.$$

On obtient ainsi la majoration suivante de $|\sum_{i+j=n} \mathbf{u}_i * \mathbf{u}_j|$:

$$\left|\sum_{i+j=n} \mathbf{u}_i * \mathbf{u}_j\right| \leq C^n \frac{|\zeta|^{\frac{n}{2}-1}}{\Gamma(\frac{n}{2})} \sum_{i+j=n} g(i)g(j).$$

Posons $\mathbf{v}_n = \sum_{i+j=n} \mathbf{u}_i * \mathbf{u}_j$, on a $\mathbf{u}_n = \frac{a_2(\varepsilon)}{\zeta+1}\zeta^{-\frac{1}{2}} * \mathbf{v}_n$. Pour majorer $|\mathbf{u}_n|$ on utilise le lemme suivant, dont la démonstration est immédiate :

Lemme 5.3 *Si on a dans S : $|\mathbf{v}_n| \leq B|\zeta|^{\frac{n}{2}-1}$, où B est une constante, alors il existe une constante K''_S, telle que* $|\frac{a_2(\varepsilon)}{\zeta+1}\zeta^{-\frac{1}{2}} * \mathbf{v}_n| \leq |a_2(\varepsilon)|K''_S B|\zeta|^{\frac{n}{2}-1}$

En appliquant ce lemme on obtient :

$$|\mathbf{u}_n(\zeta)| \leq C^n \frac{|\zeta|^{\frac{n}{2}-1}}{\Gamma(n/2)} |a_2(\varepsilon)| K''_S \sum_{i+j=n} g(i)g(j).$$

Pour avoir la majoration (24) il suffit donc de poser :

$$g(n) = |a_2(\varepsilon)| K''_S \sum_{i+j=n} g(i)g(j).$$

Avec la condition $g(1) = 1$, ceci détermine la fonction g. On a donc démontré la majoration (24) pour tout $n \geq 1$; pour en déduire la majoration (23), il ne reste plus qu'à étudier la suite $g(n)$. En introduisant la "fonction génératrice"

$$G(X) = \sum_{n \geq 1} g(n) X^n,$$

on peut calculer explicitement les $g(n)$:

$$g(n) = (2|a(\varepsilon)|K''_S)^{n-1} \frac{1.3.5.(2n-3)}{n\,!}$$

d'où l'on déduit que $g(n) \leq \frac{1}{n}(4|a_2(\varepsilon)|K''_S)^{n-1}$. Avec la valeur de C donnée en (25), ceci permet d'obtenir la majoration affirmée par le lemme (5.2). □

On déduit de ce lemme 5.2 que la série $\sum_{n=1}^{\infty} \mathbf{u}_n$ converge uniformément sur tout compact de S et définit une fonction analytique $\mathbf{u}$ sur S, majorée sur S par :

$$|\mathbf{u}(\zeta)| \leq \sum_{n \geq 1} K_S^n |a_0|^n |a_2|^{n-1} \frac{|\zeta|^{\frac{n}{2}-1}}{n\Gamma(\frac{n}{2})}.$$

En séparant les termes pairs et impairs dans cette somme, on obtient :

$$|\mathbf{u}(\zeta)| \leq \left(K_S^2 |a_0|^2 |a_2| + K_S |a_0| |\zeta|^{-1/2}\right) e^{K_S^2 |a_0|^2 |a_2|^2 |\zeta|}.$$

Le comportement de $\mathbf{u}$ par rapport à ε se déduit alors des hypothèses faites sur f_0 et f_2 ; on a $a_0(\varepsilon) \cong A_0\varepsilon^{\frac{1}{2}}$, et $a_2(\varepsilon) \cong A_2\varepsilon^{-\frac{1}{2}}$, d'où la majoration pour $|\mathbf{u}|$:

$$|\mathbf{u}(\zeta)| \leq L\varepsilon^{\frac{1}{2}}|\zeta|^{-\frac{1}{2}}e^{M|\zeta|} \tag{26}$$

avec L et M des constantes indépendantes de ζ et de ε. □

Remarques :

- Le comportement en ε de la solution ci-dessus au delà de $x = 1$ dépend de la contribution à l'intégrale de Laplace que vont apporter les singularités de $\mathbf{u}(\zeta)$ lorsque la demi-droite d'intégration franchit $\mathbf{R}_-$. Dans certains cas particuliers on peut avoir existence d'une solution infinitésimale au delà de $x = 1$, comme par exemple pour l'equation $\varepsilon\frac{du}{dx} = (x+i)u + \varepsilon + u^2$ qui possède la solution $u(x,\varepsilon) = -\varepsilon/(x+i)$, infinitésimale sur tout $\mathbf{R}$.

- On peut étendre les résultats précédents aux équations

$$\varepsilon\frac{du}{dx} = (x+i)u + f_0(\varepsilon) + f_2(\varepsilon)u^2 + f_3(\varepsilon)u^3 + \dots$$

  où l'on suppose que $f_0(\varepsilon) = c_1\varepsilon + c_2\varepsilon^2 + \dots$ et que la série $\sum_{n=1}^{\infty} f_n(\varepsilon)u^n$ a un rayon de convergence $\rho(\varepsilon)$ non infinitésimal. La construction d'une solution infiniment petite pour tout $x \ll 1$ est semblable à la précédente. Les équations définissant les $\mathbf{u}_n$ s'écrivent :

$$\begin{aligned} (\zeta+1)\mathbf{u}_1 &= a_0(\varepsilon)\zeta^{-\frac{1}{2}} \\ (\zeta+1)\mathbf{u}_n &= \zeta^{-\frac{1}{2}} * \textstyle\sum_{p\geq 2} a_p(\varepsilon) \sum_{i_1+i_2+\cdots+i_p=n} \mathbf{u}_{i_1} * \mathbf{u}_{i_2} * \cdots * \mathbf{u}_{i_p}. \end{aligned} \tag{27}$$

Ces formules faisant intervenir des produits de convolution à p termes la majoration des $|\mathbf{u}_n|$ se fait moins simplement que précédemment, mais aboutit à une formule analogue à celle du lemme 5.2.

References

[1] E. Benoit. Relation entrée-sortie. *C.R.Acad.Sci. Paris*, 293(Série 1):293–296, 1981.

[2] E. Benoit, J.L. Callot, F. Diener, et M. Diener. Chasse au canard. *Collectanea Mathematica, Barcelone*, 31(1-3):37–119, 1981.

[3] B. Candelpergher. Une introduction à la résurgence. *Gazette des mathématiciens*, 42:31–64, 1989.

[4] B. Candelpergher, J–C. Nosmas, et F. Pham. *Approche de la résurgence.* à paraître, 1990.

[5] F. Diener et M. Diener. *Canards et fleuves.* en préparation, 1990.

[6] F. Diener et G. Reeb. *Analyse Non Standard.* Hermann, 1989.

[7] J. Ecalle. *Les fonctions résurgentes* , tomes 1–3. Prépublication , Publications mathématiques d'Orsay, Université de Paris Sud, Département de Mathématiques, bât. 425, F91425 Orsay, 1985.

[8] C. Lobry et G. Wallet. La traversée de l'axe imaginaire n'a pas toujours lieu là où l'on croit l'observer. In M. Diener et G. Wallet, editeurs, *Mathématiques Finitaires et Analyse Non Standard*, pages 45–51, Publications Mathématiques de l'Université de Paris VII, 31 :2, 1989.

[9] A. I. Neishtadt. Persistence of stability loss for dynamical bifurcations I. *Differentsial'nye Uravneniya (Differential Equations)*, 23(12):2060–2067 (1385–1390), 1987 (88).

[10] A. I. Neishtadt. Persistence of stability loss for dynamical bifurcations II. *Differentsial'nye Uravneniya (Differential Equations)*, 24(2):226–233 (171–176), 1988 (88).

[11] J–P. Ramis. Dévissage gevrey. In *Journées singulières de Dijon*, pages 173–204, Astérisque 59–60, 1978.

[12] M. A. Shishkova. Examination of a system of differential equations with a small parameter in the highest derivatives. *Dokl. Akad. Nauk. SSSR*, 209(3):576–579, 1973.

[13] Y. Sibuya. *Gevrey property of formal solutions in a parameter.* Preprint, School of Mathematics, University of Minnesota, Minneapolis, Minnesota 55455 USA, 1989.

[14] G. Wallet. Entrée-sortie dans un tourbillon. *Ann. Inst. Fourier*, 36(4):157–184, 1986.

ON BIFURCATION OF LIMIT CYCLES FROM CENTERS *

CARMEN CHICONE †

Abstract. For a one parameter family of plane vector fields $\mathbf{X}(\cdot,\epsilon)$ depending analytically on a small real parameter ϵ, we determine the number and position of the local families of limit cycles which emerge from the periodic trajectories surrounding a center. Aside from the intrinsic interest in the example we choose, it serves to illustrate some techniques which are developed for treating similar bifurcation problems when the first order methods are inconclusive. Actually, we are able to treat the bifurcations of all orders.

Key Words. Limit cycles, center bifurcations, multiple Hopf bifurcation.

AMS(MOS) subject classification. 58F14, 58F21, 58F30, 34C15, 34C25.

1. Introduction. In this paper we study the bifurcations of limit cycles in the four parameter system $\mathcal{E}_\lambda$

$$\dot{x} = -y, \quad \dot{y} = x + x^3 + \lambda_1 y + \lambda_2 x^2 + \lambda_3 xy + \lambda_4 x^2 y$$

which appears in the paper of G. Danglemeyer and J. Guckenheimer [11]. The problem considered falls within the general context of a class of bifurcation problems studied by many authors (e.g., [2, 11, 14, 18, 29]). To be specific, let $(x,y) \mapsto \mathbf{X}(x,y,\epsilon)$, where $\epsilon \in \mathbf{R}$, be a one parameter family of analytic plane vector fields, where the unperturbed vector field $\mathbf{X}_0(x,y) := \mathbf{X}(x,y,0)$ has a center at the origin surrounded by a family of periodic trajectories. The problem is to determine the number and position of the local families of limit cycles which emerge from the periodic trajectories of $\mathbf{X}_0$ as the parameter ϵ is varied. A theory for the analysis of this bifurcation problem is based on the following simple idea: replace the problem of finding the periodic trajectories of the unperturbed system where a family of limit cycles emerges by an equivalent problem of finding the zeros of a suitable function. To obtain the appropriate function, one first constructs a Poincaré section Σ which intersects the family of periodic trajectories surrounding the center of the unperturbed system. Then, using the Poincaré return map $(\xi,\epsilon) \mapsto h(\xi,\epsilon)$ where ξ is a coordinate on Σ, the displacement function $d(\xi,\epsilon) := h(\xi,\epsilon) - \xi$ is defined. The zeros of $d(\cdot,\epsilon)$ correspond to the periodic trajectories of $\mathbf{X}(\cdot,\epsilon)$ which intersect Σ. Since $d(\xi,0) \equiv 0$, determining the location and multiplicity of the zeros of d requires some knowledge of its partial derivatives. For example, if $d_\epsilon(\xi_0,0) = 0$, but $d_{\xi\epsilon}(\xi_0,0) \neq 0$, then a corollary of the Implicit Function Theorem implies the existence of a smooth function $\epsilon \mapsto \beta(\epsilon)$ such that $\beta(0) = \xi_0$ and $d(\beta(\epsilon),\epsilon) \equiv 0$. The curve β corresponds to a family of limit cycles emerging from the periodic trajectory of the unperturbed system which meets Σ at ξ_0. Perhaps the most important result of the classical theory, which is presented in detail in the excellent book of A. A. Andronov *et al.* [1, Ch. XIII], is an integral representation for the partial derivative $d_\epsilon(\xi,0)$. Using

* This document was written March 13, 1990 This research was supported by The Air Force Office of Scientific Research under grant AF-AFOSR-89-0078.

† Department of Mathematics, University of Missouri, Columbia, MO 65211.

this representation it is often possible to find the simple zeros of $d_\epsilon(\xi, 0)$, and, thus, the number and position of the local families of limit cycles which emerge from the periodic trajectories of the unperturbed system. Of course, if $d_\epsilon(\xi, 0) \equiv 0$, or if one of its zeros is not simple, then higher order derivatives must be computed. Our main interest here is to show how all higher order bifurcations can often be treated after obtaining knowledge of the first few partial derivatives of the displacement function.

This paper is organized as follows. In §2 we outline the classical theory in some detail and we give a precise formulation of the main theorem which will be proved in this paper. In §3 we complete the first essential step in the proof of our main theorem, viz., a complete analysis of the first order bifurcation problem, i.e., when $d_\epsilon(\xi, 0) \not\equiv 0$. The theorems and computational techniques used, especially our methods for treating the problem of finding the zeros of linear combinations of complete elliptic integrals, should prove valuable in other circumstances. These methods are inspired by the work on the applications of Picard-Fuchs equations to bifurcation theory, e.g., [7, 10, 21, 23, 29]. In §4 we analyze the higher order bifurcations when $d_\epsilon(\xi, 0) \equiv 0$. This section sets forward our program for using the ideas in the celebrated paper of N. Bautin [3] to obtain a complete analysis of the global bifurcation of limit cycles from the periodic trajectories surrounding a center in a family of analytic perturbations. Bautin proved that in a sufficiently small neighborhood U of a stationary point, which is either a focus or a center, of a quadratic system, all sufficiently small quadratic perturbations of the given system have at most three limit cycles in U, and that three arbitrarily small amplitude limit cycles can be produced. In addition, he found a structure theorem for the power series development of the displacement function in powers of ξ, the distance coordinate from the stationary point along a Poincaré section, which gives a means of computing the zeros of the displacement function locally near the origin. While this theorem is only local, we show how the analogous result for $\mathcal{E}_\lambda$ leads to an analysis of the higher order global bifurcations. The results in §4 for the global bifurcations of limit cycles in the family $\mathcal{E}_\lambda$ provide an extension of the bifurcation theory carried out for the quadratic systems with isochronous centers in [6], to a cubic system (with a non isochronous center). However, the methods of analysis initiated in [6] and continued here can, in principle, be carried out in many other cases. To wit, if one is given an analytic family of plane vector fields depending on a finite number of parameters where the analog of Bautin's theorem can be obtained for the finite parameter system, and where, for a certain finite integer N, the partial derivatives of the displacement function with respect to ϵ can be computed in closed form up to order N, then we show how it is often possible to also obtain the number and position of the unperturbed periodic trajectories at which a family of limit cycles emerges for the bifurcations of all orders.

2. Bifurcation of Limit Cycles from Centers. Suppose $(x, y) \mapsto \mathbf{X}(x, y, \epsilon)$ is a family of plane analytic vector fields depending on a small real parameter ϵ. If Γ is a periodic trajectory of the unperturbed system $\mathbf{X}_0(x, y) := \mathbf{X}(x, y, \epsilon)$, we say a local family of periodic trajectories emerges (or bifurcates) from Γ if there is a number $\epsilon_0 > 0$ and a continuous family of closed plane curves $\epsilon \mapsto \Gamma_\epsilon$ defined for $|\epsilon| < \epsilon_0$ such that $\Gamma_0 = \Gamma$ and Γ_ϵ is a periodic trajectory of $\mathbf{X}(\cdot\,, \epsilon)$. For the remainder of this paper we assume the phase portrait of the unperturbed system $\mathbf{X}_0$, contains a *period annulus*, i.e., a continuous one parameter family of periodic trajectories. In this section we will review the theory which will allow us to analyze the following bifurcation problem: determine the number and position of the periodic trajectories in the period annulus at which a continuous family of periodic trajectories emerges. For this problem it will also be convenient to consider the differential equation corresponding to the vector field family $\mathbf{X}$. We assume this differential equation has the form

$$\dot{x} = P(x,y) + \epsilon p(x,y,\epsilon), \quad \dot{y} = Q(x,y) + \epsilon q(x,y,\epsilon).$$

Now, given Γ and $\epsilon_0 > 0$, we can always arrange the coordinates so that a certain open interval J on the horizontal line $y = y_0$ is transverse to the flow of $\mathbf{X}$ for $|\epsilon| < \epsilon_0$ and so that Γ meets J at the point $(\xi_*, 0)$. Then, there is a subinterval $\Sigma \subset J$ and a number $\epsilon_1 > 0$, with $0 < \epsilon_1 < \epsilon_0$, such that the Poincaré scalar return map $(\xi,\epsilon) \mapsto h(\xi,\epsilon)$ is defined on $\Sigma \times (-\epsilon_1, \epsilon_1)$. It assigns to (ξ,ϵ) the abscissa $h(\xi,\epsilon)$ of the point where the trajectory of $(x,y) \mapsto \mathbf{X}(x,y,\epsilon)$ starting at (ξ, y_0) first returns to Σ. We also define the *displacement function* $(\xi,\epsilon) \mapsto d(\xi,\epsilon)$ by $d(\xi,\epsilon) = h(\xi,\epsilon) - \xi$. Finally, for the purpose of obtaining some classical formulas to follow, we make one further definition. Let $\mathbf{X}_0^\perp$ denote the orthogonal vector field with components $(-Q, P)$ and $\mathbf{H}(\xi,\epsilon)$ the vector $(d(\xi,\epsilon), 0)$. Then, we define the *normalized displacement function* F by

$$F(\xi,\epsilon) := \mathbf{X}_0^\perp(\xi, y_0) \cdot \mathbf{H}(\xi,\epsilon) = -Q(\xi, y_0) d(\xi,\epsilon).$$

Of course, $F(\xi,0) \equiv 0$ and $F(\xi,\epsilon) = 0$ for $\epsilon \neq 0$ exactly when the trajectory of $(x,y) \mapsto \mathbf{X}(x,y,\epsilon)$ through (ξ, y_0) is periodic. Moreover, the existence of a continuous local family of periodic trajectories Γ_ϵ with $\Gamma_0 = \Gamma$ is equivalent to the existence of a continuous function $\epsilon \mapsto \beta(\epsilon)$ defined on $|\epsilon| < \epsilon_1 \leq \epsilon_0$ such that $F(\beta(\epsilon),\epsilon) \equiv 0$. The next result is the basic lemma which gives sufficient conditions for the existence of such local families of periodic trajectories. The lemma is proved in [6].

LEMMA 2.1 (BIFURCATION LEMMA). *Let $(\xi,\epsilon) \mapsto F(\xi,\epsilon)$ be an analytic function defined on a product neighborhood of the point $(\xi_*, 0)$, and let*

$$F(\xi,0) \equiv \partial_\epsilon F(\xi,0) \equiv \partial_\epsilon^2 F(\xi,0) \equiv \cdots \equiv \partial_\epsilon^{k-1} F(\xi,0) \equiv 0.$$

If

$$\partial_\epsilon^k F(\xi_*,0) = 0, \quad \text{with } \partial_\xi \partial_\epsilon^k F(\xi_*,0) \neq 0$$

(i.e., ξ_* is a simple root of $\partial_\epsilon^k F(\xi,0) = 0$), then there is a number $\epsilon_1 > 0$ and a unique smooth function $\epsilon \mapsto \beta(\epsilon)$, $|\epsilon| < \epsilon_1$, such that $\beta(0) = \xi_*$ and $F(\beta(\epsilon),\epsilon) \equiv 0$. Moreover, for $0 < |\epsilon| < \epsilon_1$, the point $\xi = \beta(\epsilon)$ is a simple root of the equation $F(\xi,\epsilon) = 0$. On the other hand, if

$$\partial_\epsilon^k F(\xi_*,0) = \partial_\xi \partial_\epsilon^k F(\xi_*,0) = \cdots = \partial_\xi^{l-1} \partial_\epsilon^k F(\xi_*,0) = 0, \quad \text{with } \partial_\xi^l \partial_\epsilon^k F(\xi_*,0) \neq 0$$

(i. e., ξ_* is a root of multiplicity l), then there are at most l distinct functions $\epsilon \mapsto \beta_i(\epsilon)$ such that $\beta_i(0) = \xi_*$ and $F(\beta_i(\epsilon),\epsilon) \equiv 0$, $i = 1, \ldots, l$.

In practice, when we seek to determine the number of periodic trajectories of the unperturbed system for which a continuous local family of periodic trajectories emerges, we encounter the case where Σ is a section intersecting all the periodic trajectories of the period annulus of the unperturbed system. Here we wish to consider the normalized displacement function defined globally on Σ. The next lemma summarizes some of the properties of the normalized displacement function which we require in this paper.

LEMMA 2.2 (GLOBAL BIFURCATION LEMMA). *If Σ is a horizontal line segment which is a Poincaré section intersecting the periodic trajectories of a period annulus of the analytic plane vector field $(x,y)\mapsto \mathbf{X}(x,y,\epsilon)$ and ξ is the distance coordinate along Σ, then the normalized displacement function, $(\xi,\epsilon)\mapsto F(\xi,\epsilon)$, is defined and analytic on an open neighborhood U of $\Sigma\times\{0\}$. Moreover, if*

$$F(\xi,0)\equiv \partial_\epsilon F(\xi,0)\equiv \partial_\epsilon^2 F(\xi,0)\equiv\cdots\equiv \partial_\epsilon^{k-1}F(\xi,0)\equiv 0$$

and $\partial_\epsilon^k F(\xi,0)$ has N simple zeros on Σ, then there are exactly N periodic trajectories of the unperturbed system (corresponding to the zeros of F) at which a local family of periodic trajectories emerges, while if $\partial_\epsilon^k F(\xi,0)$ has N zeros, counting multiplicities, then at most N local families of periodic trajectories emerge.

In order to apply the bifurcation lemmas just stated we require the partial derivatives at $\epsilon=0$ of the normalized displacement function. The next theorem is a version of the theorem presented by Andronov *et al.* [1], cf. [6, The Appendix]. It gives formulas for $F_\epsilon(\xi,0)$ and $F_{\xi\epsilon}(\xi,0)$. For the statement of this theorem we require one additional function associated with $\mathbf{X}(x,y,\epsilon)$. The *time map* $(\xi,\epsilon)\mapsto T(\xi,\epsilon)$ on Σ assigns the minimum positive time required for the trajectory starting at (ξ,ϵ) to return to Σ. This function is defined and analytic on the same open set where F is defined.

THEOREM 2.3. *Let $(x,y)\mapsto \mathbf{X}(x,y,\epsilon)$ denote the vector field whose corresponding differential equation is*

$$\dot{x}=P(x,y)+\epsilon p(x,y)+O(\epsilon^2),\quad \dot{y}=Q(x,y)+\epsilon q(x,y)+O(\epsilon^2),$$

and let ϕ_t^ϵ denote its flow.

(i) If $\mathbf{X}_0$ has a period annulus with Poincaré section $\Sigma:\ y=y_0$, then

$$F_\epsilon(\xi,0)=\int_0^{T(\xi,0)}(Pq-Qp)(\gamma(t))\exp\left(-\int_0^t \operatorname{div}\mathbf{X}_0(\gamma(s))\,ds\right)dt$$

where $\gamma(t):=\phi_t^0(\xi,y_0)$ is the integral curve corresponding to the periodic trajectory Γ through (ξ,y_0).

(ii) If $F_\epsilon(\xi_0,0)=0$ for some ξ_0, then

$$\begin{aligned}F_{\xi\epsilon}(\xi_0,0)= & -\ Q(\xi_0,y_0)\Big\{\operatorname{div}\mathbf{X}_0(\xi_0,y_0)T_\epsilon(\xi_0,0)\\ & +\int_0^{T(\xi_0,0)}\operatorname{div}(p,q)(\gamma(t))\,dt\\ & +\int_0^{T(\xi_0,0)}\frac{d}{d\epsilon}\operatorname{div}\mathbf{X}_0(\phi_t^\epsilon(\xi_0,y_0))\Big|_{\epsilon=0}dt\Big\}.\end{aligned}$$

(iii) In particular, if $\mathbf{X}_0$ *is Hamiltonian so that* $\operatorname{div}\mathbf{X}_0 \equiv 0$, *then*

$$F_\epsilon(\xi,0) = \int_0^{T(\xi,0)} (Pq - Qp)(\gamma(t))\,dt = -\int_\Omega \operatorname{div}(p,q)(x,y)\,dxdy,$$

where Ω *is region enclosed by the curve* Γ, *and, if* $F_\epsilon(\xi,0) \equiv 0$, *then*

$$F_{\xi\epsilon}(\xi_0,0) = -Q(\xi_0,y_0)\int_0^{T(\xi_0,0)} \operatorname{div}(p,q)(\gamma(t))\,dt.$$

(iv) If $F_\epsilon(\xi_0,0) = 0$ *and* $F_{\xi\epsilon}(\xi_0,0) \neq 0$, *then there is a continuous family* Γ_ϵ *of periodic trajectories with* $\Gamma_0 = \Gamma$ *which are hyperbolic limit cycles for sufficiently small* $\epsilon \neq 0$. *If* $\epsilon F_{\xi\epsilon}(\xi_0,0) < 0$, Γ_ϵ *is attracting while if* $\epsilon F_{\xi\epsilon}(\xi_0,0) > 0$, Γ_ϵ *is repelling.*

Remark. In practice a reasonable approach to the application of the theorem is to calculate $F_\epsilon(\xi,0)$ using the integral representation of the theorem. Then, if the integral can be evaluated, one can find the points where continuous curves of limit cycles emerge by finding the simple roots of the equation $F_\epsilon(\xi,0) = 0$. This is usually more fruitful than computing the mixed partial derivative using the representation given in the theorem.

In the present context, the unperturbed system $\mathcal{E}_0$ is in Hamiltonian form and we can always take the positive x-axis as a Poincaré section. Also, we have an analytic displacement function. So, for s, the coordinate of a point on the positive x-axis in the domain of the function $\xi \mapsto d(\xi,0)$, there are constants $\xi_0 > 0$ and $\epsilon_0 > 0$ such that, for $|\xi - s| < \xi_0$ and for $|\epsilon| < \epsilon_0$, the displacement function on $(s-\xi_0, s+\xi_0)\times(-\epsilon_0,\epsilon_0)$ can be expressed as a convergent series

$$d(\xi,\epsilon) = \sum_{k=1}^{\infty} d_k(\xi)\epsilon^k.$$

It follows that the coefficients $d_k(\xi)$ are independent of the choice of ξ_0. In particular, the first order bifurcation function, $d_\epsilon(\xi,0)$, is given by

$$d_1(\xi) = d_\epsilon(\xi,0) = -\frac{1}{Q(\xi,0)}\int_0^{T(\xi,0)} (Pq - Qp)\,dt.$$

The main results of the paper for the system $\mathcal{E}_\lambda$ are as follows.

1. *If* $\lambda_2\lambda_3 = 0$, *then there is at most one periodic trajectory surrounding the origin of* $\mathcal{E}_0$ *from which a family of limit cycles emerges in the system* $\mathcal{E}_\lambda$. *Also, the bound is sharp.*
2. *If* $d_1 \neq 0$, *then, at first order, there is at most one periodic trajectory surrounding the origin of* $\mathcal{E}_0$ *from which a family of limit cycles emerges in the system* $\mathcal{E}_\lambda$. *Also, the bound is sharp.*
3. *If* $d_1 \equiv 0$, *and if* $d_2(\xi)$ *has at most* k *zeros (counting multiplicities) for each choice of* λ_1, λ_2, λ_3 *and* λ_4, *then there are at most* k *periodic trajectories surrounding the origin of* $\mathcal{E}_0$ *from which a family of limit cycles emerges in the system* $\mathcal{E}_\lambda$.

The proofs of these statements are given in §4 where we also offer some evidence for the conjecture: *In the third statement above* $k = 2$.

3. First Order Bifurcation. We consider the first order bifurcation of limit cycles in the system

$$\dot{x} = -y, \quad \dot{y} = x + x^3 + \epsilon(\lambda_1 y + \lambda_2 x^2 + \lambda_3 xy + \lambda_4 x^2 y)$$

with ϵ the bifurcation parameter. The unperturbed system ($\epsilon = 0$) is Hamiltonian with energy function

$$H(x,y) = y^2/2 + x^2/2 + x^4/4.$$

Its phase portrait has a center at the origin surrounded by periodic trajectories which fill the entire punctured plane. The periodic trajectories are parameterized in the usual way by the distance coordinate of their intersection $\xi \in (0,\infty)$ along the positive x-axis or by their energy $h \in (0,\infty)$. These coordinates are related by the transformations

$$\xi = \sqrt{-1+\sqrt{1+4h}}, \quad h = \frac{2\xi^2+\xi^4}{4}.$$

Since the divergence of the unperturbed system vanishes, the first order bifurcation function is given by

$$d_1(\xi) = \frac{1}{\xi+\xi^3}\int_0^{T(\xi)} y(t)\left[\lambda_1 y(t) + \lambda_2 (x(t))^2 + \lambda_3 x(t)y(t) + \lambda_4 (x(t))^2 y(t)\right] dt$$

where the line integral is computed along the solution $(x(t), y(t))$ of the unperturbed system starting at $(\xi, 0)$. Using the symmetry of the unperturbed system with respect to the coordinate axes we find the bifurcation function is given by an elliptic integral:

$$d_1(\xi) = \frac{2\sqrt{2}}{\xi+\xi^3}\int_0^{\xi} (\lambda_1 + \lambda_4 x^2)\sqrt{4h-2x^2-x^4}\, dx.$$

There are a number of methods available to determine the number of zeros of functions given by elliptic integrals. We take the most elementary approach and simply express the function d_1 in terms of complete elliptic integrals. In fact, using the notation of Byrd and Friedman [4], we will express the integral in terms of complete elliptic integrals of the first and second kinds:

$$K(k) := \int_0^1 \frac{d\tau}{\sqrt{(1-\tau^2)(1-k^2\tau^2)}},$$

$$E(k) := \int_0^1 \sqrt{\frac{1-k^2\tau^2}{1-\tau^2}}\, d\tau.$$

Since the reduction involves some choices in order to obtain the precise form that we need, we give some of the significant steps in the calculation. With $v(x) := \sqrt{4h-2x^2-x^4}$ we have

$$d_1(\xi) = \frac{2\sqrt{2}}{\xi+\xi^3}\int_0^{\xi} \frac{(\lambda_1+\lambda_4 x^2)(4h-2x^2-x^4)}{v(x)}\, dx.$$

Next, the reduction formula

$$\frac{d}{dx}\left(x^N v(x)\right) = \frac{-(N+2)x^{N+3} - 2(N+1)x^{N+1} + 4Nhx^{N-1}}{v(x)}$$

is used to reduce the terms x^4/v and x^6/v in the integrand to exact terms and terms of the form x^2/v or $1/v$. After some computation using $v(\xi) = 0$ we compute

$$d_1(\xi) = \frac{2\sqrt{2}}{15(\xi+\xi^3)}\{8h(5\lambda_1 - \lambda_4))J_1 - 2(5\lambda_1 - 12\lambda_4 h - 4\lambda_4)J_2\}$$

where

$$J_1 := \int_0^\xi \frac{1}{v(x)}\,dx \text{ and } J_2 := \int_0^\xi \frac{x^2}{v(x)}\,dx.$$

To compute J_1 and J_2, first observe that

$$\begin{aligned} v(x) &= 2\sqrt{h}\left[\left(1+\frac{x^2}{1+\sqrt{1+4h}}\right)\left(1+\frac{x^2}{1-\sqrt{1+4h}}\right)\right]^{1/2} \\ &:= 2\sqrt{h}\sqrt{(1+Ax^2)(1+Bx^2)}. \end{aligned}$$

Then, with $z^2 = Bx^2 + 1$,

$$k^2 := \frac{2h}{\sqrt{1+4h}+1+4h},$$

and some straightforward manipulation, we find

$$J_1 = \frac{1}{\sqrt{2}(1+4h)^{1/4}}K(k), \quad J_2 = -\sqrt{2}(1+4h)^{1/4}\left[(1-k^2)K(k) - E(k)\right].$$

Now, define

$$\begin{aligned} f_1(k) &:= (1-k^2)k^2(K(k) - 2E(k)), \\ f_2(k) &:= E(k) - (1-k^2)K(k) \end{aligned}$$

and

$$\alpha := 5\lambda_1 - \lambda_4, \quad \beta := 5\lambda_1 - 4\lambda_4$$

to obtain

$$d_1(\xi(k)) = -\frac{4\sqrt{2}}{15k(1-2k^2)}\{2\alpha f_1(k) + \beta f_2(k)\}.$$

Since

$$k^2 = \frac{\xi^2}{2(1+\xi^2)},$$

the domain of the bifurcation function is $0 < k < 1/\sqrt{2}$. Thus, the study of the zeros of d_1 has been reduced to the study of the zeros of linear combinations (with real coefficients) of the functions f_1, and f_2 on this domain.

Next we prove two lemmas which will facilitate the proof of the bifurcation theorem which follows.

LEMMA 3.1. *Let f be a real analytic function defined on the interval $[a,b)$ with the property that there is an $\epsilon > 0$ such that $f(x)f'(x) > 0$ for $a < x < a+\epsilon$. In addition let p, q, and r be real analytic functions defined on (a,b) satisfying $p(x)r(x) > 0$ on (a,b). If f satisfies the differential equation*

$$p(x)f''(x) = q(x)f'(x) + r(x)f(x)$$

on the open interval (a,b), then f is strictly increasing on the interval $[a,b)$. If $f(x) > 0$ and $p(x)r(x) < 0$ on (a,b), and if there is an $\epsilon > 0$ such that $f'(x) < 0$ on $(a, a+\epsilon)$, then $f(x)$ is strictly decreasing on $[a,b)$.

Proof. We take the case where both $f(x)$ and $f'(x)$ are positive on $(a, a+\epsilon)$. We note that since f satisfies the differential equation, and $r(x)/q(x) > 0$ on (a,b), it follows that any critical point $\xi \in (a,b)$ with $f(\xi) > 0$ must be a proper local minimum for the function f. However, under the stated assumptions $f(x)$ is initially increasing (strictly), and so can fail to be increasing on the entire interval only if there is a stationary point $a < \xi < b$ where f has a positive relative maximum. This is a contradiction. If f and f' are both locally negative at a, then we apply the same proof as above to the functions $-f$ and $-f'$. The proof of the last assertion of the lemma is similar. □

This simple Lemma is often useful in obtaining inequalities involving expressions in the complete elliptic functions E and K, especially, when the expression defines a function of the form

$$f(k) = \mu_1(k)E(k) + \mu_2(k)K(k).$$

Since

$$E'(k) = \left[\frac{E(k) - K(k)}{k}\right], \quad K'(k) = \left[\frac{E(k) - (1-k)^2K(k)}{k(1-k^2)}\right],$$

the first step, finding a linear homogeneous second order differential equation which f satisfies, is easy.

LEMMA 3.2.

(i) The function f_2 defined above is positive and strictly increasing on $[0, 1/\sqrt{2}]$.

(ii) The function defined by $\phi_1(k) := K(k)/E(k)$ is strictly increasing on $[0,1)$. Also, $\phi_1(0) = 1$ and $\phi_1(1/\sqrt{2}) < 2$.

(iii) The function defined by

$$\phi_2(k) := 2(k^2-1)K(k) + (3-2k^2)E(k)$$

is positive on $[1/2, 1/\sqrt{2}]$.

(*iv*) *The function* $\mathcal{B}(k) := f_1(k) + 2f_2(k)$ *is positive and monotone increasing on* $(0, 1/\sqrt{2}]$.

(*v*) *The function* $Z(k) := -f_1(k)/f_2(k)$ *is positive and monotone decreasing on* $(0, 1/\sqrt{2})$. *Also, if* $0 < k < 1/\sqrt{2}$, *then* $1/2 < Z(k) < 2$.

Proof. Since

$$f_2(k) = \frac{\pi}{4}k^2 + O(k^4),$$

it follows that f_2 is positive and monotone increasing near $k = 0$. But, also

$$f_2''(k) = \frac{1}{k}f_2'(k) + \frac{1}{1-k^2}f_2(k)$$

and statement (*i*) follows from the first lemma.

The facts stated in (*ii*) are standard. However, for the sake of completeness we give a proof. Both K and E are positive on $[0, 1)$ with both functions having value $\pi/2$ at $k = 0$. Moreover, we have

$$\begin{aligned} K''(k) &= -\frac{1-3k^2}{k(1-k^2)}K'(k) + \frac{1}{1-k^2}K(k), \\ E''(k) &= -\frac{1}{k}E'(k) - \frac{1}{1-k^2}E(k) \end{aligned}$$

and

$$K(k) = \frac{\pi}{2} + \frac{1}{4}k^2 + O(k^4), \quad E(k) = \frac{\pi}{2} - \frac{1}{4}k^2 + O(k^4).$$

Applying the first lemma, we find K is monotone increasing and E is monotone decreasing on $[0, 1)$ and as a result that ϕ_1 is monotone increasing. The second statement of (*ii*),

$$\frac{K(\frac{1}{\sqrt{2}})}{E(\frac{1}{\sqrt{2}})} < 2,$$

is an easy consequence of Legendre's Relation [4, p.10]:

$$E(k)K(\sqrt{1-k^2}) + E(\sqrt{1-k^2})K(k) - K(k)K(\sqrt{1-k^2}) = \frac{\pi}{2}.$$

If

$$\phi_2'(r) = \frac{(4r^2-1)K(r) + (1-6r^2)E(r)}{r} = 0,$$

then

$$\frac{K(r)}{E(r)} = \frac{6r^2-1}{4r^2-1}.$$

But, the right side of the last equation is monotone decreasing from ∞ to 2 on the interval $(1/2, 1/\sqrt{2}]$, so $K(r)/E(r) > 2$, a contradiction by (ii) just proved. It follows that ϕ_2 is monotone on $[1/2, 1/\sqrt{2}]$. But, using (ii) again, we compute ϕ_2 to be positive and its slope to be negative at $k = 1/\sqrt{2}$. Thus, ϕ_2 is positive on I and this completes the proof of (iii).

The proof of (iv) follows from the previous lemma after computing

$$\mathcal{B}(k) = \frac{15\pi}{16}k^4 + O(k^6)$$

and

$$\mathcal{B}''(k) = \frac{3-5k^2}{k^3(1-k^2)}\mathcal{B}'(k) + \frac{5}{1-k^2}\mathcal{B}(k).$$

For (v) consider first the behavior of the graph of Z at the end points of its domain. We compute

$$Z(k) = 2 - \frac{15}{4}k^2 + O(k^4)$$

to conclude $Z(k)$ is monotone decreasing near $k = 0$ while

$$\lim_{k \to 0^+} Z(k) = 2.$$

At the right endpoint we compute $Z(1/\sqrt{2}) = 1/2$ and

$$Z'(\frac{1}{\sqrt{2}}) = \frac{\sqrt{2}K(\frac{1}{\sqrt{2}})}{K(\frac{1}{\sqrt{2}}) - 2E(\frac{1}{\sqrt{2}})}.$$

By (ii) we find $Z'(1/\sqrt{2})$ is negative and, therefore, Z is monotone decreasing near $k = 1/\sqrt{2}$. Now, if Z is not monotone decreasing on the interval $0 < k < 1/\sqrt{2}$, there must be three numbers r, s and t with $0 < r < s < t < 1/\sqrt{2}$ such that

$$Z(r) = Z(s) = Z(t).$$

An application of the Cauchy Mean Value Theorem shows there are numbers u and v with $r < u < s < v < t$ such that

$$\frac{f_1'(u)}{f_2'(u)} = \frac{f_1'(v)}{f_2'(v)}.$$

Thus, it suffices to show the function Y defined by

$$Y(k) = -\frac{f_1'(k)}{f_2'(k)}$$

is monotone on $0 < k < 1/\sqrt{2}$. For this we compute the derivative

$$Y'(k) = \frac{a(k)K(k)^2 + b(k)K(k)E(k) + c(k)E(k)^2}{(f_2'(k))^2} := \frac{Q(k)}{k(1-k^2)K(k)^2}$$

where

$$\begin{aligned} a(k) &:= -5(1-k^2)(1-4k^2), \\ b(k) &:= -2a(k), \\ c(k) &:= -5(1-2k^2). \end{aligned}$$

It suffices to show

$$Q(k) = a(k)(K(k)-E(k))^2 + (c(k)-a(k))E(k)^2$$

is negative on $0 < k < 1/\sqrt{2}$. We find

$$c(k) - a(k) = -5k^2(3-4k^2)$$

is negative on $0 < k < 1/\sqrt{2}$. Since $a(k) < 0$ on $0 < k < 1/2$, it follows that $Q(k) < 0$ on $0 < k \leq 1/2$. We will now complete the proof of (v) by showing

$$\mathcal{Q}(k) := -\frac{1}{5}Q(k) = k^2(3-4k^2)E(k)^2 + (1-k^2)(1-4k^2)(K(k)-E(k))^2 > 0$$

for k in the interval $(1/2, 1/\sqrt{2})$. Define

$$A := k^2(3-4k^2), \quad B := (1-k^2)(1-4k^2)$$

and observe that since $E(k) > 0$ it suffices to show

$$q(x) := Bx^2 - 2Bx + A + B > 0$$

when $x = K(k)/E(k)$. As $A > 0$ and $B < 0$ the quadratic q has a positive maximum at $x = 1$. Thus, $q(x)$ will be positive for x between its two real roots

$$r_1 := 1 + \frac{\sqrt{-AB}}{B}, \quad r_2 := 1 - \frac{\sqrt{-AB}}{B}.$$

Since $x = K(k)/E(k) > 1$ and $r_1 < 1 < r_2$, it suffices to show $x < r_2$. For this we first observe that

$$\frac{4k^2-1}{2} < \sqrt{-AB}.$$

Then, it suffices to show

$$x < 1 + \frac{4k^2-1}{2(-B)},$$

or, equivalently, that

$$\mathcal{S}(k) := 2(k^2-1)K(k) + (3-2k^2)E(k) > 0.$$

But, by (iii), $\mathcal{S} = \phi_2$ is positive and this completes the proof of (v). □

The bifurcation theorem for this section can now be proved.

THEOREM 3.3. *If $\lambda_1^2 + \lambda_4^2 \neq 0$, then there is at most one periodic trajectory surrounding the origin of the system*

$$\dot{x} = -y, \quad \dot{y} = x + x^3 + \epsilon(\lambda_1 y + \lambda_2 x^2 + \lambda_3 xy + \lambda_4 x^2 y)$$

at which a continuous family of limit cycles emerges. For any periodic trajectory Γ in the period annulus at the origin, there is a choice of the perturbation so that a continuous family of limit cycles emerges from Γ at $\epsilon = 0$.

Proof. From the discussion above and the expression for d_1 in terms of the elliptic modulus k, it suffices to consider the zeros of

$$\mathcal{D}(k, \alpha, \beta) := \mathcal{D}(k) := 2\alpha f_1(k) + \beta f_2(k)$$

where α, and β are real numbers. We will show the following:

(i) Given $0 < r < 1/\sqrt{2}$, there are constants α and β (not both zero) such that

$$\mathcal{D}(r) = 0.$$

(ii) Given constants α and β with $\alpha^2 + \beta^2 \neq 0$, there cannot exist r and s with $0 < r < s < 1/\sqrt{2}$ such that

$$\mathcal{D}(r) = \mathcal{D}(s) = 0.$$

(iii) When $\mathcal{D}(k)$ has exactly one zero in $(0, 1/\sqrt{2})$, then this zero is simple.

The theorem follows directly from these statements and the Global Bifurcation Lemma in §1. The statements (*i*)—(*iii*) will now be proved in the order listed.

If $\alpha = 0$ and $\beta \neq 0$ then, from the second lemma (*i*), $\mathcal{D}$ has no zeros on its domain. In view of this fact it clearly suffices to show (*i*)–(*iii*) for $\mathcal{D}(k, 1, c)$, i.e., for the function

$$f(k) := f_1(k) + cf_2(k)$$

where c is a real number. If r is chosen with $0 < r < 1/\sqrt{2}$, then we simply define $c = -f_1(r)/f_2(r)$ to obtain $f(r) = 0$. This proves (*i*). In order to prove f has at most one distinct zero on its domain, it suffices to show the function

$$Z(k) := -\frac{f_1(k)}{f_2(k)}$$

is monotone on $0 < k < 1/\sqrt{2}$. But, this is precisely statement (*v*) proved in the preceding lemma and the proof of (*ii*) is complete. Finally, suppose $f(r) = f'(r) = 0$ for some $r \in (0, 1/\sqrt{2})$ and observe this implies $Z'(r) = 0$. But, we have just proved in (*ii*) that Z is decreasing. Thus, we must have $Z'(r) = Z''(r) = 0$. This is equivalent to saying

$$\frac{f_1(r)}{f_2(r)} = \frac{f_1'(r)}{f_2'(r)} = \frac{f_1''(r)}{f_2''(r)}.$$

So, in particular, $f_2'(r)f_1''(r) - f_1'(r)f_2''(r) = 0$ and we have $Y'(r) = 0$ which is impossible by our proof of (*ii*). □

4. Higher Order Bifurcations. In this section we discuss the higher order bifurcations of limit cycles in the system $\mathcal{E}_{\lambda(\epsilon)}$ given by

$$\dot{x} = -y, \quad \dot{y} = x + x^3 + \lambda_1 y + \lambda_2 x^2 + \lambda_3 xy + \lambda_4 x^2 y$$

where the parameters are analytic functions of the bifurcation parameter ϵ. We assume each function $\epsilon \mapsto \lambda_i(\epsilon)$ is represented by a convergent power series of the form

$$\lambda_i(\epsilon) = \sum_{j=1}^{\infty} \lambda_{ij}\epsilon^j$$

and, for notational convenience, we define $\lambda(\epsilon) := (\lambda_1(\epsilon), \ldots, \lambda_4(\epsilon))$. Our goal is to show how to determine the number of continuous local families of limit cycles which emerge from the periodic trajectories of the unperturbed system $\mathcal{E}_{\lambda(0)}$ when the bifurcation analysis is carried out to arbitrary order. The example at hand, $\mathcal{E}_\lambda$, illustrates clearly how certain classes of bifurcation problems of this type can be analyzed completely after the first few low order bifurcation functions are obtained. For $\mathcal{E}_\lambda$, there are two cases: $\lambda_2\lambda_3 = 0$ and $\lambda_2\lambda_3 \neq 0$. We will show in the first case that the first order bifurcation analysis in §3 solves the problem completely, while in the second case the analysis of all higher order bifurcations depends only on a complete analysis of the zeros of the second order bifurcation function.

Our method depends on obtaining an analogue of the Bautin structure lemma [3] for the coefficients of the series expansion of the displacement function for quadratic systems. Naturally, our treatment follows closely the ideas in Bautin's paper, but the proof in the present context is easier. The first step is to identify the centers.

LEMMA 4.1 (CENTER LEMMA). *If λ is in the variety of the ideal $\mathcal{I} := (\lambda_1, \lambda_4, \lambda_2\lambda_3)$ contained in the ring $\mathbf{R}[\lambda_1, \lambda_2, \lambda_3, \lambda_4]$, then $\mathcal{E}_\lambda$ has a center at the origin*

Proof.

$$\dot{x} = -y, \quad \dot{y} = x + x^3 + \lambda_2 x^2$$

is symmetric with respect to the x-axis while the system

$$\dot{x} = -y, \quad \dot{y} = x + x^3 + \lambda_3 xy$$

is symmetric with respect to the y-axis. □

The converse of the Center Lemma is also true and is a corollary of the next result. There are more general results known which identify the centers of cubic systems [8, 20].

LEMMA 4.2 (STRUCTURE LEMMA). *If λ^* is in the variety of the ideal $\mathcal{I}$, then there are positive numbers R_1 and R_2 such that the displacement function d for $\mathcal{E}_\lambda$ has the form*

$$d(\xi, \epsilon) = \lambda_1 \sum_{k=1}^{\infty} \delta_k(\lambda)\xi^k + \bar{v}_3(\lambda)\xi^3 \left[1 + \sum_{k=1}^{\infty} \alpha_k(\lambda)\xi^k\right] + \bar{v}_5(\lambda)\xi^5 \left[1 + \sum_{k=1}^{\infty} \beta_k(\lambda)\xi^k\right]$$

for $|\xi| < R_1$ and $||\lambda - \lambda^*|| < R_2$, where

$$\delta_1(\lambda) = \frac{1}{\lambda_1}\left[\exp\left(2\pi\lambda_1/\sqrt{4 - \lambda_1^2}\right) - 1\right],$$

δ_k, α_k and β_k, for each $k = 2, 3, \ldots$, are analytic functions and where

$$\bar{v}_3(\lambda) = \frac{\pi}{4}(\lambda_4 - \lambda_2\lambda_3), \quad \bar{v}_5(\lambda) = \frac{5\pi}{24}\lambda_4.$$

Moreover, δ_k is an analytic function of λ_1, and when $\lambda_1 = 0$, both α_k and β_k are polynomials in the ring $\mathbf{R}[\lambda_2, \lambda_3, \lambda_4]$.

Proof. The displacement function is analytic for sufficiently small $|\xi|$ and for λ sufficiently near λ^*. In order to obtain the representation of the displacement function as a power series in ξ we change to polar coordinates. In fact, we compute

$$\frac{dr}{d\theta} = \frac{r\Psi \sin\theta}{1 + \Psi\cos\theta}$$

where

$$\Psi := \lambda_1 \sin\theta + r\cos\theta\,(\lambda_2\cos\theta + \lambda_3\sin\theta) + r^2\cos^2\theta\,(\cos\theta + \lambda_4\sin\theta).$$

The right hand side of this differential equation can be expanded as a convergent power series in r of the form

$$\frac{dr}{d\theta} = \sum_{k=1}^{\infty} R_k r^k$$

where each R_k is a polynomial in $\sin\theta$, $\cos\theta$ and the λ_i. The solution of the differential equation is also analytic and is represented by a power series of the form

$$r(\theta, \xi, \lambda) = \sum_{k=1}^{\infty} \nu_k(\theta, \lambda)\xi^k.$$

This series is convergent for all sufficiently small $|\xi|$, λ sufficiently close to λ^*, and for $0 \leq \theta \leq 2\pi$. Also, since $r(0, \xi, \lambda) = \xi$ we find $\nu_1(0, \lambda) = 1$, while $\nu_k(0, \lambda) = 0$ for each $k = 2, 3, \ldots$. The ν_k can be determined recursively by substitution of the series representation for $r(\theta, \xi, \lambda)$ into the differential equation and then equating the coefficients. The displacement function is then given by

$$d(\xi, \lambda) = (\nu_1(2\pi, \lambda) - 1)\xi + \sum_{k=2}^{\infty} \nu_k(2\pi, \lambda)\xi^k.$$

To compute ν_1 we do not use the recursive differential equation directly, but rather compute

$$\nu_1(2\pi, \lambda) = r_\xi(2\pi, 0, \lambda).$$

In fact, $r_\xi(\theta, 0, \lambda)$ is the solution of the variation initial value problem

$$\frac{d\rho}{d\theta} = \frac{\lambda_1 \rho \sin^2(\theta)}{1 + \lambda_1\cos\theta\sin\theta}, \quad \rho(0, 0, \lambda) = 1$$

which we see is exactly the polar coordinate representation of the linear part of $\mathcal{E}_\lambda$:

$$\dot{x} = -y, \quad \dot{y} = x + \lambda_1 y; \quad x(0) = 1, \; y(0) = 0.$$

In particular, the solution of this initial value problem returns to the positive x-axis at time $T = 4\pi/\sqrt{4-\lambda_1^2}$ and therefore

$$\nu_1(2\pi, \lambda) = x\,(T) = \xi \exp\left(2\pi\lambda_1/\sqrt{4-\lambda_1^2}\right)$$

as required.

When $\lambda_1 = 0$, one can prove by induction, using the recursively defined differential equations for the ν_k, that each ν_k, for $k = 2, 3, \ldots$, is a polynomial in $\mathcal{R} := \mathbf{R}[\lambda_2, \lambda_3, \lambda_4]$. It follows that

$$\nu_k(2\pi, \lambda) = \nu_k^0(\lambda_2, \lambda_3, \lambda_4) + \lambda_1 \nu_k^1(\lambda)$$

where $\lambda \mapsto \nu_k^1(\lambda)$ is analytic and $\nu_k^0 \in \mathcal{R}$. In order to obtain the structure of the polynomial part of these coefficients we first recall that the first nonvanishing derivative of the displacement function with respect to ξ is of odd order, cf. [1, p. 243]. It follows that $\nu_2(2\pi, \lambda) = \lambda_1 \nu_2^1(\lambda)$. To obtain the higher order ν_k it is advantageous to compute using a computer algebra system. However, the first few coefficients are published in the literature for the general case, see [1, p. 252],[15]. For the present case the important ν_k were first obtained in [11]. We find $\nu_3^0 = \bar{v}_3$. Then, we must have $\nu_4^0 = p_4\bar{v}_3$, for some $p_4 \in \mathcal{R}$. Next, we compute $\nu_5^0 = \bar{v}_5 + p_5\bar{v}_3$ where again $p_5 \in \mathcal{R}$.

For k=6,7,8,..., we will show there exit p_k, $q_k \in \mathcal{R}$ such that $\nu_k^0 = p_k\bar{v}_3 + q_k\bar{v}_5$, i.e., ν_k^0 is in the ideal generated by $\bar{v}_3$ and $\bar{v}_5$ in $\mathcal{R}$, or, equivalently, that $\nu_k \in \mathcal{I}$, the ideal defined in the Center Lemma. This step of the argument for more general situations is very difficult. For example, the most difficult part of Bautin's argument for quadratic systems is devoted to a proof of this ideal membership, cf. [3]. However, for the case at hand, the result is elementary. In fact, by the Center Lemma and the fact that the displacement function vanishes when $\mathcal{E}_\lambda$ is a center, each ν_k^0 is a polynomial in $\mathcal{R}$ which vanishes on the (real) variety of $\mathcal{I}$, i.e. ν_k^0 vanishes when $\lambda_4 = 0$ and $\lambda_2\lambda_3 = 0$. We have

$$\nu_k^0(\lambda_2, \lambda_3, \lambda_4) = a_N(\lambda_2, \lambda_3)\lambda_4^N + a_{N-1}(\lambda_2, \lambda_3)\lambda_4^{N-1} + \cdots + a_0(\lambda_2, \lambda_3)$$

where each a_j is a polynomial in the indicated variables. But, after rearrangement we have

$$\nu_k^0(\lambda_2, \lambda_3, \lambda_4) = \lambda_4 \left[a_N(\lambda_2, \lambda_3)\lambda_4^{N-1} + \cdots + a_1(\lambda_2, \lambda_3)\right] + a_0(\lambda_2, \lambda_3),$$

and we see a_0 vanishes when $\lambda_2\lambda_3 = 0$. It follows that $\lambda_2\lambda_3$ is a factor of a_0 and as a result that $\nu_k^0 \in \mathcal{I}$.

We have just shown each ν_k can be expressed in the form

$$\nu_k = \delta_k\lambda_1 + p_k\bar{v}_3 + q_k\bar{v}_5$$

where δ_k is a convergent power series in λ. Thus, in the ring of convergent power series $\mathcal{R}_s := \mathbf{R}\{\lambda_1, \lambda_2, \lambda_3, \lambda_4\}_{\lambda^*}$, the Taylor coefficients $d_k(\lambda)$ of the series expansion

$$d(\xi, \lambda) = \sum_{k=1}^{\infty} d_k(\lambda)\xi^k$$

are all in the ideal $\mathcal{J} := (\lambda_1, \bar{v}_3, \bar{v}_5) \subset \mathcal{R}_s$. By rearrangement of the series one sees that the representation of the series of the displacement function given in the statement of the Lemma is formally correct. However, it is not clear that the rearrangement is convergent. Fortunately, the convergence can be proved using some results from the theory of several complex variables, cf. [17, Th. 2 p. 13 and Th. 7 p. 32]. In outline, the Späth-Cartan Preparation Theorem gives $\nu_k = \delta_k\lambda_1 + \nu_k^0$ and the existence of a constant D not dependent on k such that $||\nu_k||_\Lambda \leq D||\nu_k||_\Lambda$, where Λ is a small ball about the origin in the λ-space. We observe $||\nu_k^0|| \leq 2||\nu_k||$ for small Λ and then apply the theorem again to ν_k^0 and $\bar{v}_3$ to obtain $\nu_k^0 = p_k\bar{v}_3 + \nu_k^1$ with a similar estimate on p_k, and, in turn, on q_k. The Cauchy inequalities for ν_k with respect to the series expansion for $d(\xi, \lambda)$, yield for sufficiently small $|\xi|$ and $|\lambda|$, estimates of the form $||\nu_k|| \leq C/\xi_0^k$ where $C > 0$ and $\xi_0 > 0$ are fixed. Combining the estimates on the coefficients δ_k, p_k and q_k with the Cauchy estimate, it follows easily that the series

$$\sum_{k=1}^{\infty} \delta_k(\lambda)\xi^k, \quad \sum_{k=1}^{\infty} p_k(\lambda)\xi^k, \quad \sum_{k=1}^{\infty} q_k(\lambda)\xi^k$$

all converge for sufficiently small $|\xi|$ and $|\lambda|$. Thus, the required representation of the displacement function is convergent on the same domain. □

The displacement function for $\mathcal{E}_{\lambda(\epsilon)}$ is simply $d(\xi, \lambda(\epsilon))$ which we abbreviate by $d(\xi, \epsilon)$. Since the origin is a center for $\epsilon = 0$, we have $d(\xi, 0) \equiv 0$ and thus, for $|\xi|$ and $|\epsilon|$ sufficiently small,

$$d(\xi, \epsilon) = \sum_{k=1}^{\infty} d_k(\xi)\epsilon^k = \sum_{k=1}^{\infty} \frac{1}{k!}\frac{\partial^k}{\partial\epsilon^k} d(\xi, 0)\epsilon^k.$$

It is important to note that while this series representation of the displacement function is only defined locally, each of the functions $\xi \mapsto d_k(\xi)$ is defined and analytic on $0 < \xi < \infty$ as proved in the Global Bifurcation Lemma.

To determine the points along the positive x-axis at which families of limit cycles emerge we must determine the zeros (and their multiplicities) of $d_k(\xi)$ under the assumption that $d_j(\xi) \equiv 0$ for $j < k$. Thus, we must identify as explicitly as possible the d_k under this hypothesis. This will be accomplished using the Structure Lemma. It is perhaps worthwhile to note that the determination of the zeros of $d_k(\xi, 0)$, which solves our bifurcation problem, does not depend on the *global* convergence of the series expansion for the displacement function.

Recall that each $\epsilon \mapsto \lambda_i(\epsilon)$ is given by a convergent power series

$$\lambda_i(\epsilon) = \sum_{j=1}^{\infty} \lambda_{ij}\epsilon^j.$$

Thus, by composition, we obtain power series representations for the functions which appear in the Structure Lemma:

$$\bar{v}_3(\lambda(\epsilon)) = \sum_{k=1}^{\infty} \bar{v}_{3k}\epsilon^k, \quad \bar{v}_5(\lambda(\epsilon)) = \sum_{k=1}^{\infty} \bar{v}_{5k}\epsilon^k,$$

$$\alpha(\lambda(\epsilon)) = \sum_{k=0}^{\infty} \alpha_{ik}\epsilon^k, \quad \beta(\lambda(\epsilon)) = \sum_{k=0}^{\infty} \beta_{ik}\epsilon^k, \quad \delta(\lambda(\epsilon)) = \sum_{k=0}^{\infty} \delta_{ik}\epsilon^k.$$

From the computation of $\delta_1(\lambda)$ in the Structure Lemma we find $\delta_{10} = \pi$. Then, by rearrangement, we see that, for $|\xi|$ and $|\epsilon|$ sufficiently small, the displacement function has the following local representation:

$$\begin{aligned} d(\xi,\epsilon) &= \sum_{i=1}^{\infty} \lambda_{ij}\epsilon^i \left[\sum_{i=0}^{\infty}\sum_{k=1}^{\infty} \delta_{ki}\xi^{k-1}\epsilon^i\right]\xi + \sum_{i=1}^{\infty} \bar{v}_{3i}\epsilon^i \left[1 + \sum_{i=0}^{\infty}\sum_{k=1}^{\infty} \alpha_{ki}\xi^k\epsilon^i\right]\xi^3 \\ &+ \sum_{i=1}^{\infty} \bar{v}_{5i}\epsilon^i \left[1 + \sum_{i=0}^{\infty}\sum_{k=1}^{\infty} \beta_{ki}\xi^k\epsilon^i\right]\xi^5. \end{aligned}$$

Finally, we define, for sufficiently small $|\xi|$, the functions

$$\begin{aligned} b_1(\xi) &:= \left(\pi + \sum_{k=2}^{\infty} \delta_{k0}\xi^{k-1}\right)\xi, \\ b_3(\xi) &:= \left(1 + \sum_{k=1}^{\infty} \alpha_{k0}\xi^k\right)\xi^3, \\ b_5(\xi) &:= \left(1 + \sum_{k=1}^{\infty} \beta_{k0}\xi^k\right)\xi^5 \end{aligned}$$

and we observe that these three functions are linearly independent for sufficiently small $|\xi|$.

THEOREM 4.3.

(i) The functions b_1, b_3 and b_5 are local power series representations at $\xi = 0$ of three corresponding functions B_1, B_3 and B_5 which are defined and analytic for $0 < \xi < \infty$.

(ii) The first order Taylor coefficient of the displacement function d for $\mathcal{E}_{\lambda(\epsilon)}$ at $\epsilon = 0$ is given by

$$d_1(\xi) = \lambda_{11}B_1(\xi) + \lambda_{41}\left(\frac{\pi}{4}B_3(\xi) + \frac{5\pi}{24}B_5(\xi)\right).$$

(iii) If $d_1(\xi) \equiv 0$, then the second order Taylor coefficient of d is

$$d_2(\xi) = \lambda_{12}B_1(\xi) + \frac{\pi}{4}\left(\lambda_{42} - \lambda_{21}\lambda_{31}\right)B_3(\xi) + \frac{5\pi}{24}\lambda_{42}B_5(\xi).$$

(iv) If the Taylor coefficients of d vanish up to order $k-1$, then the Taylor coefficient of order k is a linear combination of the functions B_1, B_3 and B_5. In particular, if $\lambda_2\lambda_3 = 0$, then

$$d_k(\xi) = \lambda_{1k}B_1(\xi) + \lambda_{4k}\left(\frac{\pi}{4}B_3(\xi) + \frac{5\pi}{24}B_5(\xi)\right).$$

Proof. Using the formulas for $\bar{v}_3(\lambda)$ and $\bar{b}_5(\lambda)$ given in the Structure Lemma we easily compute

$$\begin{aligned}\bar{v}_3(\lambda(\epsilon)) &:= \frac{\pi}{4}\left(\lambda_{41}\epsilon + (\lambda_{42} - \lambda_{21}\lambda_{31})\epsilon^2)\right) + O(\epsilon^3) \\ \bar{v}_5(\lambda(\epsilon)) &:= \frac{5\pi}{24}\left(\lambda_{41}\epsilon + \lambda_{42}\epsilon^2\right) + O(\epsilon^3).\end{aligned}$$

Thus, near $\epsilon = 0$,

$$\begin{aligned} d_1(\xi) &= \lambda_{11}\left[\pi + \sum_{k=2}^{\infty}\delta_{k0}\xi^{k-1}\right]\xi + \bar{v}_{31}\left[1 + \sum_{k=1}^{\infty}\alpha_{k0}\xi^k\right]\xi^3 + \bar{v}_{51}\left[1 + \sum_{k=1}^{\infty}\beta_{k0}\xi^k\right]\xi^5 \\ &= \lambda_{11}b_1(\xi) + \lambda_{41}\left(\frac{\pi}{4}b_3(\xi) + \frac{5\pi}{24}b_5(\xi)\right).\end{aligned}$$

If $\lambda_{41} = 0$ and $\lambda_{11} = 1$, we have the first derivative with respect to ϵ of the displacement function given by $b_1(\xi)$ for sufficiently small $|\xi|$. But, since this derivative is actually analytic on $0 < \xi < \infty$, it is an analytic extension $B_1(\xi)$ of $b_1(\xi)$. If $\lambda_{11} = 0$ and $\lambda_{41} = 0$, we find, for $|\xi|$ sufficiently small,

$$d_2(\xi) = \lambda_{12}b_1(\xi) + \frac{\pi}{4}(\lambda_{42} - \lambda_{21}\lambda_{31})\,b_3(\xi) + \frac{5\pi}{24}\lambda_{42}b_5(\xi).$$

Since, for each choice of λ_{12}, λ_{21}, λ_{31}, λ_{42}, the function $d_2(\xi)$ corresponds to a constant multiple of the second derivative of the corresponding displacement function, $d_2(\xi)$ is analytic on $0 < \xi < \infty$ and we find both b_3 and b_5 have global analytic extensions B_3 and B_5 respectively.

The proof of (iii) is by an easy induction argument.□

THEOREM 4.4.

(i) The number of periodic trajectories of $\mathcal{E}_{\lambda(0)}$ at which a continuous local family of limit cycles emerges in the system $\mathcal{E}_{\lambda(\epsilon)}$ is bounded by the number of zeros (counted with multiplicity) of an arbitrary linear combination of the functions B_1, B_3 and B_5.

(ii) If $\lambda_2(\epsilon)\lambda_3(\epsilon) \equiv 0$, then there is at most one periodic trajectory of $\mathcal{E}_{\lambda(0)}$ at which a local family of limit cycles emerges in the system $\mathcal{E}_{\lambda(\epsilon)}$. Moreover, if $0 < \xi_0 < \infty$, there is an analytic curve $\epsilon \mapsto \lambda(\epsilon)$ such that a continuous local family of limit cycles emerges at ξ_0 in the system $\mathcal{E}_{\lambda(\epsilon)}$.

(iii) If $d_1 \not\equiv 0$, then there is at most one periodic trajectory of $\mathcal{E}_{\lambda(0)}$ at which a local family of limit cycles emerges in $\mathcal{E}_{\lambda(\epsilon)}$ and, if $0 < \xi_0 < \infty$, there is a local family of limit cycles that emerges at ξ_0. If the first order bifurcation function $d_1(\xi) \equiv 0$, and, if $d_2(\xi)$ has at most k zeros for each $\lambda_1, \lambda_2, \lambda_3, \lambda_4$, then there are at most k periodic trajectories of $\mathcal{E}_{\lambda(0)}$ at which a continuous families of limit cycles emerges in the system $\mathcal{E}_{\lambda(\epsilon)}$.

Proof. The first statement of the theorem is an immediate corollary of part (*iv*) of the preceding theorem. The second statement is an immediate corollary of this same (*iv*) and the first order bifurcation analysis of §3. The first statement of (*iii*) is just a restatement of the results in §3. The rest follows from (*iii*) and (*iv*) of the preceding theorem after the observation that the coefficients of the B_i in the expression for d_2 are independent.□

The theorem shows how to make use of the analyticity of the bifurcation problem in the analysis of the higher order bifurcations by reducing the problem to the analysis of the zeros of the first few bifurcation functions $d_i(\xi)$. For the system $\mathcal{E}_\lambda$, with $\lambda_2\lambda_3$ arbitrary, the third statement of the previous theorem reduces the problem to counting the zeros of d_2. Or, what is the same thing, finding the zeros of a linear combination of the functions B_1, B_3 and B_5. From the first order analysis we know the function B_1 and the linear combination $\pi/4B_3 + 5\pi/24B_5$. Unfortunately, the formulas computed below for B_3 and B_5 are very complicated and, as a result, we are not able to give a complete analysis of the zeros of d_2. But, we do have the following conjecture: *There are at most two periodic trajectories of $\mathcal{E}_{\lambda(0)}$ at which a local continuous families of limit cycles can emerge in the system $\mathcal{E}_{\lambda(\epsilon)}$. Moreover, given $0 < \xi_1 < \xi_2 < \infty$, there are local continuous families which emerge simultaneously at both ξ_1 and ξ_2.* The first statement of the conjecture is equivalent to the conjecture: *An arbitrary linear combination of the functions B_1, B_3 and B_5 has at most two zeros counting multiplicities.* The remainder of the section is devoted to a discussion of this conjecture.

We have the following result.

THEOREM 4.5. *If $d_1(\xi) \equiv 0$ and $0 < \xi_1 < \xi_2 < \infty$, then there is an analytic function $\epsilon \mapsto \lambda(\epsilon)$ such that the corresponding second order bifurcation function d_2 has zeros at both ξ_1 and ξ_2.*

Proof. Recall from §3 that

$$d_1(\xi(k)) = -\frac{4\sqrt{2}}{15k(1-2k^2)}\left\{2(5\lambda_{11}-\lambda_{41})f_1(k) + (5\lambda_{11}-4\lambda_{41})f_2(k)\right\}.$$

Since f_1 and f_2 are globally defined analytic functions we find

$$B_1(\xi(k)) = \frac{-20\sqrt{2}}{15k(1-2k^2)}\mathcal{B}_1$$

and

$$B(\xi(k)) := \frac{\pi}{4}B_3(\xi(k)) + \frac{5\pi}{24}B_5(\xi(k)) = \frac{8\sqrt{2}}{15k(1-2k^2)}\mathcal{B}$$

where $\mathcal{B}_1(k) := 2f_1(k) + f_2(k)$ and $\mathcal{B}(k) = f_1(k) + 2f_2(k)$ ($\mathcal{B}$ is defined in Lemma 3.1 (*iv*)). Thus, we can express d_2 as follows:

$$\begin{aligned} d_2(\xi(k)) &= \lambda_{12}B_1(\xi(k)) + \frac{\pi}{4}(\lambda_{42}-\lambda_{21}\lambda_{31})B_3(\xi(k)) + \lambda_{42}\left(B(\xi(k)) - \frac{\pi}{4}B_3(\xi(k))\right) \\ &= \lambda_{12}B_1(\xi(k)) - \frac{\pi}{4}\lambda_{21}\lambda_{31}B_3(\xi(k)) + \lambda_{42}B(\xi(k)). \end{aligned}$$

Now, given $0 < r < s < 1/\sqrt{2}$, we set $\lambda_{21} = \lambda_{31} = 1$. Then, d_2 will vanish at both r and s if there is a simultaneous solution of the equations

$$\begin{aligned} \lambda_{12}B_1(\xi(r)) + \lambda_{42}B(\xi(r)) &= \frac{\pi}{4}B_3(\xi(r)) \\ \lambda_{12}B_1(\xi(s)) + \lambda_{42}B(\xi(s)) &= \frac{\pi}{4}B_3(\xi(s)). \end{aligned}$$

It suffices to show the determinant

$$\Delta(r,s) := B_1(\xi(r))B(\xi(s)) - B(\xi(r))B_1(\xi(s))$$

does not vanish. In view of the formulas for B and B_1 given above, Δ does not vanish if

$$\mathcal{B}_1(r)\mathcal{B}(s) - \mathcal{B}(r)\mathcal{B}_1(s) \neq 0.$$

But, by Lemma 3.2 (*iv*), we see both $\mathcal{B}(r)$ and $\mathcal{B}(s)$ are not zero, so it suffices to show

$$\frac{\mathcal{B}_1(r)}{\mathcal{B}(r)} - \frac{\mathcal{B}_1(s)}{\mathcal{B}_1(s)} \neq 0.$$

For this we consider

$$W(k) := \frac{\mathcal{B}_1(k)}{\mathcal{B}(k)} = \frac{2f_1(k) + f_2(k)}{f_1(k) + 2f_2(k)} = \frac{2Z(k) - 1}{Z(k) - 2}$$

where $Z = -f_1/f_2$ was defined in Lemma 3.2. Finally, using the Lemma 3.2 (*v*), we have

$$W'(k) = -\frac{3}{(Z(k) - 2)^2}Z'(k) > 0.$$

It follows that W is monotone and the determinant does not vanish. □

The theorem shows d_2 can have two zeros, but it does not show these zeros are simple. Thus, we can not be certain that a local family of limit cycles emerges from each of these points. It seems that one must compute B_3 and B_5 in order to make further progress. However, since we know a formula for B, which is a linear combination of these functions, it suffices to compute just one of them, say B_3. To this end, note that B_3 can be computed from the displacement function of the system $\mathcal{E}_\epsilon$ given by

$$\dot{x} = -y, \quad \dot{y} = x + x^3 + \epsilon x^2 + \epsilon xy.$$

Indeed, from Theorem 4.3, the series expansion of the displacement function for $\mathcal{E}_\epsilon$ will have $d_1(\xi) \equiv 0$ and $d_2(\xi) = -\pi/4B_3(\xi)$. Unfortunately, the computation of d_2 for $\mathcal{E}_\epsilon$ is rather lengthy. Thus, we will only give an outline of the intermediate steps.

We use polar coordinates (r, θ) and represent r as a series

$$r(\theta, \xi, \epsilon) = \sum_{k=0}^{\infty} r_k(\theta, \xi)\epsilon^k.$$

Here, $d(\xi,\epsilon) = r(2\pi,\xi,\epsilon) - \xi$. Thus, $r_1(2\pi,\xi) = d_1(\xi)$ and

$$r_2(2\pi,\xi) = d_2(\xi) = -\frac{\pi}{4}B_3(\xi).$$

Now, from the Hamiltonian for the unperturbed system and with

$$w := \sqrt{1+4h\cos^4\theta},$$

we compute

$$r_0^2 = \frac{4h}{1+w}$$

where h is the energy. In fact,

$$h = \frac{2\xi^2+\xi^4}{4}.$$

After substitution of the series for r into the polar form of $\mathcal{E}_\epsilon$ and equating coefficients, we have recursively defined differential equations for the coefficients r_j of the form

$$\frac{dr_j}{d\theta} = \rho(\theta)r_j + g_j(\theta)$$

where the initial conditions are $r_0(0) = \xi$ and $r_j(0) = 0$, for $j > 0$. We compute

$$\rho(\theta) = -\frac{w+2}{2w(w+1)}\frac{dw}{d\theta}$$

and from this the integrating factor for each of these linear differential equations. This integrating factor has a simple form, namely,

$$\mu(\theta) := \frac{\sqrt{1+w(0)}}{w(0)}\frac{w}{\sqrt{1+w}}.$$

After computing g_1, we have

$$\begin{aligned} r_1(\theta) &= \frac{1}{\mu(\theta)}\int_0^\theta \mu(s)g_1(s)\,ds \\ &= \frac{4h\sqrt{1+w}}{w}\int_0^\theta \frac{\sin s\cos s(\sin s+\cos s)}{w(1+w)^{3/2}}\,ds \end{aligned}$$

and, after computing g_2, we have

$$\frac{\pi}{4}B_3(\xi(k)) = -M(h)\int_0^{2\pi}\frac{\cos s}{w^2(1+w)^2}\left(A_0(s)+A_1(s)r_1(s)+A_2(s)r_1^2(s)\right)ds,$$

where

$$\begin{aligned} A_0(s) &:= 4h\cos^2 s(2\cos^3 s - 2\cos s - \sin s), \\ A_1(s) &:= -2(2-w)(1+w)(\cos^2 s - \sin s\cos s - 1), \\ A_2(s) &:= (4-w)(1+w)\sin s\cos^2 s, \\ M(h) &:= \frac{\sqrt{4h}\sqrt{1+\sqrt{1+4h}}}{\sqrt{1+4h}} \end{aligned}$$

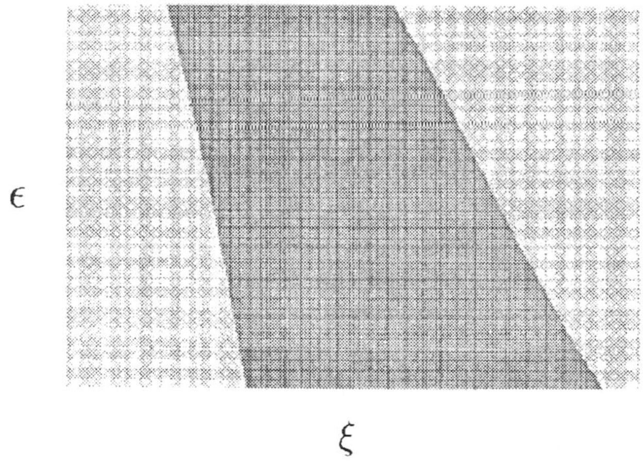

FIG. 1. *Bifurcation diagram of the displacement function* $d(\xi,\epsilon)$ *of* $\mathcal{E}^{num}$ *for* $25 < \xi < 1.25$ *and* $-1 < \epsilon < 1$ *showing two limit cycle families. The dark grey tone corresponds to* $d(\xi,\epsilon) < 0$ *and the light grey tone corresponds to* $d(\xi,\epsilon) \geq 0$.

and

$$h = \frac{k^2(1-k^2)}{(1-2k^2)^2}.$$

At present these expressions seem to be too complex to analyze analytically. However, in order to supply some evidence for the conjecture, a numerical experiment was performed which suggests that two continuous families of limit cycles can bifurcate simultaneously. For this we fix two values of ξ, in this case $\xi = 1/2$ and $\xi = 1$, set $\lambda_{21} = \lambda_{31} = 1$, $\lambda_{11} = \lambda_{41} = 0$, and then solve for λ_{12} and λ_{42} using the system of linear equations for these parameters given in the proof of Theorem 4.5. The chosen values of ξ correspond to $k = 1/\sqrt{10}$ and $k = 1/2$ respectively. After computing the values for B, B_1 and B_3 at these two values using the formulas developed above and after solving the resulting system of linear equations for the parameters, we find

$$\lambda_{12} = .0235473593, \quad \lambda_{42} = .4561286509.$$

Thus, the system $\mathcal{E}^{num}$ defined by

$$\dot{x} = -y, \quad \dot{y} = x + x^3 + \lambda_{12}\epsilon^2 y + \epsilon x^2 + \epsilon xy + \lambda_{42}\epsilon^2 x^2 y$$

has a first order bifurcation function which vanishes identically and a second order bifurcation function with two zeros. The results of the numerical experiment are shown in Figure 1. In the figure the horizontal coordinate is the distance ξ along the positive x-axis while the vertical coordinate is the value of the parameter ϵ. For a grid point $(\xi,\epsilon) \in (.25, 1.25) \times (-1,1)$, the corresponding differential system (in polar coordinates) was integrated numerically to obtain the displacement $d(\xi,\epsilon)$. Then, the point was plotted in the dark grey tone if the displacement was negative and in the light grey tone if the displacement was nonnegative. Thus, the boundaries between the light and dark regions of the graph indicate the approximate positions of limit cycles. The experiment clearly suggests that two families of limit cycles bifurcate as $|\epsilon|$ increases from zero.

REFERENCES

[1] A. A. Andronov et. al., *Theory of bifurcations of dynamical systems on a plane*, John Wiley and Sons, New York, 1973.

[2] V. I. Arnold, *Geometric methods in the theory of ordinary differential equations*, Springer–Verlag, New York, 1982.

[3] N. N. Bautin, *On the number of limit cycles which appear with the variation of coefficients from an equilibrium position of focus or center type*, Amer. Math. Soc. Transl , 100(1954), 1–19.

[4] P. F. Byrd and M. D. Friedman, *Handbook of elliptic integrals for engineers and physicists*, Springer–Verlag, Berlin, 1954.

[5] C. Chicone and M. Jacobs, *Bifurcation of critical periods for plane vector fields*, Trans. Amer. Math. Soc., 312(1989), 433–486

[6] C. Chicone and M. Jacobs, *Bifurcation of limit cycles from quadratic isochrones*, to appear in J. Differential Equations.

[7] S. N. Chow and J Sanders, *On the number of critical points of the period*, J Differential Equations, 64(1986), 51–66.

[8] C. J. Christopher and N. G. Lloyd, *On the paper of Jin and Wang concerning the conditions for a center in certain cubic systems*, Bull. London Math. Soc. 22(1990) 5–12

[9] W A. Coppel, *The limit cycle configurations of quadratic systems*, Ninth Conference on Ordinary and Partial Differential Equations, University of Dundee.

[10] R. Cushman and J. Sanders, *A codimension two bifurcation with a third order Picard–Fuchs equation*, J Differential Equations, 59(1985), 243–256.

[11] G. Dangelmeyer and J. Guckenheimer, *On a four parameter family of planar vector fields*, Arch. Rat. Mech. Anal., 97(1987) 321–352

[12] S. I. Diliberto, *On systems of ordinary differential equations*, in Contributions to the Theory of Nonlinear Oscillations, Annals of Mathematics Studies, Vol. 20, Princeton University Press, Princeton, 1950.

[13] W. Farr, C. Li, I. Labouriau, and W. Langford, *Degenerate Hopf bifurcation formulas and Hilbert's 16th problem*, SIAM J. Math. Anal., 20(1989) 13–30.

[14] J. Guckenheimer and P. Holmes, *Nonlinear oscillations, dynamical systems, and bifurcations of vector fields*, second ed., Springer–Verlag, New York, 1986.

[15] F. Göbber and K. D. Willamowski, *Lyuapunov approach to multiple Hopf bifurcation*, J. Math. Anal. Appl., 71(1979),333–350 .

[16] P. Henrici, *Applied and computational complex analysis*, Vol. I, John Wiley and Sons, New York, 1974.

[17] M. Herve, *Several complex variables*, Oxford Univ. Press, 1963.

[18] S Lefschetz, *Differential equations. geometric theory*, second ed., Dover, New York, 1977.

[19] N G. Lloyd, *Limit cycles of polynomial systems, some recent developments*, New Directions in Dynamical Systems, LMS Lecture Notes, Series No. 127, Cambridge University Press, 1988, 192–234.

[20] N. G. Lloyd and J. M. Pearson, *Conditions for a centre and the bifurcation of limit cycles in a*

class of cubic systems, Preprint, The University College of Wales, 1990.

[21] C. Li and C. Rousseau, *Codimension 2 symmetric homoclinic bifurcations and applications to 1:2 resonance*, preprint, Université de Montréal, 1988.

[22] W. S. Loud, *Behavior of the periods of solutions of certain plane autonomous systems near centers*, Contributions to Differential Equations, 3(1964), 21–36.

[23] G. S. Petrov, *Elliptic integrals and their nonoscillation*, Func. Anal. Appl., 20(1986), 37–40.

[24] J. W. Reijn, *A bibliography of the qualitative theory of quadratic systems of differential equations in the plane*, Technical Report 89-71, Faculty of Tecnical Mathematics and Informatics, Delft University of Technology, The Netherlands, 1989.

[25] R. Roussarie, *A note on finite cyclicity property and Hilbert's* 16th *problem*, Lecture Notes in Mathematics, 1331(1986), 161–168.

[26] C. Rousseau, *Bifurcation methods in quadratic systems*, Canadian Mathematical Society Conference Proceedings, 8(1987) 637–653.

[27] K. S. Sibirsky, *Introduction to the algebraic theory of invariants of differential equations*, Manchester University Press, New York, 1988.

[28] C. K. Siegel and J. K. Moser, *Lectures on celestial mechanics*, Springer–Verlag, New York, 1971.

[29] H. Żołądek, *Abelian integrals in non–symmetric perturbation of symmetric Hamiltonian vector fields*, preprint, Warsaw University, 1988.

On the saddle loop bifurcation

by F. Dumortier and R. Roussarie

Abstract

It is shown that the set of C^∞ (generic) saddle loop bifurcations has a unique modulus of stability $\gamma \in]0,1[\cup]1,\infty[$ for (C^0, C^r)-equivalence, with $1 \le r \le \infty$. We mean for an equivalence $(x,\mu) \mapsto (h(x,\mu), \varphi(\mu))$ with h continuous and φ of class C^r. The modulus γ is the ratio of hyperbolicity at the saddle point of the connection. Already asking φ to be a lipeomorphism forces two saddle loop bifurcations to have the same modulus, while two such bifurcations with the same modulus are $(C^0, \pm Identity)$-equivalent.

A side result states that the Poincaré map of the connection is C^1-conjugate to the mapping $x \mapsto x^\gamma$.

In the last part of the paper is shown how to finish the proof that the Bogdanov-Takens bifurcation has exactly two models for (C^0, C^∞)-equivalence.

Freddy Dumortier, Limburgs Universitair Centrum, Universitaire Campus, B-3610 Diepenbeek, Belgium.

Robert Roussarie, Département de Mathématiques, Université de Bourgogne, UFR de Sciences et Techniques, Laboratoire de Topologie, B.P. 138 (U.A. n 755 de CNRS), F-21004 Dijon Cedex, France

Introduction

Given X_μ and Y_μ two families of vectorfields, we say that they are **equivalent** if there exists a homeomorphism $\phi(\mu)$ in the parameter space and a parameter dependent family $h_\mu(x)$ of homeomorphisms in the phase space such that, for each μ, h_μ is an equivalence between X_μ and $Y_{\phi(\mu)}$.

Depending on various restrictions on ϕ and h we say that the equivalence (h_μ, ϕ) is :

- a (fibre C^k, C^l)-equivalence : if ϕ is C^l and for each μ, h_μ is C^k but no assumption is made on the continuity of h in function of μ,
- a (C^0, C^0)-equivalence : if $h(x,\mu) = h_\mu(x)$ is continuous in function of (μ, x).
- a $(C^0, C^k), (C^0, C^\omega)$-equivalence : if moreover ϕ is C^k, C^ω.
- a $(C^0$, Lip)-equivalence : if ϕ is a lipeomorphism.
- a (C^0, Id)-equivalence : if h is continuous in (μ, x) and if $\phi = Id$.

In case of a study of local families all these notions have of course restrictions to germs of families at some value μ_0 of the parameter and along some compact invariant subset Γ of X_{μ_0}.We call such a germ an **unfolding of the germ of X_{μ_0} along Γ**. Let us take $\mu_0 = 0$. Here, we are interested in **the generic saddle loop bifurcation** for vector fields in $\mathbb{R}^2$, with $\mu \in \mathbb{R}$.
It is an unfolding X_μ of the germ of X_0 along a saddle loop.
We suppose that X_0 has a homoclinic connection (or "saddle loop") Γ at a hyperbolic saddle point.
Let $-\lambda_1, \lambda_2$, with $\lambda_i > 0$,be the eigenvalues at the saddle point s of Γ, and let $\gamma = \lambda_1/\lambda_2$. We call it : the **ratio of hyperbolicity**. Generic means that :

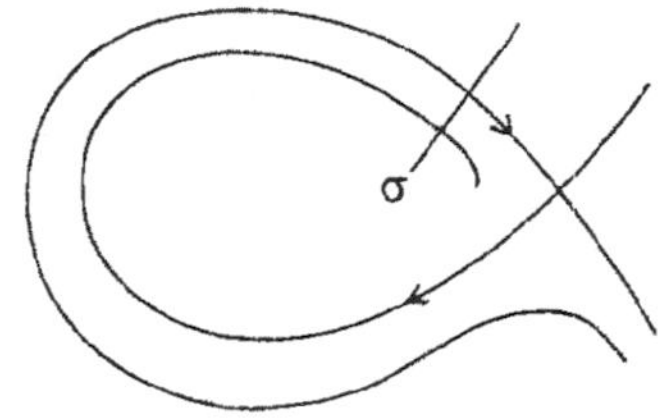

fig. 1

1) The ratio of hyperbolicity is different from 1.

2) Let σ be a segment transverse to Γ, oriented toward the inside of Γ, at some regular point. Let $\alpha(\mu)$ be the shift function equal to the distance along σ from the first intersection point of the stable manifold to the first intersection point of the unstable manifold of the saddle $s(\mu)$ of X_μ near s. Then $\alpha(0) = 0$ and $\frac{d\alpha}{d\mu}(0) \neq 0$.

This generic bifurcation was described in detail by Andronov and others in [A.L.]. They show that the type of bifurcation, up to (fibre C^0, C^∞) equivalence is determined by the sign of $\gamma - 1$. Of course, it suffices to consider the case where $\gamma < 1$ and $\frac{d\alpha}{d\mu} > 0$. We suppose these conditions to be fulfilled throughout the paper. In this case the saddle loop is repelling. There exists a repelling limit cycle for $\mu < 0$ which becomes the saddle loop for $\mu = 0$ and disappears when $\mu > 0$. This description characterizes uniquely the (fibre C^0, Id)-equivalence type of the family. In [M.P.], it was proved that two such families are (C^0, C^0) equivalent and in [A.A.D.] it was observed in the appendix that the exact value of γ was important in order to improve this result to a (C^0, C^k) equivalence with $k \geq 1$. In the present article we want to concretisize this result in the following theorems :

Theorem 1
Let X_μ and Y_μ be two generic C^∞ saddle loop bifurcations which are (C^0, Lip)-equivalent. Let γ and γ' be their respective ratios of hyperbolicity at $\mu = 0$. Then $\gamma = \gamma'$.

Theorem 2
Let X_μ and Y_μ be two generic C^∞ saddle loop bifurcations with equal ratio of hyperbolicity at $\mu = 0$. Then X_μ and Y_μ are (C^0, Id)-equivalent.

Because of the theorems 1 and 2 we obtain that $\gamma \in]0,1[\cup]1,\infty[$ is a modulus of stability for (C^0, C^r)-equivalence, with $1 \leq r \leq \infty$. By this we mean that the (C^0, C^r)-equivalence classes are completely determined by γ.
A C^1 assumption is sufficient to formulate the genericity of the family. In the proofs of the theorems we will use the C^∞ assumption for technical reasons. In particular we use it to establish a C^1-linearisation of the Poincaré-map of the saddle loop at the value $\mu = 0$:

Theorem 3
Let X be a C^∞ vector field with a saddle loop Γ at a saddle point s with ratio of hyperbolicity $\gamma \neq 1$. Let σ be a transverse segment at some regular point of Γ. Then we can choose a C^1 parametrization of σ by a parameter x, such that $\{x = 0\} = \Gamma \cap \sigma$ and such that the Poincaré map defined inside Γ (cf. Fig. 1) is equal to :

$$P(x) = x^\gamma \quad \text{for } x \geq 0$$

($\{x > 0\}$ corresponds to the inside part of Γ).

This "linearization" result, proved in paragraph 1, will be essential to obtain a simplified form for the family $f_\mu(x)$ of the Poincaré maps of a C^∞ generic family X_μ. This simplified form is given in the paragraph 2, where we also reduce the proof of theorems 1, 2 to proving equivalent theorems for the families of Poincaré maps. These theorems will be proved in the paragraphs 3 and 4. Next we apply theorem 2 to the generic 2-parameter unfoldings of nilpotent singular points, known as the Bogdanov-Takens unfoldings.

In his paper [B.], Bogdanov anounced but did not prove completely that any C^∞ 2-parameter generic unfolding of X_0, with $j^2X_0 = y\frac{\partial}{\partial x} + (x^2 \pm xy)\frac{\partial}{\partial y}$ is (C^0, C^∞)–equivalent to the polynomial family :

$$X^N_{(\mu,\nu)} = y\frac{\partial}{\partial x} + (x^2 + \mu + y(\nu \pm x))\frac{\partial}{\partial y} \qquad (N)$$

In fact Bogdanov just proved a (fibre C^0, C^∞) equivalence because there remain some hidden difficulties in order to obtain the continuity of the family of homeomorphisms along the line C of saddle saddle loops of the family $X_{(\mu,\nu)}$. Here, in the paragraph 5, we will use a 2-parameter version of the theorem 2 in order to complete the proof. A preliminar (C^0, C^0) proof was obtained in [A.A.D.], with a C^∞ change of parameter outside the line C.

Theorem 4

Any generic C^∞ unfolding of X_0, with $j^2X_0 \cong y\frac{\partial}{\partial x} + (x^2 \pm xy)\frac{\partial}{\partial y}$ is (C^0, C^∞)-equivalent to the polynomial one (N).

Remark

Concerning theorem 1 we want to observe that the statement remains true if we change $(C^0$, Lip)-equivalence by (weak C^∞,Lip)-equivalence. A weak-C^∞-equivalence (see [T2]) in this case is given by a C^∞ diffeomorphism respecting the singularities and their stable and unstable manifolds. As we will see in paragraph 2, the whole obstruction to construct a $(C^0$, Lip)-equivalence comes from the movement of the invariant manifolds of the saddle point.

Acknowledgments

Both authors thank the IMPA of Rio de Janeiro where this work was initiated. They thank the Université de Bourgogne (Dijon) and the L.U.C. (Diepenbeek) where this work was prepared and completed. The NFWO (Belgium) is thanked for its financial aid.

1 C^1-linearisation of the Poincaré map of a saddle loop (Theorem 3)

In this paragraph, we want to prove theorem 3 of the introduction.
We begin with a C^∞ transversal σ with regular parameter x, where $x > 0$ corresponds to the inside part of Γ (the part where the Poincaré map $P(x)$ is defined). We want to find a C^1 change of coordinate on σ in order to write $P(x)$ as : $x \to x^\gamma$. More precisely we will prove :

Theorem 1.1.
Under the hypothesis of theorem 3, there exists a C^1 diffeomorphism U of some interval $[0, c]$ in σ, with $U'(0) = 1$, such that $U^{-1}PU(x) = x^\gamma$.

In what follows, we suppose that $\gamma < 1$. The proof uses the existence of an asymptotic expansion of the transition map near the saddle point, as obtained by Dulac in [D]. Let (x, y) be local coordinates near the saddle point s, such that Ox and Oy are respectively the local unstable and stable manifold ($s = (0,0)$). We suppose that this system of coordinates contains the rectangle $\{|x| \leq 2, |y| \leq 2\}$. Let D be the transition map of X from $\sigma_0 = \{y = 1, x \geq 0\}$ to $\tau_0 = \{x = 1\}$ (The inside part of Γ is supposed to correspond to $\{x \geq 0, y \geq 0\}$). The asymptotic expansion of Dulac for D is a formal series :

$$\tilde{D}(x) = \sum_{\substack{n, m \in \mathbb{N} \\ n \geq 1}} x^{n\gamma+m} P_{n,m}(\log x), \quad P_{n,m} \text{ a polynomial of degree } \leq n\gamma + m$$

which gives a C^∞-approximation of D in the sense that for any K :

$$D(x) - \sum_{n\gamma+m \leq K+1} x^{n\gamma+m} P_{n,m}(\log x) \quad \text{is a } C^K, K - \text{flat function}$$

The function D is C^∞ outside 0. The successive derivatives of D also admit a Dulac expansion which is obtained from the D-expansion by termswise derivation. For what follows we need some precisation about the first terms of the D-expansion. Because they are not easily accessible in the Dulac proof we calculate them below, using the Dulac method :

Lemma 1.2.
$D(x) = x^\gamma(1+\Delta(x))$ with :

$$\Delta(x) = O(x|\log x|),\ \Delta'(x) = O(|\log x|),\ \Delta''(x) = O(\frac{1}{x})$$

Proof

As said above, D and hence also Δ may be expanded in Dulac series. So the different estimates on $\Delta, \Delta', \Delta''$ follow from the first one on Δ by a termswise derivation. As it follows easily from properties of Dulac expansions we just need to prove that :

$$D(x) = x^\gamma(1 + \alpha x \log x + o(x \log x)) \text{ for some } \alpha \in \mathbf{R}$$

We establish this, by means of a C^∞ normal form for X : if γ is irrational, then X is C^∞ linearisable and the Dulac series $\tilde{D}$ reduces to $\tilde{D}(x) = x^\gamma$. In this case $\alpha = 0$. Suppose now that $\gamma = p/q \in \mathbf{Q}$. Then, X has a C^∞ normal form in terms of the resonant monomial $u = x^p y^q$. This normal form is equal to : $X = x\frac{\partial}{\partial x} - \gamma(1+f(u))y\frac{\partial}{\partial y}$ up to a C^∞ equivalence (here f is a C^∞ function, such that $f(0) = 0$, and $j^\infty f(0) \neq 0$ if $f \neq 0$). Following the method of Dulac we look for a first integral of the dual 1-form :

$$\Omega = xdy + \gamma(1+f(u))ydx.$$

We will find the first integral $I(x,y)$ for Ω by resolving the equation $\Omega = 0$. It is equivalent to :

$$\frac{dy}{y} + \gamma(1+f(u))\frac{dx}{x} = 0 \tag{1}$$

Using $\frac{du}{u} = p\frac{dx}{x} + q\frac{dy}{y}$, (1) gets :

$$\frac{1}{pf(u)u}du + \frac{dx}{x} = 0 \tag{2}$$

If $j^\infty f(x)$ is equal to zero, X is again linearisable. So we suppose that $j^\infty f(0) \neq 0$.

We can write : $puf(u) = \alpha u^k + \ldots$ for some $k \geq 2$ and $\alpha \neq 0$.
From (2), we obtain a first integral I :

$$I(x,y) = -\frac{1}{\alpha(k-1)u^{k-1}}(1+\psi(u)) + \log x + c \log u$$

(where c is some constant and ψ a C^∞ function).
Now, the transition map $y = D(x)$ is obtained from the equation :

$$I(x,1) = I(1,y) \tag{3}$$

$$\begin{aligned} I(x,1) &= -\frac{1}{\alpha(k-1)x^{(k-1)p}}(1+\psi(x^p)) + \log x + c\log x^p \\ I(1,y) &= -\frac{1}{\alpha(k-1)y^{(k-1)q}}(1+\psi(y^q)) + c\log y^q \end{aligned}$$

From (3) it easily follows that $y = D(x) \sim x^\gamma$. To look for the second term of $D(x)$, we put $y = x^\gamma(1+\Delta(x))$ in (3) :

$$\begin{aligned} &-\frac{1}{\alpha(k-1)x^{(k-1)p}}(1+\psi(x^p)) + \log x = \\ &\quad -\frac{1}{\alpha(k-1)x^{(k-1)p}}\frac{(1+\psi(x^p(1+\Delta)^q))}{(1+\Delta)^{(k-1)q}} + c\log(1+\Delta)^q \end{aligned} \tag{4}$$

From (4) we obtain $\Delta(x)$ in an implicit way :

$$1+\psi(x^p) - \frac{1+\psi(x^p(1+\Delta)^q)}{(1+\Delta)^{(k-1)q}} = (k-1)\alpha x^{(k-1)p}[\log x + c\log(1+\Delta)^q]$$

This implicit equation is at least C^{k-1} in (x,Δ) (with $k \geq 2$), and can be written as :

$$\frac{1}{\alpha(k-1)x^{(k-1)p}}.((k-1)q\Delta + \ldots) = \log x + \ldots$$

$$\text{or} \quad \Delta(x) = \frac{\alpha}{q}x^{(k-1)p}\log x + \ldots \tag{5}$$

where $+\ldots$ stands for terms of higher order.

From this it follows that $\Delta(x) = O(x|\log x|)$ in any case ($\Delta \sim x|\log x|$ if $k = 2$ and $p = 1$; if not $\Delta = o(x|\log x|)$).

We now return to the proof of theorem 1.1. Of course it suffices to prove the theorem for any σ. So we prove it for the segment τ_0 used to define the transition map D. Recall that D is defined from σ_0 to a second transversal τ_0 of the local unstable manifold. Call now $R(x)$ the transition map along the regular part of Γ from $\sigma = \tau_0$

to σ_0. (We now denote by x the parameter on $\sigma = \tau_0$).

$$R(x) = ax(1+\varphi(x)) \text{ for some } a > 0 \text{ and a } C^\infty\varphi.$$

The Poincaré map relative to σ is the composition :

$$P(x) = D \circ R(x) = R(x)^\gamma(1+\Delta(R(x))) \tag{6}$$

$R(x)^\gamma = a^\gamma x^\gamma(1+\varphi(x))^\gamma$ and $\overline{\Delta}(x) = \Delta \circ R(x)$ has the same expansion as Δ. So from the Lemma 1.2. it follows that :

$$\overline{\Delta}(x) = O(x|\log x|), \overline{\Delta}'(x) = O(|\log x|), \overline{\Delta}''(x) = O\left(\frac{1}{x}\right) \tag{7}$$

Using this in (6) we finally obtain :

$$P(x) = a^\gamma x^\gamma(1+\psi(x)) \quad (7)$$

with $\psi(x) = O(x|\log x|), \psi'(x) = O(|\log x|), \psi''(x) = O\left(\frac{1}{x}\right)$.
Now, changing the coordinate x by the linear scaling $x \to a^{\gamma/1+\gamma}x$ we can eliminate the term a^γ.
So, at this point, we have proved that on any transverse segment σ, we can choose a C^∞ parameter x such that the Poincaré map P(x) has the expansion :

$$P(x) = x^\gamma(1+\psi(x))$$

with ψ of class C^∞ for $x > 0$ and : (8)

$$\psi(x) = O(x|\log x|)\ ,\ \psi'(x) = O(|\log x|), \psi''(x) = O\left(\frac{1}{x}\right)$$

Let us now look for a diffeomorphism $U(x) = x(1+u(x))$ satisfying $U^{-1}PU(x) = x^\gamma$. Write $P_0(x) = x^\gamma$. Such a U is a fixed point of the map : $U \to P \circ U \circ P_0^{-1}$.
It is easy to see that necessarily $u(x) = O(x^{1/\gamma})$. Since we want U to be C^1, it would be natural to look for u of class C^1 and verifying $u'(x) = O(x^\delta)$ with $\delta = \frac{1}{\gamma} - 1$.
But for reasons which will become clear later in the proof, we weaken somewhat this condition. Take any $\delta' : 0 < \delta' < \delta$, and consider the following functional space :

$$\mathcal{E}_c : \{u : [0,c] \to \mathbf{R} | u\ C^1, u(0) = 0 \text{ and } u'(x) = O(x^{\delta'})\}$$

On $\mathcal{E}_c$, we introduce the following norm :

$$\|u\|_1 = \sup_{x\in[0,c]} \left|\frac{u'(x)}{x^{\delta'}}\right|$$

Note that $\mathcal{E}_c$ with this norm is a Banach space. We define an operator T on $\mathcal{E}_c$ by the condition :

$$x(1+Tu(x)) = P \circ U \circ P_0^{-1}(x) \text{ when } U(x) = x(1+u(x)) \tag{9}$$

Clearly, there exists a $c_0 > 0$ such that if $c \leq c_0, Tu$ is defined on $[0,c]$ for $u \in \mathcal{E}_c$. The fixed point problem on U is replaced by a fixed point problem for T. Using the formula (8) we have :

$$Tu(x) = [1+u(x^{1/\gamma})]^\gamma[1+\psi(x^{1/\gamma}(1+u(x^{1/\gamma}))] - 1 \tag{10}$$

Theorem 1.1 will be a consequence of the following proposition :

Proposition 1.3
Let any $c \leq c_0$. The operator T defined above acts on $\mathcal{E}_c$. Let $B_c(K) = \{v \in \mathcal{E}_c \mid \|v\|_1 \leq K\}$. Then for any $K > 0$ and for any c sufficiently small (depending on K), T is a Lipschitz-contraction on $B_c(K)$, with a Lipschitz constant going to zero when $c \to 0$.

Proof
Let $v = Tu$. We calculate $v' = \dfrac{dv}{dx}$:

$$\begin{aligned} v'(x) &= \gamma[1+u(x^{1/\gamma})]^{\gamma-1}\frac{1}{\gamma}.u'(x^{1/\gamma})x^\delta[1+\psi(x^{1/\gamma}(1+u(x^{1/\gamma})))] \\ &\quad +[1+u(x^{1/\gamma})]^\gamma\psi'(x^{1/\gamma}(1+u(x^{1/\gamma})))[\frac{1}{\gamma}x^\delta(1+u(x^{1/\gamma}))+\frac{1}{\gamma}x^{1/\gamma}u'(x^{1/\gamma})x^\delta] \end{aligned} \tag{11}$$

Let $\alpha = \delta - \delta' > 0$. From (11) it follows :

$$\begin{aligned} \frac{v'(x)}{x^{\delta'}} &= [1+u(x^{1/\gamma})]^{\gamma-1}[1+\psi(x^{1/\gamma}(1+u(x^{1/\gamma})))]u'(x^{1/\gamma})x^\alpha \\ &\quad \frac{1}{\gamma}[1+u(x^{1/\gamma})]^\gamma[1+u(x^{1/\gamma})+x^{1/\gamma}u'(x^{1/\gamma})]\psi'(x^{1/\gamma}(1+u(x^{1/\gamma})))x^\alpha \end{aligned} \tag{12}$$

Note that $|u(x)| \leq \int_0^x |u'(\zeta)|d\zeta \leq \|u\|_1 \int_0^x \zeta^{\delta'}d\zeta$,
so that :

$$|u(x)| \leq \frac{1}{\delta'+1}\|u\|_1 x^{\delta'+1} \tag{13}$$

Next, using estimates (8) we have for some constant M and for $y = x^{1/\gamma}\left(1 + \frac{\gamma}{\delta'+1}\|u\|_1 x^{(\delta'+1)1/\gamma}\right)$:

$$\left.\begin{array}{l} |\psi(x^{1/\gamma}(1+u(x^{1/\gamma}))| \leq My|\log y| \\ |\psi(x^{1/\gamma}(1+u(x^{1/\gamma}))| \leq M|\log y| \end{array}\right\} \tag{14}$$

These estimates (13), (14) together with $|u'(x^{1/\gamma})| \leq \|u\|_1 x^{\delta'/\gamma}$, imply that for a fixed u : $\frac{v'(x)}{x^{\delta'}} = O(x^\alpha|\log x|)$ and so that $v \in \mathcal{E}_c$. Let us now fix some $K > 0$ and take $u \in B_c(K)$.
In the estimates on $|u(x)|$ and $|u'(x)|$ we can take $\|u\|_1 \leq K$. We deduce from (12) that $\|v\|_1 \leq O(c^\alpha|\log c|)$ (in a way independent on u, if $\|u\|_1 \leq K$). So, we find a $c_1, 0 < c_1 \leq c_0$, such that for any $c \leq c_1, T(B_c(K)) \subset B_c(K)$.
Again with the same fixed value of K, and $c \leq c_1$, we show that T is a Lipschitz map with a constant $L(c) \to 0$ when $c \to 0$. The result will be established when $0 < c \leq c_2$ for some new $c_2 : 0 < c_2 \leq c_1$.

To prove this Lipschitz property take $u_1, u_2 \in B_c(K)$ and let $v_1 = Tu_1, v_2 = Tu_2$. We need to calculate : $\dfrac{v_1'(x)}{x^{\delta'}} - \dfrac{v_2'(x)}{x^{\delta'}}$.
Therefore, consider the expression (12) in function of the independent variables (x, U, U') with $U = u(x^{1/\gamma})$ and $U' = u'(x^{1/\gamma})$: $\dfrac{v'(x)}{x^{\delta'}} = F(x, u(x^{1/\gamma}), u'(x^{1/\gamma}))$ with

$$\begin{aligned} F(x,U,U') &= (1+U)^{\gamma-1}(1+\psi(x^{1/\gamma}(1+U))U'x^\alpha \\ &\quad + \frac{x^\alpha}{\gamma}(1+U)^\gamma(1+U+x^{1/\gamma}U')\psi'(x^{1/\gamma}(1+U)) \end{aligned} \tag{15}$$

Now :
$\dfrac{v_2'(x)}{x^{\delta'}} - \dfrac{v_1'(x)}{x^{\delta'}} = F(x, U_2, U_2') - F(x, U_1, U_1')$ may be estimated by the mean value theorem. There exists $U \in [U_1, U_2], U' \in [U_1', U_2']$ such that :

$$\frac{v_2'(x)}{x^{\delta'}} - \frac{v_1'(x)}{x^{\delta'}} = \frac{\partial F}{\partial U}(x,U,U')(U_2 - U_1) + \frac{\partial F}{\partial U'}(x,U,U')(U_2' - U_1') \tag{16}$$

We know that :

$$\begin{array}{c} |U_2 - U_1| = |u_2(x^{1/\gamma}) - u_1(x^{1/\gamma})| \leq \frac{1}{\delta'+1}\|u_2 - u_1\|_1 x^{(1+\delta')/\gamma} \\ \text{and } |U_2' - U_1'| \leq \|u_2 - u_1\|_1 x^{\delta'/\gamma} \end{array} \tag{17}$$

So, it suffices to prove that $\frac{\partial F}{\partial U}(x, U, U')$ and $\frac{\partial F}{\partial U'}(x, U, U')$, with (U, U') as above, are uniformly bounded for $u_1, u_2 \in B_c(K), c \leq c_1$. Now, for $i = 1, 2$:

$$\begin{aligned} |U_i| &\leq \frac{1}{\delta'+1}\|u_i\|_1 x^{(1+\delta')\gamma} \leq \frac{K}{\delta'+1}x^{(1+\delta')/\gamma}, \text{ and hence} \\ |U| &\leq \frac{K}{\delta'+1}x^{(1+\delta')/\gamma}, \ |U'| \leq Kx^{\delta'/\gamma} \end{aligned} \tag{18}$$

The boundedness of $\frac{\partial F}{\partial U}$ and $\frac{\partial F}{\partial U'}$ follow from (18) and the estimates (8) on ψ, ψ', ψ'' which enter in the computation of $\frac{\partial F}{\partial U}, \frac{\partial F}{\partial U'}$.
For instance from :
$\psi''(x^{1/\gamma}(1+U)) = O(x^{-1/\gamma})$ follows that :
$\frac{\partial}{\partial U}[\psi''(x^{1/\gamma}(1+U))] = O(1)$.
In all the cases the term x^{α} can be used to overcome the logarithmic contributions given by ψ'.
Finally, we obtain :

$$\|v_2 - v_1\|_1 \leq L(c)\|u_1 - u_2\|_1$$

with a constant $L(c) = O(c^{\delta'/\gamma})$ for a fixed K. This achieves the proof of Proposition 1.3.

Remark 1 : Theorem 3 remains correct for X of class C^r with r sufficiently big.
Remark 2 : The use of lemma 1.2 can be avoided for vector fields X which are C^2-linearisable at the saddle point.
Remark 3 : It is clear from the proof that the C^1-"linearization" depends in a continuous way (for the C^1-topology) on extra parameters, if the vector field does (for the C^r-topology, with r sufficiently big).

2 One parameter family of Poincaré maps

2.1 A simple form for the return maps

Let X_μ represent a generic saddle loop bifurcation at $\mu = 0$. Like in the previous section we suppose that the saddle point has a fixed position s and that $\gamma(0) = \gamma < 1$ where $\gamma(\mu)$ is the ratio $\dfrac{\lambda_1(\mu)}{\lambda_2(\mu)}$, with $-\lambda_1(\mu)$ and $\lambda_2(\mu)$ the negative and positive eigenvalues of X_μ at s.
Also, as in introduction, we call $\alpha(\mu)$ the shift function of the invariant manifolds, calculated on some transversal σ with respect to the stable manifold and with a positive orientation toward the inside of the saddle loop Γ. We assume $\alpha'(0) = \dfrac{d\alpha}{d\mu}(0) > 0$. Let x denote any C^∞ parameter on σ.
To calculate the return map $P_\mu(x)$ on σ we can replace σ by any other transversal. Like in paragraph 1 we introduce local coordinates x, y at s such that the axes $0x, 0y$ are respectively the local invariant unstable and stable manifold, but now **for any** μ, and we take $\sigma = \sigma_\Gamma \subset \{y = 1\}$. ($\sigma_0, \mu_0$ defined as in paragraph 1).
If the 1-jet of X_μ at $s = 0$ is equal to : $j^1X_\mu(0) = x\dfrac{\partial}{\partial x} - \gamma(\mu)y\dfrac{\partial}{\partial y}$, it is trivial to

prove that the transition map $D_\mu(x)$ from σ_0 to μ_0 has the form :

$$D_\mu(x) = x^{\gamma(\mu)}[1 + \phi_1(x,\mu)] \tag{19}$$

where ϕ_1 is a continuous function and $\phi_1(0,\mu) \equiv 0$.
The transition map $R_\mu(x)$ from τ_0 to σ_0 is C^∞ and can be expanded :

$$R_\mu(y) = \alpha(\mu) + \beta(\mu)y[1 + \psi(y,\mu)] \tag{20}$$

where ψ, α, β are C^∞, $\beta(\mu) > 0$ and $\phi(y,\mu) = O(y)$.
The Poincaré map relative to $\sigma = \sigma_0$ is equal to $R_\mu \circ D_\mu$.
From (19), (20) we obtain :

$$P_\mu(x) = \alpha(\mu) + \beta(\mu)x^{\gamma(\mu)}(1 + \phi_1(x,\mu))(1 + \psi(x^\gamma(\mu)(1 + \phi_1(x,\mu)),\mu))$$
$$\text{So :} P_\mu(x) = \alpha(\mu) + \beta(\mu)x^{\gamma(\mu)}(1 + \phi_2(x,\mu)) \tag{21}$$

with again ϕ_2 a continuous function such that $\phi_2(0,\mu) \equiv 0$.

We can reduce $\beta(\mu)$ to the value 1 by the following μ-dependent C^∞ change of coordinates on $\sigma : x \to x.\beta^{\frac{1}{1-\gamma}}$. In this new coordinate :

$$P_\mu(x) = \alpha(\mu)(\beta(\mu))^{-\frac{1}{1-\gamma(\mu)}} + x^{\gamma(\mu)}(1 + \phi(x,\mu)) \tag{22}$$
$$\text{where } \phi(x,\mu) = \phi_2(\beta^{\frac{1}{1-\gamma}}x,\mu)$$

We still have $\phi(x,\mu)$ continuous and $\phi(0,\mu) \equiv 0$, and the new shift function is $\alpha\beta^{-\frac{1}{1-\gamma}}$, having the same properties as α.
So, without restriction, we can suppose that $P_\mu(x)$ has the form (21) with $\beta \equiv 1$. Write ϕ for ϕ_2.
We now use the result of paragraph 1 : there exists a C^1 diffeomorphism $U(x)$ of σ, with $U'(0) = 1$, such that

$$U^{-1}P_0U(x) = x^\gamma \tag{23}$$

With U we introduce a new x-coordinate along σ; it is just a C^1-coordinate. In the new coordinate x, we obtain :

$$\widetilde{P}_\mu(x) = U^{-1}P_\mu U(x) = U^{-1}[\alpha(\mu) + U(x)^{\gamma(\mu)}(1 + \phi(U(x),\mu))] \tag{24}$$

We develop this formula, using the mean value theorem for $V = U^{-1}$:
$U^{-1}(\alpha(\mu) + U^{\gamma(\mu)}(1 + \phi(U,\mu))) = U^{-1}\alpha(\mu) + K(x,\mu)U(x)^{\gamma(\mu)}(1 + \phi(U,\mu))$ where

$K(x,\mu) = \int_0^1 \frac{\partial V}{\partial x}(\alpha(\mu) + tU^{\gamma(\mu)}(1 + \phi(U,\mu)))dt.$

The function K is continuous, $k_0(\mu) = K(0,\mu) = \frac{\partial V}{\partial x}(\alpha(\mu))$ is differentiable at 0 with $k_0'(0) \neq 0$. Moreover, $\frac{\partial V}{\partial x}(0) = 1$ implies that $K(0,0) = 1$. We can write $K(x,\mu) = (1 + k(x,\mu)).k_0(\mu)$ with $k(0,\mu) \equiv 0$ and $k_0(0) = 1$. Using this and the μ-dependent coordinate change $x \to (k_0(\mu))^{\frac{1}{1-\gamma(\mu)}}x$.

$$\begin{aligned} \widetilde{P}_\mu(x) &= (k_0(\mu))^{-\frac{1}{1-\gamma(\mu)}}U^{-1}\alpha(\mu) + (1 + k(x,\mu)x^\gamma(1 + u(x))^\gamma(1 + \phi(U,x)) \\ &= (k_0(\mu)^{-\frac{1}{1-\gamma(\mu)}}U^{-1}\alpha(\mu) + x^\gamma(1 + \widetilde{\phi}(x,\mu)) \end{aligned}$$

We again have : $\widetilde{\phi}$ continuous and $\widetilde{\phi}(0,\mu) \equiv 0$. (Since $k(0,\mu) \equiv u(0) \equiv \phi(U(0,\mu),\mu) \equiv 0$). But now, for $\mu = 0, U^{-1}P_0U = \widetilde{P}_0(x) = x^\gamma$.

So, we have : $\widetilde{\phi}(x,0) \equiv 0$.

Finally, we have proved :

Proposition 2.1.
Let X_μ be a generic C^∞ homoclinic saddle loop bifurcation as above and σ a segment transverse to Γ. Then there exists a C^1 μ-dependent choice of the coordinate x on σ such that the return map $P_\mu(x)$ has the following form :

$$P_\mu(x) = \widetilde{\alpha}(\mu) + x^{\gamma(\mu)}(1 + \widetilde{\phi}(x,\mu)) \tag{25}$$

where $\widetilde{\alpha}(0) = 0$, $\widetilde{\alpha}$ is differentiable at 0, $\widetilde{\alpha}'(0) > 0$ and $\widetilde{\phi}$ is a continuous function such that $\widetilde{\phi}(x,0) \equiv \widetilde{\phi}(0,\mu) \equiv 0$.

2.2 Reduction of the theorems 1, 2 to propositions relative to Poincaré maps

Let X_μ, Y_μ be two generic families and $f_\mu(x), g_\mu(x)$ the associated Poincaré maps relative to transverse sections σ_1, σ_2 with C^1-coordinate x as obtained in the proposition 2.1. :

$$\begin{aligned} f_\mu(x) &= \alpha_1(\mu) + x^{\gamma_1(\mu)}(1 + \phi_1(x,\mu)) \\ \text{and } g_\mu(x) &= \alpha_2(\mu) + x^{\gamma_2(\mu)}(1 + \phi_2(x,\mu)) \end{aligned} \tag{26}$$

where $\alpha_i, \gamma_i, \phi_i$ have the properties given in proposition 2.1.

Now if X_μ, Y_μ are (C^0, Lip)-equivalent by means of $(H_\mu(x), \varphi(x))$, then the families

f_μ and g_μ are seen to be (C^0, Lip)-conjugate as follows : the conjugacy $h_\mu(x)$ between f_μ and $g_{\varphi(\mu)}$ is obtained by the composition of $H_\mu|\sigma_1$ from σ_1 to $\sigma_1' = H_\mu(\sigma_1)$ and the transition map from σ_1' to σ_2, induced by the flow of $Y_{\varphi(\mu)}$.
Now to say that f_μ and g_μ are (C^0, Lip)-conjugate by (h_μ, φ) is the same as to say that f_μ and $g_{\varphi(\mu)}$ are (C^0, Id)-conjugate. Moreover, up to the mapping $\alpha_1(\mu)$, we can replace μ by $\tilde{\mu} = \alpha_1(\mu)$.
Let $f_{\tilde{\mu}} = \tilde{\mu} + x^{\delta(\tilde{\mu})}(1 + h(x,\tilde{\mu}))$ with $\delta(\tilde{\mu}) = \gamma_1(\alpha_1^{-1}(\tilde{\mu}))$, $h(x,\tilde{\mu}) = \phi_1(x, \alpha_1^{-1}(\tilde{\mu}))$ and $g_{\tilde{\mu}} = \alpha(\tilde{\mu}) + x^{\overline{\delta}(\tilde{\mu})}(1 + \overline{h}(x,\tilde{\mu}))$ with $\alpha(\tilde{\mu}) = \alpha_2 \circ \varphi \circ \alpha_1^{-1}(\tilde{\mu})$, $\overline{\delta}(\tilde{\mu}) = \gamma_2 \circ \varphi \circ \alpha_1^{-1}(\tilde{\mu}), \overline{h}(x,\tilde{\mu}) = \phi_2(x, \varphi \circ \alpha_1^{-1}(\tilde{\mu}))$. The families f_μ and $g_{\varphi(\mu)}$ are (C^0, Id)-conjugate if and only if this property holds for $f_{\tilde{\mu}}, g_{\tilde{\mu}}$. Because α_i is differentiable at 0 with $\alpha_i'(0) \neq 0$ and φ is a lipeomorphism, there exist $A, B : 0 < A < B$ such that on the domain of definition $[0,\mu_1]$ we have : $A\mu \leq \alpha(\mu) \leq B\mu$. Finally we denote $\tilde{\mu}$ by μ, and we see that the theorem 1 is a consequence of the following theorem which we will prove in chapter 3.

Theorem 2.2
Take $f_\mu(x) = \mu + x^{\delta(\mu)}(1 + h(x,\mu))$ and $g_\mu(x) = \alpha(\mu) + x^{\overline{\delta}(\mu)}(1 + \overline{h}(x,\mu))$ both defined on $[0,x_1] \times [0,\mu_1]$. Suppose that there exist $0 < A < B$ with $A\mu \leq \alpha(\mu) \leq B\mu$ on $[0,\mu_1]$, that $\delta, \overline{\delta}$ are continuous at 0 with $\overline{\delta}(0) < \delta(0) < 1$, and that $h, \overline{h}$ are continuous at $(0,0)$. Then f_μ, g_μ are not (C^0, Id)-conjugate for $\mu \geq 0, x \geq 0$.

Take now two generic families X_μ, Y_μ as in theorem 2 with $\gamma(0) < 1$. Let f_μ, g_μ be the Poincaré maps associated to them as in proposition 2.1.
As it is well-known X_μ and Y_μ are (C^0, Id)-equivalent when f_μ and g_μ are (C^0, Id)-conjugate. For $\mu \leq 0$, these maps have a fixed point and it is relatively easy to establish the conjugacy (see [A.A.D]). So we can restrict to the side $\mu \geq 0$ and to families $f_\mu(x) = \alpha(\mu) + x^{\gamma(\mu)}(1 + \phi(x,\mu))$ like in the proposition 2.1., with $\gamma(0) = \gamma$ a fixed value $0 < \gamma < 1$. Among such families, the simplest one is :

$$F_\mu(x) = \mu + x^\gamma$$

In order to show that two families like f_μ are (C^0, Id)-conjugate, it suffices to prove that any family f_μ is (C^0, Id)-conjugate to F_μ, and the theorem 2 reduces to :

Theorem 2.3
Let $\alpha(\mu), \gamma(\mu)$ be differentiable at $0, \alpha(0) = 0, \alpha'(0) > 0, 0 < \gamma(0) = \gamma < 1$, let $\phi(x,\mu)$ be continuous on $[0,x_1] \times [0,\mu_0]$ such that $\phi(0,\mu) \equiv \phi(x,0) \equiv 0$. Then $f_\mu(x) = \alpha(\mu) + x^{\gamma(\mu)}(1 + \phi(x,\mu))$ is (C^0, Id)-conjugate to $F_\mu(x) = \mu + x^\gamma$.

3 The necessary condition

We prove here theorem 2.2 which implies theorem 1, as explained in paragraph 2. We use the notations introduced in §2.

Proof of theorem 2.2
Choose $x_1 \geq x_0 > 0,\ 1 > \gamma > \overline{\gamma} > 0$ such that for all $(x, \mu) \in]0, x_0] \times]0, \mu_0]$ we have :

$$x + x^{\delta(\mu)}(1 + h(x, \mu)) < x^{\gamma} \tag{27}$$

$$\text{and} \qquad x^{\overline{\gamma}} < x^{\overline{\delta}(\mu)}(1 + \overline{h}(x, \mu)) \tag{28}$$

By induction on $k \in \mathbb{N}$, we can prove that :

$$f_x^k(x) < x^{\gamma^k} \qquad \text{for } \forall x \in]0, x_0^{\gamma^{1-k}}] \tag{29}$$

Indeed, if one has the result for $k = l$, then :

$$\begin{aligned} f_x^{l+1}(x) &= f_x(f_x^l(x)) = x + (f_x^l(x))^{\delta(\mu)}(1 + h(f_x^l(x), \mu)) \\ &\underset{(1)}{<} f_x^l(x) + (f_x^l(x))^{\delta(\mu)}(1 + h(f_x^l(x), \mu)) \\ &\underset{(2)}{<} (f_x^l(x))^{\gamma} \underset{(3)}{<} (x^{\gamma^l})^{\gamma} = x^{\gamma^{l+1}} \end{aligned}$$

The inequality (1) is valid on $]0, 1[$, and (2) on $]0, f_x^{-l}(x_0)]$ which contains $]0, x_0^{\gamma^{-l}}]$, and (3) on $]0, x_0^{\gamma^{1-l}}]$.
We suppose to have :

$$\begin{cases} h_\mu \circ f_\mu = g_\mu \circ h_\mu \\ h_\mu(0) = 0 \end{cases} \tag{30}$$

on $[0, \mu_0] \times [0, x_0]$ and with $h(x, \mu) = h_\mu(x)$ continuous on $[0, \mu_0] \times [0, x_0]$.
Now for any $\mu \in [0, \mu_0]$:

$$h_\mu(\mu) = h_\mu(f_\mu(0)) = g_\mu(0) = \alpha(\mu) \tag{31}$$

By induction on $i \in \mathbf{N}$, this gives :

$$h_\mu(f^i_\mu(0)) = g^i_\mu(0) \tag{32}$$

For any x with $0 < x \leq x_0$, we choose :

$$\mu_1 > \mu_2 > \mu_3 > \ldots \qquad \text{with } f^i_{\mu_i}(\mu_i) = x \tag{33}$$

The existence of these μ_i follows from the fact that for all $i \in \mathbf{N}^*$, $f^i_\mu(\mu)$ is continuous in μ, with $f^i_0(0) = 0$ and $f^i_{\mu_0}(\mu_0) > \mu_0$. Moreover $f^{i+1}_\mu(\mu) > f^i_\mu(\mu)$ when $\mu > 0$. Because of the continuity of h; we need to have :

$$h_0(x) = \lim_{i\to\infty} h_{\mu_i}(f^i_{\mu_i}(\mu_i)) = \lim_{i\to\infty} g^i_{\mu_i}(\alpha(\mu_i)) \tag{34}$$

We will now contradict this fact by showing that :

$$\liminf_{i\to\infty} g^i_{\mu_i}(\alpha(\mu_i)) \geq 1 \tag{35}$$

Because of (27), we can find a strictly decreasing sequence

$$(\nu_i)_{i\in\mathbf{N}^*} \text{ with } \nu_i < \mu_i \text{ and } \nu_i^{\gamma'} = x \tag{36}$$

Because of (28) : $g_{\mu_i}(\alpha(\mu_i)) > (\alpha(\mu_i))^{\bar\gamma} \geq A^{\bar\gamma}\mu_i^{\bar\gamma} > A^{\bar\gamma}\nu_i^{\bar\gamma}$ and by induction :

$$g^i_{\mu_i}(\alpha(\mu_i)) > A^{\bar\gamma'}\nu_i^{\overline{\gamma}'} \tag{37}$$

Using (36) and (37) we find :

$$g^i_{\mu_i}(\alpha(\mu_i)) > A^{\bar\gamma'}\nu_i^{\bar\gamma'} = A^{\bar\gamma'}(\nu_i^{\gamma'})^{(\frac{\bar\gamma}{\gamma})'} = A^{\bar\gamma'}x^{(\frac{\bar\gamma}{\gamma})'} \tag{38}$$

As $\bar\gamma < 1$ and $\overline{\gamma}/\gamma < 1$, this last expression tends to 1 for $i \to \infty$, from which we obtain (35).

4 The sufficient condition

We prove here the theorem 2.3 which implies theorem 2 as explained in paragraph 2. The notations are those introduced in §2. We will decompose the proof in several steps.

4.1 Reduction to two fundamental cases

Proposition 4.1
The theorem 2.3 reduces to prove that the two following pairs of families are (C^0, Id)-conjugate : Pair A : $F_\mu(x) = \mu + x^\gamma$ and $f_\mu(x) = \mu + x^{\delta(\mu)}(1 + h(x,\mu))$ where $\delta(\mu) = \gamma + \varepsilon(\mu)$, ε is continuous, there exist $a < b$ such that $a\mu \le \varepsilon(\mu) \le b\mu$ for $\mu \in [0, \mu_1]$, $h(x,\mu)$ is continuous on $[0, x_1] \times [0, \mu_1]$ with $h(0,\mu) \equiv h(x,0) \equiv 0$.
Pair B : $F_\mu(x)$ and $\alpha(\mu) + x^\gamma$ where there exist $0 < A < B$ such that $A\mu \le \alpha(\mu) \le B\mu$.

Proof

Take any family $f_\mu(x) = \alpha(\mu) + x^{\gamma(\mu)}(1 + \phi(x,\mu))$ as in theorem 2.3. If we introduce the parameter $\widetilde{\mu} = \alpha(\mu)$, knowing that α is differentiable at $\mu = 0$ with $\alpha'(0) > 0$, the family f_μ becomes : $\widetilde{f}_{\widetilde{\mu}}(x) = \widetilde{\mu} + x^{\widetilde{\gamma}(\widetilde{\mu})}(1 + \widetilde{\phi})$
with $\widetilde{\gamma}(\widetilde{\mu}) = \gamma\alpha^{-1}(\widetilde{\mu}), \widetilde{\phi}(x,\widetilde{\mu}) = \phi(x, \alpha^{-1}(\widetilde{\mu}))$
Clearly, $\widetilde{\gamma}$ is differentiable at 0 and $\widetilde{\phi}(x,\widetilde{\mu})$ verifies $\widetilde{\phi}(0,\widetilde{\mu}) \equiv \widetilde{\phi}(x,0) \equiv 0$. We can choose domains $[0,\mu_0]$, $[0,x_0]$ to apply the "Case A" : $\widetilde{f}_{\widetilde{\mu}}$ is (C^0, Id)-conjugate to $\widetilde{\mu} + x^{\widetilde{\gamma}(0)} = \widetilde{\mu} + x^\gamma$.
If we return to the parameter μ : f_μ is (C^0, Id)-conjugate to $\alpha(\mu) + x^\gamma$.
Now, because α has the necessary properties, we can apply the "Case B" to $\alpha(\mu) + x^\gamma$ and F_μ. Finally we have that f_μ is (C^0, Id)-conjugate to F_μ.

4.2 The property of equal convergence

Definition
Let two families $f_\mu(x) = \alpha_1(\mu) + x^{\gamma_1(\mu)}(1 + \phi_1(x,\mu))$ and $g_\mu(x) = \alpha_2(\mu) + x^{\gamma_2(\mu)}(1 + \phi_2(x,\mu))$, with α_i, γ_i continuous on $[0,\mu_1]$, ϕ_i continuous on $[0,x_1] \times [0,\mu_1]$, $\phi_i(0,\mu) \equiv 0$ and $\alpha_i(0) = 0$. We say that this pair of families has the **property of equal convergence** (or **e.c.-property**) if and only if :
For any strictly decreasing sequence $(\mu_i)_{i \in \mathbb{N}^*} \to 0$ in $[0,\mu_1]$: $f^i_{\mu_i}(0) \to a \in]0, x_1]$ if and only if $g^i_{\mu_i}(0) \to a$.

Remark
The e.c.-property is equivalent to the similar property for any subsequences (i_j) of $\mathbb{N}$; i.e. : For any $(i_j) \to \infty$, increasing subsequence of $\mathbb{N}$, and any strictly decreasing

sequence $(\mu_{i_j})_j \to 0$ in $[0, \mu_1]$:

$$F_{i_j}(\mu_{i_j}) = f^{i_j}_{\mu_{i_j}}(0) \to a \in]0, x_1] \Leftrightarrow G_{i_j}(\mu_{i_j}) = g^{i_j}_{\mu_{i_j}}(0) \to a \in]0, x_1]$$

To see this, take sequences (i_j) and (μ_{i_j}) as above. It suffices to show that we can complete the sequence (μ_{i_j}) to a sequence $(\mu_i) \to 0$, strictly decreasing and such that $F_i(\mu_i) \to a$. Take any j with $i_{j+1} - i_j \geq 2$. Using the ordering between the F_i we have that $F_{i_j+1}(\mu_{i_j}) > F_{i_j}(\mu_{i_j})$ and $F_{i_j+1}(\mu_{i_{j+1}}) < F_{i_{j+1}}(\mu_{i_{j+1}})$.
So, from the continuity of $F_{i_j+1}(\mu)$ there exists a μ_{i_j+1} : $\mu_{i_{j+1}} < \mu_{i_j+1} < \mu_{i_j}$ such that $F_{i_j+1}(\mu_{i_j+1})$ lies between $F_{i_j}(\mu_{i_j})$ and $F_{i_{j+1}}(\mu_{i_{j+1}})$.
Continuing in the same way, we can complete the sequence with μ_{i_j+l} such that $F_{i_j+l}(\mu_{i_j+l})$ lies between $F_{i_j}(\mu_{i_j})$ and $F_{i_{j+1}}(\mu_{i_{j+1}})$, and consequently complete every gap $\{i_j, i_{j+1}\}$ in the same way.

It is easy to see that the existence of a $(C^0$,Id)-conjugacy by some $h_\mu(x)$ with $h_0(x) \equiv x$ implies the e.c.-property. Here we are interested in the converse :

Proposition 4.2
Let f_μ and g_μ be two families as above with $f_0(x) = g_0(x) = x^\gamma(1 + \phi(x))$ and $\phi(0) = 0$. Suppose that this pair (f_μ, g_μ) verifies the e.c.-property. Then there exists a continuous family of homeomorphisms $h_\mu(x) = h(x, \mu)$ on $[0, x_0] \times [0, \mu_0]$ (for some $0 < x_0 \leq x_1$, $0 \leq \mu_0 \leq \mu_1$), with $h_0(x) \equiv x$, representing a $(C^0$, Id)-conjugacy between f_μ and g_μ : for all $\mu \in [0, \mu_0]$, $g_\mu \circ h_\mu = h_\mu \circ f_\mu$.

Proof

We choose $\mu_0 > 0$ and $x_0 > 0$ small enough such that f_μ, g_μ are increasing homeomorphisms on $[0, x_0]$ for $\mu \in [0, \mu_0]$.
We restrict to the rectangle $T = [0, x_0] \times [0, \mu_0]$. For the required conjugacy $h_\mu(x)$ we must have for each $\mu \in [0, \mu_0]$ that $h_\mu(0) = 0$. And, by induction :

$$h_\mu(f^i_\mu(0)) = g^i_\mu(0) \text{ for all } i \in \mathbb{N}^* \tag{39}$$

Let $H(x, \mu) = (h_\mu(x), \mu)$ be the global homeomorphism on T. Let also $F_i(\mu) = f^i_\mu(0)$ and $G_i(\mu) = g^i_\mu(0)$. The condition (39) is equivalent to the following one. If $\mathcal{F} = \bigcup_{i=0}^{\infty}$ Graph $(F_i) \bigcap T$ and $\mathcal{G} = \bigcup_{i=0}^{\infty}$ Graph $(G_i) \bigcap T$, H must send $\mathcal{F}$ onto $\mathcal{G}$.

The restriction H_0 of H to $\mathcal{F}$ is given by : $H_0(F_i(\mu),\mu) = (G_i(\mu),\mu)$. The problem consists in extending H_0 from $\mathcal{F}$ to T. The e.c.-property will permit it. It is equivalent to the fact that the map H_0 can be extended continuously by taking the identity on $[0,x_0] \times \{0\}$.

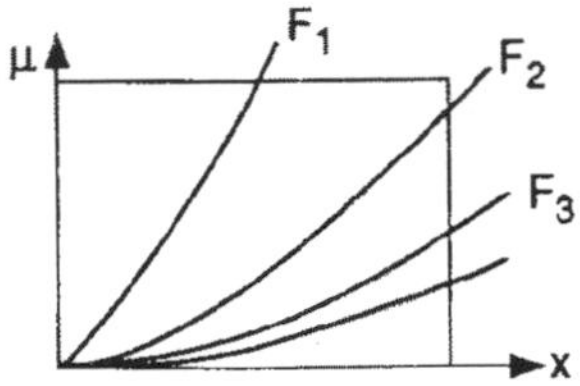

fig. 2

We now first extend H_0 on a fundamental domain D of f_μ. Take a, $0 < a < x_0$, such that $f_\mu(a) < x_0$ for $\forall \mu \in [0,\mu_0]$, and define D by $D = \{(x,\mu) | a \leq x \leq f_\mu(a)\}$. As $f_\mu(x)$ is increasing there exists a strictly decreasing sequence $(\mu_i)_i \to 0$ with $f^i_{\mu_i}(0) = F_i(\mu_i) = a$ (for i sufficiently big : $i \geq i_0 \geq 1$). The e.c.-property implies that $g^i_{\mu_i}(0) \to a$. Let $\sigma'(\mu)$ be a continuous function such that for $\forall i, \sigma'(\mu_i) = g^i_{\mu_i}(0) = G_i(\mu_i)$. The graph of σ' is a segment topologically transverse to $0x$ at the point $(a,0)$ and which contains all the points $(G_i(\mu_i),\mu_i)$.
A fundamental domain D' for g_μ is defined by :

$$D' = \{(x,\mu) | \sigma'(\mu) \leq x \leq g_\mu(\sigma'(\mu))\}$$

Now remark that for all $i \in \mathbb{N}$ and all μ, $F_i(\mu) < F_{i+1}(\mu)$ and $G_i(\mu) < G_{i+1}(\mu)$. So the graphs of the F_i cut D along segments l_i : the part of the graph over $[\mu_i, \mu_{i-1}]$. The same is true for the G_i and D'. The graphs of G_i cut D' along a segment l'_i. We want to define H from D onto D'. Necessarily, H must send each l_i onto l'_i. This means that for each $\mu > 0, \mu_i \leq \mu \leq \mu_{i-1}$ H must send :

- (a,μ) to $(\sigma'(\mu),\mu)$
- $(F_i(\mu),\mu) \in l_i$ to $(G_i(\mu),\mu) \in l'_i$
- $(f_\mu(a),\mu)$ to $(g_\mu(\sigma'(\mu)),\mu)$

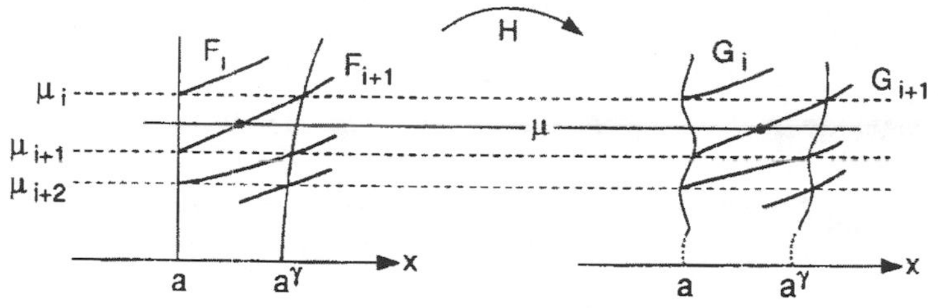

fig. 3

We extend H_0 in a unique way on the segment $[a, f_\mu(a)] \times \{\mu\}, \mu > 0$, by taking it piecewise linear. We now extend H on the complement of D by saturation : if $x = f_\mu^j(x_0)$ with $(x_0, \mu) \in D$ we take $H(x, \mu) = (g_\mu^j(h_\mu(x_0)), \mu)$. In this way we define H for all (x, μ) with $0 \leq x \leq x_0,\ 0 < \mu \leq \mu_0$.
Finally we take $H(x, 0) = (x, 0)$ for all $0 \leq x \leq x_0$.
It is clear that H is a conjugacy which is continuous for $\mu > 0$. We have to show the continuity at the points $(x, 0)$:
a)$x \in]a, f_0(a)[$
Let a sequence $(M_i, x_i) \to (0, x)$. We can assume that this sequence belongs to D. We want to prove that $h_{M_i}(x_i) \to x$.
Suppose that $h_{M_i}(x_i) \not\to x$. Since $h_{M_i}(x_i) \in [\sigma'(M_i), g_{M_i}(\sigma'(M_i))]$, $\sigma'(M_i) \to a$ and $g_{M_i}(\sigma'(M_i)) \to f_0(a) = g_0(a)$, we can find a subsequence $(M_{i_j})_j$ such that $h_{M_{i_j}}(x_{i_j}) \to y \neq x$ where $y \in [a, f_0(a)]$.
Now, for each i_j, there exists a k_j such that $M_{i_j} \in [\mu_{k_j}, \mu_{k_j - 1}]$ and
$x_{i_j} \in [a, F_{k_j}(M_{i_j})]$ or $x_{i_j} \in [F_{k_j}(M_{i_j}), f_{M_{i_j}}(a)]$
By extracting a new subsequence of $(i_j)_j$, again called i_j we can suppose, for instance, that for $\forall j, x_{i_j} \in [a, F_{k_j}(M_{i_j})]$ and also that $F_{k_j}(M_{i_j}) \to z \in [a, f_0(a)]$.
Let t_{i_j} be such that :

$$x_{i_j} = t_{i_j} a + (1 - t_{i_j}) F_{k_j}(M_{i_j}).$$

Because $x_{i_j} \to x$ and $F_{k_j}(M_{i_j}) \to z$ we have that $(t_{i_j}) \to t \in [0, 1]$ such that $x = ta + (1 - t)z$.
Now we have defined $h_{M_{i_j}}$ in a piecewise linear way :

$$h_{M_{i_j}}(x_{i_j}) = t_{i_j} h_{M_{i_j}}(a) + (1 - t_{i_j}) h_{M_{i_j}}(F_{k_j}(M_{i_j}))$$

Because $F_{k_j}(M_{i_j}) \to z$, we have, thanks to the e.c.-property, that :

$$h_{M_{i_j}}(F_{k_j}(M_{i_j})) = G_{k_j}(M_{i_j}) \to z.$$

So $h_{M_{i_j}}(x_{i_j}) \to ta + (1-t)z = x$
But this is in contradiction with $h_{M_{i_j}}(x_{i_j}) \to y \neq x$.
So we can conclude that H is continuous on $]a, t_0(a)[\times\{0\}$.
b)$x = a, f_0(a)$
If $(x_i, \mu_i) \to (a, 0)$ for instance, extracting subsequences we can assume that all the points belong to D or to $f^{-1}(D)$ where $f = (f_\mu, \mu)$. In the first case we have the same proof as in case a). In the second case ($(x_i, \mu_i) \in f^{-1}(D)$) we apply f to return to D, and prove a convergence towards $f_0(a)$, in the same way as we did in case a).
c)For $x > 0$
By iteration, using the conjugacy, we can send x to $[a, f_0(a)]$ and apply the preceeding steps.
d) $x = 0$
Let $(x_i, \mu_i)_i \in [0, x_1] \times [0, \mu_1]$ be a sequence converging to $(0,0)$. There exists a map $j(i) : \mathbb{N} \to \mathbb{N}$ (not necessarily monotone) such that :

$$x_i \in [F_{j(i)-1}(\mu_i), F_{j(i)}(\mu_i)].$$

Now, since $|F_{j(i)-1}(\mu_i) - F_{j(i)}(\mu_i)| \to 0$ with $i \to \infty$, $(x_i, \mu_i) \to (0,0)$ is equivalent to $(F_{j(i)}(\mu_i), \mu_i) \to (0,0)$. On the other hand $H(x_i, \mu_i) \to 0$ if $(G_{j(i)}(\mu_i), \mu_i) \to 0$ because, by construction, if $H(x_i, \mu_i) = (y_i, \mu_i)$, we have :

$$y_i \leq G_{j(i)}(\mu_i).$$

So it suffices to prove that $G_{j(i)}(\mu_i) \to 0$. Suppose this is not the case. Then $j(i)$ is unbounded (otherwise it is trivial that $G_{j(i)}(\mu_i) = g_{\mu_i}^{j(i)}(0) \to 0$). So, we can extract a subsequence (i_k) with $j(i_k)$ strictly increasing such that :
$\mu_{i_k} \to 0$ in a strictly descreasing way and $G_{j(i_k)}(\mu_{i_k}) \to a \in]0, x_1]$.
But, applying the e.c.-property to this subsequence $(j(i_k))_k$, we obtain that $F_{j(i_k)}(\mu_{i_k}) \to a \neq 0$, which contradicts $F_{j(i)}(\mu_i) \to 0$.

4.3 The property of equal convergence for the pairs A,B

To prove the propostion 2.3, it suffices now to establish that the pairs A,B as introduced in 4.1 have the property of equal convergence.

4.3.1 Equal convergence of the pair A

It is the pair $F_\mu(x) = \mu + x^\gamma, \gamma < 1, f_\mu(x) = \mu + x^{\delta(\mu)}(1 + h(x,\mu))$ where $\delta(\mu) = \gamma + \varepsilon(\mu), a\mu \leq \varepsilon(\mu) \leq b\mu$ with $a < b$ for $\mu \in [0, \mu_1]$ and $h(x,\mu)$ continuous on $[0,x_1]\times[0,\mu_1]$ with $h(0,\mu)\equiv h(x,0)\equiv 0$.

Let $A(\mu)=\sup_{x\in[0,x_0]}|h(x,\mu)|$. This function is continuous on $[0,\mu_1]$ with $A(0)=0$. Take a sequence $\mu_1>\mu_2>\mu_3>\dots$ tending to zero such that $F^i_{\mu_i}(\mu_i)\to x$ for some $0<x\le\min(x_0,\mu_0)$. We need to show that $\lim_{i\to\infty} f^i_{\mu_i}(\mu_i)=x$. The way to do this will be to look at the ratio $f^i_{\mu_i}(\mu_i)/F^i_{\mu_i}(\mu_i)$, estimating it from above and from below. For $i=1$:

$$\frac{f_y(y)}{F_y(y)}=\frac{y+y^{\gamma+\varepsilon(y)}(1+h(y,y))}{y+y^\gamma}$$

$$\le\frac{y+y^{\gamma+ay}(1+A(y))}{y+y^\gamma}\underset{(+)}{\le}\frac{y^{\gamma+ay}(1+A(y))}{y^\gamma}=y^{ay}(1+A(y))$$

Inequality (+) comes from $\beta\ge\alpha>0\Rightarrow\frac{\gamma+\beta}{\gamma+\alpha}\le\frac{\beta}{\alpha}$.

Analogously : $\dfrac{f_y(y)}{F_y(y)}\ge y^{by}(1-A(y))$

By induction, we may now prove that :

$$\prod_{i=0}^{k-1}y^{b\gamma^i y}(1-A(y))^{\gamma^i}\le\frac{f_y^k(y)}{F_y^k(y)}\le\prod_{i=0}^{k-1}y^{a\gamma^i y}(1+A(y))^{\gamma^i}\qquad(40)$$

Indeed :

$$\begin{aligned}\frac{f_y^k(y)}{F_y^k(y)}&=\frac{y+(f_y^{k-1}(y))^{\gamma+\varepsilon(y)}(1+h(f_y^{k-1}(y),y))}{y+(F_y^{k-1}(y))^\gamma}\\&\le\frac{(f_y^{k-1}(y))^{\gamma+\varepsilon(y)}(1+h(f_y^{k-1}(y),y))}{(F_y^{k-1}(y))^\gamma}\\&\le\left(\frac{f_y^{k-1}(y)}{F_y^{k-1}(y)}\right)^\gamma(f_y^{k-1}(y))^{ay}(1+A(y))\\&\le\left(\prod_{i=0}^{k-2}y^{a\gamma^i y}(1+A(y))^{\gamma^i}\right)^\gamma y^{ay}(1+A(y))\\&\le\prod_{i=0}^{k-1}y^{a\gamma^i y}(1+A(y))^{\gamma^i}\end{aligned}$$

and similarly from below.

As $\prod_{i=0}^{k-1}y^{a\gamma^i y}(1+A(y))^{\gamma^i}=y^{ay\left(\sum_{i=0}^{k-1}\gamma^i\right)}(1+A(y))^{\left(\sum_{i=0}^{k-1}\gamma^i\right)}$.

(40) can be read as :

$$y^{by\frac{1-\gamma^k}{1-\gamma}}(1-A(y))^{\frac{1-\gamma^k}{1-\gamma}} \leq \frac{f_y^k(y)}{F_y^k(y)} \leq y^{ay\frac{1-\gamma^k}{1-\gamma}}(1+A(y))^{\frac{1-\gamma^k}{1-\gamma}}$$

which for $k \to \infty$ gives :

$$y^{\frac{b}{1-\gamma}y}(1-A(y))^{\frac{1}{1-\gamma}} \leq \frac{f_y^k(y)}{F_y^k(y)} \leq y^{\frac{a}{1-\gamma}y}(1+A(y))^{\frac{1}{1-\gamma}} \tag{41}$$

Now clearly both left and right hand side of (41) tend to 1 when $y = \mu_i \to 0$.

4.3.2 Equal convergence of the pair B

The pair B is the pair $F_\mu(x) = \mu + x^\gamma$ and $f_\mu(x) = \alpha(\mu) + x^\gamma$, $0 < \gamma < 1$, α continuous on $[0, \mu_1]$ and $B\mu \leq \alpha(\mu) \leq A\mu$ for some $A, B, 0 < B < 1 < A$.

First, we will show that :

$$BF_\mu^{k+1}(0) \leq f_\mu^{k+1}(0) \leq AF_\mu^{k+1}(0) \quad \text{for } \forall k \in \mathbb{N} \tag{42}$$

Let $G_k(\mu) = F_\mu^k(\mu) = F_\mu^{k+1}(0)$ and $G_0(\mu) = \mu$.
Then $F_\mu(\mu) = G_1(\mu) = \mu + \mu^\gamma$, $F_\mu^2(\mu) = G_2(\mu) = \mu + (\mu + \mu^\gamma)^\gamma, \ldots$
On the other hand :

$$f_\mu(0) = \alpha(\mu),\ f_\mu(\alpha(\mu)) = f_\mu^2(0) = \alpha(\mu) + \alpha(\mu)^\gamma = G_1(\alpha(\mu)), \ldots$$

So, (42) is equivalent to :

$$BG_k(\mu) \leq G_k(\alpha(\mu)) \leq AG_k(\mu) \tag{43}$$

We prove (43) by induction on k. For $k = 0$, we have : $B\mu \leq \alpha(\mu) \leq A\mu$.
Suppose (43) to be established up to $k - 1$:

$$BG_{k-1}(\mu) \leq G_{k-1}(\alpha(\mu)) \leq AG_{k-1}(\mu).$$

By definition we have :

$$G_k(\mu) = \mu + G_{k-1}(\mu)^\gamma \text{ and } G_k(\alpha) = \alpha + G_{k-1}(\alpha)^\gamma$$

So :

$$G_k(\alpha) \le \alpha + A^\gamma G_{k-1}(\mu)^\gamma \le A\mu + A^\gamma G_{k-1}(\mu)^\gamma.$$

Because $A > 1$ and $\gamma < 1$ we have $A^\gamma < A$ and hence :

$$G_k(\alpha) \le A(\mu + G_{k-1}(\mu)^\gamma) = AG_k(\mu).$$

Similarly

$$G_k(\alpha) = \alpha + G_{k-1}(\alpha)^\gamma \ge B\mu + B^\gamma G_{k-1}(\mu)^\gamma.$$

And hence :

$$G_k(\alpha) \ge B(\mu + G_{k-1}(\mu)^\gamma) = BG_k(\mu)$$

(Here : $B < 1$ and $\gamma < 1$ imply that $B^\gamma > B$).

Suppose now that $F_{\mu_i}^{i+1}(0) \to x, 0 < x \le x_0$, where $(\mu_i)_i$ is a decreasing sequence tending to 0.
From (42) we see that $\mathrm{Lim}(f_{\mu_i}^{i+1}(0))$, the set of limit points of $(f_{\mu_i}^{i+1}(0))_i$, for $i \to \infty$, satisfies :

$$\mathrm{Lim}(f_{\mu_i}^{i+1}(0)) \subset [Bx, Ax] \tag{44}$$

Note that this estimation is established for any x, $0 < x \le x_0$. Take some $x_1, 0 < x_1 \le x_0$. We want to show that the validity of (44) for any x and for any sequence (μ_i) as above, implies that if $F_{\mu_i}^{i+1}(0) \to x_1$ for some sequence $(\mu_i)_i$, that then also $f_{\mu_i}^{i+1}(0) \to x_1$ (i.e. the e.c.-property).
Suppose that $f_{\mu_i}^{i+1}(0) \not\to x_1$. Since $\mathrm{Lim}(f_{\mu_i}^{i+1}(0)) \subset [Bx_1, Ax_1]$, we can find some $r \ne 1$ with $r \in [B, A]$ and a subsequence $(\mu_{i_j})_j$ such that : $f_{\mu_{i_j}}^{i_j+1}(0) \to rx_1$. But this implies that : $f_{\mu_{i_j}}^{i_j+2}(0) = f_{\mu_{i_j}}(f_{\mu_{i_j}}^{i_j+1}(0)) \to f_0(rx_1) = r^\gamma x_1^\gamma$.
On the other hand : $(F_{\mu_{i_j}}^{i_j+2})_j \to x_1^\gamma$.
By induction, for any $k \in \mathbb{N}$: $(f_{\mu_{i_j}}^{i_j+k}(0))_j \to r^{\gamma^k} x_1^{\gamma^k}$ with $F_{\mu_{i_j}}^{i_j+k}(0))_j \to x_1^{\gamma^k}$.
As $\gamma < 1$ and $r \ne 1$, we can choose k such that :

$$r^{\gamma^k} \notin [B, A]$$

which contradicts (42).

5 (C^0, C^∞)-equivalence of Bogdanov-Takens bifurcations

A Bogdanov-Takens bifurcation is a generic 2-parameter unfolding of a germ X_0 with 2-jet at zero C^∞-equivalent to :

$$y\frac{\partial}{\partial x} + (x^2 \pm xy)\frac{\partial}{\partial y}$$

Of course, we can restrict to the repelling case (sign +). We briefly recall that Bogdanov in [B], and Takens partly in [T1], proved that any such generic 2-parameter unfolding is (fibre C^0, C^∞)-equivalent to the polynomial model :

$$X^N_{(\mu,\nu)} = y\frac{\partial}{\partial x} + (x^2 + \mu + y(\nu + x))\frac{\partial}{\partial y}$$

In fact, Bogdanov shows the existence of a C^∞ change in parameter space of the type $(\mu, \nu) \to (\mu, \overline{\nu}(\mu, \nu))$ sending the bifurcation diagram of any given generic unfolding onto the following set of 3 lines :

- 0ν : a line of saddle-node bifurcations
- a half-parabola $C = \{\mu = -\frac{49}{25}\nu^2 | \nu \geq 0\}$ which is a line of generic saddle loop bifurcations
- a half-parabola $H = \{\mu = -\nu^2 | \nu \geq 0\}$ which is a line of *Hopf* bifurcations.

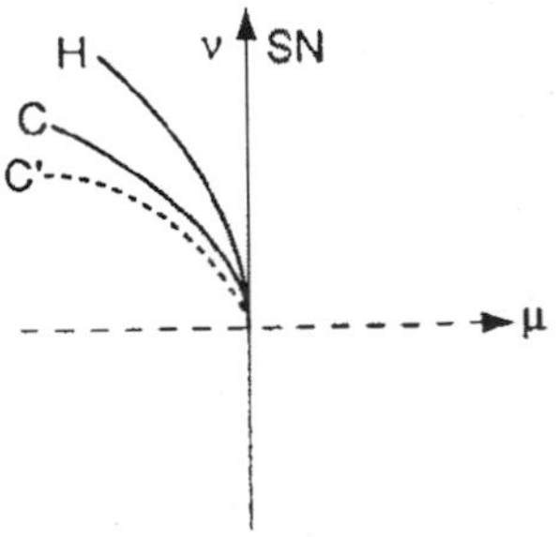

fig. 4

Next, in [A.A.D], it was proved that any two Bogdanov-Takens unfoldings are (C^0, C^0)-equivalent. More precisely, the authors managed to prove the (C^0, C^∞) equivalence for (μ, ν) outside a small cone CC' between C and any graph C' above C, as in figure 3. This cone can be chosen arbitrarily flat at 0. Inside this cone there are difficulties as explained in the introduction.
To overcome these, $[A.A.D]$ used a continuous change of parameters in CC'.
Here, we explain how to obtain a (C^0, C^∞) equivalence everywhere. For this, we need a 2-parameter version of proposition 4.2 :

Proposition 5.1
Suppose that f_μ, g_μ as in proposition 4.2 depend continuously on some extra parameter $\delta \in E$, with E some topological space : $f_{\mu,\delta}(x), g_{\mu,\delta}(x)$ are given by $\alpha_i(\mu,\delta)+x^{\gamma_i(\mu,\delta)}(1+\phi_i(x,\mu,\delta)), i = 1,2$ respectively, with α_i, γ_i continuous functions on $[0,\mu_1] \times E$ and ϕ_i continuous on $[0,x_1] \times [0,\mu_1] \times E$. For every $\delta \in E$ we suppose that $\gamma_1(0,\delta) = \gamma_2(0,\delta) = \gamma(\delta) < 1$ and that : $f_{0,\delta}(x) = g_{0,\delta}(x) = x^{\gamma(\delta)}(1+\phi(x,\delta))$ with $\phi(0,\delta) \equiv 0$. Suppose that for each δ, the pair $(f_{\mu,\delta}, g_{\mu,\delta})$ verifies the property of equal convergence. Then, there exist continuous functions $x_0(\delta), \mu_0(\delta)$ on E, with $0 < x_0(\delta) \leq x_1$ and $0 < \mu_0(\delta) \leq \mu_1$, and a continuous family of homeomorphisms $h_{\mu,\delta}(x)$, continuous on $\bigcup_{\delta\in E}[0,x_0(\delta)] \times [0,\mu_0(\delta)] \times \{\delta\}$, giving a ($C^0$, Id)-conjugacy between $f_{\mu,\delta}$ and $g_{\mu,\delta}$ on $[0,x_0(\delta)]$, for any $(\mu,\delta) \in \bigcup_{\delta\in E}[0,\mu_0(\delta)] \times \{\delta\}$.

Proof
If we return to the proof of 4.2, we can see that the construction of H is completely determined by the choice of a point $a \in [0,\mu_0]$. Now, we can choose in a continuous way $\mu_0(\delta), x_0(\delta)$ and next $a(\delta)$. The resulting construction of H is clearly continuously depending on δ. It gives as a family $H_\delta = (h_{\mu,\delta},\mu)$, which is continuous on $\bigcup_{\delta\in E}[0,x_0(\delta)] \times [0,\mu_0(\delta)] \times \{\delta\}$.

We now return to our problem. Take any repelling Bogdanov-Takens unfolding. As it was shown by Bogdanov, this unfolding has, up to a C^∞- equivalence, the following normal form :

$$X_\lambda = y\frac{\partial}{\partial x} + (x^2 + \mu + y(\nu + x + x^2h(x,\lambda)) + y^2Q(x,y,\lambda))\frac{\partial}{\partial y} \tag{45}$$

with $\lambda = (\mu,\nu), h, Q$ are C^∞ functions and $Q = O((|x|+|y|+|\lambda|)^N)$, N may be chosen arbitrarily large.

As recalled above, we can find a C^∞ change of parameters $\varphi(\mu,\nu) = (\mu,\overline{\nu}(\mu,\nu))$ sending the bifurcation lines on the triple $(C,H,0\nu)$. We still call (μ,ν) the new parameters. The form (45) is preserved with ν replaced by the function $\overline{\nu}(\mu,\nu)$ (the old parameter). In fact : $\frac{\partial\overline{\nu}}{\partial\nu}(0,0) = 1$.
We take ν as a regular C^∞ parameter along C, and we have to calculate the ratio-function $\gamma(\nu)$ along C.
The saddle point of X_λ, for $\mu < 0$ is $s_\lambda = (\sqrt{-\mu},0)$. Along C, its first coordinate is equal to : $x(\nu) = \sqrt{-\mu(\nu)} = \frac{7}{5}\nu$. (Recall that the equation of C is : $\mu = -\left(\frac{7}{5}\right)^2\nu^2$ for $\nu \geq 0$).
To compute the eigenvalues at the saddle point $(x(\nu),0)$ we change the coordinates

by : $x = X + \frac{7}{5}\nu, y = y$. The 1-jet of X_λ at this point is equal to :

$$y\frac{\partial}{\partial x} + \left(\frac{14}{5}\nu X + T(\nu)y\right)\frac{\partial}{\partial y} \tag{46}$$

where $T(\nu)$ is the trace at $(x(\nu), 0)$. It is a C^∞ function of ν, equal to :

$$\begin{aligned} T(\nu) &= \overline{\nu} + x(\nu) + x(\nu)^2 h(x(\nu), \mu(\nu), \nu) \\ T(\nu) &= \frac{12}{5}\nu + O(\nu^2) \end{aligned} \tag{47}$$

The characteristic polynomial at $(x(\nu), 0)$ is : $\Lambda^2 - T\Lambda - \frac{14}{5}\nu$.
Let $\Delta(\nu) = T^2 + \frac{56}{5}\nu = \frac{56}{5}\nu + 0(\nu^2)$ its discriminant.
The eigenvalues $-\lambda_1(\nu), \lambda_2(\nu)$ are equal to :

$$-\lambda_1(\nu) = \frac{T - \sqrt{\Delta}}{2} \quad \text{and} \quad \lambda_2 = \frac{T + \sqrt{\Delta}}{2} \tag{48}$$

So, the ratio of hyperbolicity $\gamma(\nu) = \frac{\sqrt{\Delta} - T}{\sqrt{\Delta} + T}$ has the following expression :

$$\gamma(\nu) = \frac{\sqrt{\nu}.A(\nu) - T(\nu)}{\sqrt{\nu}.A(\nu) + T(\nu)} \tag{49}$$

where $A(\nu)$ is a C^∞ function of ν, with $A(0) = \sqrt{\frac{56}{5}}$.

Now, take two different Bogdanov-Takens unfoldings, with the same bifurcation diagram $(C, H, 0\nu)$. Let $\gamma_i(\nu), \nu = 1,2$ be their respective ratio-functions along $C : \gamma_i(\nu) = \frac{\sqrt{\nu}A_i(\nu) - T_i(\nu)}{\sqrt{\nu}A_i(\nu) + T_i(\nu)}$, $i = 1,2$ where A_i, T_i are C^∞ and as above : $A_i(0) = \sqrt{\frac{56}{5}}$, $T_i(\nu) = \frac{12}{5}\nu + 0(\nu^2)$.

We are going to make a new C^∞ change in parameter space in order to ajust the ratio function of the first unfolding to the second one, without changing the bifurcation diagram. We use a transformation of the following type :

$$(\mu, \nu) \rightarrow \left(\left(\frac{\varphi(\nu)}{\nu}\right)^2 \mu, \varphi(\nu)\right) \tag{50}$$

where $\varphi(\nu)$ is a C^∞ diffeomorphism of the line, of the form $\varphi(\nu) = \nu + O(\nu^2)$.
The transformation (50) preserves each parabola $\{\mu = k\nu^2\}$ for any $k \in \mathbf{R}$, so

preserves the bifurcation diagram $(C, H, 0\nu)$. If we apply it to the first unfolding, the ratio-function $\gamma_1(\nu)$ is replaced by $\gamma_1 \circ \varphi(\nu)$. So to ajust this new ratio to γ_2, we have to solve the following equation in φ :

$$\gamma_1 \circ \varphi = \gamma_2 \tag{51}$$

The equation may be expressed as :

$$\frac{\sqrt{\varphi}A_1(\varphi) - T_1(\varphi)}{\sqrt{\varphi}A_1(\varphi) + T_1(\varphi)} = \frac{\sqrt{\nu}A_2(\nu) - T_2(\nu)}{\sqrt{\nu}A_2(\nu) + T_2(\nu)} \tag{52}$$

which reduces to :

$$\frac{1}{(A_1(\varphi))^2}\frac{(T_1(\varphi))^2}{\varphi} = \frac{1}{(A_2(\nu))^2}\cdot\frac{(T_2(\nu))^2}{\nu} \tag{53}$$

This equation is of the form :

$$F \circ \varphi(\nu) = G(\nu) \tag{54}$$

where F, G are C^∞ germs of diffeomorphism of the real line at 0, such that

$$F(\nu), G(\nu) = \frac{18}{35}\nu + 0(\nu^2)$$

As a solution φ we have the germ of diffeomorphism $\varphi(\nu) = F^{-1} \circ G(\nu)$.

At this point, we have two Bogdanov-Takens unfoldings X_λ, Y_λ with the same bifurcation diagram $(C, H, O\nu)$ and the same ratio-function $\gamma(\nu)$ along C. Using (49) we have :

$$\gamma(\nu)) = 1 - 2\frac{T(\nu)}{A(\nu)\sqrt{\nu}} + O(\nu) = 1 - \frac{6}{35}\sqrt{70}\sqrt{\nu} + O(\nu) \tag{55}$$

This expansion shows that $\gamma(\nu) < 1$ if $\nu > 0$.

To apply the proposition 5.1 in the cone CC' (but outside 0) we take $E =]0, \mu_0]$ for a small value $\mu_0 > 0$ and $\delta = \mu$. We choose some parameter $\bar{\mu}$ transverse to the line C such that for instance the segments $\{\delta = C^t\}$ are vertical : this new parameter $\bar{\mu}$ is some rescaling of $\nu - \frac{5}{7}\sqrt{-\mu}$ such that its domain of definition is $[0, \nu_1]$ for some $\nu_1 > 0$ and any $\delta \in E$. We also choose some family of C^∞ segments σ, transverse to the saddle connections Γ, with parameter $x \in [0, x_1]$ corresponding to the inside of Γ. (Of course in the phase space, the length of this segment goes to zero when $\delta \to 0$. So the new parameter x does not coincide with the x-coordinate in the phase space, but with some x-coordinate obtained by rescaling; see [B] or [D.R.S.] for instance for more details). These choices of $(\bar{\mu}, \delta)$ give a parametrization of the cone CC'.

Let $f_{\bar{\mu},\delta}(x), g_{\bar{\mu},\delta}(x)$ be the resulting Poincaré maps. Because of remark 3 at the end of §1, we may suppose (after taking a C^1-"linearization" which depend in a continuous way on δ) that $f_{0,\delta}(x) = g_{0,\delta}(x)$. Applying proposition 5.1 we obtain a (C^0, Id)-conjugacy $h_{\bar{\mu},\delta}$ of these 2 families and from it a (C^0, Id)-equivalence of the 2 unfoldings X_λ, Y_λ for $\lambda \in CC' \setminus \{(0,0)\}$, on the inner part of Γ.

Domain covered by Proposition 5.1

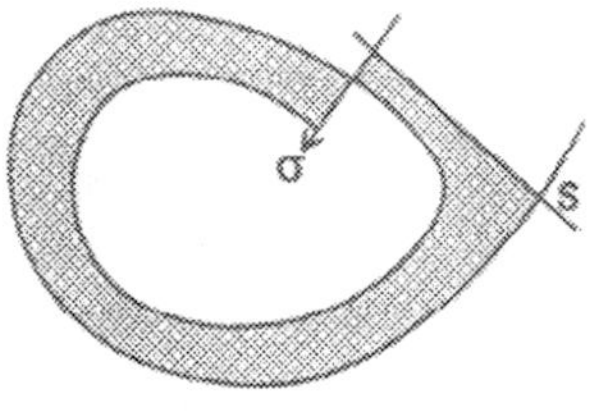

fig. 5

It was shown in [A.A.D.] that this equivalence may be extended in a (C^0, Id) way everywhere. Remark in particular that we do not know if $h_{\bar{\mu},\delta}$ has a limit for $\delta \to 0$. But the domain in phase space covered by this equivalence goes to zero and this observation allows to obtain the continuity of the equivalence at the point $(0,0)$ in parameter space. (See [A.A.D] for the details).

References

[A.L] A. Andronov, E. Leontovich, et al.
Theory of Bifurcations of Dynamical Systems on a Plane
I.P.S.T., Jerusalem, 1971.

[A.A.D.] H. Annabi, M.L. Annabi, F. Dumortier.
Continuous dependence on parameters in the Bogdanov-Takens bifurcation.
To appear in the proceedings of the workshop on Chaotic Dynamics and Bifurcations, Longman Research Notes.

[B.] R.I. Bogdanov.
Versal deformation of a singularity of a vector field on the Plane in the Case of Zero Eigenvalues
(R) Seminar Petrovski, 1976, (E) Selecta Mathematica Sovietica, Vol. 1, 4, 389-421, 1981.

[D.] M.H. Dulac.
Sur les cycles limites.
Bull. Soc. Math. France 51, 45-188, 1923.

[D.R.S.] F. Dumortier, R. Roussarie, J. Sotomayor.
Generic 3-parameter families of vector fields on the plane, unfolding a singularity with nilpotent linear part. The cusp case.
Erg. Theor. and Dyn. Sys. 7, 375-413, 1987.

[M.P.] I.P. Malta, J. Palis
Families of vector fields with finite modulus of stability.
Lecture Notes in Mathematics 898, Dyn. Systems and Turbulence, Warwick 1980, Springer-Verlag, 212-229, 1981.

[T1] F. Takens.
Forced oscillations and bifurcations.
Applications of Global Analysis 1, Communications of Math. Inst. Rijksuniv. Utrecht, 3, 1974.

[T2] F. Takens.
Unfoldings of Certain Singularities of Vector fields - Generalized Hopf Bifurcations.
Journal of Diff. Equations 14, 476-493, 1973.

FINITUDE DES CYCLES-LIMITES ET ACCELERO-SOMMATION DE L'APPLICATION DE RETOUR

Par J. Ecalle

IV - Appendices.

IV-1 Analysabilité de Borel-Laplace et formules cryptolinéaires.
IV-2 Accélérations faibles et fonctions cohésives. Constructibilité du prolongement quasi-analytique et du contournement des singularités quasianalytiques.
IV-3 Notion de douceur pour les fonctions modérées ou très lentes. Equivalences universelles.
IV-4 Types de croissance (TC) et types de croissance différentiables (TCD). Itération d'ordre transfini et échelle naturelle de croissance. Le Grand Cantor.

V - Références.

Avant-Propos.

Cet article développe deux exposés qui furent prononcés à Luminy, en septembre 89, lors d'un colloque sur les équations différentielles, et qui présentaient une preuve constructive de la finitude des cycles-limites (pour un champ de vecteurs analytique sur $\mathbf{R}^2$). Il condense également un livre, [E.5], consacré au même sujet et dont le "tapuscrit" fut aussi distribué lors du colloque de Luminy.

La *partie locale* de la preuve a été considérablement abrégée, car ce sont là des résultats assez anciens, qui sont exposés très en détail dans [E.5]. En revanche, les articulations essentielles de la *partie globale* ont été maintenues, de sorte que cet article, bien que conçu pour servir d'introduction à [E.5], se suffit presque à lui-même.

La plupart des mathématiciens qui s'intéressaient au problème de Dulac (finitude des cycles-limites) semblaient et semblent toujours n'y voir qu'une étape vers le 16ème problème de Hilbert (majoration du nombre des cycles-limites pour un champ polynomial de degré n). C'est leur droit, mais je tiens à dire que cet article et le livre qu'il résume ont été écrits dans une toute autre optique. Ils cherchent avant tout à illustrer, à l'occasion d'une application particulière, un ensemble de méthodes nouvelles (d'analyse et surtout de resommation) qui tournent autour de plusieurs idées-clef comme : ***résurgence ; dérivées étrangères ; compensation ; fonction médiane ; accélération ; transséries ; fonctions analysables ; fonctions cohésives.*** Ce parti-pris explique que le présent article, et le livre [E.5] plus encore, contiennent beaucoup de développements qui ne sont pas strictement indispensables à la preuve de la conjecture de Dulac ou qui même n'ont aucun rapport avec elle. Tel est le cas en particulier des quatres Appendices qui concluent l'article et le livre. Malgré la "gratuité" de ces Appendices, je me permets de les recommander à l'attention du lecteur, car ils abordent d'assez curieuses questions.

Tous mes remerciements vont à mes collègues Jean Martinet, Robert Moussu et Jean-Pierre Ramis, avec qui j'ai eu de nombreux échanges. Sans leur amicale insistance, je ne me serais d'ailleurs jamais occupé du problème de Dulac.

Je dois également signaler que Y.S. Ilyashenko a trouvé, indépendamment, une autre preuve de la finitude des cycles-limites. Cette preuve [Il-3] diffère totalement de celle qui est exposée ici et n'a pas, semble-t-il, le même caractère "constructif" et "réel". En un mot, alors que ma méthode consiste à associer à l'application de retour F d'un polycycle un objet formel et réel $\tilde{F}$, avec une correspondance $F \leftrightarrow \tilde{F}$ constructive et explicite dans les deux sens, la méthode de Y. Ilyashenko, si je la comprends bien, revient à associer à F un autre objet formel, disons $F^{\#}$, qui n'est pas véritablement réel et qui, tout en caractérisant l'application F, ne permet pas de la reconstituer effectivement.

I - SURVOL DES RESULTATS ET OUTILS DE BASE

I-1. L'application de retour et ses propriétés de régularité : non-oscillation, analysabilité, douceur.

Le présent article condense mon livre [E.5] où sont exposées deux démonstrations de l'énoncé de Dulac : pour tout champ de vecteurs X analytique sur un domaine de $\mathbf{R}^2$, les cycles-limites (c'est-à-dire les courbes intégrales de X analytiques, fermées et isolées) ne s'accumulent pas(*). Comme l'accumulation redoutée ne pourrait se produire que sur un *polycycle*(**) (courbe intégrale réunion finie d'arcs analytiques), il suffit de former les *applications de passage* g_i en chaque sommet S_i du polycycle ainsi que leur composée f, dite *application de retour* :

$$f = g_r o \ldots g_2 o\, g_1 \quad ; \quad g_i : x_i \rightarrow x_{i+1} \quad ; \quad f : x = x_1 \rightarrow x' = x_{r+1}$$

puis de montrer qu'il existe un intervalle $]0, \varepsilon]$ sur lequel l'application f est soit l'identité ($f(x) \equiv x$) soit sans points fixes ($f(x) < x$ ou $x < f(x)$).

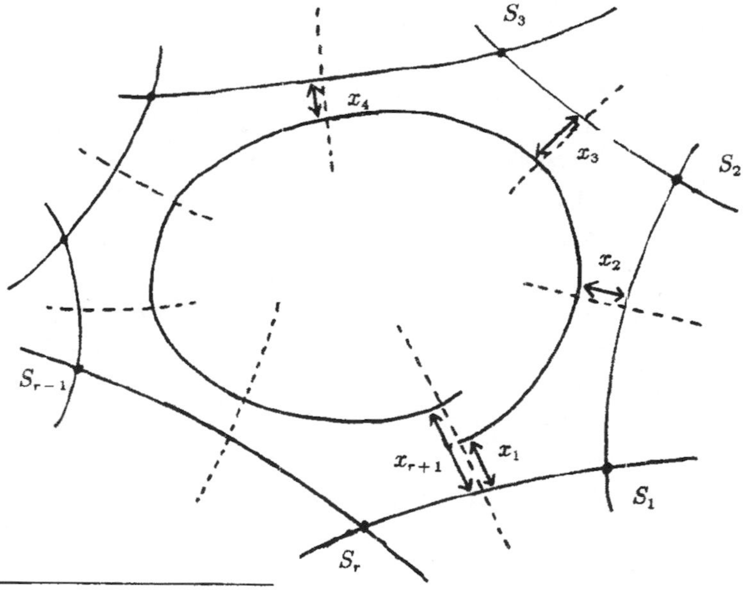

(*) L'énoncé original (1923) portait en fait sur les seuls champs polynomiaux. Dulac en donna une démonstration qui passa longtemps pour valide, mais dont l'insuffisance fut soupçonnée par Dumortier dans les années 70, puis mise à jour d'une façon indiscutable par Y.S. Il'yashenko.

(**) ou sur un point singulier de X qu'on peut ramener, par éclatement, à un polycycle.

Il est en fait plus commode de manier des infiniment grands, autrement dit de passer aux variables $z_i = 1/x_i$ ($x_i > 0$ petit ; $z_i > 0$ grand) et aux applications G_i et F :

$$F = G_r o \ldots G_2 o\, G_1 \;;\; G_i : z_i \to z_{i+1} \;;\; F : z = z_1 \to z' = z_{r+1}$$

La méthode consiste à associer à l'objet géométrique F une *transsérie* $\tilde{F}$, c'est-à-dire un objet formel qui généralise la notion de *série asymptotique* ; qui permet de *reconstituer* F ; et sur lesquel se peuvent lire *toutes* les propriétés de F.

A vrai dire, mon propos dans [E.5] n'est pas seulement d'établir le théorème de finitude, mais aussi et surtout d'obtenir le maximum d'informations sur l'application de retour. Celle-ci s'avère posséder de nombreuses propriétés de régularité : *non-oscillation, analysabilité, douceur.*

i) *Régularité minimale : non-oscillation de F.*
Nous verrons que l'application de retour est non-oscillante. En particulier, l'expression $F(z) - z$ est soit identiquement nulle, soit de signe constant pour z assez grand. Cette non-oscillation de F implique évidemment le théorème de finitude des cycles-limites, mais elle n'épuise pas, tant s'en faut, la régularité de F.

ii) *Régularité maximale : analysabilité de F.*
Nous verrons que l'application de retour F est entièrement "formalisable" en ce sens qu'on peut, d'une manière unique, lui associer une transsérie formelle $\tilde{F} = \Sigma a_n \square_n$, qui se présente comme une suite *bien ordonnée* de transmonômes $\square_n$, qui eux-mêmes sont des échafaudages *irréductibles* fabriqués avec la variable z, les symboles (+, -, ×, /, log, exp) et un nombre fini ou transfini de coefficients réels ; et qu'inversement F peut être reconstituée à partir de $\tilde{F}$. Plus précisément, les transmonômes $\square_n$ et la transsérie $\tilde{F}$ elle-même, bien que généralement divergents, possèdent toujours un nombre fini de "*temps critiques*" $\{\{z_i\}\}$ et peuvent être *sommés* selon le processus d'accéléro-sommation, qui débute par une transformation de *Borel formelle* relativement au temps critique $\{\{z_1\}\}$ le plus lent ; est suivi d'une cascade *d'accélérations* (cf.§.I.2 ci-après) faisant passer successivement par tous les temps critiques $\{\{z_i\}\}$; et se termine par une transformation de *Laplace* relativement au temps critique le plus rapide $\{\{z_r\}\}$.
Les sommes $\square_n$ des transmonômes formels $\square_n$ sont des fonctions positives, non-oscillantes et *comparables deux à deux* (comme germes infiniment petits ou grands) selon un ordre que reflète leur indexation n. De plus, pour tout indice n_0 fini ou transfini, la transsérie partielle $\sum_{n<n_0} a_n \square_n$ peut être sommée exactement ou modulo un idéal de fonctions arbitrairement petites, et la différence $F(z) - \sum_{n<n_0} a_n \square_n(z)$ est équivalente à la somme (automatiquement non-oscillante) du premier transmonôme négligé, soit $a_{n_0} \square_{n_0}(z)$. De plus, toutes

les opérations usuelles (+, ×, composition, dérivation, intégration, etc...) applicables à F peuvent être effectuées sur sa transsérie $\tilde{F}$ et elles *commutent avec l'accéléro-sommation* $\tilde{F} \to F$. C'est tout cela qu'on exprime en disant que F est *analysable*. Outre l'intérêt conceptuel que présente la réduction complète d'une fonction F à un objet formel $\tilde{F}$, l'analysabilité implique une régularité optimale. Dans le cas qui nous occupe, elle signifie que F est non-oscillante, ainsi que toutes ses dérivées $F^{(n)}$ et, plus généralement, toute fonction appartenant à l'algèbre engendrée par :

$$\{F, F', F'' \dots \quad ; A, A', A'' \dots \quad ; B, B', B'' \dots \quad ; C, C', C'' \dots\}$$

où A, B, C... désignent des monômes ou transmonômes quelconques ou même d'autres fonction analysables - par exemple les applications de retour de plusieurs polycycles différents.

(iii) *Régularité intermédiaire : douceur de F*
Y-a-t'il un moyen terme entre la *non-oscillation,* qui ne donne qu'une très pâle idée de la prodigieuse régularité de F, et *l'analysabilité,* qui va au fond des choses mais qui ne peut pas s'étudier ni même s'énoncer sans recours aux transmonômes, aux transséries et à tout l'appareil accéléro-sommatoire ? La réponse est OUI et elle tient dans la remarquable notion de *douceur* d'une fonction. La chose est expliquée en fin d'article, dans l'appendice IV.3, mais voici en deux mots de quoi il s'agit. Les fonctions $\mathcal{L}$ à croissance *très lente* (i.e. plus lente que tout logarithme itéré L_q) possèdent une propriété étonnante : *ou bien* les *dérivées posthomogènes* de $\mathcal{L}$ (c'est-à-dire les $D\mathcal{L} = D^* H$ dérivées homogènes de $H = log(1/\mathcal{L}')$) ne présentent aucune régularité à l'infini ; *ou bien* elles possèdent des *séries asymptotiques universelles* as $(D\mathcal{L}) = S^D$ qui sont totalement indépendantes de $\mathcal{L}$ et fonction du seul opérateur différentiel D. Dans le second cas, on dit que la fonction très lente $\mathcal{L}$ est *douce*. Quant aux fonctions F à croissance modérée, i.e. ni très lentes ni très rapides (ne sortant pas de l'échelle des logarithmes et exponentielles itérées) on dit qu'elles sont *douces* si, composées avec les fonctions très lentes, elles préservent leur douceur ; autrement dit si, pour toute fonction $\mathcal{L}$ très lente et douce, les fonctions très lentes $Fo\mathcal{L}$ et $\mathcal{L}oF$ sont douces elles-aussi. Dans ce cas, en écrivant que les dérivées posthomogènes de $Fo\mathcal{L}$ et $\mathcal{L}oF$ possèdent les mêmes séries asymptotiques que celles de $\mathcal{L}$, on obtient pour les fonctions F modérées des *équivalences universelles :*

$$as(D(Fo\mathcal{L})) = as(D(\mathcal{L}oF)) = as(D(\mathcal{L})) = S^D$$

qui expriment une assez forte régularité et ceci sans aucun recours aux transséries. Il se trouve que l'application de retour F et plus généralement toutes les fonctions analysables (qui par construction sont toujours à croissance modérée) vérifient aussi la condition de *douceur* et toutes les équivalences universelles qui s'ensuivent. L'inverse n'est pas vrai : il y a beaucoup de fonctions modérées et douces qui ne sont pas analysables.

I-2. Les opérateurs d'accélération

Désignons par $\mathbf{L}^{int}(\mathbf{R}^+)$ l'algèbre des fonctions localement intégrables sur $\mathbf{R}^+$ (y compris en 0) avec le produit de convolution :

(I.2.1) $$\hat{\varphi}_1 * \hat{\varphi}_2 = \hat{\varphi}_3 \text{ avec } \hat{\varphi}_3(\varsigma) = \int_0^\varsigma \hat{\varphi}_1(\varsigma_1)\hat{\varphi}_2(\varsigma - \varsigma_1)\, d\varsigma_1 \quad (\varsigma \in \mathbf{R}^+)$$

Opérateur de Laplace-Borel

La transformation de Laplace :

(I.2.2) $$\mathcal{L} : \hat{\varphi}(\varsigma) \to \varphi(z) = \int_0^{+\infty} e^{-\varsigma z} \hat{\varphi}(\varsigma)\, d\varsigma$$

est un homomorphisme de la sous-algèbre convolutive $\mathbf{L}^{int}_{exp}(\mathbf{R}^+) \subset \mathbf{L}^{int}(\mathbf{R}^+)$ formée des $\hat{\varphi}$ à croissance au plus exponentielle, dans l'algèbre multiplicative $\mathbf{B}$ formée des germes $\varphi(z)$ de fonctions définies, bornées, holomorphes dans un demi-plan $Re z \geq x_0$(*). L'inverse de $\mathcal{L}$ est la transformation de Borel :

(I.2.3) $$B : \varphi(z) \to \hat{\varphi}(\varsigma) = \frac{1}{2\pi i} \int_{c-i\infty}^{c+i\infty} e^{z\varsigma} \varphi(z) dz$$

Pour toute série formelle $\tilde{\varphi}(z) = \Sigma \varepsilon_n(z)$ dont le terme général $\varepsilon_n(z) \in \mathbf{B}$ possède une transformée de Borel $\hat{\varepsilon}_n(\varsigma)$, on a une notion de transformation de Borel formelle :

(I.2.4) $$\tilde{B} : \tilde{\varphi}(z) \to \hat{\varphi}(\varsigma) = \Sigma\, \hat{\varepsilon}_n(\varsigma)$$

Par exemple :

(I.2.4 bis) $$\tilde{B} : \tilde{\varphi}(z) = \Sigma\, a_n z^{-n} \to \hat{\varphi}(\varsigma) = \Sigma\, a_n \varsigma^{n-1}/\Gamma(n) \quad (n > 0)$$

Opérateurs d'accélération

Pour toute fonction F définie holomorphe sur un voisinage ramifié de l'infini, réelle positive sur $\mathbf{R}^+$ et vérifiant pour $z \to \infty$:

(I.2.5) $$x^{-1} F(x) \to 0 \;;\; \delta F(z) \sim \delta F(x) \;;\; \delta^2 F(z) \sim \delta^2 F(x) \quad \text{avec :}$$

(*) x_0 dépend de φ .

(I.2.6) $$0 < x \to +\infty \; ; \; z = x.e^{i\theta} (\theta \text{ réel fixé}) \; ;$$

$$\delta F(z) = \frac{zF'(z)}{F(z)} \; ; \; \delta^2 F(z) = 1 + \frac{zF''(z)}{F'(z)} - \frac{zF'(z)}{F(z)}$$

uniformément sur tout secteur, l'opérateur C_F de changement de variable $z_1 = F(z_2)$:

(I.2.7) $$C_F : \mathbf{B} \to \mathbf{B}, \varphi_1 \to \varphi_2 \text{ avec } \varphi_1(z_1) \equiv \varphi_2(z_2) \equiv \varphi_1(F(z_2))$$

est transmuté par Borel-Laplace en un opérateur $\hat{C}_F$, dit *opérateur d'accélération :*

(I.2.8) $$\hat{C}_F \begin{cases} \mathbf{L}^{int}_{exp}(\mathbf{R}) \to \mathbf{L}^{int}_{exp}(\mathbf{R}) \\ \hat{\varphi}_1(\varsigma_1) \to \hat{\varphi}_2(\varsigma_2) = \int_0^{+\infty} C_F(\varsigma_2, \varsigma_1)\hat{\varphi}_1(\varsigma_1) d\varsigma_1 \end{cases}$$

de noyau intégral égal à la transformée de Borel (en z_2) de exp $(-\varsigma_1 F(z_2))$:

(I.2.9) $$C_F(\varsigma_2, \varsigma_1) = \frac{1}{2\pi i} \int_{c-i\infty}^{c+i\infty} \exp\,(\varsigma_2 z_2 - \varsigma_1 F(z_2)).\; dz_2$$

Pour $\varsigma_1 > 0$ fixe et $0 < \varsigma_2 \to 0, C_F(\varsigma_2, \varsigma_1)$ tend vers 0. Pour $\varsigma_2 > 0$ fixe et $0 < \varsigma_1 \to +\infty, C_F(\varsigma_2, \varsigma_1)$ tend vers 0 *surexponentiellement* en ς_1.

Par suite, *l'opérateur d'accélération* $\hat{C}_F$ *a un domaine naturel de définition plus grand que celui de l'opérateur de Laplace* (c'est-à-dire plus grand que $\mathbf{L}^{int}_{exp}(\mathbf{R}^+)$). Une $\hat{\varphi}_1$ de $\mathbf{L}^{int}(\mathbf{R}^+)$ est dite *accélérable* pour F si l'intégrale (I-2-9) converge absolument pour tout $\varsigma_2 \in]0, c[$. Le plus grand des c convenables est dit *abscisse d'accélération* de $\hat{\varphi}_1$.

Accélérations fortes, moyennes, faibles. Accélérées analytiques ou quasianalytiques.

On est conduit à classer les accélérations $z_1 \to z_2$ selon leur force.
On distingue ainsi :
(i) les accélérations fortes : $log z_2 / log z_1 \to +\infty$
(ii) les accélérations moyennes : $log z_2 / log z_1 \to 1/\alpha > 1$
(iii) les accélérations faibles : $log z_2 / log z_1 \to 1$ mais $z_2 / z_1 \to +\infty$
La décroissance en ς_1, du noyau $C_F(\varsigma_2, \varsigma_1)$ est toujours décrite par la formule :

(I.2.10) $$\log C_F(\varsigma_2, \varsigma_1) \sim -\varsigma_1 G(\varsigma_2/\varsigma_1) \to -\infty (0 < \varsigma_2 \text{ fixe } ; 0 < \varsigma_1 \to +\infty)$$

pour une fonction $G(\varsigma)$, dite *co-accélératrice* (cf.§.IV-2), définie par :

(I.2.11) $$G(f(z)) \equiv F(z) - zf(z) \quad (\text{avec } f(z) \equiv F'(z), z > 0)$$

et tendant vers $+\infty$ quand ς tend vers $+0$ par valeurs positives.

Les *accélérations fortes* ont des noyaux à décroissance relativement faible (i.e. à peine surexponentielle) en ς_1 et les accélérées $\hat{\varphi}_2(\varsigma_2)$ correspondantes sont automatiquement définies holomorphes sur un voisinage de 0 ramifié et d'ouverture infinie.

Les *accélérations moyennes* ont des noyaux à décroissance plus forte (exp $(-c_\alpha . \varsigma_1^{1/\beta} . \varsigma_2^{-\alpha/\beta}$ avec $\beta = 1 - \alpha$) et les accélérées $\hat{\varphi}_2(\varsigma_2)$ correspondantes sont automatiquement définies holomorphes sur un voisinage sectoriel de 0 d'ouverture $\{-\frac{\pi}{2}.\frac{\beta}{\alpha} < \arg \varsigma < \frac{\pi}{2}.\frac{\beta}{\alpha}\}$.

Les *accélérations faibles* ont un noyau très fortement décroissant en ς_1. Les accélérées $\hat{\varphi}_2(\varsigma_2)$ correspondantes sont définies à la racine de l'axe $\mathbf{R}^+$, au voisinage de 0. Elles sont toujours quasianalytiques au sens de Denjoy-Carleman, mais généralement non analytiques (cf.§.IV-2).

I-3. Notion de dérivée étrangère et de fonction médiane

Mineurs et majeurs. Les algèbres $\mathcal{S}(0^+)$ et $\mathcal{S}^{int}(0^+)$.

Appelons *majeur* $\check{\varphi}$ un germe de fonction définie holomorphe à la racine du secteur $-2\pi < arg\varsigma < 0$ et possédant des valeurs-limite holomorphes sur les bords de ce secteur. Appelons *mineur* de $\check{\varphi}$ le germe $\hat{\varphi}$ défini holomorphe à la racine de $\mathbf{R}^+$ (mais pas forcément en 0) et donné par :

$$\hat{\varphi}(\varsigma) = \check{\varphi}(\varsigma) - \check{\varphi}(\varsigma.e^{-2\pi i}) \qquad (\varsigma > 0 \text{ petit}) \tag{I.3.1}$$

Désignons par $\overset{\nabla}{\varphi}$ la classe de $\check{\varphi}$ modulo les fonctions régulières (i.e. holomorphes) *en* 0. Notons $\mathcal{S}(0^+)$ l'espace de toutes ces classes $\overset{\nabla}{\varphi}$ et notons $\mathcal{S}^{int}(0^+)$ la partie de $\mathcal{S}(0^+)$ formée des classes $\overset{\nabla}{\varphi}$ dont *les* majeurs $\check{\varphi}$ et *le* mineur $\hat{\varphi}$ vérifient :

$$\operatorname{Lim}_{\varsigma \to 0} \varsigma\, \check{\varphi}(\varsigma) = 0 \quad \text{et} \quad \int_0^\epsilon |\hat{\varphi}(\varsigma)|\, |d\varsigma| < +\infty \qquad (\epsilon > 0 \text{ petit}) \tag{I.3.2}$$

Contrairement aux éléments de $\mathcal{S}(0^+)$, ceux de $\mathcal{S}^{int}(0^+)$ sont entièrement déterminés par *leur* mineur.

Pour deux classes $\overset{\nabla}{\varphi}_1, \overset{\nabla}{\varphi}_2 \in \mathcal{S}(0^+)$ et u proche de 0, la classe $\overset{\nabla}{\varphi}_3$ du majeur $\check{\varphi}_{3.u}$ défini par :

$$\check{\varphi}_{3.u}(\varsigma) = \int_u^{\varsigma - u} \check{\varphi}_1(\varsigma_1)\, \check{\varphi}_2(\varsigma - \varsigma_1)\, d\varsigma_1 \tag{I.3.3}$$

est indépendante de u et du choix des majeurs $\check{\varphi}_1$ et $\check{\varphi}_2$ dans $\overset{\triangledown}{\varphi}_1$ et $\overset{\triangledown}{\varphi}_2$. La loi $\overset{\triangledown}{\varphi}_1 * \overset{\triangledown}{\varphi}_2 = \overset{\triangledown}{\varphi}_3$ définit une *convolution* commutative et associative sur $S(0^+)$. Sur $S^{\text{int}}(0^+)$, elle induit une *convolution des mineurs :*

(I.3.4) $$\hat{\varphi}_1 * \hat{\varphi}_2(\zeta) = \hat{\varphi}_3(\zeta) = \int_0^{\zeta} \hat{\varphi}_1(\zeta_1)\, \hat{\varphi}_2(\zeta - \zeta_1)\, d\zeta \qquad (\zeta \text{ proche de } 0)$$

Fonctions résurgentes au dessus de $\mathbf{R}^+$. Les algèbres $\mathcal{R}(\mathbf{R}^+)$ et $\mathcal{R}^{\text{int}}(\mathbf{R}^+)$.

Notons $\mathcal{R}(\mathbf{R}^+)$ la sous-algèbre de $S(0^+)$ formée des classes $\overset{\triangledown}{\varphi}$ dont le mineur $\hat{\varphi}$ se prolonge analytiquement le long de tout chemin qui longe $\mathbf{R}^+$ (sans retour en arrière) en contournant, à droite ou à gauche, une suite arbitraire de points singuliers $0 < \omega_1 < \omega_2 < \omega_3 \ldots$ (*). Pour toute séquence de signes $\varepsilon_i = \pm$ et $\zeta \in]\omega_r, \omega_{r+1}[$ notons $\hat{\varphi}^{\varepsilon_1, \ldots, \varepsilon_r}_{\omega_1, \ldots, \omega_r}(\zeta)$ la détermination de $\hat{\varphi}(\zeta)$ obtenue à partir de 0^+ en contournant chaque ω_i à droite si $\varepsilon_i = +$ et à gauche si $\varepsilon_i = -$. L'espace $\mathcal{R}(\mathbf{R}^+)$ est stable pour la convolution $*$. C'est l'algèbre des *fonctions résurgentes au-dessus* de $\mathbf{R}^+$. Les éléments $\hat{\varphi}$ de S^{int} dont tous les prolongés $\hat{\varphi}^{\varepsilon_1 \ldots \varepsilon_r}_{\omega_1 \ldots \omega_r}$ sont *intégrables* sur leur intervalle de définition $]\omega_r, \omega_{r+1}[$ forment une sous-algèbre $\mathcal{R}^{\text{int}}(\mathbf{R}^+) \subset \mathcal{R}(\mathbf{R}^+)$. C'est l'algèbre des fonctions résurgentes *intégrables au-dessus* de $\mathbf{R}^+$.

Dérivations étrangères et fonction médiane.

Pour toute séquence finie de signes $\varepsilon_i = \pm$, notons p (resp. q) le nombre de signes $+$(*resp*$-$) et introduisons les poids :

(I.3.5) $$\delta^{\varepsilon_1, \ldots, \varepsilon_{r-1}} = \delta_{p,q} = \frac{p!q!}{(p+q+1)!} \qquad (p+q = r-1)$$

(I.3.6) $$\lambda^{\varepsilon_1, \ldots, \varepsilon_r} = \lambda_{p.q} = \frac{(2p)!(2q)!}{4^{p+q}.p!q!(p+q)!} \qquad (p+q = r)$$

Pour tout $\omega \in \mathbf{R}^+$ et tout $\hat{\varphi} \in \mathcal{R}^{\text{int}}(\mathbf{R}^+)$, la relation :

(I.3.7) $$\Delta_\omega \hat{\varphi}(\zeta) = \sum_{\varepsilon_i = \pm} \delta^{\varepsilon_1, \ldots, \varepsilon_{r-1}} . \left\{ \hat{\varphi}^{\varepsilon_1, \ldots, \varepsilon_{r-1}, +}_{\omega_1, \ldots, \omega_{r-1}, \omega}(\zeta + \omega) - \hat{\varphi}^{\varepsilon_1, \ldots, \varepsilon_{r-1}, -}_{\omega_1, \ldots, \omega_{r-1}, \omega}(\zeta + \omega) \right\}$$

valable pour $\zeta > 0$ *petit* puis *étendue par prolongation analytique,* définit un nouvel élément $\Delta_\omega \hat{\varphi}$ de $\mathcal{R}^{\text{int}}(\mathbf{R}^+)$. L'opérateur Δ_ω est une *dérivation* de l'algèbre de résurgence $\mathcal{R}^{\text{int}}(\mathbf{R}^+)$, dite *dérivation étrangère d'indice* ω.

(I.3.8) $$\Delta_\omega(\hat{\varphi}_1 * \hat{\varphi}_2) = (\Delta_\omega \hat{\varphi}_1) * \hat{\varphi}_2 + \hat{\varphi}_1 * (\Delta_\omega\, \hat{\varphi}_2)$$

L'action de Δ_ω s'étend d'une manière unique à l'algèbre $\mathcal{R}(\mathbf{R}^+)$.

(*) qui ne sont pas fixés, mais dépendent de $\hat{\varphi}$.

Pareillement, l'application $\hat{\varphi} \to \text{med}\, \hat{\varphi}$ définie pour tout $\varsigma > 0$ par :

$$\text{(I.3.9)} \qquad \text{med}\, \hat{\varphi}(\varsigma) = \sum_{\epsilon_i} \lambda^{\epsilon_1 \cdots \epsilon_r} \hat{\varphi}^{\epsilon_1 \cdots \epsilon_r}_{\omega_1 \cdots \omega_r}(\varsigma) \quad (\text{si } \omega_r < \varsigma < \omega_{r+1})$$

est un *homomorphisme* de l'algèbre convolutive $\mathcal{R}^{\text{int}}(\mathbf{R}^+)$ dans l'algèbre convolutive $\mathbf{L}^{\text{int}}(\mathbf{R}^+)$ des fonctions localement intégrables sur $\mathbf{R}^+$:

$$\text{(I.3.10)} \qquad \text{med}\,(\hat{\varphi}_1 * \hat{\varphi}_2) = (\text{med}\, \hat{\varphi}_1) * (\text{med}\, \hat{\varphi}_2)(^*).$$

La fonction *uniforme* med $\hat{\varphi}$ est dite *médiane* de la fonction *multiforme* $\hat{\varphi}$. L'application med (prise de la médiane) s'étend à $\mathcal{R}(\mathbf{R}^+)$ à condition de remplacer $\mathbf{L}^{\text{int}}(\mathbf{R}^+)$ par une algèbre convenable de fonctionnelles analytiques.

On montre que les poids $\delta_{p,q}$ et $\lambda_{p,q}$ sont les seuls qui dépendent uniquement de p et q et qui assurent les propriétés (I-3-8) et (I-3-10). On vérifie aussi que les définitions (I-3-7) et (I-3-9) sont cohérentes, en ce sens qu'elles sont indépendantes de la séquence ω_i choisie, pourvu que cette séquence contienne toutes les singularités de la fonction envisagée $\hat{\varphi}$.

I-4. Articulation générale de la démonstration.

Le principe analytique.

La démarche suivie s'inspire du *Principe Analytique*, qui cherche à pousser aussi loin que possible la correspondance entre objets géométriques et objets formels. L'étude se scinde tout naturellement en deux étapes, l'une locale, l'autre globale.

L'*étude locale* part de cette constatation : moyennant éclatement, tout polycycle d'un champ de vecteurs analytique sur $\mathbf{R}^2$ (ou un domaine de $\mathbf{R}^2$) se ramène à un polycycle élémentaire, ne comportant au plus que trois types de sommets. On étudie alors l'application de passage G_i associée à chaque type de sommet ainsi que sa contrepartie formelle $\widetilde{G_i}$. Pour les sommets de type I ou II (cols hyperboliques sans ou avec résonance) $\widetilde{G_i}$ est une série asymptotique ordinaire. Pour les sommets de type III (cols semi-hyperboliques), $\widetilde{G_i}$ est un cas élémentaire de *transsérie*, car elle mêle deux sortes

(*) Ici $*$ désigne bien sûr la convolution (I-3-4) des mineurs, mais elle n'a pas tout à fait le même sens aux deux membres : *au premier membre*, on convole deux germes en 0 puis on considère le prolongement *multiforme* de ce convolé et on en prend la médiane (fonction *uniforme* sur $\mathbf{R}^+$) ; *au second membre*, on convole directement les médianes comme fonctions *uniformes* sur $\mathbf{R}^+$.

d'infiniment petits différents - les puissances et les exponentielles. Dans tous les cas, le procédé de resommation de $\widetilde{G}_i$ consiste à effectuer Borel formel puis Laplace, relativement à une variable $z_i = h_i(z)$ bien choisie, dite *temps critique*. Néanmoins, les causes de divergence varient selon les types de sommets, tout comme varient les remarquables phénomènes de *compensation* qui s'y produisent.

L'*étude globale* consiste à reconstituer l'application de retour $F = G_r o ... G_1$ à partir de sa contrepartie formelle $\tilde{F} = \tilde{G}_r o ... \tilde{G}_1$. Cette dernière se présente comme une transsérie réelle, de forme parfois très compliquée, mais elle admet *toujours* une écriture réduite unique et peut *toujours* être resommée par le procédé général d'accéléro-sommation, qui comporte une cascade d'accélérations et fait passer par un nombre fini de temps critiques z_i provenant des divers sommets. Les temps critiques z_i ne sont définis univoquement que modulo une relation d'équivalence $z_i \approx z_i'$ qui signifie essentiellement que $z_i' = k(z_i)$ avec $z_i k'(z_i)/k(z_i) \to 1$ quand $z_i \to +\infty$. Il est donc plus pertinent de parler des *classes critiques* $\{\{z_i\}\}$ relatives à $\approx$. Fait crucial, les classes critiques $\{\{z_i\}\}$ sont comparables *deux à deux*, ce qui permet d'accéléro-sommer en commençant par les plus lentes et en terminant par les plus rapides. Techniquement, il est commode de choisir dans *chaque* classe critique $\{\{z_i\}\}$ un temps z_i "plutôt lent", relativement auquel toutes les transformées de Borel ne présentent que des singularités intégrables (au-dessus de $\mathbf{R}^+$), ce qui permet de travailler avec les seuls mineurs $\hat{\varphi}_i(\zeta_i)$ sans s'embarrasser des majeurs $\check{\varphi}_i(\zeta_i)$. Autre fait remarquable : lorsque deux temps critiques consécutifs, mais de classes distinctes, sont proches, autrement dit lorsque :

$$\log z_i / \log z_{i-1} \to 1 \quad \text{et} \quad z_i / z_{i-1} \to +\infty$$

l'accéléro-sommation oblige à passer par des fonctions $\hat{\varphi}_i(\zeta_i)$ qui s'avèrent quasi-analytiques au sens de Denjoy-Carleman et qu'il s'agit alors de prolonger quasi-analytiquement en *contournant* leurs éventuelles singularités, et ceci sans quitter l'axe $\mathbf{R}^+$ en dehors duquel ces fonctions ne sont pas définies !

II - ETUDE LOCALE

II-1. Les trois types de sommets d'un polycycle réduit.

Soit C une courbe simple, fermée, inscrite sur une surface analytique réelle S et X un champ de vecteurs analytique réel, défini au voisinage de C et tel que C soit une réunion finie de trajectoires de X, avec r "sommets" $S_1, S_2, ..., S_r$ correspondant à des points singuliers de X et r arcs analytiques C_i joignant S_{i-1} à S_i. On fixe sur chaque arc C_i un point P_i et une transversale analytique Γ_i coupant C_i en P_i. On munit Γ_i d'une abscisse analytique x_i nulle en P_i et positive du "bon côté", c'est-à-dire vers "l'intérieur" du polycycle C (voir figure au § I-1). On suppose que la trajectoire de X coupant Γ_i en $x_i = 1/z_i (x_i \sim +0 \; ; \; z_i \sim +\infty)$ coupe Γ_{i+1} en $x_{i+1} = 1/z_{i+1} (x_{i+1} \sim +0 \; ; \; z_{i+1} \sim +\infty)$. On pose $z_{i+1} = G_i(z_i)$. Les G_i sont les ***applications de passage*** associées aux différents sommets et leur composée $F = G_r o ... G_2 o \, G_1$ n'est autre que ***l'application de retour*** (parfois dite de "premier retour") du polycycle C de X (relativement à la transversale Γ_1 et à son abscisse $x_1 = 1/z_1$).

Par le théorème de résolution des singularités des champs de vecteurs sur $\mathbf{R}^2$ (cf [S]) on peut se limiter au cas où les sommets S_i de C correspondent à des singularités ***hyperboliques*** (X admet en S_i deux valeurs propres $\neq 0$) ou ***semi-hyperboliques*** (X admet en S_i une valeur propre $\neq 0$ et une autre qui est nulle). C'est ce que nous supposerons dans toute la suite.

Proposition II.1.1. Les trois types de sommets

En tout sommet S_i du polycycle (réduit) C, il existe des cartes locales C^∞ qui normalisent la forme différentielle $\bar{\omega}$ associée au champ X, en la ramenant à l'une des trois expressions suivantes :

Type I

(II.1.1)
$$\bar{\omega}_I = \frac{dx_{i+1}}{x_{i+1}} + \lambda_i \frac{dx_i}{x_i} \qquad (\lambda_i > 0)$$

Type II

(II.1.2)
$$\bar{\omega}_{II} = \left(1 + \rho_i^+ . x_i^{p_i^-} . x_{i+1}^{p_i^+}\right) . p_i^+ . \frac{dx_{i+1}}{x_{i+1}}$$
$$+ (1 + \rho_i^- . x_i^{p_i^-} . x_{i+1}^{p_i^+}) . p_i^- . \frac{dx_i}{x_i}$$

avec

(II.1.2 Bis) $$\begin{cases} p_i^+ \text{ et } p_i^- \in \mathbf{N}^* \ ; \ p_i = (p_i^+, p_i^-) = p.g.c.d. \text{ de } p_i^+ \text{ et } p_i^- \\ (\rho_i^+ - \rho_i^-).p_i^+ \ p_i^- \ = \ \varepsilon \ p_i \ \text{ avec } \ \varepsilon \in \{+,-\} \end{cases}$$

Type III⁺

(II.1.3) $$\bar{\omega}_{III}^+ = \frac{dx_{i+1}}{x_{i+1}} + (1 + \rho_i . x_i^{p_i}).p_i \frac{dx_i}{(x_i)^{1+p_i}} \quad (p_i \in \mathbf{N}^*, \rho_i \in \mathbf{R})$$

Type III⁻

(II.1.3 bis) Même chose avec x_i et x_{i+1} échangés. ∎

La normalisabilité formelle de $\bar{\omega}$ aux types $\bar{\omega}_I, \bar{\omega}_{II}, \bar{\omega}_{III}$ est élémentaire. La normalisabilité C^∞ est classique (cf [Mou],[S]) et n'est signalée ici que pour mémoire, car elle ne servira pas directement. On donnera en effet des résultats plus précis, qui impliqueront, pour chaque sommet, la *normalisabilité "quasianalytique" et constructive* de $\bar{\omega}$ et qui, surtout, donneront la solution de ce qu'on peut appeler le problème de Dulac local : pour chaque sommet de type I ou II (resp III) on associera univoquement à l'*application de passage locale* $G_i : z_i \to z_{i+1}$, une série (resp. transsérie) formelle $\widetilde{G}_i$, puis on indiquera comment reconstituer G_i à partir de $\widetilde{G}_i$.

II-2. Sommets de type I. Compensation des petits diviseurs liouvilliens

Il existe des cartes analytiques donnant à l'équation d'un col de type I la forme *préparée* suivante :

(II.2.1) $$\frac{dx_{i+1}}{dx_i} = -\frac{x_{i+1}}{x_i} . \left\{ \lambda_i + \sum_{N \in \mathbf{N}^* \times \mathbf{N}^*} \beta_N . x^N \right\}$$

$$= -\frac{x_{i+1}}{x_i} . \left\{ \lambda_i + \sum_{m,n \geq 1} \beta_{m,n} x_i^m x_{i+1}^n \right\}$$

Proposition II-2-1. Compensation des petits diviseurs liouvilliens

En tout sommet de type I, l'application de passage G_i possède une série asymptotique de la forme :

(II.2.2) $$\widetilde{G}_i = \widetilde{K}_i o P_{\lambda_i} o \widetilde{H}_i \quad \text{avec} \quad P_{\lambda_i}(z) = z^{\lambda_i} \quad (\lambda_i > 0)$$

Les facteurs $\widetilde{H}_i$ et $\widetilde{K}_i$ sont de la forme

(II.2.3) $\widetilde{H}_i(z) = z.\{a_i + \Sigma\, a_{i,n} z^{-n}\}$; $\widetilde{K}_i(z) = z.\{b_i + \Sigma\, b_{i,n} z^{-n}\}$ $(a_i, b_i > 0; a_{i,n}, b_{i,n} \in \mathbf{R})$

et peuvent diverger quand λ_i est liouvillien, c'est-à-dire quand ce nombre ne vérifie pas la condition diophantienne de Brjuno [Br] :

(II.2.4) $\Sigma\, 2^{-n}.\log\,(1/\sigma(2^n)) < +\infty$ avec $\sigma(k) = inf\ \mid \lambda_i.q - p \mid$ pour $q \leq k, p \in \mathbf{N}$

Toutefois, la série composée $\widetilde{G}_i$ est le siège de *compensations* et peut s'écrire :

$$\text{(II.2.5)} \qquad \widetilde{G}_i(z) = z^{\lambda_i}.\{1 + \sum_r \sum_{N_j} C_{N_1 \ldots N_r}(1/z)^{<\omega_1 \ldots \omega_r <}(1/z)^{m_1 + \ldots m_r}\}$$

avec $r \geq 1$, $N_j = (m_j, n_j) \in \mathbf{N}^* \times \mathbf{N}^*$, $\omega_j = -m_j + n_j \lambda_i$ et avec

$$\text{(II.2.6)} \quad C_{N_1 \ldots N_r} = (-1 + n_1)\,(-1 + n_1 + n_2)\ldots(-1 + n_1 + n_2 + \ldots n_{r-1}).\beta_{N_1}\ldots\beta_{N_r}$$

$$\text{(II.2.7)} \quad (1/z)^{<\omega_1 \ldots \omega_r <} = \sum_{i=1}^{r}(-1)^{r-i}.(1/z)^{\omega_1 + \ldots \omega_i}[(\omega_1 + \ldots\omega_i)(\omega_2 + \ldots\omega_i)\ldots\omega_i]^{-1}$$
$$[\omega_{i+1}(\omega_{i+1} + \omega_{i+2})\ldots(\omega_{i+1} + \ldots\omega_r)]^{-1}$$

Les $(1/z)^{< \ \ <}$ sont appelés *compensateurs* parce qu'ils restent petits même lorsque certains de leurs termes sont très grands. Ils vérifient des majorations (*)qui assurent la convergence uniforme (**)de $\widetilde{G}_i(z)$ sur un voisinage ramifié de ∞ défini par :

$$\text{(II.2.8)} \qquad \mid z \mid^{-1} \mid \log \mid z \mid\mid^{c_0} < \varepsilon_0 \qquad (\varepsilon_0, c_0 \text{ constantes } > 0)$$ ■

Proposition II.2.2. Sériabilité de $\widetilde{G}_i$.
La fonction $G_i(z)$ peut se calculer à partir de sa série asymptotique $\widetilde{G}_i(z)$ directement et sans compensation (***)grâce au procédé de *sériation* (****)qui consiste à effectuer une transformation de Borel formelle relativement à la classe critique :

$$\text{(II.2.9)} \qquad \{\{\log z\}\} = \{\{\log G_i(z)\}\}$$

(*) Cf [E.5] et [E.7].

(**) Pourvu que $\mid \beta_{m,n} \mid^{1/n} < Cste$, ce à quoi on peut toujours se ramener par des dilatations sur les variables x_i. Le meilleur voisinage de convergence uniforme (pour chaque λ_i) est donné dans [E.5].

(***) c'est-à-dire sans regrouper ses termes en paquets finis (compensateurs).

(****) Ainsi nommé parce qu'il permet de "sérier" les contributions des différents termes de $\widetilde{G}_i(z)$. Nous verrons au §.II.3 en quoi ce procédé diffère de la "sommation".

avec comme choix possible de temps critique *lent* z_0 :

(II.2.10) $z = k(z_0) = e^{z_0}(z_0)^{K_0} \Leftrightarrow z_0 = h(z) = \log z - K_0 \log \log z + (...)$ (pour tout $K_0 > 0$)

La série formelle :

(II.2.11) $$\widetilde{G}_1(z) = \widetilde{G}_1(k(z_0)) = \Sigma\, c_\sigma . \varepsilon_\sigma(z_0) \text{ avec}$$

$$\varepsilon_\sigma(z_0) = (k(z_0))^{-\sigma} = e^{-\sigma z_0} . (z_0)^{-\sigma K_0}$$

soumise à Borel formel $z_0 \to \varsigma_0$, livre une série :

(II.2.12) $$\hat{\varphi}_0(\varsigma_0) = \Sigma\, c_\sigma . \hat{\varepsilon}_\sigma(\varsigma_0) \qquad (\sigma \in \mathbf{N}^* + \lambda_1 \mathbf{N}^*)$$

avec des fonctions $\hat{\varepsilon}_\sigma(\varsigma_0)$ continues sur $\mathbf{R}^+$ et nulles sur $[0,\sigma]$. En chaque point $\varsigma_0 > 0$ donné, la fonction $\hat{\varphi}_0(\varsigma_0)$ est donc calculable comme somme finie et cette fonction s'avère être de croissance au plus exponentielle sur $\mathbf{R}^+$, ce qui permet de lui appliquer Laplace et de reconstituer ainsi $G_1(z)$. ■

Preuve succincte :
La validité formelle de (II.2.5) peut se vérifier directement sur (II.2.1) et la proposition II.2.1 en résulte, moyennant les majorations des compensateurs établies dans [E.7] et [E.8]. Quant à la validité du procédé de sériation, elle tient à ce que le $G_1(z)$ obtenu comme somme convergente de la série de compensateurs (II.2.5) est borné et holomorphe sur le domaine (II.2.8). D'où la proposition II.2.2. Mais on peut aussi déduire ces deux énoncés d'un résultat plus général concernant la linéarisation des champs de vecteurs quasi-résonnants (en toute dimension) par des changements de variables non entiers mais "compensables", i.e. séries convergentes de compensateurs. Voir [E.7] et [E.5].

Remarque 1. Compensation et sériation
Bien que la mise de la série $\widetilde{G}_1$ sous forme compensée (II.2.5) soit indispensable pour l'obtention de résultats optimaux sur la complexifiée de G_1 (cf [E.7]) il est commode de pouvoir sommer $\widetilde{G}_1$ directement, sans recours aux compensateurs, par le procédé de sériation (II.2.11,12), surtout lorsque, comme ce sera le cas pour nous, on est conduit à soumettre $\tilde{G}_1$ à des opérations (composition etc...) qui affecteraient profondément sa forme compensée.

Remarque 2. Divergence surmontable et divergence insurmontable
Il existe non pas un, mais une infinité de changements de variables H qui sont C^∞ et qui conjuguent la forme différentielle (II.2.1) à la forme normale :

(II.2.13)
$$\frac{dy_{i+1}}{dy_i} = -\lambda_i \frac{y_{i+1}}{y_i}$$

et, dans cette infinité, aucun H ne se distingue vraiment des autres. C'est ce fait qui explique qu'en cas de divergence, l'unique(*)changement de variable normalisant *formel* $\tilde{H}$ ne puisse être sommé d'aucune manière canonique. Au contraire, la correspondance $G_i \leftrightarrow \widetilde{G_i}$ étant biunivoque, le Principe Analytique "garantit" l'existence d'une resommation constructive $\widetilde{G_i} \to G_i$(**). Bien entendu, l'absence d'une correspondance biunivoque entre objet géométrique et objet formel n'*interdit pas* l'existence d'une resommation canonique. Nous venons de signaler que, pour les sommets irréguliers de type I, la resommation $\tilde{H} \to H$ n'existe pas pour les changements de variables (entiers) normalisants, mais nous verrons, grâce à la résurgence, qu'elle existe pour les sommets de type II et III.

II-3. Sommets de type II. Compensation de la résurgence

Proposition II.3.1. Difféomorphismes unitaires
Un difféomorphisme U (local à l'infini et holomorphe) de C est dit *unitaire* si son inverse coïncide avec son conjugué complexe : $Uo\bar{U} =$ id. Si U n'est pas l'identité, il s'écrit :

(II.3.1)
$$U(z) = z\{1 + i\ \varepsilon\ a\ z^{-p} + o(z^{-p})\} \text{ avec } p \in \mathbf{N}^*, \varepsilon = \pm, a > 0$$

et sa série associée $\tilde{U}$ admet une factorisation :

(II.3.2)
$$\tilde{U} = {}^*\tilde{U} o T_{2\pi i \varepsilon} o \tilde{U}^* \quad ({}^*\tilde{U} o \tilde{U}^* = id)$$

avec

(II.3.2 bis)
$$\begin{cases} \tilde{U}^* = Q_\rho^* o P_p o \tilde{H} = \alpha\ z^p + o(z^p) \text{ pour } \alpha = \frac{2\pi}{pa}, p \in \mathbf{N}^*, \alpha > 0, \rho \in \mathbf{R} \\ Q_\rho^*(z) = z + \rho \log z\ ;\ P_p(z) = z^p\ ;\ \tilde{H}(z) = c.z.\{1 + \Sigma\ c_n z^{-n}\} \quad (c > 0, c_n \in \mathbf{R}) \end{cases}$$

Les séries formelles réelles $\tilde{U}^*$ et ${}^*\tilde{U}$ sont définies modulo une même translation (resp. à gauche et à droite). Elles sont dites *itérateurs* (resp. direct et inverse) de U. Elles sont

(*) unique modulo une dilatation $x_i \to c_i x_i$ des variables.
(**) ou plus exactement : suggère fortement cette existence, sans pour autant nous dispenser du soin de l'établir directement dans chaque cas.

généralement divergentes, mais *toujours* résurgentes de classe critique $\{\{z\}\}$ et $\{\{\alpha z^p\}\}$ respectivement, avec des transformées de Borel à croissance au plus exponentielles et de points singuliers tous situés au dessus de $\mathbf{Z}$. Leurs équations de résurgence s'écrivent (dans le modèle formel) :

(II.3.3) $\Delta_n \,{}^\bullet\tilde{U} = +A_n.\partial_z \,{}^\bullet\tilde{U} \quad (A_n \in i\mathbf{R}\ ;\ n \in \mathbf{Z}^\bullet\ ;\ \Delta_n = \text{dér. étr. en } z)$

(II.3.4) $\Delta_n \tilde{U}^\bullet = -A_n exp\,(-n(\tilde{U}^\bullet(z) - \alpha z^p))(A_n \in i\mathbf{R}\ ;\ n \in \mathbf{Z}^\bullet\ ;\ \Delta_n = \text{dér. étr. en } \alpha z^p)$

(II.3.4 Bis) $\dot{\Delta}_n\, \tilde{U}^\bullet = -A_n.exp(-n\,\tilde{U}^\bullet) \quad (\dot{\Delta}_n = \text{dér. étr. pointée ; voir}[E.5])$ ■

Les scalaires p, ρ sont les invariants formels de U. Les A_n sont ses invariants analytiques.

Proposition II.3.2. Compensation de la résurgence
En tout sommet de type II, l'application G_i admet une série asymptotique :

(II.3.5) $$\widetilde{G_i} = {}^\bullet\tilde{V}_i \; o \; \tilde{U}_i^\bullet \qquad \text{avec}$$

$$\widetilde{G_i}(z) = a.z^{\lambda_i}\,\{1 + \sum_{n < Cste\, m} a_{m.n} z^{-m}(log\ z)^n\} \quad (\lambda_i = p_i^+/p_i^-)$$

avec des facteurs ${}^\bullet\tilde{V}_i$ et $\tilde{U}_i^\bullet$ qui sont des *itérateurs* inverse et direct *de types formels* (p_i^+, ρ_i^+) *et* (p_i^-, ρ_i^-) *différents* mais *d'invariants holomorphes* A_n *identiques*. Par suite, bien que les facteurs ${}^\bullet\tilde{V}_i$ et $\tilde{U}_i^\bullet$ soient chacun résurgents, leur composé $\widetilde{G_i}$ a toutes ses dérivées étrangères nulles et sa transformée de Borel de temps critique $\sim \alpha z^{p_i^-}$ n'a pas de singularités(*). C'est la "compensation" de la résurgence. ■

Proposition II.3.3. Sériabilité de $\widetilde{G_i}$.
Tout comme pour les sommets de type I, la série asymptotique $\widetilde{G_i}$ associée à un sommet de type II est *sériable* et la fonction G_i peut encore se calculer à partir de sa série asymptotique $\widetilde{G_i}$ grâce au procédé de *sériation*, expliqué à la section précédente, avec comme *classe critique :*

(II.3.6) $$\{\{\log z\}\} = \{\{\log G_i(z)\}\}$$

et avec le z_0 de (II.2.10) comme choix possible de *temps critique lent*, mais pour un K_0 assez grand (alors que pour les sommets de type I, n'importe quel K_0 convenait). ■

(*) Ni au-dessus de $\mathbf{R}$, ni bien sûr ailleurs.

Idée de la preuve.
La proposition II.3.1 résulte de [E.2]. La proposition II.3.2 résulte des propriétés de l'intégrale formelle du champ X au cols de type II ; plus exactement : de sa nature résurgente et de la forme des équations de résurgence qu'elle vérifie. Voir [E.3] et [E.4].

La *compensabilité* de $\widetilde{G_t}$ et par voie de conséquence sa *sériabilité* (proposition II.3.3) tiennent à ce que l'expression compensée (II.2.5) de $\widetilde{G_t}$ reste valide même lorsque λ_t est rationnel. Le champ X est alors résonnant en S_t et généralement non linéarisable, même formellement, par un changement de variables *entier*, mais il est *normalisable* par des changements de variables entiers résurgents (cf [E.3][E.4][E.5]) et demeure *linéarisable* par des changements de variables (non entiers) *compensables*, i.e. se présentant comme séries de compensateurs du type (II.2.5). Simplement, dans ce cas, les compensateurs en question ont certains de leurs indices ω_t qui peuvent s'annuler, d'où l'apparition de *termes logarithmiques* (cf [E.5],[E.7]).

Remarque. Sériation et sommation
En raison de sa factorisation (II.3.5), le $\widetilde{G_t}(z)$ d'un sommet de type II peut *aussi* être sommé par Borel-Laplace relativement à la classe $\{\{z^p\}\}$ critique pour la résurgence des facteurs ${}^{\bullet}\tilde{V}_t$ et $\tilde{U}_t^{\bullet}$. Profitons de ce double caractère de $\widetilde{G_t}$ pour dégager les différences entre la *sériation* d'une série compensable $\tilde{\varphi}(z)$ et la *sommation* d'une série résurgente $\tilde{\varphi}(z)$.

(*A : sommation*) La transformée de Borel globale $\hat{\varphi}(\varsigma)$ s'obtient par prolongement analytique à partir de 0. Elle est d'un "seul tenant" et se prolonge (sauf dans le cas quasi-analytique, cf §.IV.2) en dehors de l'axe réel.

(*A' : sériation*) $\hat{\varphi}(\varsigma)$ n'est défini que sur l'axe réel positif et n'est pas d'un "seul tenant" en ce sens qu'elle n'est nullement déterminée par sa restriction à un intervalle $]0,\omega[$.

(*B : sommation*) Le transit par $\hat{\varphi}(\varsigma)$ est indispensable pour le calcul de $\varphi(z)$. On peut tout au plus remplacer z par un temps équivalent $z' \approx z$. La classe $\{\{z\}\}$ est donc proprement "critique".

(*B' : sériation*) Le passage par $\hat{\varphi}(\varsigma)$ n'est pas strictement indispensable. C'est simplement le choix le plus commode, mais on peut toujours remplacer z par un temps z' de classe $\{\{z'\}\}$ un peu plus rapide (mais pas plus lente). La classe $\{\{z\}\}$ n'est donc "critique" que d'un "seul côté".

(*C : sommation*) La sommation est *polarisante* en ce sens que, lorsque $\hat{\varphi}(\varsigma)$ présente des points singuliers sur l'axe d'intégration $\mathbf{R}^+$, on doit les contourner à droite ou à gauche

(Laplace latéral) ou bien "passer à travers" (Laplace médian) mais en tout cas faire un choix.

(*C' : sériation*) La sériation n'est *jamais polarisante*. $\hat{\varphi}(\varsigma)$ est certes analytique par morceaux et possède bien des points singuliers sur $\mathbf{R}^+$, mais sa valeur en chaque point de $\mathbf{R}^+$ est donnée univoquement et il n'y a qu'une seule manière de lui appliquer Laplace.

Disons en conclusion que la *sommation des fonctions résurgentes* surmonte une divergence beaucoup plus sérieuse que celle dont vient à bout la *sériation des fonctions compensables*. Dans le premier cas, il y a d'ailleurs une riche structure algébrique sous-jacente (dérivation étrangères) mais rien de tel (plus exactement : rien de constructible) dans le second cas.

II-4. Sommets de type III. Transséries latérales et compensation des imaginaires.

Posons $E = \exp$ et $L = \log$

Proposition II.4.1. Transséries latérales et compensation des imaginaires
En tout sommet de type III$^+$, la fonction $L \circ G_\iota$ possède une série asymptotique $\tilde{U}^*_\iota$ qui est l'itérateur d'un difféomorphisme unitaire, de type formel (p_ι, ρ_ι) et "sesquilatéral", en ce sens que les invariants holomorphes $\dot{A}_n$ figurant dans les équations de résurgence (II.3.3) (II.3.4) sont nuls pour $n = -2, -3, -4...$ et que seuls subsistent les $\dot{A}_n$ correspondant à $n = -1, 1, 2, 3...$

Si on note $U^{*\pm}_\iota$ et ${}^*U^{\pm}_\iota$ les fonctions obtenues en sommant $\tilde{U}^*_\iota$ et ${}^*\tilde{U}_\iota$ par les opérateurs de Laplace latéraux $\mathcal{L}^{\pm}$ (*), les *fonctions* :

$$(\text{II.4.1}) \qquad {}^+K_\iota = (G_\iota) \circ ({}^*U^+_\iota) \circ L \quad \text{et} \quad {}^-K_\iota = (G_\iota) \circ ({}^*U^-_\iota) \circ L$$

sont méromorphes à l'infini, avec pour *séries* associées :

(*) c'est-à-dire en soumettant chacun de ces itérateurs à $\tilde{B}$ (Borel formel) relativement à un temps z pris dans sa classe critique, puis en soumettant la fonction résurgente $\hat{\varphi}(\varsigma)$ ainsi obtenue à $\mathcal{L}^{\pm}$ (Laplace latéral supérieur ou inférieur) qui consiste à calculer l'intégrale de Laplace $\varphi^{\pm}(z) = \int \hat{\varphi}(\varsigma) \exp(-\varsigma z) d\varsigma$ sur l'axe $\arg \varsigma = +0$ ou -0, i.e. en contournant toutes les singularités à gauche ou à droite respectivement.

(II.4.2) $\quad {}^+\widetilde{K_i}(z) = z\{a_i + \Sigma\ a^+_{i,n} z^{-n}\};^- \widetilde{K_i}(z) = z\{a_i + \Sigma\ a^-_{i,n} z^{-n}\} \quad (a_i > 0; a^\pm_{i,n} \in \mathbf{C})$

Cela permet d'associer à la fonction G_i les transséries ${}^+\widetilde{G_i}$ et ${}^-\widetilde{G_i}$ ainsi définies :

(II.4.3) $\quad {}^+\widetilde{G_i} = ({}^+\widetilde{K_i})\ o\ E\ o\ (\widetilde{U_i^*}) \qquad$ (factorisation latérale supérieure)

(II.4.4) $\quad {}^-\widetilde{G_i} = ({}^-\widetilde{K_i})\ o\ E\ o\ (\widetilde{U_i^*}) \qquad$ (factorisation latérale inférieure)

La *série* $\widetilde{U_i^*}$ est généralement divergente, mais à coefficients toujours réels, tandis que les *séries* ${}^+\widetilde{K_i}$ et ${}^-\widetilde{K_i}$ sont toujours convergentes, mais à coefficients généralement complexes. Quant aux *fonctions* ${}^+K_i$ et U_i^{*+} (resp ${}^-K_i$ et U_i^{*-}), elles sont généralement complexes, mais leur composée G_i est évidemment réelle. C'est la "compensation des imaginaires" ■

Comme pour les sommets de type II, la preuve repose sur les propriétés de résurgence de l'intégrale formelle du champ X en un sommet de type III. Voir [E.3] et [E.5]. Mais la différence est qu'ici les coefficients de Taylor $a^\pm_{i,n}$ des séries ${}^+\widetilde{K_i}$ et ${}^-\widetilde{K_i}$ sont des fonctions résurgentes de l'abscisse σ_{i+1} (sur $\mathcal{C}_{i+1}$) du point P_{i+1} où Γ_{i+1} coupe $\mathcal{C}_{i+1}$ (si on se met à considérer ce paramètre σ_{i+1} comme variable), ce qui explique qu'il y ait ici, pour le calcul de ces coefficients $a^\pm_{i,n}$, deux déterminations possibles, correspondant aux deux Laplaces latéraux $\mathcal{L}^\pm$. Les sommets de type III$^-$ s'étudient exactement de la même manière, mais en prenant tout "à l'envers".

Remarque 1 : Sommation des transséries $^\pm\widetilde{G_i}$.

Les $^\pm\widetilde{G_i}$ associées aux G_i des sommets de type III mêlent des infiniment petits ***non commensurables***, à savoir des puissances et des exponentielles (sommets de type III$^+$) ou des puissances et des logarithmes (sommets de type III$^-$). Ces $^\pm\widetilde{G_i}$ sont des cas particuliers très élémentaires de ce que j'appelle des *transséries* (cf. § III.1 et [E.5]). Si, pour simplifier, on se place dans le cas-type $(p_i, \rho_i) = (1, 0)$ en un sommet S_i de type III$^+$, les transséries $^\pm\widetilde{G_i}$ s'écrivent :

$$(\text{II.4.5}) \qquad {}^\pm\widetilde{G_i}(z) = \sum_{m \geq 0} a_m\ z^{-m} e^{\alpha z} + \sum_{m \geq 0} \sum_{n \geq 1} a^\pm_{m.n}\ z^{-m} e^{-(n-1)\alpha z}$$

avec $\alpha > 0, a_0 > 0, a_m \in \mathbf{R}, a^\pm_{m.n} \in \mathbf{C}$

Classer les infiniments grands ou petits du second membre dans l'ordre naturel (décroissant) revient à les indexer sur l'intervalle $[1, \omega^2[$ des ordinaux transfinis $< \omega^2$, où ω désigne le premier ordinal transfini. Il vient :

$$
(\text{II.4.6}) \quad \begin{cases} {}^{\pm}\tilde{G}_{\imath}(z) = \sum_{m\geq 0} a_m\ z^{-m} e^{\alpha z} + \sum_{m\geq 0} \sum_{n\geq 1} a^{\pm}_{\omega n+m}\ z^{-m}\ e^{-(n-1)\alpha z} \\ = \tilde{\varphi}_0(z).\ e^{\alpha z} + \sum_{n\geq 1}\ {}^{\pm}\tilde{\varphi}_n(z).e^{-(n-1)\alpha z} (\tilde{\varphi}_0(z) \in \mathbf{R}[[z^{-1}]], {}^{\pm}\tilde{\varphi}_n \in \mathbf{C}[[z^{-1}]]) \end{cases}
$$

Seule la ***partie accessible*** de la ***transsérie***, à savoir la ***série*** :

$$
(\text{II.4.7}) \qquad \sum_{m\geq 0} a_m .z^{-m}.e^{\alpha z} = \widetilde{\varphi_0}(z).e^{\alpha z}
$$

s'interprête comme développement asymptotique de $G_\imath$ et elle seule est astreinte à posséder des coefficients a_m réels. Quant aux coefficients *inaccessibles* $a^{\pm}_{\omega n+m}$ $(n \geq 1)$ portés par la partie proprement *transasymptotique* de ${}^{\pm}\widetilde{G_\imath}$, ils peuvent très bien comporter une partie imaginaire. Cela n'empêche pas la somme de ${}^{\pm}\widetilde{G_\imath}(z)$ par Borel-Laplace

$$
(\text{II.4.8}) \quad \begin{cases} {}^{\pm}G^{\pm}_{\imath}(z) = \varphi^{\pm}_0(z)e^{\alpha z} + \sum_{n\geq 1}\ {}^{\pm}\varphi^{\pm}_n(z).e^{-(n-1)\alpha z} \\ \text{avec } \varphi^{\pm}_0(z) = \mathcal{L}^{\pm}.\tilde{B}.\varphi \ \text{ et } \ {}^{\pm}\varphi^{\pm}_n(z) = \mathcal{L}^{\pm}.\tilde{\mathrm{B}}.{}^{\pm}\varphi \end{cases}
$$

d'être réelle :

$$
(\text{II.4.9}) \qquad {}^{+}G^{+}_{\imath}(z) = {}^{-}\ G^{-}_{\imath}(z) = G_{\imath}(z)
$$

En effet, la partie accessible $\widetilde{\varphi_0}(z)$ est réelle et sa transformée de Borel $\widehat{\varphi_0}(\varsigma)$ l'est aussi (à la racine de l'axe $\mathbb{R}^+$). Si $\widehat{\varphi_0}(\varsigma)$ comporte des singularités au-dessus de $\mathbf{R}^+$ (ce qui est le cas en général, compte tenu des équations de résurgence (II.3.3) et (II.3.4) vérifiées par les itérateurs), ses sommes de Laplace latérales $\varphi^{\pm}_0(z)$ ne sont pas réelles, mais leur partie imaginaire est exponentiellement petite et, dans la somme (III.4.8), elle se trouve exactement *compensée par les parties imaginaires de termes suivantes* ${}^{\pm}\varphi^{\pm}_1(z), {}^{\pm}\varphi^{\pm}_2(z)$...

Remarque 2. On peut démontrer l'énoncé de Dulac (finitude des cycles-limites) à partir des *transséries latérales* ${}^{+}\widetilde{G_\imath}$ et ${}^{-}\widetilde{G_\imath}$ (cf [E.5], deuxième démonstration, §.IV.4) mais celles-ci ont l'inconvenient d'être *deux* –d'où un certain arbitraire– et surtout d'introduire des coefficients complexes dans un problème foncièrement réel. On va donc les remplacer par la ***transsérie médiane*** $\widetilde{G_\imath}$, unique et réelle.

II-5. Sommets de type III. (suite). Transsérie médiane et compensation de la résurgence émanée.

Considérons toujours un sommet de type III$^+$ et reprenons les notations de la section précédente.

Proposition II.5.1. Transsérie médiane et compensation de la résurgence émanée
Les séries convergentes ${}^{+}\widetilde{K}_{1}$ et ${}^{-}\widetilde{K}_{1}$ diffèrent d'un facteur $\widetilde{V}_{1}$:

$$\text{(II.5.1)} \quad {}^{+}\widetilde{K}_{1} = {}^{-}\widetilde{K}_{1} o \widetilde{V}_{1} \quad \text{avec} \quad \widetilde{V}_{1}(z) = z.\{1 + i\,\varepsilon\, a\, z^{-p'_1} + \ldots\}\ \{\varepsilon = \pm 1, a > 0, p'_1 \leq +\infty\}$$

La série convergente $\widetilde{V}_{1}$ définit un difféomorphisme unitaire V_1 (local à l'infini) de $\mathbf{C}$, dit *premier émané* du difféomorphisme U_1 et représentant l'holonomie de X au sommet considéré. Les itérateurs ${}^{*}\widetilde{V}_{1}$ et $\widetilde{V}_{1}^{*}$ de V_1 ont des coefficients de Taylor réels, très simplement reliés aux invariants holomorphes de U_1, lesquels sont des imaginaires purs A_n (*). Si on note $\widetilde{V}_{1}$ la racine carrée itérative formelle de V_1, définie par :

$$\text{(II.5.2)} \qquad \widetilde{V}_{1} = {}^{*}\widetilde{V}_{1}\ o\ T_{\pi\underline{i}\varepsilon}\, o\ \widetilde{V}_{1}^{*} \qquad (T_{\pi\underline{i}\varepsilon}(z) \equiv z + \pi\underline{i}\varepsilon\ ;\ \underline{i} = \sqrt{-1})$$

et si l'on pose :

$$\text{(II.5.3)} \qquad \widetilde{K}_{1} = {}^{+}\widetilde{K}_{1}\ o\ \overset{-1/2}{\widetilde{V}_{1}} = {}^{-}\widetilde{K}_{1}\ o\ \overset{+1/2}{\widetilde{V}_{1}} \in z\,\mathbf{R}[[z^{-1}]]$$

alors la série formelle $\widetilde{K}_{1}$ ainsi définie est réelle-résurgente alors que ${}^{+}\widetilde{K}_{1}$ et ${}^{-}\widetilde{K}_{1}$ étaient *complexes convergentes*. On pose alors :

$$\text{(II.5.4)} \qquad \widetilde{G}_{1} = \widetilde{K}_{1}\ o\ E\ o\ \widetilde{U}_{1}^{*} \qquad \text{(factorisation médiane)}$$

L'objet formel $\widetilde{G}_{1}$ ainsi défini est dit *transsérie médiane* de la fonction G_1. Ses deux facteurs $\widetilde{K}_{1}$ et $\widetilde{U}_{1}^{*}$ sont chacun *réels* mais *généralement divergents* et *toujours résurgents*. Plus précisément, $\widetilde{U}_{1}^{*}(z)$ est résurgent de classe critique $\{\{z_0\}\} = \{\{z^{p_1}\}\}$ et $\widetilde{K}_{1}(z_*)$ est résurgent de classe critique $\{\{z_1\}\} = \{\{z_*^{p'_1}\}\}$. Toutefois, dans la factorisation médiane (II.5.4), la *première résurgence émanée* de $\tilde{U}^{*}$, de classe critique $\{\{z_1\}\} = \{\{z_*^{p'_1}\}\}$ avec $z_* = \exp \overline{\overline{U}}_{1}^{*}(z)$ où $\overline{\overline{U}}_{1}^{*}$ désigne la partie "infiniment grande" de $\widetilde{U}_{1}^{*}$:

$$\text{(II.5.5)} \qquad \overline{\overline{U}}_{1}^{*}(z) = \alpha_{1}.z^{p_1} + \alpha_{1\,1}.z^{p_1-1} + \ldots \alpha_{1.p_1-1}.z + \rho_1 \log z$$

détruit ou, si l'on préfère, *compense* la résurgence "vraie" ou "première" venant du facteur $\widetilde{K}_{1}$ et correspondant à la même classe critique $\{\{z_1\}\}$, si bien qu'on peut calculer la fonction G_1 à partir de sa transsérie $\widetilde{G}_{1}$ par $\mathcal{L}^{\mathrm{med}}.\tilde{B}$ (Borel formel suivi de Laplace médian) en passant par le seul temps critique $\{\{z_0\}\} = \{\{z^{p_1}\}\}$ ■

(*) cf. les équations de résurgence (II.3.3) et (II.3.4) ci-dessus et surtout [E.5] §.III.5, où on introduit, interprête et utilise la "procession" infinie des émanés de tous ordres. A signaler toutefois que les notations employées ici s'écartent légèrement de celle de [E.5].

Cela veut dire qu'on peut reconstituer G, en appliquant à $\widetilde{G}$, le même traitement qu'à $^+\widetilde{G}$, et $^-\widetilde{G}$, (voir section précédente, Remarque 1) avec cette seule différence que, dans les intégrales de Laplace, les déterminations latérales lat$^\pm \hat{\varphi}(\varsigma)$ etc... doivent être remplacées par les déterminations médianes med $\hat{\varphi}(\varsigma)$ etc... Le point important à noter est que, même si le passage de $^\pm\widetilde{G}$, à $\widetilde{G}$, rétablit la réalité de la transsérie au prix de l'introduction d'une divergence supplémentaire, à savoir celle du facteur $\widetilde{K}$, cette complication n'est heureusement qu'apparente et il ne s'introduit pas de seconde classe critique : il se trouve en effet que les singularités des transformées de Borel relatives à la seconde classe critique (attachée à $\widetilde{K}$,) sont détruites par les accélérées médianes des transforées de Borel relatives à la première classe critique (attachée à $\tilde{U}^{\cdot\cdot}$).

Il n'en irait plus de même si l'on voulait sommer *séparément* les facteurs $\widetilde{U}^{\cdot}$ et $\widetilde{K}$, de $\widetilde{G}^{\cdot}$ par le procédé $\mathcal{L}^{\mathrm{med}}\widetilde{B}$. Cela ne serait pas possible *exactement* (en général) car, contrairement aux *déterminations latérales* des transformées de Borel, leur *détermination médiane* possède (génériquement) une croissance (légèrement) *surexponentielle*. Mais il est possible, par un processus accéléro-sommatoire :

$$\mathcal{L}^{\pm} \quad \hat{\mathcal{C}}^{\mathrm{med}}_{F_s} . \hat{\mathcal{C}}^{\mathrm{med}}_{F_{s-1}} ... \hat{\mathcal{C}}^{\mathrm{med}}_{F_2} \; \hat{\mathcal{C}}^{\mathrm{med}}_{F_1} \; \widetilde{B}$$

de longueur s arbitraire, d'assigner à chacun de ces deux facteurs $\widetilde{U}^{\cdot\cdot}$ et $\widetilde{K}$, une "somme s-affinée", *définie et réelle modulo des infiniment petits d'ordre* $1/E_s(z)$ avec $E_s =\exp o \exp...\exp$ (s fois). Ce dernier point, sans être aucunement nécessaire à la preuve du théorème de finitude ni même au processus d'accéléro-sommation de la transsérie $\tilde{F}$ totale (i.e. non tronquée), n'en est pas moins très intéressant et très éclairant car il montre qu'un sens géométrique intrinsèque peut-être attaché à tout tronçon de la transsérie $\tilde{F}$. Toutes ces choses sont expliquées avec un grand luxe de détails au §.III.5 de [E.5] ("la sommation médiane des itérateurs réels et la procession sans fin des émanés").

II-6. Tableau récapitulatif

Formes normales
Sommets de type I :

$$\bar{\omega}_I = \frac{dx_{i+1}}{x_{i+1}} + \lambda_i \frac{dx_i}{x_i} \qquad (\lambda_i > 0)$$

Sommets de type II :

$$\bar{\omega}_{II} = (1 + \rho_i^+ x_i^{p_i^-} x_{i+1}^{p_i^+})p_i^+ \frac{dx_{i+1}}{x_{i+1}} + (1 + \rho_i^- x_i^{p_i^-} x_{i+1}^{p_i^+})p_i^- \frac{dx_i}{x_i}$$

avec $p_i^{\pm} \in \mathbf{N}^*, \rho_i^{\pm} \in \mathbf{R}; (\rho_i^+ - \rho_i^-)p_i^+ p_i^- = \varepsilon p_i$ où $\varepsilon = \pm$ et $p_i = \varepsilon.(p_i^+, p_i^-)$

Sommets de type III⁺ :

$$\bar{\omega}_{III}^+ = \frac{dx_{i+1}}{x_{i+1}} + (1 + \rho_i\, x_i^{p_i})p_i.\frac{dx_i}{(x_i)^{1+p_i}} \quad (p_i \in \mathbf{N}^*, \rho_i \in \mathbf{R})$$

Sommets de type III⁻ :

$$\bar{\omega}_{III}^- = \text{même chose avec } x_i \text{ et } x_{i+1} \text{ permutés}$$

Factorisation de $\tilde{G}_i$ et phénomènes de compensation :

Type I : $$\tilde{G}_i = \tilde{K}_i \; o \; P_{\lambda_i} o \; \tilde{H}_i \qquad (P_{\lambda_i}(z) \equiv z^{\lambda_i})$$

Compensation des petits diviseurs liouvilliens : Si λ_i est liouvillien, $\widetilde{H}_i$ et $\widetilde{K}_i$ divergent en général, mais leur composée $\widetilde{G}_i$ est toujours *compensable* et a fortiori *sériable.*

Type II : $$\tilde{G}_i = {}^*\tilde{V}_i o \tilde{U}_i^* \; ({}^*\tilde{V}_i \text{ et } \tilde{U}_i^* \text{ itérateurs bilatéraux})$$

Compensation de la résurgence : $\tilde{U}_i^*$ et ${}^*\tilde{V}_i$ sont résurgentes réelles, mais leur composée $\tilde{G}_i$ a toutes ses dérivées étrangères nulles. C'est même une fonction *compensable* et a fortiori *sériable.*

Type III⁺ : $${}^{\pm}\tilde{G}_i = {}^{\pm}\tilde{K}_i o E o \tilde{U}_i^* \quad (E = \exp, \tilde{U}_i^* \text{ itérateur sesquilatéral})$$

Transséries latérales ${}^+\tilde{G}$ = et ${}^-\tilde{G}$; compensation des imaginaires : $\tilde{U}_i^*$ est résurgente réelle et ${}^{\pm}\tilde{K}_i$ convergente complexe. Les sommes latérales $(U_i^*)^{\pm}$ de $\tilde{U}_i^*$ sont complexes, tout comme la somme naturelle ${}^{\pm}K_i$ de ${}^{\pm}\tilde{K}_i$, mais le germe composé :

$$G_i = {}^+K_i o \; E \; o \; (U_i^*)^+ = {}^-K_i o \; E \; o \; (U_i^*)^-$$

est réel.

Type III⁺ : $$\tilde{G}_i = \tilde{K}_i o \; E \; o \; \tilde{U}_i^* \text{ avec } \tilde{K}_i = {}^+\tilde{K}_i \; o \; \overset{-1/2}{\tilde{V}_i} = {}^-\tilde{K}_i \; o \; \overset{+1/2}{\tilde{V}_i}$$

Transsérie médiane et compensation de la résurgence émanée : $\tilde{V}_i$ est le premier émané de $\widetilde{U}_i$. Les séries $\widetilde{K}_i$ et $\tilde{U}_i^*$ sont réelles résurgentes, mais généralement pas sommables par Laplace médian. Toutefois, la composée $\widetilde{G}_i$ est, elle, sommable par Laplace médian.

Type III⁻ : Même chose à l'envers :

$${}^{\pm}\tilde{G}_i = {}^*\tilde{U}_i o \; L \; o \; {}^{\pm}\tilde{H}_i \quad \text{et} \quad \tilde{G}_i = {}^*\tilde{U}_i \; o \; L \; o \; \tilde{H}_i \qquad (L = \log)$$

Procédés sommatoires et temps critiques :

Types I et II : $G_\iota(z)$ est calculable à partir de $\widetilde{G_\iota}\,(z)$ par *sériation* de classe critique :

$$\{\{z_\iota\}\} = \{\{\log z\}\} = \{\{\log G_\iota(z)\}\}$$

Type III$^+$: $G_\iota(z)$ est calculable à partir de ${}^{\pm}\tilde{G}_\iota(z)$ (resp. $\widetilde{G_\iota}(z)$) par *sommation latérale* (resp. *médiane*) de classe critique :

$$\{\{z_\iota^*\}\} = \{\{z^{p_\iota}\}\} = \{\{\log G_\iota(z)\}\}$$

Type III$^-$: $G_\iota(z)$ est calculable à partir de ${}^{\pm}\tilde{G}_\iota(z)$ (resp. $\widetilde{G_\iota}\,(z)$) par *sommation latérale* (resp. *médiane*) de classe critique :

$$\{\{{}^*z_\iota\}\} = \{\{\log z\}\} = \{\{(G_\iota(z))^{p_\iota}\}\}$$

Pour des choix possibles, dans chaque *classe critique*, de *temps critiques lents*, voir [E.5], §§ III-2, III-3, III-4.

III - ETUDE GLOBALE

III-1. Opérations sur les transséries.

Aperçu sur les fonctions analysables.

Nous allons ici construire la *trigèbre* $\mathbf{R}[[[x]]]$ des transséries (formelles, réelles). C'est une construction au fond très naturelle puisqu'elle revient essentiellement à *clore* l'algèbre $\mathbf{R}[[x^{-1}]]$ des séries formelles ordinaires pour les opérations $+, \times, 0, \partial$ et leurs *inverses*(*). Avant même de définir par induction les transséries et les transmonômes dont elles sont formées, fixons des symboles très commodes pour les désigner.

	infiniment grands	infiniment petits	infiniment grands ou petits
transmonôme	$\sqcap$	$\sqcup$	$\square$
transmonôme d'exponentialité m	${}^{m}\sqcap$	${}^{m}\sqcup$	${}^{m}\square$
transsérie	$\sqcap\!\sqcap$	$\sqcup\!\sqcup$	$\square\!\square$
transsérie homogene d'exponentialité m	${}^{m}\sqcap\!\sqcap$	${}^{m}\sqcup\!\sqcup$	${}^{m}\square\!\square$

Commençons par introduire les *transmonômes* et *transséries alogarithmiques* (i.e. exprimables sans le symbole $L = log$). Pour ceux-ci, l'exponentialité m ne prend que des valeurs entières ≥ 0. Dans toute la suite, la variable x (variable formelle des transmonômes et transséries) désigne un infiniment grand *réel positif* ; ω désigne le premier ordinal transfini ; et n un ordinal fini ou transfini $< \omega^{\omega}$. Chaque transsérie se présentera comme une somme de transmonômes indexés sur un intervalles $\underline{n} < \underline{n}_0$ plus ou moins long.

Départ de l'induction. Exponentialité 0

Un *transmonôme alogarithmique* ${}^{0}\square$ d'exponentialité 0 est un monôme ordinaire :

(III.1.1) $${}^{0}\sqcap(x) = x^{\sigma} \quad (\sigma > 0) \quad \text{ou} \quad {}^{0}\sqcup(x) = x^{-\sigma} \quad (\sigma > 0)$$

(*) L'enrichissement viendra surtout de l'*intégration* ∂^{-1} et des *compositions* directes et inverses 0 et 0^{-1}.

avec l'ordre naturel : $\{x^\sigma < x^\tau\}$ ssi $\{\sigma < \tau\}$. Une transsérie alogarithmique ${}^0\square\!\square$ homogène d'exponentialité 0 est une expression de la forme :

(III.1.2) $\quad {}^0\sqcap\!\sqcap(x) = \Sigma a_{\underline{n}}\, x^{\sigma(\underline{n})} \quad (a_{\underline{n}} \in \mathbf{R}^* ; \sigma(\underline{n}) > 0\ ;\ \sigma(\underline{n})$ décroissant en $\underline{n})$

(III.1.3) $\quad {}^0\sqcup\!\sqcup(x) = \Sigma a_{\underline{n}}\, x^{-\sigma(\underline{n})} \quad (a_{\underline{n}} \in \mathbf{R}^* ; \sigma(\underline{n}) > 0\ ;\ \sigma(\underline{n})$ croissant en $\underline{n})$

avec les axiomes de finitude (où les σ_i dépendent évidemment de ${}^0\square\!\square$) :

(III.1.4) $\quad {}^0\sqcap\!\sqcap(x)/x^{\sigma(0)} \in \mathbf{R}[x^{-\sigma_1}, ..., x^{-\sigma_r}]\ ;\ {}^0\sqcup\!\sqcup(x) \in \mathbf{R}[[x^{-\sigma_1}, ..., x^{-\sigma_r}]]$

Suite de l'induction. Exponentialité m.

Un *transmonôme alogarithmique* ${}^m\square$ d'exponentialité m est une expression de la forme :

(III.1.5) $\quad {}^m\sqcap(x) = x^\sigma.\exp\,(+\sqcap\!\sqcap(x))$ ou ${}^m\sqcup(x) = x^\sigma.\exp\,(-\sqcap\!\sqcap(x)) \quad (\sigma \in \mathbf{R})$

avec $\sqcap\!\sqcap = {}^{m-1}\sqcap\!\sqcap + {}^{m-2}\sqcap\!\sqcap + \ldots {}^0\sqcap\!\sqcap$ et $\sqcap\!\sqcap(x) > 0$ (*) et avec l'ordre :

(III.1.6) $\quad \{\square_1 < \square_2\}$ ssi $\{\square_1/\square_2 = \sqcup_3\}$

Une *transsérie alogarithmique* ${}^m\sqcap\!\sqcap$ homogène d'exponentialité m est une expression de la forme :

(III.1.7) $\quad {}^m\sqcap\!\sqcap = \Sigma a_{\underline{n}}\,.\,{}^m\sqcap_{\underline{n}} \quad ({}^m\sqcap_{\underline{n}}$ décroissant en $\underline{n})$

(III.1.8) $\quad {}^m\sqcup\!\sqcup = \Sigma a_{\underline{n}}\,.\,{}^m\sqcup_{\underline{n}} \quad ({}^m\sqcup_{\underline{n}}$ décroissant en $\underline{n})$

avec les axiomes de finitude :

(III.1.9) $\quad {}^m\sqcap\!\sqcap/{}^m\sqcap_0 \in \mathbf{R}[[\sqcup_1, ..., \sqcup_r]],\ {}^m\sqcup\!\sqcup \in \mathbf{R}[[\sqcup_1, ..., \sqcup_r]] \quad (\text{expo}\ \sqcup_i \le m)$

qui disent que tous les transmonômes figurant dans le développement de ${}^m\square\!\square$ sont produit d'un nombre fini de transmonômes fixes $\sqcup_1, ..., \sqcup_r$ (dépendant évidemment de ${}^m\square\!\square$).

L'espace $\mathbf{R}^{\mathrm{alog}}[[[x]]]$ des transséries alogarithmiques de la forme :

(III.1.10) $\quad \square\!\square = {}^m\sqcap\!\sqcap + {}^{m-1}\sqcap\!\sqcap + \ldots {}^0\sqcap\!\sqcap + \text{cste} + {}^0\sqcup\!\sqcup + \ldots {}^{m'-1}\sqcup\!\sqcup + {}^{m'}\sqcup\!\sqcup \quad (m, m' \in \mathbf{N})$

(*) i.e. $\sqcap\!\sqcap(x) = a_0 \sqcap_0(x) + \ldots$ avec $a_0 > 0$.

est un corps stable pour la dérivation et la composition(*) mais pas pour l'intégration ni pour la prise de l'inverse de composition. Chaque transsérie algorithmique s'écrit d'une manière unique sous la forme :

(III.1.11) $\square\!\square = \Sigma a_{\underline{n}} \, \square_{\underline{n}} \; (\underline{n} < \underline{n}_0 < \omega^{\omega} \, ; a_{\underline{n}} \in \mathbf{R}^{*}, \; \square_{\underline{n}}$ décroissant en $\underline{n})$

(III.1.12) $\square\!\square = a_0 \; \square_0 . \{1 + \sqcup\!\sqcup_0\} \quad (a_0 \in \mathbf{R}^{*})$

Transmonômes et transséries généraux. La trigèbre $\mathbf{R}[[[x]]]$.
Posons L_m = log...log (m fois) et E_m = exp...exp (m fois). La limite inductive :

(III.1.13) $$\mathbf{R}[[[x]]] = \lim_{m \to \infty} \text{ind} \; \mathbf{R}^{\text{alog}}[[[L_m(x)]]] \qquad (^{**})$$

est dite *trigèbre* des transséries générales. Celles-ci admettent encore des développements (III.1.11), mais avec des transmonômes $\square$ de la forme :

(III.1.14) $\square = {}^{m}\square_{\bullet} \; o \; L_{m'} \quad (m, m' \in \mathbf{N} \; ; \; {}^{m}\square_{\bullet}$ alogarithmique d'exponentialité $m)$

Le plus petit entier m' pour lequel la décomposition (III.1.4) est possible est dit *profondeur* de $\square$ et le m correspondant est la *hauteur* de $\square$. On attribue évidemment à $\square$ une exponentialité égale à $m - m'$:

(III.1.15) $$\text{expo}\,(\square) = \text{haut}\,(\square) - \text{prof}\,(\square) \in \mathbf{Z}$$

L'exponentialité k d'un transmonôme Π se lit sur la relation :

(III.1.16) $L_m \, o \, \Pi\,(x) \sim L_{m-k}(x)$ (au sens formel ; m fixe assez grand ; $x \to +\infty$)

La trigèbre $\mathbf{R}[[[x]]]$ des transséries possède un ordre naturel $>$ et elle est stable pour les opérations $+, \times, 0, \partial$ ainsi que pour les opérations inverses.

Transséries portées par une autre.
Une transsérie $B = \Sigma \, b_{\underline{n}} \; B_{\underline{n}}$ est dite *contenue* dans une transsérie :

(III.1.17) $$A = \Sigma a_{\underline{n}} A_{\underline{n}} = \Sigma a_{\underline{n}} \; \exp\,(\sqcap\!\sqcap_{\underline{n}}) \qquad (n \in I = [0, n_0[)$$

(*) pour la seule composition qui ait un sens, c'est-à-dire pour la post-composition par un infiniment grand > 0 : $\square\!\square_1, \sqcap\!\sqcap_2 \to \square\!\square_1 \, o \, \sqcap\!\sqcap_2 \quad (\sqcap\!\sqcap_2 > 0)$.
(**) pour le plongement naturel de $\mathbf{R}^{\text{alog}}[[[L_m(x)]]]$ dans $\mathbf{R}^{\text{alog}}[[[L_{m+1}(x)]]]$.

s'il existe un sous-intervalle $J \subset I$ tel que $B = \Sigma a_{\underline{n}} A_{\underline{n}}$ pour $n \in J$, ou encore s'il existe un transmonôme $\square$ qui *domine* chacun des transmonômes $B_n (n \in J)$:

$$\log \square = \Sigma \alpha_{\underline{r}} \ \Pi_{\underline{r}} \text{ et } \log A_{\underline{n}} = \Pi\!\!\Pi_{\underline{n}} = \Sigma \alpha_{\underline{n}.\underline{m}} \ \Pi_{\underline{n}.\underline{m}} \tag{III.1.18}$$

$$\text{avec } \Pi_{\underline{r}} > \Pi_{\underline{n}.\underline{m}} (\forall \underline{r}, \underline{n}, \underline{m})$$

et qui soit en facteur dans tous les $A_{\underline{n}}$:

$$\square . B = \Sigma a_{\underline{n}} \ A_{\underline{n}} \quad (\underline{n} \in J) \tag{III.1.19}$$

Une transsérie B est dite *portée par A à l'étage* 1 (resp $n \geq 2$) si elle est *contenue* (cf supra) dans le logarithme $\Pi\!\!\Pi_{\underline{n}}$ d'un transmonôme $A_{\underline{n}}$ figurant effectivement (i.e. $a_{\underline{n}} \neq 0$) dans le développement de A (resp. dans celui d'une transsérie *portée par A à l'étage* $n-1$). Pour unifier la terminologie, on dit des B contenues dans A qu'elles sont portées par A à l'étage 0.

Transséries subexponentielles ou exponentielles.

Une transsérie $A(x)$ d'expression (III.1.17) est dite *exponentielle* en x si :

$$\lim_{x \to \infty} \quad x^{-1} . \Pi\!\!\Pi_{\underline{n}}(x) = -\omega_{\underline{n}} \in \mathbf{R} \quad \text{(au sens formel ; } \forall_{\underline{n}}) \tag{III.1.20}$$

Par les axiomes de finitude (III.1.4) (III.1.9), la suite décroissante des ω_n parcourt une partie discrète Ω de $\mathbf{R}$ et $A(x)$ s'écrit d'une manière unique sous la forme :

$$A(x) = \Sigma e^{-\omega x} \mathcal{A}_\omega (x) \qquad (x \in \Omega) \tag{III.1.21}$$

Si la somme (III.1.22) se réduit au seul terme $\mathcal{A}_0(x)$, on dit que $A(x)$ est *subexponentielle* en x. Enfin, un transmonôme $\square = \exp\ (\Sigma\ a_{\underline{n}} \Pi_{\underline{n}})$ est dit *purement surexponentiel* si :

$$\lim x^{-1} \Pi_{\underline{r}}(x) = +\infty \qquad \text{(au sens formel ; } \forall \underline{r}) \tag{III.1.22}$$

Transséries finies et transséries convergentes.

Il existe une notion naturelle de *transsérie finie* (sommes finies de transmonômes dont les logarithmes sont eux-mêmes des sommes finies de transmonômes dont etc...) et de *transséries convergentes* (définissables par induction sur la hauteur : A de hauteur m est dite convergente si $A.\sqcup_0 \in \mathbf{R}\{\{\sqcup_1, ..., \sqcup_r\}\}$ pour des transmonômes $\sqcup_i$ tels que $\log \sqcup_i$ soit convergente de hauteur $< m$) mais la première notion n'est pas stable par composition et ni l'une ni l'autre ne sont stables par intégration ou résolution d'équations différentielles.

Transséries accéléro-sommables. Notion de fonction analysable.

Il existe aussi, englobant les transséries convergentes, une notion très naturelle de transsérie *accéléro-sommable*. Ce sont les transséries qu'on peut sommer par une accéléro-sommation médiane :

$$\mathcal{L}^{\mathrm{med}}\ \hat{\mathcal{C}}^{\mathrm{med}}_{F_{r-1}} \ldots \hat{\mathcal{C}}^{\mathrm{med}}_{F_2}\ \hat{\mathcal{C}}^{\mathrm{med}}_{F_1}\ \tilde{\mathcal{B}}\ ;\ x \to x_1 \to \zeta_1 \to \zeta_2 \to \ldots \zeta_r \to x_r \to x$$

faisant passer par un nombre fini de classes critiques $\{\{x_i\}\}$ comparables, avec un *second axiome de finitude*(*) qui dit ceci : on peut trouver dans chaque classe $\{\{x_i\}\}$ un temps critique x_i tel que $x_i = F_i(x_{i+1})$ pour une F_i elle-même accéléro-sommable de classes critiques $\{\{x_{i,j}\}\} \ni x_{i,j}$ avec $x_{i,j} = F_{i,j}(x_{i,j+1})$... etc ... avec des temps critiques successifs $x_i, x_{i,j}, x_{i,j,k}, \ldots$ formant un *arbre fini*.

Nous avons donc deux notions de transséries sommables : les convergentes et les accéléro-sommables. Le fait central à bien saisir est que la seconde notion ("accéléro-sommable") est incomparablement plus *vaste*, plus *stable* et, tout bien considéré, plus *naturelle* que la première ("convergente"). Les transséries convergentes, en effet, ne résistent pas aux opérations les plus simples (pas même à l'intégration, voir [E.5]) tandis que la trigèbre $\mathbf{R}\{\{\{x\}\}\} \subset \mathbf{R}[[[x]]]$ des transséries accéléro-sommables possède une stabilité à toute épreuve : elle résiste en particulier aux opérations $+, \times, \circ, \partial$ et aux opérations inverses, y compris l'intégration (et plus généralement, semble-t-il, à la résolution d'équations différentielles(**)).

Je propose d'appeler *fonctions analysables* les (germes de) fonctions qui sont sommes de transséries de $\mathbf{R}\{\{\{x\}\}\}$, c'est-à-dire, encore une fois, de transséries accéléro-sommables avec finitude de l'arbre des temps critiques. Tout comme les germes analytiques réels, les fonctions analysables sont comparables deux à deux : on a soit $f_1(x) \equiv f_2(x)$ soit $f_1(x) > f_2(x)$ soit $f_1(x) < f_2(x)$ pour x assez grand. De plus, ces inégalités (entre infiniment grands ou infiniment petits de même signe) sont stables par dérivation. Enfin, de même que toute *fonction analytique* est déterminée par la suite infinie de ses coefficients de Taylor, toute *fonction analysable* est déterminée par *l'arbre* infini de ses coefficients (i.e. des coefficients de sa transsérie). Cette notion de fonction analysable marque probablement l'extension ultime de la notion de fonction analytique (réelle) et elle paraît inclusive et stable à un degré inouï.

(*) qui est à rapprocher du *premier axiôme de finitude* (cf. (III.1.4) et (III.1.9)) et qui assure automatiquement la comparabilité des classes critiques.

(**) i.e. si une équation différentielle à coefficients dans $\mathbf{R}\{\{\{x\}\}\}$ admet une solution formelle dans $\mathbf{R}[[[x]]]$, alors celle-ci est automatiquement dans $\mathbf{R}\{\{\{x\}\}\}$.

Il est clair que la non-oscillation de l'application de retour F (et donc la finitude des cycles-limites) résulte immédiatement des propriétés de stabilité des fonctions analysables, puisque $F = G_r o \ldots G_2 o G_1$ avec des G_i analysables. Cependant, comme nous n'avons besoin ici que d'une stabilité particulière (la stabilité par composition) et ceci pour des facteurs eux-mêmes très particuliers (les facteurs G_i), je préfère ne pas invoquer la théorie des fonctions analysables dans toute sa force et démontrer "à la main" l'énoncé de Dulac (cf §§ III.2, III.3, III.4).

III-2. Recensement des temps critiques.

Factorisation naturelle et factorisation réduite de l'application de retour.

Reprenons la situation du § II-1. Soit donc un polycycle $\mathcal{C}$ à r sommets et des courbes transversales Γ_i à chacun des arcs $\mathcal{C}_i$ de $\mathcal{C}$. Soit sur $\mathcal{C}_i$ une abscisse $x_i = 1/z_i$ et la factorisation correspondante de l'application de retour :

(III.2.1) $$F = G_r o \ldots G_2 o\ G_1 \qquad (F : z = z_1 \to z' = z_{r+1})$$

Cette factorisation de F possède une contrepartie formelle "immédiate" :

(III.2.2) $$\tilde{F} = \tilde{G}_r o \ldots \tilde{G}_2 o\ \tilde{G}_1$$

avec des "facteurs de passage" définis comme au § II, c'est-à-dire :

aux sommets de type I : $\tilde{G}_i = \tilde{K}_i o\ P_{\lambda_i} o\ \tilde{H}_i$
aux sommets de type II : $\tilde{G}_i = {}^*\tilde{V}_i o\ \tilde{U}_i^*$
aux sommets de type III⁺ : $\tilde{G}_i = \tilde{K}_i o\ E o\ \tilde{U}_i^*$
aux sommets de type III⁻ : $\tilde{G}_i = {}^*\tilde{V}_i o\ L o\ \tilde{H}_i$

En regroupant dans (III.2.2) tous les facteurs consécutifs appartenant au groupe $\mathbf{K}^{\text{quasi}}$ (engendré par les ${}^*\tilde{U}_i$ et les ${}^*\tilde{V}_i$ itérateurs de difféomorphismes unitaires et par les $\tilde{G}_i$ de type I ou II(*)) on aboutit à la *factorisation naturelle :*

(III.2.3) $$\tilde{F}_{\text{nat}} = \tilde{A}_{r_3} o\ E_{\varepsilon_{r_3}} o \ldots E_{\varepsilon_2} o\ \tilde{A}_1 o\ E_{\varepsilon_1} o\ \tilde{A}_0 \qquad (\varepsilon_i = \pm 1\ ;\ \tilde{A}_i \in \mathbf{K}^{\text{quasi}})$$

avec r_3 = nombre de sommets de type III, $\varepsilon_i = \pm 1$, $E_1 = E =$ exp, $E_{-1} = L =$ log.

Cette factorisation, considérée comme définie modulo les substitutions élémentaires :

(*) Le groupe "quasianalytique" $\mathbf{K}^{\text{quasi}}$ est étudié en détail dans [E.5] § III.5.

$$\text{(III.2.4)} \qquad \begin{cases} EoT_\alpha = D_a oE \\ T_\alpha oL = LoD_a \end{cases} \quad \text{avec} \quad \begin{cases} a = \exp(\alpha) \\ T_\alpha(z) \equiv \alpha + z \; ; \; D_a(z) \equiv a\, z \end{cases}$$

ne dépend que du choix de la transversale Γ_1 et de son abscisse analytique. En effectuant dans la factorisation naturelle toutes les simplifications possibles du type :

$$\text{(III.2.5)} \qquad EoT_\beta oD_c oL = D_b oP_c$$

$$\text{(III.2.6)} \qquad LoD_b oP_c oE = T_\beta oD_c$$

$$\begin{cases} \beta \in \mathbf{R}, \quad b = \exp \beta, \quad c > 0 \\ P_c(z) \equiv z^c \end{cases}$$

on aboutit à la *factorisation réduite* :

$$\text{(III.2.7)} \qquad \tilde{F}_{red} = \tilde{B}_{r_3'} o\ E_{\eta_{r_3'}} o \ldots E_{\eta_2} o\ \tilde{B}_1 o\ E_{\eta_1} o\ \tilde{B}_0 \qquad (\eta_i = \pm 1 \; ; \; r_3' \leq r_3)$$

qui n'est pas rigoureusement unique(*) mais dont le nombre de facteurs est bien défini(**).

Altitude des facteurs.

Pour une indexation des sommets S_i du polycycle commençant au bon endroit et pour un bon choix du sens de parcours, on peut faire en sorte que :

$$\text{(III.2.8)} \qquad 0 \leq \varepsilon_1 + \ldots \varepsilon_i = \nu_i \qquad \text{pour } i = 1, 2, \ldots, r_3$$

$$\text{(III.2.9)} \qquad 0 \leq \eta_1 + \ldots \eta_i = \nu_i' \qquad \text{pour } i = 1, 2, \ldots, r_3'$$

L'entier ν_i (resp ν_i') est dit *altitude* du facteur $\tilde{A}_i$ (resp $\tilde{B}_i$). Le maximum ν (resp ν') atteint par les ν_i (resp ν_i') est dit *altitude* du polycycle (resp. du polycycle réduit). L'altitude du premier facteur ($\tilde{A}_0$ ou $\tilde{B}_0$) est toujours 0 et l'énoncé de Dulac est trivialement vrai lorsque l'altitude du dernier facteur est > 0. Nous nous concentrerons donc sur le cas où celle-ci est nulle.

Classes critiques pour la sériation et classes critiques pour la sommation.

Pour tout $i \leq r_3$ désignons par $\underline{\tilde{A}}_i$ (resp $\underline{\underline{\tilde{A}}}_i$) le produit des facteurs de (III.2.3) situés à droite de $\tilde{A}_i$ exclu (resp. inclus). Autrement dit :

$$\text{(III.2.10)} \qquad \underline{\tilde{A}}_i = E_{\varepsilon_i} o\ \tilde{A}_{i-1} o\ E_{\varepsilon_{i-1}} \ldots \tilde{A}_0 \; ; \; \underline{\underline{\tilde{A}}}_i = \tilde{A}_i\ o\ \underline{\tilde{A}}_i$$

Puis introduisons les (germes de) fonction $\underline{A}_i$ et $\underline{\underline{A}}_i$ obtenus en remplaçant dans (III.2.10) chacun des facteurs $\tilde{A}_j$ par sa somme affinée d'ordre $q - j$ pour q grand ($q \geq \nu + 2$ suffit).

(*) elle l'est modulo les substitutions (III.2.4).
(**) quant à la possibilité de "réductions médiates", voir § III-5, lemmes d'immiscibilité.

Les fonctions $\underline{A}_i$ et $\underline{\underline{A}}_i$ sont définies modulo des infiniments petits d'exponentialité ≥ 2 (i.e. inférieurs à 1/exp exp z) et leurs *classes* $\{\{\underline{A}_i\}\}$ et $\{\{\underline{\underline{A}}_i\}\}$ pour la relation d'équivalence $\approx$

(III.2.10) $\quad f \approx g$ ssi $f = h \circ g$ avec $\text{cste} < \overset{\circ}{\delta} h = x h'(x)/h(x) < \text{cste}$

sont, elles, parfaitement définies(*).

On a vu au chapitre II (voir par exemple la fin de § II.6) qu'en chaque sommet irrégulier de type I ou II la série $\tilde{G}_j(z)$ possédait une *classe sério-critique* (critique pour la sériation) :

(III.2.12) $$\{\{z'_j\}\} = \{\{\log z\}\} = \{\{\log G_j(z)\}\}$$

et qu'en chaque sommet irrégulier de type III$^+$ (resp III$^-$) la transsérie médiane $\tilde{G}_j(z)$ possédait une *classe sommo-critique* (critique pour la sommation) :

(III.2.13) $$\{\{z^*_j\}\} = \{\{U^*_j(z)\}\} \quad (\text{resp } \{\{{}^*z_j\}\} = \{\{\log z\}\})$$

Vu la manière dont les $\tilde{A}_i$ s'obtiennent à partir des $\tilde{G}_j$, ceci suggère (voir [E.5]) d'associer à chaque facteur $\tilde{A}_i$ de $\tilde{F}_{\text{nat}}$ *une classe sério-critique* (au plus), notée $\{\{z'_i\}\}$, et *deux classes sommo-critiques* (au plus), notées $\{\{z^*_i\}\}$ et $\{\{{}^*z_i\}\}$, conformément au tableau ci-dessous :

Facteur $\tilde{A}_i$	inséré dans une séquence	au plus une classe sério-critique (d'exponentialité $\nu_i - 1$ en z)	au plus deux classes sommo-critiques (d'exponentialité ν_i en z)
bas	$E \circ \tilde{A}_i \circ L$	$\{\{z'_i\}\} = \{\{\log A_i(z)\}\} = \{\{\log A_i(z)\}\}$	$\{\{z^*_i\}\} = \{\{A_i(z)\}\}$ et $\{\{{}^*z_i\}\} = \{\{A_i(z)\}\}$
haut	$L \circ \tilde{A}_i \circ E$	$\{\{z'_i\}\} =$ idem	rien
ascendant	$E \circ \tilde{A}_i \circ E$	$\{\{z'_i\}\} =$ idem	$\{\{z^*_i\}\} = \{\{A_i(z)\}\}$
descendant	$L \circ \tilde{A}_i \circ L$	$\{\{z'_i\}\} =$ idem	$\{\{{}^*z_i\}\} = \{\{A_i(z)\}\}$

(*) on peut d'ailleurs calculer des *représentants* des classes $\{\{A_i\}\}$ et $\{\{A_i\}\}$ *directement*, par simple composition de *fonctions* G_i, sans aucune "sommation affinée" dans $\mathbf{K}^{\text{quasi}}$. Voir [E.5].

Le cas "tout-analytique".

Dans le cas "tout-analytique", c'est-à-dire dans le cas où l'équation différentielle associée au champ X est analytiquement normalisable en chaque sommet S_i, les factorisations naturelle et réduite, $\tilde{F}_{nat}$ et $\tilde{F}_{red}$, mises sous forme irréductible "descendue" (*), définissent une même transsérie $\tilde{F}$ qui est *convergente* (cf § III.1) *et donc "naïvement" sommable. Par suite, ou bien* $\tilde{F}(z) = z$ et alors $F(z) = z$, *ou bien* $\tilde{F}(z) = z + a\,\tilde{\square}(z) + o(\tilde{\square}(z))$ avec $a \neq 0$ et avec un *transmonôme convergent* $\tilde{\square}$ de somme naïve $\square$ et alors
$F(z) = z + a\,\square(z) + o(\square(z))$,
si bien que dans le cas "tout-analytique" l'énoncé de Dulac est trivialement vrai. La méthode suivie consistera à étendre ce résultat au cas général, où la transsérie $\tilde{F}$ *diverge*, en ressommant celle-ci non plus "naïvement" mais par accéléro-sommation et en montrant, qu'à ce détail près, rien ne change.

Mais, pour accéléro-sommer, il faut commencer par *ordonner* les classes critiques, des plus lentes aux plus rapides. *Dans toute la suite, on surmontera d'une tilde les séries ou transséries formelles et on réservera les lettres simples à leurs sommes, qui sont des germes de fonctions réelles en* $[\ldots, +\infty[$.

III-3. Accéléro-sommation de la transsérie médiane. Non-oscillation de l'application de retour et théorème de finitude.

Classes critiques et classes critiques formelles. Représentants canoniques.

Répertorions toutes les classes sério-critiques $\{\{z'_i\}\}$ et toutes les classes sommo-critiques $\{\{z_i^*\}\}$ et $\{\{{}^*z_i\}\}$ intoduites à la fin de la section précédente. Chacune de ces classes $\{\{R(z)\}\}$ est définie à partir d'un représentant $R(z)$ qui possède une factorisation qu'on peut écrire d'une façon unique sous la forme :

$$\text{(III.3.1)} \qquad R = E_\kappa \,o\, (L_{\nu_j} \,o\, A_j \,o\, E_{\nu_j}) \,o\, \ldots (L_{\nu_2} \,o\, A_2 \,o\, E_{\nu_2}) \,o\, (L_{\nu_1} \,o\, A_1 \,o\, E_{\nu_1}) \,o\, A_0$$

où $\nu_i = \varepsilon_1 + \ldots \varepsilon_i \geq 0$ est l'*altitude* du facteur A_i et κ l'*exponentialité* de R. Ce représentant R possède une contrepartie formelle :

$$\text{(III.3.2)} \qquad \tilde{R} = E_\kappa \,o\, (L_{\nu_j} \,o\, \tilde{A}_j \,o\, E_{\nu_j}) \,o\, \ldots (L_{\nu_2} \,o\, \tilde{A}_2 \,o\, E_{\nu_2}) \,o\, (L_{\nu_1} \,o\, \tilde{A}_1 \,o\, E_{\nu_1}) \,o\, \tilde{A}_0$$

Cette factorisation définit une transsérie qui, mise sous forme descendue, s'écrit :

(*) c'est-à-dire de la forme (III.1.11) à laquelle on se ramène en expulsant (ou, si l'on préfère, en "faisant descendre") des exponentielles la partie *constante* et la partie *infiniment* petite. Ainsi : $\exp\,(z + \alpha + z^{-1}) \to a.\,(\exp z)\,.(\Sigma\, z^{-n}/n!)$ avec $a = \exp\,\alpha$.

(III.3.3) $$\tilde{R} = a\,\tilde{\Pi} + \Sigma\, a_{\underline{n}}\, \tilde{\square}_{\underline{n}} \qquad (a > 0\ ;\ a_{\underline{n}} \in \mathbf{R}\ ;\ \tilde{\square}_{\underline{n}} << \tilde{\Pi})$$

Notons deb $\tilde{R}$ (lire : "début de $\tilde{R}$") le premier transmonôme $\tilde{\Pi}$ de $\tilde{R}$ privé de son coefficient a. On a ainsi une application :

(III.3.4) $$\{\{R\}\} \to \{\{\tilde{R}\}\} = \{\widetilde{\{R\}}\}$$

de l'ensemble des classes critiques du polycycle dans l'ensemble de ce qu'on appelera ses *classes critiques formelles. Pour l'instant, on ne sait pas encore* si cette application est *injective ; on ne sait pas* si les $\{\{R\}\}$ peuvent être *reconstruites* à partir des $\{\{\tilde{R}\}\}$ et *on ne sait pas* si les $\{\{R\}\}$ sont *comparables.* Mais cela n'empêche pas d'envisager les classes critiques formelles $\{\{\tilde{R}\}\}$ qui sont, elles, manifestement comparables (les transséries le sont toujours) et qui en outre possèdent des *représentants canoniques* deb $\tilde{R} = \tilde{\Pi}$ qui sont de simplicité maximale puisque *transmonomiaux.*

Ingrédients des représentants transmonomiaux. Troncation des facteurs.

Indexons ces classes critiques formelles dans l'ordre croissant :

(III.3.5) $$\{\{\tilde{\Pi}_1\}\} << \{\{\tilde{\Pi}_2\}\} << \ldots << \{\{\tilde{\Pi}_N\}\} \qquad (N \leq r\ ;\ \tilde{\Pi}_{i+1}/\tilde{\Pi}_i >> 1)$$

Pour chaque i, notons $\kappa(i)$ l'exponentialité de $\tilde{\Pi}_i$ et i' le plus petit j tel que $\tilde{\Pi}_i$ = deb $(\tilde{R}_j)$ avec $\tilde{R} = \tilde{R}_j$ de factorisation (III.3.2) et avec $1 + j = 1 + i'$ facteurs $\tilde{A}_0, \tilde{A}_1, \ldots, \tilde{A}_j$. Soit $\tilde{A}_m\ (m \leq i')$ le facteur générique de (III.3.2) et soit $\tilde{C}_m$ le produit des divers facteurs situés à gauche de $\tilde{A}_m$ (*).

Lemme III-1. On peut, sans changer deb $(\tilde{R}_{i'}) = \tilde{\Pi}_i$, remplacer (*séparément* ou *simultanément*) chaque facteur $\tilde{A}_m$ de (III.3.2) par une *troncation* trq $\tilde{A}_m$ dont la force dépend de l'altitude ν_m du facteur en question :

si $\nu_m \geq \kappa(i) + 2$ alors trq $\tilde{A}_m \equiv z$

si $\nu_m = \kappa(i) + 1$ alors trq $\tilde{A}_m = z$ ou z^{λ_m} selon que :

(III.3.6) $$\log \tilde{C}_m(z) >> \log\log z \quad \text{ou non}$$

si $\nu_m = \kappa(i)$ alors trq $\tilde{A}_m = z^{\lambda_m}$ ou $a_m\, z^{\lambda_m}$ selon que :

(III.3.7) $$\log \tilde{C}_m(z) >> \log z \quad \text{ou non}$$

(*) on exclut $\tilde{A}_m$ lui-même mais on inclut tous les facteurs situés à gauche, y compris les E et L.

si $\nu_m = \kappa(i)-1$ alors $\text{trq}\ \tilde{A}_m = a_m\, z^{\lambda_m}\{1+ \text{somme finie}\}$ ou $\tilde{A}_m$ (tel quel) selon que :

(III.3.8) $$\log\log \tilde{C}_m(z) >> \log z \quad \text{ou non}$$

si $\nu_m \leq \kappa(i)-2$ alors $\text{trq}\ \tilde{A}_m = \tilde{A}_m$ ■

Les conditions (III.3.6,7,8) sont faciles à vérifier à partir de la formule de Taylor appliquée à la transsérie $\tilde{C}_m o \tilde{A}_m$. Il ne vaut d'ailleurs pas la peine de s'y arrêter car, au fond, la force des troncations effectuées importe peu : il nous suffira, pour enclencher la récurrence (voir ci-après) de savoir que tous les facteurs $\tilde{A}_m$ d'altitude $\nu_m \geq \kappa(i)$ sont *effectivement tronquables*, ce qui est immédiat à vérifier.

Le lemme ci-dessus dit que les seuls vrais ingrédients du transmonôme $\tilde{\Pi}_i$ d'exponentialité $\kappa(i)$ sont *les* facteurs $\tilde{A}_m$ d'altitude $\nu_m \leq \kappa(i)-2$ et *certains* des facteurs $\tilde{A}_m$ d'altitude $\nu_m = \kappa(i)-1$. Ces facteurs-là interviennent tout entiers (avec toute l'infinité de leurs coefficients) dans la définition de $\tilde{\Pi}_i$ et, s'ils sont divergents, ils introduisent de la divergence au sein même de $\tilde{\Pi}_i$.

Les autres $\tilde{A}_m$, au contraire, sont *effectivement tronquables :* seul un nombre fini de leurs coefficients (peu importe combien, en pratique) interviennent dans la fabrication de $\tilde{\Pi}_i$. Notons enfin que *certains* des facteurs d'altitude $\nu_m = \kappa(i)+1$ et *tous* les facteurs d'altitude $\nu_m \geq \kappa(i)+2$ disparaissent purement et simplement, puisque pour eux on a :

(III.3.9) $$\text{trq}\ \tilde{A}_m = L_{\nu_m} o\ (\text{trq}\ \tilde{A}_m)\ o\ E_{\nu_m} = id$$

Pour les autres facteurs $\tilde{A}_m$ d'altitude $\kappa(i)+1$ qui ne sont pas complètement éliminables, on a :

(III.3.10) $$\text{trq}\ \tilde{A}_m = P_{\lambda_m} \quad \text{et} \quad L_{\kappa(i)+1} o\ \tilde{A}_m\ o\ E_{\kappa(i)+1} = L_{\kappa(i)} o\ D_{\lambda_m}\ o\ E_{\kappa(i)}$$

$$\text{avec} \quad P_\lambda(z) \equiv z^\lambda \quad \text{et} \quad D_\lambda(z) = \lambda z$$

Temps critiques accéléro-sommables. Leurs ingrédients.

Nous verrons plus loin (par récurrence sur l'exponentialité $\kappa(i)$ des $\tilde{\Pi}_i$) que les représentants canoniques $\tilde{\Pi}_i$ des classes critiques formelles sont indéfiniment accélérables mais pas toujours accéléro-sommables. Aussi est-il commode (sans que ce soit vraiment indispensable ; voir [E.5], § IV.3, remarques finales) de leur substituer des représentants $\tilde{S}_i \approx \tilde{\Pi}_i$ qui sont des transséries accéléro-sommables (mais qui restent suffisamment élémentaires pour permettre la récurrence à venir).

Pour ce faire, on repère dans la factorisation (III-3-1) de $\tilde{R} = \tilde{R}_{i'}$, tous les sommets de type III$^+$ (resp III$^-$) dont l'application de passage

(III.3.11) $$\tilde{G}_m = \tilde{K}_m\ o\ E\ o\ \tilde{U}_m^* \qquad (\text{resp}\ \tilde{G}_m = {}^*\tilde{U}_m\ o\ L\ o\ \tilde{H}_m)$$

possède un terme $\tilde{U}^{\bullet}_m$ (*resp* ${}^{\bullet}\tilde{U}_m$) qui entre dans la définition d'un facteur non tronquable (i.e. trq $\tilde{A}_m = \tilde{A}_m$) mais dont le terme $\tilde{K}_m$ (resp $\tilde{H}_m$) entre dans la définition d'un facteur $\tilde{A}_{m-1}$ (resp $\tilde{A}_{m+1}$) tronquable.

Alors que ces $\tilde{K}_m$ et $\tilde{H}_m$ *frontaliers* (qui sont généralement divergents et qui, rappelons-le, servent dans le processus de sommation de $\tilde{G}_m$ à compenser la résurgence émanée de $\tilde{U}^{\bullet}_m$ ou ${}^{\bullet}\tilde{U}_m$; cf §§ II-4, II-5 et surtout [E.5], §§ III.5 et III.6) disparaissent essentiellement de la factorisation :

(III.3.12) $\quad \text{trq } \tilde{R}_{\iota'} = E_{\kappa(\iota)} o\ (L_{\nu_{\iota'}} o\ (\text{trq } \tilde{A}_{\iota'})\ o\ E_{\nu_{\iota'}})\ldots\ o\ (L_{\nu_1} o\ (\text{trq } \tilde{A}_1)\ o\ E_{\nu_1})\ o\ (\text{trq } \tilde{A}_0)$

où ils ne figurent que tronqués, le procédé de fabrication des $\tilde{S}_\iota$ consiste justement à maintenir ces $\tilde{K}_m$ et $\tilde{H}_m$ frontaliers pour préserver la compensabilité, autrement dit à poser :

(III.3.13) $\quad \tilde{S}_\iota = \text{trq}^* \tilde{R}_{\iota'} = E_{\kappa(\iota)} o\ (L_{\nu_{\iota'}} o\ (\text{trq}^* \tilde{A}_{\iota'})\ o\ E_{\nu_{\iota'}})\ldots\ o\ (L_{\nu_1} o\ (\text{trq}^* \tilde{A}_1)\ o\ E_{\nu_1})\ o\ (\text{trq}^* \tilde{A}_0)$

avec

(III.3.14) $\quad \text{trq}^* \tilde{A} = \text{trq } \tilde{A}$ $\quad$ si $\tilde{A}$ ne contient pas de $\tilde{K}$ ou $\tilde{H}$ frontaliers.

(III.3.15) $\quad \text{trq}^* \tilde{A} = \tilde{H}\ o\ (\text{trq } \tilde{D})\ o\ \tilde{K}$ $\quad$ si $\tilde{A} = \tilde{H}\ o\ \tilde{D}\ o\ \tilde{K}$ avec $\tilde{H}$ et $\tilde{K}$ frontaliers.

(III.3.16) $\quad \text{trq}^* \tilde{A} = (\text{trq } \tilde{D})\ o\ \tilde{K}$ $\quad$ si $\tilde{A} = \tilde{D}\ o\ \tilde{K}$ avec $\tilde{K}$ frontalier.

(III.3.17) $\quad \text{trq}^* \tilde{A} = \tilde{H}\ o\ \text{trq } \tilde{D}$ $\quad$ si $\tilde{A} = \tilde{H}\ o\ \tilde{D}$ avec $\tilde{H}$ frontalier.

Ici, trq $\tilde{D}$ désigne le facteur $\tilde{D}$ tronqué au maximum (c'est-à-dire autant qu'on peut le faire sans affecter deb $\tilde{S}_\iota$, qui doit rester égal à $\tilde{\Pi}_\iota$).

Temps critiques accéléro-sommables lents.

Pour obtenir des temps critiques encore plus commodes, nous allons *ralentir* les temps formels $\tilde{S}_\iota(z)$ et les remplacer par des temps formels $\tilde{F}_\iota(z)$:

(III.3.18) $$\tilde{F}_\iota = \mathcal{F}_\iota^{o(-1)} o\ \tilde{S}_\iota \quad ; \quad \tilde{S}_\iota = \mathcal{F}_\iota o\ \tilde{F}_\iota$$

obtenus en soumettant $\tilde{S}_\iota$ à un *ralentissement* (cf. [E.5] § II-8) défini par une transsérie *convergente* $\mathcal{F}_\iota$ avec $\mathcal{F}_\iota(z) \sim z$ et $(z^{-1}\mathcal{F}_\iota(z) - 1)$ à décroissance *très* lente. On a d'ailleurs l'embarras du choix pour les $\mathcal{F}_\iota$. On peut prendre par exemple :

(III.3.19) $\quad \mathcal{F}_\iota(z) = z + z/L_{2+\kappa(\iota)}(z)$ et $\mathcal{F}_\iota^{o(-1)}$ = inverse de composition (converge !) de $\mathcal{F}_\iota$.

Pour d'autres choix possible, voir [E.5].

Accéléro-sommation des temps critiques.

Lemme III-2. Arbre des temps critiques.
Pour un polycycle fixe :
a) Les classes critiques $\{\{R\}\}$ sont en bijection avec les classes critiques formelles $\{\{\tilde{R}\}\}$.

b) Les représentants canoniques $\tilde{\Pi}_i$ des classes critiques formelles $\{\{\tilde{R}_i\}\}$ sont toujours indéfiniment accélérables mais généralement non accéléro-sommables.

c) Les représentants $\tilde{S}_i$ et $\tilde{F}_i$ des $\{\{\tilde{R}_i\}\}$ sont toujours accéléro-sommables.

d) Plus précisément, si l'on note $F_i(z)$ la somme de $\tilde{F}_i(z)$ et si on pose $z_i = F_i(z)$, alors la transsérie $\tilde{F}_i$ est accéléro-sommable avec une séquence S_i de classes critiques $\{\{z_k\}\}$ correspondant à tous les k tels que :

(III.3.19 bis) $$k' < i' \ (*) \text{ et } \log z_k << \log\log z_i .$$

e) Plus généralement, pour tous $i < j$, la transsérie $\tilde{F}_{i,j} = \tilde{F}_i \, o \, \tilde{F}_j^{0(-1)}$ est accéléro-sommable avec une séquence $S_{i,j}$ de classes critiques (*rapportées à z*) qui vérifient :

(III.3.20) $$S_{i,j} \subset S_i \cup S_j$$

et qui par suite sont toutes de la forme $\{\{z_k\}\}$ avec

(III.3.21) $$k' < \sup(i', j') \text{ et } \log z_k << \log\log z_j$$

Bien entendu, la fonction $F_{i,j}$ somme de $\widetilde{F_{i,j}}$ vérifie $z_i = F_{i,j}(z_j)$

f) Chaque classe critique $\{\{z_k\}\}$ de $\tilde{F}_i$ (resp $\widetilde{F_{i,j}}$) provient d'un temps critique d'un facteur $\tilde{A}_i$ *non tronqué* figurant dans la factorisation de $\tilde{S}_i$ (resp. à un facteur $\tilde{A}_m$ *non tronqué* subsistant dans la factorisation de $\tilde{S}_i \, o \, \tilde{S}_j^{0(-1)}$ après qu'on ait procédé à toutes les *simplifications immédiates*(**)). ■

Nous *admettrons provisoirement* ce lemme, qui nous servira à *décrire* le processus d'accéléro-sommation de $\tilde{F}$. Ensuite, une même récurrence sur l'exponentialité nous permettra de *prouver simultanément* l'accéléro-sommabilité de $\widetilde{F}$, des $\tilde{F}_i$ et des $\widetilde{F_{i,j}}$, et ceci

(*) avec la même correspondance $i \to i'$ que précédemment.
(**) c'est-à-dire du type (III.2.5),(III.2.6).

grâce au fait que ces trois types de transséries possèdent des factorisations de même nature, faisant intervenir E,L et les facteurs $\tilde{A}_m$.

Signalons encore que, dans ce problème particulier, toutes les classes critiques d'exponentialité $\kappa = -1, 0, 1$ ont toujours pour représentant canonique $\tilde{\Pi}$ un transmonôme *élémentaire*, c'est-à-dire ne dépendant que d'un nombre fini de paramètres (voir § III.4 et [E.5] § IV.1). A partir de l'exponentialité 2, les $\tilde{\Pi}$ sont (génériquement) non élémentaires.

Accéléro-sommation de la transsérie $\tilde{F}$.

Nous allons voir que l'application de retour F peut être reconstituée à partir de sa transsérie médiane $\tilde{F}$ par un processus d'accéléro-sommation qui consiste :
(i) à soumettre toute transsérie $\tilde{A}$ portée par $\tilde{F}$ et exponentielle en $z_1 = \tilde{F}_1(z)$ à la transformation de Borel formelle $z_1 \to \varsigma_1$
(ii) à effectuer successivement les accélérations médianes :

$$(z_1, \varsigma_1) \to (z_2, \varsigma_2) \to \ \dots\ (z_N, \varsigma_N)$$

la i-ème accélération ne portant que sur les transséries portées par $\tilde{F}$ et exponentielles en z_i
(iii) à soumettre la dernière accélération à la transformation de Laplace médiane : $\varsigma_N \to z_N$

Rappelons que toute transsérie exponentielle en $\tilde{\Pi}(z)$ possède une écriture unique.

$$\tilde{A}(z) = \Sigma\ e^{-\omega\ \tilde{\Pi}(z)}.\tilde{A}_\omega(z) \qquad \text{(III.3.22)}$$

avec des ω parcourant une suite discrète $\omega_0 < \omega_1 < \omega_2 < \dots$ de $\mathbf{R}$ et des transséries $\tilde{A}_\omega(z)$ telles que

$$\log\ |\ \tilde{A}_\omega(z)\ | \sim \log \operatorname{deb} \tilde{A}_\omega(z) << \tilde{\Pi}(z) \quad \text{(formellement)} \qquad \text{(III.3.23)}$$

Dans la suite, nous supposerons toujours que $\omega_0 = 0$, par simple commodité, afin de n'avoir à envisager que des facteurs exp $(-\omega\ \tilde{\Pi}(z))$ infiniment petits. On peut toujours se ramener à ce cas soit par multiplication de (III.3.22) par exp $(-a\ \tilde{\Pi}(z))$ soit par mise à l'écart d'un *nombre fini* de termes au second membre.

Première étape. La transformation de Borel $z_1 \to \varsigma_1$
Toute transsérie $\tilde{A}(z)$ portée par $\tilde{F}$ et exponentielle en $\tilde{\Pi}_1(z) = \log z$ se présente sous la forme :

$$\tilde{A}(z) = \Sigma\ e^{-\omega \tilde{\Pi}_1(z)}\ \tilde{A}_\omega(z) \quad (\omega \in \Omega_1 \text{ discret dans } \mathbf{R}^+) \qquad \text{(III.3.24)}$$

avec des $\tilde{A}_\omega(z)$ polynomiaux en $\Pi_1(z) = \log z$. Par rapport au temps lent $z_1 = F_1(z)$ cela s'écrit :

(III.3.25) $\tilde{A}(z) = \Sigma\, e^{-\omega\, z_1}\, \tilde{\tilde{A}}_\omega(z)$ avec $\tilde{\tilde{A}}_\omega(z) = e^{-\omega(\tilde{\Pi}_1(z)-\tilde{F}_1(z))}.\tilde{A}_\omega(z)$ $(\omega \geq 0)$

Comme $\Pi_1(z) - F_1(z) \sim (\log z)/(\log\log z)$, la somme *naturelle* $A_\omega(z)$ de $\tilde{\tilde{A}}_\omega(z)$ est une fonction subexponentielle de $z_1 = F_1(z)$ (si $\omega > 0$), uniformément décroissante dans les demi-plans Re $z_1 \geq$ cste et possédant une transformée de Borel $z_1 \to \zeta_1$:

(III.3.26)
$$\hat{A}^1_\omega(\zeta_1) = \int_{-i\infty}^{+i\infty} A_\omega(z).e^{-z_1\zeta_1}.dz_1 \quad (z_1 = F(z))$$

qui est définie C^∞ sur $\mathbf{R}^+$ (y compris en 0) et analytique en dehors de 0.(*).

Proposition III.3.3
Pour chaque transsérie (III.3.29) portée par $\tilde{F}(z)$, la fonction uniforme sur $\mathbf{R}^+$ définie par :

(III.3.27)
$$\hat{A}^1(\zeta_1) = \sum_{0\leq\omega<\zeta_1} \hat{A}^1_\omega(\zeta_1-\omega) \qquad (\forall\, \zeta_1 > 0)$$

(et complétée au besoin par la donnée du majeur $\check{A}^1_0$) est continue et même C^∞ sur $\mathbf{R}^+$ (sauf au plus en 0) ; analytique en dehors de l'ensemble discret Ω_1 et de croissance accélérable pour l'accélération $z_1 \to z_2$. ■

Etape i-ème $(2 \leq i \leq N)$. *L'accélération* $(z_{i-1}, \zeta_{i-1}) \to (z_i, \zeta_i)$

Toute transsérie $\tilde{A}(z)$ portée par $\tilde{F}$ et exponentielle en $\tilde{\Pi}_i(z)$ s'écrit d'une façon unique sous la forme :

(III.3.28)
$$\tilde{A}(z) = \sum_\omega e^{-\omega\,\tilde{\Pi}_i(z)}.\tilde{A}_\omega(z) \quad (\omega \in \Omega_i \text{ discret dans } \mathbf{R}^+)$$

(III.3.29) $\tilde{A}_\omega(z) = \sum_{\underline{n}} \tilde{B}_{\omega.\underline{n}}\tilde{D}_{\omega.\underline{n}} = \sum_{\underline{n}} e^{\tilde{C}_{\omega.\underline{n}}}.\tilde{D}_{\omega.\underline{n}}$ $(\underline{n} < \underline{n}(\omega)$ fini ou transfini)

avec des transséries $\tilde{D}_{\omega.\underline{n}}(z)$ exponentielles en $\tilde{\Pi}_{i-1}(z)$ et des transmonômes $\tilde{B}_{\omega.\underline{n}}$ subexponentiels en $\tilde{\Pi}_i$ et strictement surexponentiels en $\tilde{\Pi}_{i-1}$. Comme la transsérie $\tilde{D}_{\omega.\underline{n}}$ est exponentielle en $\tilde{\Pi}_{i-1}$ et qu'on peut toujours (en intercalant entre les classes critiques vraies des classes critiques fictives(**)) faire en sorte que $\log \tilde{\Pi}_i << \tilde{\Pi}_{i-1}$, la transsérie $\tilde{C}_{\omega.\underline{n}}$ est elle aussi exponentielle (et même subexponentielle) en $\tilde{\Pi}_{i-1}$. Par conséquent,

(*) Seul le coefficient A_0 peut faire exception. Sa transformée de Borel doit alors être caractérisée par l'ensemble mineur-majeur : $(\hat{A}^1_0, \check{A}^1_0)$

(**) c'est une simple commodité pour uniformiser l'exposition du processus, mais ce n'est aucunement nécessaire pour les calculs !

d'après l'induction(*), les transséries $\tilde{D}_{\omega.\underline{n}}$ et $\tilde{C}_{\omega.\underline{n}}$, après multiplication éventuelle par des facteurs régularisants :

(III.3.30) $$\tilde{\tilde{D}}_{\omega.\underline{n}} = e^{-\eta_{\omega.\underline{n}}\tilde{\Pi}_{i-1}}.\tilde{D}_{\omega.\underline{n}} \; ; \; \tilde{\tilde{C}}_{\omega.\underline{n}} = e^{-\eta'_{\omega.\underline{n}}\tilde{\Pi}_{i-1}}.\tilde{C}_{\omega.\underline{n}}$$

possèdent des "transformées de Borel médianes" (ce sont en fait des accélérées !) en $z_{i-1} = F_{i-1}(z)$, que l'on note :

(III.3.31) $$\text{med } \hat{D}^{i-1}_{\omega.\underline{n}}(\zeta_{i-1}) \quad \text{et} \quad \text{med } \hat{C}^{i-1}_{\omega.\underline{n}}(\zeta_{i-1}) \quad (\zeta_{i-1} \in \mathbf{R}^+)$$

et qui ont une *croissance accélérable* pour l'accélération $z_{i-1} \to z_i$. On a même un peu plus :

Proposition III.3.4.
Il existe $\varepsilon = \varepsilon_i > 0$ indépendant de ω et $\underline{n}$, tel que les intégrales

(III.3.32) $$\hat{D}^{i}_{\omega.\underline{n}}(\zeta_i) = \int_0^\infty C\,(\zeta_i, \zeta_{i-1}) \text{ med } \hat{D}^{i-1}_{\omega.\underline{n}}(\zeta_{i-1}).d\,\zeta_{i-1}$$

(III.3.33) $$\hat{C}^{i}_{\omega.\underline{n}}(\zeta_i) = \int_0^\infty C\,(\zeta_i, \zeta_{i-1}) \,.\, \text{med } \hat{C}^{i-1}_{\omega.\underline{n}}\,(\zeta_{i-1}).d\,\zeta_{i-1}$$

relatives au noyau d'accélération :

(III.3.34) $$C(\zeta_i, \zeta_{i-1}) = C_{F_{i-1,i}}(\zeta_i, \zeta_{i-1}) \text{ pour } z_{i-1} = F_{i-1,i}(z_i)$$

convergent uniformément sur tout l'intervalle $0 < \zeta_i \leq \varepsilon$.

Rapportée au temps $z_i = F_i(z)$, la transsérie (III.3.28) s'écrit alors :

(III.3.35) $$\tilde{\tilde{A}}(z) = \Sigma\, e^{-\omega\, z_i}\, \tilde{\tilde{A}}_\omega(z) \quad (\omega \in \Omega_i) \quad \text{avec :}$$

(III.3.36) $$\tilde{\tilde{A}}_\omega(z) = \Sigma\, \tilde{\tilde{B}}_{\omega.\underline{n}} \tilde{\tilde{D}}_{\omega.\underline{n}} = \Sigma\, \tilde{\tilde{A}}_{\omega.\underline{n}}$$

(III.3.37) $$\log \tilde{\tilde{B}}_{\omega.\underline{n}} = -\omega(\tilde{\Pi}_i - \tilde{F}_i) + \tilde{C}_{\omega.\underline{n}} + (\eta_{\omega.\underline{n}} + \eta'_{\omega.\underline{n}})\, \tilde{\Pi}_{i-1}$$

On peut alors calculer l'exponentielle de convolution qui donnera la "transformée de Borel" en z_i de $\tilde{\tilde{B}}_{\omega.\underline{n}}$ (cf [E.5]) puis le produit de convolution $\hat{B}^{i}_{\omega.\underline{n}} * \hat{D}^{i}_{\omega.\underline{n}}$ qui donnera la "transformée de Borel" en z_i de $\tilde{\tilde{A}}_{\omega.\underline{n}}$, c'est-à-dire $\hat{A}^{i}_{\omega.\underline{n}}$. ■

(*) c'est-à-dire d'après la Proposition III.3.3 si on est à l'étape $i = 2$ et, si on est à l'étape $i \geq 3$, d'après la Proposition III.3.4 relative à l'étape précédente.

Proposition III.3.5

Pour chaque $\omega \in \Omega_i$, la somme

$$(\text{III}.3.38) \qquad \hat{\mathcal{A}}^{i}_{\omega}(\varsigma_i) = \Sigma \hat{\mathcal{A}}^{i}_{\omega,\underline{n}}(\varsigma_i) \qquad (0 \le \underline{n} < \underline{n}(\omega))$$

converge uniformément sur $0 < \varsigma_i \le \varepsilon$ et y définit une fonction $\mathcal{A}^{i}_{\omega}(\varsigma_i)$ qui est analytique (resp. Denjoy quasianalytique de classe finie, cf. § IV.2, si l'accélération $z_{i-1} \to z_i$ est faible) et qui possède *par dessus* $\mathbf{R}^{+}$ (cf §§ I,1 et I,2) un prolongement analytique (resp. Denjoy quasianalytique si l'accélération $z_{i-1} \to z_i$ est faible) et continu (et même $\mathcal{C}^{\infty}$ pour les temps z_i très lents qu'on a pris soin de choisir) et de *singularités* (au sens ordinaire ou au sens quasianalytique ; cf § IV.2) situées au dessus d'un ensemble discret $\Omega_i \subset \mathbf{R}^{+}$ et indépendant de ω. ∎

Proposition III.3.6.

Si on note med $\hat{\mathcal{A}}^{i}_{\omega}(\varsigma_i)$ la *fonction médiane* (cf (I.3.9)) de $\hat{\mathcal{A}}^{i}_{\omega}(\varsigma_i)$, alors la fonction *uniforme* sur $\mathbf{R}^{+}$ définie par :

$$(\text{III}.3.39) \qquad \text{med } \hat{\mathcal{A}}^{i}(\varsigma_i) = \sum_{0 \le \omega < \varsigma_i} (\text{med } \hat{\mathcal{A}}^{i}_{\omega})(\varsigma_i - \omega) \qquad (\varsigma_i > 0)$$

est continue (et même $\mathcal{C}^{\infty}$ pour notre choix de z_i) et analytique (resp. Denjoy *q.a* si l'accélération $z_{i-1} \to z_i$ était *faible*) en dehors d'un ensemble discret Ω_i. Cette fonction *med* $\hat{\mathcal{A}}^{i}(\varsigma_i)$ est dite "transformée de Borel médiane" en z_i de la transsérie $\tilde{\mathcal{A}}(z)$(*) ∎

Proposition III.3.7.

Si $\{\{z_i\}\}$ n'est pas la plus rapide des classes critiques, la fonction médiane med $\hat{\mathcal{A}}^{i}(\varsigma_i)$ est de croissance accélérable pour l'accélération $z_i \to z_{i+1}$(**).

Pour $\{\{z_N\}\}$ la plus rapide des classes critiques, la fonction médiane med $\hat{\mathcal{A}}^{N}(\varsigma_N)$ est de croissance au plus exponentielle. ∎

Dernière étape. La transformation de Laplace $\varsigma_N \to z_N$.

Tous les énoncés de l'étape i restent en vigueur pour $i = N$ avec cette addition que, si la dernière classe $\{\{z_N\}\}$ est assez rapide(***), la transsérie $\tilde{F}(z)$ toute entière se présente

---

(*) car, dans le cas trivial "tout-analytique" (cf fin de §.III.2), où la transformée de Borel en z_i est *directement calculable*, elle coïncide bien avec la définition (*indirecte* mais *toujours valable*) que nous venons de donner.

(**) et ceci pour tout représentant z_i lent de la classe $\{\{z_i\}\}$.

(***) ce à quoi on peut toujours se ramener par l'introduction de classes "critiques" fictives (i.e. théoriquement inutiles !). Ici encore, ce n'est qu'une simple commodité d'exposition, nullement nécessaire aux calculs.

maintenant sous la forme :

(III.3.40)
$$\tilde{F}(z) = \tilde{C}_0(z_N).\{a_0 + \sum_{\underline{n}} \tilde{B}_{\underline{n}}\, \tilde{D}_{\underline{n}}\} \qquad (a_0 > 0)$$

avec des transséries $\tilde{D}_n$ exponentielles en $\tilde{\Pi}_{N-1}(z)$ et des transmonômes $\tilde{B}_n$ subexponentiels en $\tilde{\Pi}_N$ et strictement surexponentiels en $\tilde{\Pi}_{N-1}$, ce qui permet de prendre $\tilde{F}$ (rapportée à z_N) telle quelle et de la traiter comme une transsérie (III.3.29) de l'étape i.

Variante : La démarche d'accéléro-sommation qu'on vient de décrire est ***d'exécution très commode*** en ce sens qu'elle ramène pratiquement tout à des manipulation sur des transséries infiniment petites $\widetilde{\sqcup}$, ce qui, du côté "convolutif", permet de ne manipuler que des *mineurs* (avec une toute petite réserve pour les $\tilde{C}_{\omega.\underline{n}}$ et $\tilde{C}_{\omega.\underline{n}}$). La contrepartie est que les transséries $\widetilde{\sqcap}$ *portées* par $\tilde{F}$ et infiniment grandes doivent être multipliées par un facteur qui les rende infiniment petites. D'où un processus d'accéléro-sommation assez long à décrire. Si au contraire on accepte de manipuler telles quelles les transséries $\widetilde{\sqcap}$, l'accéléro-sommation devient très simple à décrire(*) (**) mais il faut, à chaque pas, du côté "convolutif", travailler sur les paires majeurs-mineurs (voir [E.5]).

Conclusion du processus d'accéléro-sommation :

Proposition III.3.7. Deux cas seulement sont possibles :
ou bien $\tilde{F}(z) \equiv z$ et alors $F(z) \equiv z$
ou bien $\tilde{F}(z) \equiv z + a\,\tilde{\square}(z) + o(\tilde{\square}(z))$ et alors $F(z) = z + a\,\square(z) + o(\square(z))$
avec $a \in \mathbf{R}^*$ et avec un transmonôme $\tilde{\square}(z)$ de somme $\square(z) > 0$. ■

L'application $z \to F(z)$ ne peut donc pas avoir une infinité de points fixes s'accumulant en $+\infty$. Cela revient à dire que le champ X ne peut pas avoir de cycles-limites s'accumulant sur le polycycle C et ceci établit l'énoncé de Dulac.

Preuve : Il s'agit de montrer (par induction sur l'exponentialité) l'accéléro-sommabilité des transséries :

(III.3.41)
$$\tilde{H}_q = \tilde{G}_q \;o\; \tilde{G}_{q-1} \;o \dots \tilde{G}_2 \;o\; \tilde{G}_1$$

(*) surtout si on convient (convention qu'il *faut* faire et que l'avenir imposera sans nul doute) de dire que *la* "transformée de Borel" en z_1 d'une transsérie analysable $\tilde{F}(z) \equiv \tilde{F}_1(z_1)$ ***est la collection*** des "transformées de Borel" en z_1 (calculées indirectement, par accéléro-sommation) ***de toutes les transséries*** portées par $\tilde{F}_1(z_1)$ et exponentielles en z_1.
(**) Il n'est même plus nécessaire de choisir des *temps lents* dans les classes sommo-critiques ou sério-critiques (encore que, pour ces dernières, la renonciation aux temps lents soit assez coûteuse ; cf [E.5]).

c'est-à-dire la possibilité de les sommer par un processus $\mathcal{L}\ \hat{C}_{\iota_q} \dots \hat{C}_1\ \tilde{\mathcal{B}}$, avec pour somme :

$$H_q = G_q \ o\ G_{q-1}\ o \dots G_2\ o\ G_1 \tag{III.3.42}$$

et avec des accélérations $C_\iota\ :\ z_\iota \rightarrow z_{\iota+1}$ faisant passer par des temps critiques z_ι appartenant à l'ensemble des $z_\iota = F_\iota(z)$ répertoriés ci-avant (avec une autre indexation, éventuellement). Comme on passe de z_ι à $z_{\iota+1}$ par une accélération $z_\iota = F_{\iota,\iota+1}(z_{\iota+1})$ de transsérie $\tilde{F}_{\iota,\iota+1} = \tilde{F}_\iota^{0(-1)} o\ \tilde{F}_{\iota+1}$ admettant (à des facteurs élémentaires près, venant des troncations et des $\tilde{F}_\iota$) une factorisation qui est encore de la forme (III.3.41) (quoique pour des $\tilde{G}_j$ plus forcément consécutifs) mais avec un nombre de facteurs inférieur (d'au moins une unité) et surtout avec une *exponentialité inférieure* (d'au moins une unité), ***une même induction*** (sur q et sur l'exponentialité) va permettre de prouver l'accéléro-sommabilité des $\tilde{H}_q$ (et donc de $\tilde{F} = \tilde{H}_r$) et celle des $\tilde{F}_\iota$ et $\tilde{F}_{\iota,\iota+1}$(*).

Raisonnons donc sur les $\tilde{H}_q$ de la forme (III.3.41) et choisissons (parmi les deux inductions possibles) celle qui consiste à *précomposer* $\tilde{H}_q$ par un facteur $\tilde{G}_{q+1}$, c'est-à-dire par une application de passage $\tilde{G}(z)$ correspondant à un sommet de type I ou II ou III et donc de la forme :

$$\tilde{G}(z) = z^\lambda \{a + \Sigma\ a_\sigma\ z^{-\sigma}\} \qquad (a > 0\ ;\ \lambda > 0\ ;\ \sigma \in \mathbf{N} + \lambda\,\mathbf{N}) \tag{III.3.43}$$

$$\tilde{G}(z) = z^{p/q}.\{a + \Sigma\ a_{m\ n}\ z^{-m/q}(\log z)^n\}\ (a > 0\ ;\ n/m\ <\ \text{cste}) \tag{III.3.44}$$

$$\tilde{G}(z) = e^{P(z)}.z^\rho.\{a + \Sigma\ a_{m.n} e^{-mP(z)} z^{-m\rho - n}\} \tag{III.3.45}$$

$$(P \text{ polynome en } z\ ;\ P(z) > 0; a > 0)$$

$$\tilde{G}(z) = (\log z)^{1/p}.\{a + \Sigma\ a_{m\ n}\ z^{-m}\,(\log z)^{m-n}\}\ \ (a > 0) \tag{III.3.46}$$

avec dans chaque cas une classe critique unique et des temps critiques lents $R_\bullet(z)$ indiqués au § II.

Désignons par COMP l'opération qui consiste à composer les facteurs $\tilde{G}_{q+1}$ et $\tilde{H}_q$ (ou ce qui, dans chaque modèle, leur correspond) au moyen de :

(i) la composition formelle 0 dans le modèle formel z

(ii) la composition des germes 0 dans le modèle sommé z

(*) L'accéléro-sommabilité de $\tilde{F}_i$ est indispensable pour assurer la ***comparabilité*** des classes critiques et non pas, bien sûr, pour "calculer" des représentants de ces classes, car il y a des représentants directement "donnés" par la géométrie.

(iii) la composition-convolution $\hat{0}$ dans chaque modèle convolutif ς_i

Il s'agit de montrer la commutativité du diagramme de gauche :

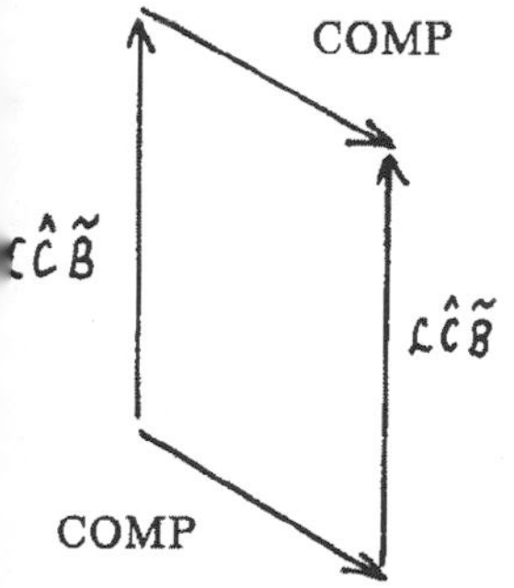

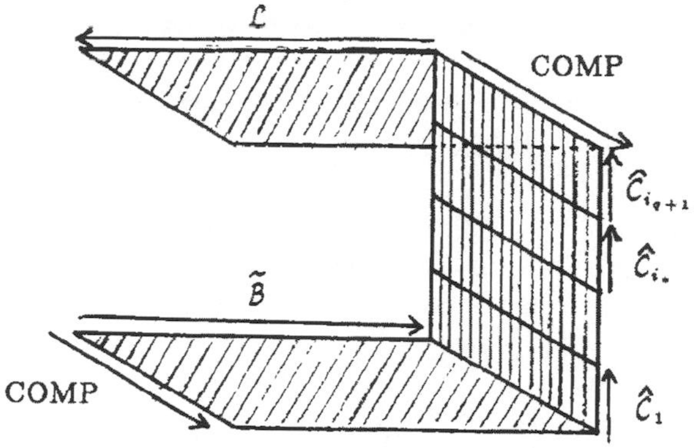

qui exprime que la composition formelle des transséries $\tilde{G}_{q+1}(z)$ de classe critique $\{\{R_\bullet(z)\}\}$ et $\tilde{H}_p(z)$ de classes critiques formelles $\{\{\tilde{R}_{q,i}(z)\}\}$ $(1 \le i \le i_q)$ livre une transsérie $\tilde{H}_{q+1}$ accéléro-sommable de classes critiques formelles :

$$\text{(III.3.47)} \qquad \{\{R_\bullet \; o \; \tilde{H}_p(z)\}\}, \{\{\tilde{R}_{q.1}(z)\}\}, ..., \{\{\tilde{R}_{q.i_q}(z)\}\} \qquad (\textit{une} \text{ répétition possible})$$

et de somme H_{q+1} égale à la *composée* $G_{q+1} \; o \; H_q$ des sommes G_{q+1} et H_q de $\tilde{G}_{q+1}$ et $\tilde{H}_q$. Tout se ramène donc à montrer la commutativité des parallélépipèdes hachurés dans le diagramme de droite. C'est effectivement la méthode à suivre pour montrer que deux transséries accéléro-sommables générales $\widetilde{\boxbar}_2$ et $\widetilde{\Pi}_1$ de temps critiques multiples ont une composée $\widetilde{\boxbar}_2 \; o \; \widetilde{\Pi}_1$ accéléro-sommable mais, dans le cas très particulier qui nous occupe ($\widetilde{\boxbar}_2 = \tilde{G}_{q+1}$, $\widetilde{\Pi}_1 = \tilde{H}_q$), nous allons voir que la commutativité *de tous* les diagrammes élémentaires est *triviale* sauf (et encore !) pour le diagramme correspondant à l'accélération $C_{i_\bullet}$, et ayant pour "base" le modèle convolutif $\varsigma_{i_\bullet}$ correspondant à la "nouvelle" classe critique.

A ce propos, deux cas peuvent se présenter : ou bien la "nouvelle" classe critique formelle est égale à une classe "ancienne", ou bien elle s'insère entre deux telles classes ("s'insère" au sens large : elle peut les précéder toutes ou les suivre toutes). On numérote donc $\{\{z_1\}\}, ..., \{\{z_{i_{q+1}}\}\}$ l'ensemble total des classes critiques $(i_{q+1} = i_q$ ou $1 + i_q)$ et on note $i_\bullet$ l'indice de la "nouvelle" (qui peut coïncider avec une ancienne). Ces classes sont *comparables*, non seulement formellement mais effectivement (car, par *l'induction* sur l'exponentialité, elles possèdent des représentants non seulement accéléro-sommables mais *déjà* accéléro-sommés !) et on peut choisir dedans des représentants "lents" $z_1, z_2, ..., z_{i_{q+1}}$.

Supposons pour commencer que $\tilde{G}_{q+1}$ *est de type I et donc de la forme* (III.3.45) *et* *que* $\imath_{\bullet} < i_{q+1}$ (i.e. la "nouvelle" classe n'est par la dernière (*)).

Rapportée à la variable d'origine, la transsérie $\tilde{H}_q(z)$ s'écrit :

(III.3.48) $$\tilde{H}_q(z) = a.\tilde{\Pi}(z).\{1 + \Sigma\, a_{\underline{n}}\, \tilde{\sqcup}_{\underline{n}}(z)\}$$

Rapportée à chaque temps critique $z_\imath$, la transsérie $\tilde{\Pi}_q(z)$ s'écrit :

(III.3.49) $$\tilde{\Pi}_{q,\imath}(z_\imath) = a.\tilde{\Pi}_\imath(z_\imath).\{1 + \Sigma\, a_{\underline{n},\imath}\, \tilde{\sqcup}_{\underline{n},\imath}(z_\imath)\}\ (**)$$

Notons $C_\imath$ l'accélération $z_\imath \to z_{\imath+1}$. Pour toute valeur $i < \imath_{\bullet}$ (quand il y en a), la commutation :

(III.3.50) $$\text{COMP}.\ \hat{C}_\imath = \hat{C}_\imath.\ \text{COMP}$$

est essentiellement triviale, car chaque monôme $P_\sigma(z) = z^{-\sigma}$ de $\tilde{G}_{q+1}(z)$, quand on y effectue la substitution $z = \tilde{H}_{q,\imath}(z_\imath)$, débute par un transmonôme $a^{-\sigma}(\tilde{\Pi}_\imath(z_\imath))^{-\sigma}$ qui est un infiniment petit *surexponentiel* (non strictement, en général) en $z_\imath$. Par suite, la transsérie $P_\sigma\ o\ \tilde{H}_{q,\imath}(z_\imath)$ n'est pas "réalisée" en totalité dans le modèle $\zeta_\imath$. Seules, certaines des transséries *portées* par elle seront réalisées comme fonctions ; mais le début (transmonôme dominant) de chaque $P_\sigma\ o\ \tilde{H}_{q,\imath}(z_\imath)$ reste symbole. Par suite, pour un tel i, la commutation (III.3.50) équivaut à un certain nombre de permutations.

(III.3.51) $$\int .\Sigma = \Sigma.\int$$

où $\int$ représente l'intégrale (I.2.8) d'accélération $z_\imath \to z_{\imath+1}$(***) et où Σ représente la somme finie $\Sigma_{\square}$ qui correspond au décompte des parties (au plus) exponentielles en $z_\imath$ qu'on trouve en facteur de chaque transmonôme $\tilde{\square}$ *strictement surexponentiel* en $z_\imath$ lors de la mise de $\tilde{G}_{q+1} o \tilde{H}_{q,\imath}(z_\imath)$ sous forme "réduite" ou "descendue" (voir § III.1). Il y a en général une infinité de permutations (III.3.51) à considérer (autant que de $\tilde{\square}$) mais pour chacune d'elles $\Sigma_{\square}$ est finie(****).

(*) à vrai dire, c'est automatiquement réalisé pour $\tilde{H}_{q+1}$ quand $\tilde{G}_{q+1}$ est de type I ou II, étant donné l'arc-origine C_1 choisi sur le polycycle C; mais ce n'est pas automatiquement réalisé pour les $\tilde{F}_\imath$ et $\tilde{F}_{\imath,\imath+1}$ dont nous avons *aussi* besoin pour faire marcher l'induction.

(**) Bien entendu, chaque $\tilde{\sqcup}_{\underline{n},\imath}$ se calcule à partir de la *totalité* des $\tilde{\sqcup}_{\underline{m}}$ pour $\underline{m} \leq \underline{n}$ et non pas (sauf exception) par simple changement de variable $z_\imath = \tilde{F}_\imath(z)$ dans $\tilde{\sqcup}_{\underline{n}}$.

(***) avec intervention des majeurs quand c'est nécessaire.

(****) c'est une conséquence immédiate du "premier axiome" de finitude des transséries qui est évidemment vérifié ici.

Pour les i qui sont $> i_*$ (quand il y en a) la commutation (III.3.50) est encore triviale, mais pour une raison *exactement opposée* : parce que les transséries P_σ o $\tilde{H}_{q,i}(z_i)$ ont un transmonôme dominant, à savoir deb P_σ o $\tilde{H}_{q,i}$, qui est subexponentiel en z_i et qui donc a *déjà* été sommé (à partir du modèle ς_{i_*} très exactement) et parce que la somme Σ a_σ P_σ o $\tilde{H}_{q,i}(z_i)$ dans le modèle ς_i (c'est-à-dire en fait chacunne de ses parties "réalisables" comme fonctions ; autrement dit, tout ce qu'elle *porte* - voir § III.1 - d'exponentiel en z_i en ses différents étages ≥ 0) s'obtient essentiellement par une accélération de la somme correspondante du modèle ς_{i-1}. Evidemment, il peut très bien y avoir dans le "début" de P_σ o $\tilde{H}_{q,i}(z_i)$, des termes surexponentiels en z_i ; on doit alors considérer les *sommes finies* (de fonctions *réalisables*) $\Sigma_\square$ qui sont en facteur de chacun de ces transmonômes $\square$ strictement surexponentiels en z_i et, ici encore, la commutativité (III.3.50) équivaut à des permutations (III.3.51) pour des sommes $\Sigma_\square$ finies (mais en nombre peut-être infini !) et un $\int$ qui représente toujours l'intégrale d'accélération $\hat{\mathcal{C}}_i : z_i \to z_{i+1}$.

Seule pourrait donc faire difficulté l'étape $i = i_*$, c'est-à-dire le passage du modèle ς_{i_*} au modèle ς_{i_*+1}, car c'est dans le modèle ς_{i_*} que les $a_\sigma P_\sigma$ vont être *réalisés* pour la première fois (i.e. qu'ils vont perdre leur statut de *symbole formel* pour être actualisés comme *fonctions*) et que ces $a_\sigma P_\sigma$ vont être *sommés en* σ. Par suite, c'est à ce moment précis que la divergence de $\tilde{G}_{q+1}$ (si $\tilde{G}_{q+1}$ correspond à un sommet irrégulier) pourrait se faire sentir. Mais il n'en est rien, comme nous allons voir.

Choisissons en effet, dans la classe critique $\{\{\log z\}\}$ de $\tilde{G}_{q+1}(z)$, un représentant "lent" z_0 de la forme :

(III.3.52)
$$\begin{cases} z_0 = A(z) = \log z - K_0 \log\log z & \text{(exactement)} \\ z = B(z_0) = \exp(z_0 + K_0 \log z_0 + \ldots) & (K_0 > 0) \end{cases}$$

(Techniquement, il est commode ici de prendre A "simple" plutôt que B). D'après le § II.2, la transsérie tout à fait élémentaire $\tilde{S}(z_0) = \tilde{G}_{q+1}$ o $\tilde{B}(z_0)$ admet une transformée de Borel $z_0 \to \varsigma_0$ qu'on notera $\hat{S}(\varsigma_0)$ et qui est définie, continue et analytique par morceaux sur $\mathbf{R}^+$ et, surtout, à croissance (au plus) exponentielle en ς_0 sur $\mathbf{R}^+$.

D'autre part, si $\tilde{H}_q$ est accéléro-sommable, L o $\tilde{H}_q$ et A o $\tilde{H}$ le sont aussi (c'est élémentaire ; voir [E.5]). De plus, pour tout représentant z_{i_*} de la "nouvelle" classe critique, on a :

(III.3.53) $$L\ o\ \tilde{H}_{q,i_*}(z_{i_*}) \sim A\ o\ \tilde{H}_{q,i_*}(z_{i_*}) = a.z_{i_*} + b\,\tilde{\square}_1(z_{i_*}) + \tilde{o}(\ldots)$$

avec $a > 0, b \neq 0$ et $\tilde{\square}_1(z_{i_*}) = \tilde{o}(z_{i_*})$. On peut toujours (grâce au jeu sur les $\tilde{\mathcal{F}}_i$) choisir un temps z_{i_*} du type (III.3.18) tel que $a = 1$ et $b > 0$(*).

(*) la condition $b > 0$ n'est essentielle que si $\tilde{\square}_1 = \tilde{\Pi}_1$ = infiniment grand.

D'où :

(III.3.54) $A \circ \tilde{H}_{q.i_*}(z_{i_*}) = z_{i_*} + \tilde{Q}(z_{i_*})$ avec $\tilde{Q}(z_{i_*}) > 0$ (formellement)

(III.3.55) $\tilde{Q}(z_{i_*}) = \tilde{Q}_0(z_{i_*}) + \sum_{0<\underline{n}<\underline{n}_0} \tilde{B}_{\underline{n}}(z_{i_*})\tilde{Q}_{\underline{n}}(z_{i_*})$ $(\tilde{Q}_0(z_{i_*}) = \tilde{Q}(z_{i_*}))$

avec des *transséries* $\tilde{Q}_{\underline{n}}(z_{i_*})$ exponentielles (au plus) en z_{i_*} et des *transmonômes* $\tilde{B}_{\underline{n}}(z_{i_*})$ *strictement surexponentiels* en z_{i_*}. Bien sûr, le second membre de (III.3.55) peut se réduire à $\tilde{Q}_0$).

Considérons du "côté formel" la transsérie composée :

$$\text{(III.3.56)} \quad \begin{cases} \tilde{H}_{q+1.i_*}(z_{i_*}) = \tilde{G}_{q+1} \circ \tilde{H}_{q.i_*}(z_{i_*}) = \tilde{S}(z_{i_*} + \tilde{Q}(z_{i_*})) \\ = \tilde{S}(z_{i_*}) + \sum_{m\geq 1} \frac{1}{m!} (\tilde{Q}(z_{i_*}))^m . \partial^m . \tilde{S}(z_{i_*}) \text{ avec } \partial = \partial/\partial z_{i_*} \end{cases}$$

et ce qui lui correspond du "côté convolutif", c'est-à-dire :

$$\text{(III.3.57)} \quad \hat{H}_{q+1.i_*}(\zeta_{i_*}) = \hat{S}(\zeta_{i_*}) + \sum_{m\geq 1} \frac{1}{m!} \left[(\hat{Q}(\zeta_{i_*}))^{*m} * ((-\zeta_{i_*})^m . \hat{S}(\zeta_{i_*}))\right]$$

car $\partial \to \hat{\partial}$ = multiplication par $(-\zeta_{i_*})$. En fait, l'identité (III.3.57) est ici écrite sous forme "compacte", car elle mêle des *parties réalisées* (fonctions) et des *parties encore virtuelles* (transmonômes formels). Si on veut interpréter (III.3.57) comme identité entre *fonctions*, on doit "projeter" cette identité en une infinité d'identités [III.3.57.$\underline{n}$] qui expriment l'identité des *fonctions* qui apparaissent aux deux membres de (III.3.57) en facteur de chaque transmonôme $\sqcup_{\underline{n}}(z_{i_*})$ qui est *strictement surexponentiel* en z_{i_*} (comme *infiniment petit,* car s'en est forcément un) et qui est, de ce fait même, condamné à rester formel dans le modèle ζ_{i_*}. Pour $\underline{n} = 0$, l'identité (III.3.57. 0) est un peu particulière : elle s'obtient en ne "retenant" de (III.3.57) que la partie antérieure au premier transmonôme $\sqcup_1(z_{i_*})$, c'est-à-dire tout simplement en remplaçant $\hat{Q}$ par $\hat{Q}_0$:

$$\text{(III.3.57.0))} \quad \hat{H}_{q+1.i_*.0}(\zeta_{i_*}) = \hat{S}(\zeta_{i_*}) + \sum_{m\geq 1} \frac{1}{m!} \left[(\hat{Q}_0(\zeta_{i_*}))^{*m} * ((-\zeta_{i_*})^n . \hat{S}(\zeta_{i_*}))\right]$$

Les identités (III.3.57.$\underline{n}$) générales ($\underline{n} \geq 1$) se présentent sous une forme très analogue (voir [E.5]) avec toujours, comme ingrédient essentiel, la suite infinie des *puissances de convolution de* $\hat{Q}_0(\zeta_{i_*})$. On voit donc que, pour $i = i_*$, la commutation (III.3.50) ne va plus de soi, car elle équivaut, pour chaque $\underline{n}$, à la permutabilité (III.3.51) de l'intégrale d'accélération $\int$ et d'une somme Σ qui est cette fois *infinie.*

Toutefois, par *l'induction,* on sait que $\tilde{Q}_0$ est accélérable de z_{i_*} à z_{i_*+1} (on le sait soit *directement,* parce que la "nouvelle" classe critique $\{\{z_{i_*}\}\}$ coïncide avec une ancienne, soit *indirectement*, parce que $\tilde{Q}_0$, est accélérable de $z_{\bullet-1}$ à $z_{\bullet+1}$ et qu'on peut toujours décomposer cette accélération en deux accélérations $z_{\bullet-1} \to z_{\bullet} \to z_{\bullet+1}$). Par suite, $\hat{Q}_0(\zeta_{\bullet})$ possède la *croissance accélérable* $z_{\bullet} \to z_{\bullet+1}$, autrement dit vérifie :

(III.3.58) $$|\hat{Q}_0(\zeta_{\bullet})| \leq \text{cste} \,/\, C_{\bullet}(r, \zeta_{\bullet}) \quad (\text{pour } \zeta_{\bullet} \to +\infty)$$

où $C_{\bullet}(\bullet, \bullet)$ désigne le noyau de l'accélération $z_{\bullet} \to z_{\bullet+1}$ et r un réel > 0 assez petit (pour l'asymptotique des noyaux d'accélération, voir § I.2, § IV.2 et surtout [E.5] et [E.6]). D'où, par des estimations classiques en théorie de l'accélération, une majoration, uniforme en m, des puissances de convolution de $\hat{Q}_0$:

(III.3.59) $$|\hat{Q}_0^{*m}(\zeta_{\bullet})| \leq C_{m,\epsilon}/C_{\bullet}(r', \zeta_{\bullet}) \quad (\varepsilon > 0, 0 < r' < r)$$

uniformémant sur $\zeta_{\bullet} \in [\varepsilon, +\infty]$ et avec des constantes $C_{m,\epsilon}$ telles que :

(III.3.60) $$\frac{1}{m}\, |C_{m,\epsilon}|^{1/m} \to 0 \text{ quand } m \to +\infty.(^*)$$

Moyennant quelques précautions à l'origine (où il faut travailler sur les *majeurs*) les majorations (III.3.59) et (III.3.60) assurent *l'absolue accélérabilité* (pour *tout* $\zeta_{\bullet+1} < r'$) du second membre de (III.3.57. 0), ou, si l'on préfère, la *possibilité d'intégrer en* $\zeta_{\bullet}$ sous le signe somme (encore une fois, pour $\zeta_{\bullet+1}$ assez petit) c'est-à-dire de faire (III.3.51). Pour chaque identité (III.3.57. $\underline{n}$), les majorations qui assurent l'absolue accélérabilité sont tout à fait du même type et il existe d'ailleurs une présentation de la question qui évite à peu près tout calcul (cf [E.5]).

Il ne reste plus qu'à lever les deux hypothèses restrictives que nous avons faites, à savoir $\{i_{\bullet} < i_{q+1}\}$ et $\{\tilde{G}_{q+1}$ de type I$\}$.

Lorsque $i_{\bullet} = i_{q+1}$ (i.e. lorsque la nouvelle classe critique est la dernière - quant à la possibilité de cette éventualité pour les $\tilde{H}_q$ et les $\tilde{F}_{\bullet}$, voir note antérieure dans cette même section), c'est la *dernière étape* de l'accéléro-sommation, à savoir *Laplace* (case supérieure du diagramme "de droite" ci-avant) qui est non triviale et il s'agit alors de montrer la commutation :

(III.3.61) $$\text{COMP}.\mathcal{L} = \mathcal{L}.\text{COMP}$$

ce qui revient à prendre l'intégrale de Laplace *sous* le signe somme de (III.3.57.0)(**), permutation qui, ici encore, est justifiée par les majorations (III.3.59) (III.3.60), mais avec

---

(*) Par exemple, si la transsérie $\tilde{Q}_0(z_{\bullet})$ est du type $\tilde{\mathbb{W}}$ (i.e. infiniment petite) on peut prendre $C_{m,\epsilon} \equiv c_{\epsilon}(\forall\varepsilon)$. Les constants $C_{m,\epsilon}$ sont d'autant moins bonnes que deb $(\tilde{Q}_0)$ est plus grand.

(**) ici, il n'y a plus que $\underline{n} = 0$

le *noyau exponentiel* à la place de l'inverse du noyau d'accélération :

(III.3.62) $\qquad \exp(\varsigma_{\iota_\bullet}/r')$ à la place de $1/C_{\iota_\bullet}(r'/\varsigma_{\iota_\bullet})$

En effet, comme on travaille à la dernière étape, on sait, en vertu de l'*induction*, que $\hat{Q}_0(\varsigma_{\iota_\bullet})$ est "laplaçable" et donc à croissance exponentielle (au plus), ainsi que ses puissances de convolution.

Lorsque $\tilde{G}_{q+1}$ *est de type* II et donc de la forme (III.3.44), l'argument est exactement le même, à ceci près qu'il faut prendre K_0 assez grand (alors que pour le type I tout $K_0 > 0$ convenait).

Lorsqu'enfin $\tilde{G}_{q+1}$ *est de type* III$^+$ *ou* III$^-$ et donc de la forme (III.3.45) ou (III.3.46), c'est encore le même argument, mais avec une "nouvelle" classe critique formelle $\{\{R_\bullet o\tilde{H}_q(z)\}\}$ avec $R_\bullet(z) = z^p$ ou z (et A, B modifiés en conséquence) si bien qu'ici, la "nouvelle" classe critique peut effectivement être la "dernière", même pour les $\tilde{H}_{q+1}$ et l'origine du polycycle choisie comme nous l'avons fait.

C.Q.F.D.

Remarques : On donne dans [E.5] des informations complémentaires et des indications sur les variantes possibles de la preuve. Telle est en effet la *simplicité de principe* du processus d'accéléro-sommation (passage obligatoire par un nombre déterminé de "modèles critiques") et sa *souplesse d'exécution* (dans le choix des temps critiques lents ; dans l'actualisation progressive des parties symboliques, etc...) que toute variante de ce processus qui conserve un sens conduit automatiquement au bon résultat. D'ailleurs, à ceux qui trouveraient compliquée la preuve ci-dessus, on peut répondre ceci :

Premièrement, cette méthode livre beaucoup plus que la preuve de l'énoncé de Dulac (finitude des cycles-limites) pour lequel il existe des raccourcis. Voir §.III.5 ci-après.

Deuxièmement, cette méthode (modulo les variantes) est la seule (du moins est-ce mon absolue conviction) qui puisse jamais faire passer constructivement de l'objet géométrique F à l'objet formel $\tilde{F}$, et vice versa, et par suite conduire à une *connaissance totale* du premier.

Troisièmement, cette méthode possède un champ d'application naturel (celui des *fonctions analysables*) infiniment plus vaste que le problème de Dulac, qui n'en est qu'une illustration.

Quatrièmement, la complication du procédé tel qu'il est exposé ici n'est qu'*apparente*. Elle est surtout de l'ordre des notations et du langage. Comme ces choses sont nouvelles, certains raccourcis, abréviations et syncopes, qui par la suite entreront dans l'usage et ne soulèveront aucune objection, ne sont pas encore acceptables et doivent être remplacés pas des "développements" assez pénibles (*). Mais dès qu'on s'est bien familiarisé avec la démarche, celle-ci se résoud en une suite d'évidences d'où pratiquement toute difficulté est évacuée. On serait presque tenté de dire que, dans le problème de Dulac, il n'y avait à peu près rien à prouver, car il fait partie de ces problèmes qui *se dissolvent* (dans une vision juste) *plutôt qu'ils ne se résolvent.* Quoi qu'il en soit, dans la plupart des questions de géométrie locale (voire de physique) où surgit de la divergence, l'avenir est assurément du côté de ces méthodes. Cf [E.3] [E.6] [E.8]. Aux sceptiques : rendez-vous dans dix ans !

III.4. Exemples d'accéléro-sommation

Au §.III.2, nous avons introduit deux invariants des polycycles, à savoir leur altitude ν et leur altitude réduite ν', qui s'interprêtent de la manière suivante. Si on figure la factorisation naturelle $\tilde{F}_{nat}$ (resp. la factorisation réduite $\tilde{F}_{red}$) de l'application de retour par un graphique où les facteurs $\tilde{A}_i$ (resp $\tilde{B}_i$) sont représentés par des points (•), les facteurs $E = \exp$ par des segments ascendants (↖) et les facteurs $L = \log$ par des segments descendants (↙), on obtient une ligne brisée qui s'inscrit dans une bande horizontale dont la hauteur est précisément ν(resp ν').

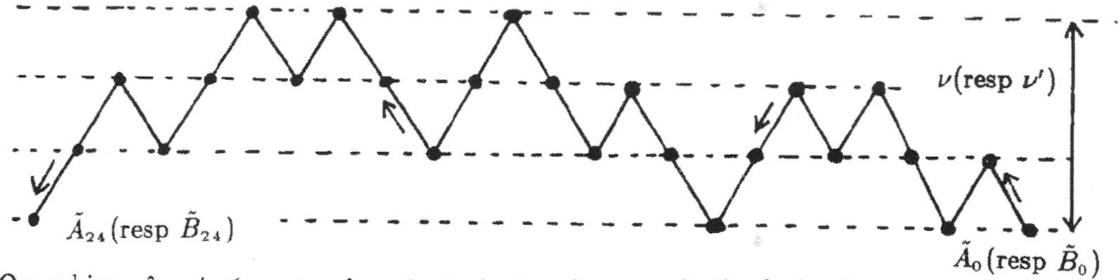

On a bien sûr $\nu' \leq \nu$ et même $0 \leq \nu' \leq \nu$, à cause de l'endroit où nous avons choisi l'origine sur le polycycle. La transsérie $\tilde{F}$ se calcule indifféramment à partir de $\tilde{F}_{red}$ ou $\tilde{F}_{nat}$. Raisonnons sur $\tilde{F}_{red}$, qui est plus courte, et regardons ce que représente le processus d'accéléro-sommation pour les premières valeurs de ν'.

Altitude $\nu' = 0$ ($\Rightarrow$ *Une seule classe critique, d'exponentialité* -1)

(*) Parlant des *fonctions analysables*, on se trouve un peu dans la position de quelqu'un qui parlerait pour la première fois des *fonctions analytiques* et n'aurait pas le droit de dire "je prolonge le long du chemin Γ" mais devrait à chaque fois parler de points intermédiaires, d'éléments analytiques, de disques de convergence se chevauchant, etc, puis justifier l'invariance du résultat par rapport à l'arbitraire des étapes intermédiaires.

Il n'y a alors qu'un seul facteur, soit $\tilde{B}_0$. Celui-ci peut "résumer" beaucoup de facteurs $\tilde{G}_i$ correspondant à des sommets irréguliers de type I ou II, mais il y a toujours (au plus) une *classe critique* (sério-critique) *unique*, à savoir la classe $\{\{\log z\}\}$ avec le z_0 de (II.2.10) comme choix possible de *temps lent*, pour K_0 assez grand.

Altitude $\nu' = 1$ ($\Rightarrow$ *Une seule classe critique d'exponentialité -1 ; plusieurs classes d'exponentialité* 0)

Pour la forme que revêt la transsérie $\tilde{F}$ dans ce cas, voir [E.5]. En plus de la classe critique $\{\{\log z\}\}$ déjà rencontrrée, qui correspond à l'*ensemble* des sommets irréguliers de type I ou II et d'altitude (réduite) nulle, peuvent venir s'ajouter un nombre fini de *classes critiques* $\{\{z^{\sigma_i}\}\}$, qui correspondent soit à des sommets irréguliers de type III et d'altitude (réduite) égale à $1/2$ (*) (*sommo-criticité*) ou à des sommets irréguliers de type I ou II et d'altitude (réduite) égale à 1 (*sério-criticité*). Dans ces classes, on peut prendre comme temps lents les $z_i = z^{\sigma_i} - z^{\sigma_i}/\log z$, qui conviennent "largement". Il n'y a jamais, ici, de *vraie* classe critique d'exponentialité 1 mais, pour la commodité de l'accéléro-sommation, on peut passer par un nombre fini de "temps critiques fictifs" du type $z_i' = \exp(z^{\alpha_i} - z^{\beta_i})$ $(\alpha_i > \beta_i > 0)$.

Altitude $\nu' = 2$ ($\Rightarrow$ *Une seule classe critique d'exponentialité -1 ; plusieurs classes d'exponentialité* 0 *ou* 1)

Pour la forme de $\tilde{F}$, voir [E.5]. Aux classes critiques d'exponentialité -1 ou 0 déja vues, peuvent s'ajouter des classes critiques d'exponentialité 1, qui correspondent à des sommets irréguliers d'altitude 2 (type I ou II) ou d'altitude $3/2$ (type III). Ces nouvelles classes admettent des représentants canoniques (transmonomiaux, mais ici en fait "monomiaux") de la forme :

$$\text{(III.4.1)} \qquad \tilde{\Pi}_i(z) = z^{\sigma_i} . \exp P_i(z) \quad \text{avec} \quad P_i(z) = \Sigma\, a_{i,n} z^{\sigma_{i,n}} > 0 \quad (\Sigma \text{ fini})$$

Altitude $\nu' = 3$ ($\Rightarrow$ *Une seule classe critique d'exponentialité* 0 ; *plusieurs classes critiques d'exponentialité* 0 *ou* 1 *ou* 2)

C'est le *premier cas* où *peuvent* intervenir des classes critiques d'exponentialité 2, dont les représentants canonique $\tilde{\Pi}_i$ *peuvent* ne pas être élémentaires (**), mais comporter un nombre infini de coefficients :

Par exemple :

(*) c'est-à-dire à des sommets figurés sur le graphique par un trait oblique joignant un point d'altitude 0 à un point d'altitude 1.

(**) cela dépend des conditions (III.3.8)

(III.4.2) $\quad \tilde{\Pi}_{\iota}(z) = \exp\left((\exp(z^5 + z^2)) \bullet (1 + \sum_n a_n z^{-n})\right) \quad (\Sigma \text{ infini})$

A partir de l'altitude $\nu' = 4$, les classes critiques d'exponentialité ≥ 3 qui peuvent apparaître, possèdent *génériquement* des représentants canoniques $\tilde{\Pi}_{\iota}$ non-élémentaires (de "vrais" transmonômes).

III.5. Informations complémentaires

Variante de la démonstration : par accéléro-sommation des transséries latérales.

On donne aussi dans [E.5] une variante de la précédente démonstration, qui consiste à associer à l'application de retour F non plus sa transsérie médiane $\tilde{F}$ mais ses transséries latérales ${}^{+}\tilde{F}$ et ${}^{-}\tilde{F}$.

(III.5.1) $\quad {}^{+}\tilde{F} = {}^{+}\tilde{G}_{r} o \ldots {}^{+}\tilde{G}_{1} \;\; ; \;\; {}^{-}\tilde{F} = {}^{-}\tilde{G}_{r} o \ldots {}^{-}\tilde{G}_{1}$ avec :

$$\begin{cases} \text{pour les sommets de type I ou II :} \quad {}^{+}\tilde{G}_{\iota} = {}^{-}\tilde{G}_{\iota} = \tilde{G}_{\iota} \quad (\text{cf } \S II.2 \text{ et } \S II.6) \\ \text{pour les sommets de type III :} \quad {}^{+}\tilde{G}_{\iota} \text{ et } {}^{-}\tilde{G}_{\iota} \text{ comme en } (II.4.3) \text{ et } (II.4.4) \end{cases}$$

Prenons ${}^{+}\tilde{F}$ par exemple. La *première différence* vient évidemment de ce que, du fait des coefficients complexes qui peuvent figurer dans les facteurs ${}^{+}\tilde{G}_{\iota}$ correspondant à des sommets de type III, la transsérie ${}^{+}\tilde{F}$ appartient à une trigèbre complexe ${}^{+}\mathbf{C}[[[z]]]$ construite d'une façon analogue à $\mathbf{R}[[[x]]]$ mais avec des coefficients complexes et une convention spécifique de "descente" (ou d'expulsion des infiniment petits des exponentielles) conduisant à une écriture unique. La *deuxième différence* tient à ce que les temps critiques qu'on est conduit à prendre ne sont plus exactement réels, mais seulement équivalents à des temps réels z_{ι}. La *troisième différence* tient, bien sûr, à ce qu'il faut appliquer le processus d'accéléro-sommation *latéral*, c'est-à-dire prendre, dans les intégrales d'accélération ou dans la dernière intégrale de Laplace, les déterminations *latérales* (ici, supérieures) des $\hat{\varphi}(\varsigma_{\iota})$. Cette méthode présente un léger avantage pratique en ceci, justement, qu'elle dispense d'envisager cette moyenne "compliquée" qu'est la fonction médiane, mais elle a l'inconvénient d'introduire une *dissymétrie* (latéral supérieur ou inférieur) et surtout des *coefficients imaginaires* dans un problème foncièrement réel. Avec cette méthode, il faut vérifier (par induction sur la hauteur des transséries) que les parties imaginaires des transséries portées par les exponentielles sont toujours infiniment petites, et l'*absence de termes oscillatoires* perd ainsi le caractère "d'évidence" qu'elle possède quand on envisage la transsérie médiane. Un autre inconvéniant de la méthode est qu'elle ne permet pas de donner un sens "géométrique" à tout *tronçon* de la transsérie, car elle conduit à attribuer aux tronçons "non-commençants" des sommes qui peuvent n'être ni réelles ni même équivalentes à des fonctions réelles.

Autre type de démonstration : par les lemmes d'immiscibilité.

Les démonstrations par accéléro-sommation ne se justifient pleinement que par rapport au but que nous nous sommes assigné, à savoir la "connaissance totale" de F grâce à une *bijection constructive* entre les F et les $\tilde{F}$. Mais si on vise seulement l'énoncé de Dulac (non oscillation de $F(z) - z$), on peut donner des preuves simplifiées, qui reposent sur des lemmes *d'immiscibilité purement formels*, selon lesquels il ne peut exister d'identités mixtes non triviales du type :

(III.5.2) $$Q^{-1} o \tilde{K}_r o Q o \tilde{H}_r o \ldots Q^{-1} o \tilde{K}_1 o Q o \tilde{H}_1 \equiv \text{identité (*)}$$

avec des facteurs $\tilde{H}_i$ et $\tilde{K}_i$ de la forme $z\{a_0 + \Sigma\, a_n\, z^{-n}\}$ $(a_0 > 0)$ et avec par exemple : (*cas 0*) $Q(z) = z^\lambda$; λ irrationnel ; (*cas 1*) $Q(z) = E(z) =$ exp z; (*cas n*) $Q(z) = E_n(z) =$ exp ... exp z $(n fois)$.

Ces lemmes interdisent (resp. limitent) la possibilité de simplifications "médiates" (**) dans la factorisation réduite $\tilde{F}_{red}$(cf (III.2.7)) et conduisent à diverses démonstrations simplifiées dans lesquelles la part de l'accélération est d'autant plus faible que les lemmes d'immiscibilité invoqués sont plus forts. Ils donnent également des renseignements complémentaires sur les *symétries* que présente nécessairement le polycycle lorsque l'application de retour est l'identité.

Par exemple, les lemmes d'immiscibilité, pour tout $n \geq 2$,admettent des preuves *formelles* (plus n est grand, plus c'est facile, si fortes deviennent les contraintes) et ils permettent de conclure à la non-oscillation à partir des seules accélérations élémentaires du type :

(III.5.3) $z \to z^\alpha$; $z \to z^\alpha \exp A(z)$; $z^\alpha \exp A(z) \to z^\beta \exp B(z)$ $(A, B$ sommes finies $\Sigma a_i z^{\sigma_i})$

Les lemmes d'immiscibilité pour *tout* $n \geq 1$ permettraient même de *tout* ramener aux seules accélérations archi-élémentaires $z \to z^\sigma$ (voir [E.5]). Malheureusement, je ne possède pas de preuve *formelle* du lemme d'immiscibilité pour $n = 1$ (c'est le seul cas difficile) et, *pour l'instant, la preuve la plus "courte" de la non-oscillation, est celle qui passe par les accélérations* (III.5.3) *quand même très élémentaires !*

(*) au sens des transséries.
(**) c'est-à-dire autres que les simplifications "immédiates" (III.2.5) et (III.2.6).

IV - APPENDICES

Les quatre Appendices qui suivent abordent des questions fort intéressantes en elles-mêmes, mais ils n'ont qu'un lien très ténu avec le problème de Dulac et ne sont nullement nécessaires à sa solution. Aussi l'exposé sera-t-il très rapide et les démonstrations quasiment absentes. Détails et preuves figurent dans [E.5] et surtout [E.8].

Le *premier Appendice* explicite une transformation purement formelle qui est sous-jacente aux transformations fonctionnelles de Borel-Laplace $\varphi(z) \leftrightarrow \hat{\varphi}(\varsigma)$ et qui échange les comportements (trans)-asymptotiques (en $z \sim +\infty$ et $\varsigma \sim +0$) les plus généraux, lesquels sont décrits par des (trans)séries. Chose curieuse, on se trouve dans l'obligation de factoriser ces transformations foncièrement linéaires que sont Borel-Laplace en un produit de deux applications non linéaires, la composition inverse et le *quartage*, le tout conduisant aux *formules cryptolinéaires*.

Le *second Appendice* montre l'identité de deux notions d'origine fort différente : les *fonctions cohésives* (qui sont les plus régulières des fonctions quasi-analytiques au sens de Denjoy-Carleman) et les *fonctions accélérées* (pour des accélérations lentes), et en tire une méthode très constructive pour *prolonger* quasianalytiquement ces fonctions et, surtout, pour *contourner* leurs singularités sans quitter l'axe réel.

Le *troisième Appendice* répond à la question suivante : quelle est la "régularité maximale" que peuvent présenter les difféomorphismes $\mathcal{L}$ (resp $\mathcal{E}$) de $[..., +\infty]$ qui sont *très lents* (resp. *très rapides*), c'est-à-dire qui vérifient $\mathcal{L}(x) << L_q(x)$ (resp. $\mathcal{E}(x) >> E_q(x)$) pour tout q. Il s'avère que cette régularité maximale - la *douceur* - s'exprime par l'existence, pour les *dérivées posthomogènes* de $\mathcal{L}$, de certaines *séries asymptotiques* qui sont *universelles*, i.e. totalement indépendantes de la fonction considérée. Les fonctions très lentes ou très rapides se rencontrent rarement en mathématiques, mais leur *douceur* induit une notion analogue chez les fonctions à croissance modérée, pour qui la douceur équivaut à l'existence d'une infinité d'*équivalences universelles*. Il est beau de voir une structure aussi riche jaillir d'une interrogation aussi simple.

Le *quatrième Appendice* aborde la question de *"l'échelle naturelle de croissance"*. Il utilise la notion de *douceur* pour réduire au maximum l'indétermination inhérente au problème et pour tout ramener à deux sortes de classes d'équivalence, à savoir les TC et ZC (types et zones de croissance) et les TCD et ZCD (types et zones de croissance différentiables). Sur les *types* est définie une *itération transfinie*, qui reflète la structure discrète des ordinaux (ici, on s'arrête à ω^ω). Quant aux *zones*, elles s'ordonnent selon une structure *fractale*, qui se signale à la fois par son origine inhabituelle (un principe d'indiscernabilité) et par son net caractère universel.

IV-1. Analysabilité de Borel-Laplace et formules cryptolinéaires.

On sait que, dans les cas simples, les transformations mutuellement inverses de Borel et de Laplace :

(IV.1.1) $$\mathcal{B} : \varphi(z) \to \hat{\varphi}(\varsigma) = \frac{1}{2\pi i} \int_{-i\infty}^{+i\infty} \varphi(z).e^{z\varsigma} dz$$

(IV.1.2) $$\mathcal{L} : \hat{\varphi}(\varsigma) \to \varphi(z) = \int_{0}^{+\infty} \hat{\varphi}(\varsigma).e^{-z\varsigma} d\varsigma$$

échangent le comportement asymptotique de $\varphi(z)$ en $z = +\infty$ et $\hat{\varphi}(\varsigma)$ en $\varsigma = +0$.
Par exemple :

(IV.1.3) $$\varphi(z) = z^{-\sigma} \quad \leftrightarrow \quad \hat{\varphi}(\varsigma) = \varsigma^{\sigma-1}/\Gamma(\sigma) \qquad (\forall \sigma > 0)$$

Mais est-ce absolument général ? Est-ce encore vrai lorsque $\varphi(z)$ ou $\hat{\varphi}(\varsigma)$ s'exprime par un échafaudage de puissances, d'exponentielles et de logarithmes itérés, conduisant à des séries irréductiblement empilées sur des millions d'étages ? Et si oui, quelle est la forme exacte de cette correspondance ? Quel est l'algorithme qui permet de traduire le comportement asymptotique de $\varphi(z)$ en celui de $\hat{\varphi}(\varsigma)$ et vice versa ?

Borel formel et Laplace formel. Les algèbres $\mathcal{E}(z)$ et $\hat{\mathcal{E}}(\varsigma)$.

La théorie des *transséries* formelles et des *fonctions analysables* permet de donner à ces questions une forme précise et, mieux encore, de leur apporter une réponse totalement positive. Elle permet surtout de montrer qu'il existe, sous-jacentes aux transformations fonctionnelles $\mathcal{B}$ et $\mathcal{L}$ (Borel et Laplace), deux transformations *purement formelles* $\tilde{\mathcal{B}}$ et $\tilde{\mathcal{L}}$, qui échangent l'algèbre multiplicative $\mathcal{E}(z)$ des transséries formelles en z (*) qui sont formellement à décroissance subexponentielle en $z \sim +\infty$:

(IV.1.4) $\varphi(z) = ⨆(z) = \Sigma\, a_{\underline{n}} \sqcup_{\underline{n}}(z)$ avec $-\log \sqcup_{\underline{n}}(z) << z$ (formellement)
et l'algèbre convolutive(**) des transséries formelles en ς^{-1} : (***)

(*) Signalons toutefois qu'il convient d'affaiblir quelque peu l'axiome de finitude (III.1.9). Voir [E.5] et aussi quelques indications en fin de section.
(**) Pour une convolution, également formelle et susceptible d'une définition directe, mais dont nous nous contenterons de dire qu'elle est la transmutée par Borel de la multiplication de $\mathcal{E}(z)$.
(***) Avec affaiblissement du premier axiome de finitude. Voir en fin de section.

(IV.1.5) $\quad \hat{\varphi}(\varsigma) = \boxplus(\varsigma^{-1}) = \Sigma\, a_{\underline{n}}\, \square_{\underline{n}}(\varsigma^{-1}) = \Sigma\, a_{\underline{n}}\, \sqcap_{\underline{n}}(\varsigma^{-1}) + \mathrm{cste} + \Sigma\, a_{\underline{n}}\, \sqcup_{\underline{n}}(\varsigma^{-1})$

décomposables en transmonômes $\square_{\underline{n}}(\varsigma^{-1})$ qui sont "formellement" intégrables en 0, ce qui, dans ce contexte, équivaut à l'existence d'un q et d'un $\varepsilon > 0$ tels que le transmonôme dominant vérifie :

(IV.1.6) $\quad \varsigma\, L_1(\varsigma^{-1})\, L_2(\varsigma^{-1}) \ldots L_{q-1}(\varsigma^{-1})(L_q(\varsigma^{-1}))^{1+\varepsilon}\, \square_0(\varsigma^{-1}) \to 0$ (formellement)

Curieusement, on est obligé de factoriser ces transformations foncièrement linéaires que sont Borel-Laplace en un produit de deux transformations non-linéaires, à savoir la prise de l'inverse de composition et le "quartage" ; et aussi de décomposer les algèbres $\mathcal{E}$ et $\hat{\mathcal{E}}$ en cellules "maximales" $\mathcal{E}_M$ et $\hat{\mathcal{E}}_N$, qui se correspondent deux à deux.
(**N.B.** Comme toutes les transséries considérées ici sont formelles, nous ometterons systématiquement les tildes)

Décomposition de $\mathcal{E}(z)$ et $\hat{\mathcal{E}}(\varsigma)$ en cellules maximales.
Commençons par introduire les espaces $\mathcal{F}$ et $\mathcal{G}$ formées des transséries f et g telles que :

(IV.1.7) $\quad z^{-1} << f(z) << 1$ et $1 << \varsigma^{-1} << g(\varsigma)$ (formellement)

Tout $f \in \mathcal{F}$ est la réciproque formelle d'un $g \in \mathcal{G}$ et on a (formellement) :

(IV.1.8) $\quad f : +\infty \to +0\ ;\ g : +0 \to +\infty\ ;\ g(f(z)) \equiv z\ ;\ -f'(z) > 0\ ;\ -g'(\varsigma) > 0$

Introduisons aussi les dérivées homogènes $f^{((m))}$ et $g^{((m))}$ pour $m \geq -2$:

(IV.1.9) $\quad f^{((-2))} = \grave{f}\ ;\ f^{((0))} = \log(-f')$ et $g^{((-2))} = \grave{g}\ ;\ g^{((0))} = \log(-g')$ (*)
($m = -2$ ou 0).

(IV.1.10) $\quad f^{((m))} = f^{(1+m)}.(-f')^{-1-\frac{m}{2}}/(1+m)!$ et

$$g^{((m))} = g^{(1+m)}(-g')^{-1-\frac{m}{2}}/(1+m)! \qquad (m = -1 \text{ ou } m \geq 1)$$

On dit que deux transmonômes e^{-F_1}, e^{-F_2} de $\mathcal{E}(z)$ ou e^{-G_1}, e^{-G_2} de $\hat{\mathcal{E}}(\varsigma)$ appartiennent à une même cellule ssi on a (au sens formel) :

(IV.1.11) $\quad \lim_{z \to +\infty} (f_1^{((-1))}(z) - f_2^{((-1))}(z)) = 0$ ou $\lim_{\varsigma \to +0} (g_1^{((-1))}(\varsigma) - g_2^{((-1))}(\varsigma)) = 0$

(*) où $\grave{f}$ et $\grave{g}$ désignent deux primitives de f et g liées par $\grave{f}(g(\varsigma)) + \grave{g}(\varsigma) = \varsigma g(\varsigma)$

avec :

(IV.1.12) $$f_1 = F_1' \; ; \; f_1^{((-1))} = (f_1).(-f_1')^{1/2} = (F_1').(-F_1'')^{-1/2}$$

$$(F_1 > 0 \; ; \; f_1 > 0 \; ; \; f_1^{((-1))} > 0)$$

(IV.1.13) $$g_1 = -G_1' \; ; \; g_1^{((-1))} = (g_1).(-g_1')^{-1/2} = (-G_1').(G_1'')^{-1/2}$$

$$(G_1 > 0 \; ; \; g_1 > 0 \; ; \; g_1^{((-1))} > 0)$$

Les cellules $\mathcal{E}_M$ de $\mathcal{E}$ (resp $\hat{\mathcal{E}}_N$ de $\hat{\mathcal{E}}$) sont donc paramétrées par les classes $[f]$ des $f \in \mathcal{F}$ (resp par les classes $[g]$ des $g \in \mathcal{G}$) caractérisées par la valeur M (resp N) de $f^{((-1))}$ (resp $g^{((-1))}$) modulo les infiniment petits :

(IV.1.14) $\sqcup = e^{-F} \in \mathcal{E}_M$ ssi $f^{((-1))} - M \to 0$ avec $M(z) = \Sigma\, a_{\underline{n}}\, \Pi_{\underline{n}}(z) +$ cste

(IV.1.15) $\square = e^{-G} \in \mathcal{E}_N$ ssi $g^{((-1))} - N \to 0$ avec $N(\varsigma) = \Sigma\, b_{\underline{n}}\, \Pi_{\underline{n}}(\varsigma^{-1}) +$ cste

$\mathcal{E}$ et $\hat{\mathcal{E}}$ sont sommes directes de deux sortes de cellules, à savoir les *cellules lentes* $\mathcal{E}_\sigma$ et $\hat{\mathcal{E}}_\sigma$:

(IV.1.16) $\sqcup = e^{-F} \in \mathcal{E}_\sigma$ ssi $\lim\limits_{z \to +\infty} (f_{(z)}^{((-1))})^2 = \sigma$ (formellement ; $\sigma > 0$)

(IV.1.17) $\square = e^{-G} \in \hat{\mathcal{E}}_\sigma$ ssi $\lim\limits_{\varsigma \to +0} (g_{(\varsigma)}^{((-1))})^2 = \sigma - 1$ (formellement ; $\sigma > 0$)

et les *cellules rapides* $\mathcal{E}_M$, $\hat{\mathcal{E}}_N$, pour $M(z)$ et $N(\varsigma)$ infiniment grands.

On dit que les cellules lentes $\mathcal{E}_\sigma$ et $\hat{\mathcal{E}}_\sigma$ de même indice σ sont *conjuguées* et on dit qu'une cellule rapide $\mathcal{E}_M = \mathcal{E}_{[f_0]}$ est *conjuguée* à une cellule rapide $\hat{\mathcal{E}}_N = \hat{\mathcal{E}}_{[g_0]}$ ssi pour *un* (et donc *tout*) $f \in [f_0]$ on a $g \in [g_0]$, où g désigne la réciproque formelle de f (*).

Regroupement des termes cocellulaires :

Toute transsérie φ ou $\hat{\varphi}$ appartenant à une cellule rapide :

(IV.1.18) $$\varphi(z) = \Sigma\, a_{\underline{n}} \sqcup_{\underline{n}}(z) \in \mathcal{E}_M \; ; \; \hat{\varphi}(\varsigma) = \Sigma\, b_{\underline{n}} \sqcup_{\underline{n}}(\varsigma^{-1}) \in \hat{\mathcal{E}}_N$$

(*) i.e. $g(f(z)) \equiv z$

peut d'une manière unique s'écrire sous la forme :

(IV.1.19) $\varphi(z) = a_0.e^{-F} \quad (F > 0)\ ;\ \hat{\varphi}(\varsigma) = b_0.e^{-G}(G > 0)$

avec $F' = f \in \mathcal{F}$ et $-G' = g \in \mathcal{G}$ et avec les propriétés de "*décroissance homogène*" :

(IV.1.20) $| f^{((m))}(z) | << (C_1(z))^m << 1$ et $| g^{((m))}(\varsigma) | << (C_2(\varsigma))^m << 1$ $(\forall m \geq 1)$(*)

(IV.1.21) $\lim_{m \to \infty} f^{((m))} = 0\ ;\ \lim_{m \to +\infty} g^{((m))} = 0$ (au sens des transséries ; i.e. terme à terme)

S'il s'agissait de *séries*, (IV.1.21) serait une simple conséquence de (IV.1.20) mais, s'agissant de transséries, c'est une condition tout à fait indépendante.

Quant aux transséries appartenant aux cellules lentes, elles s'écrivent :

(IV.1.22) $\varphi(z) = z^{-\sigma} A(z)$ avec $\log | A(z) | = o(\log z)$ (formellement) si $\varphi \in \mathcal{E}_\sigma$

(IV.1.23) $\hat{\varphi}(\varsigma) = \varsigma^{\sigma-1} B(\varsigma)$ avec $\log | B(\varsigma) | = o(\log \varsigma)$ (formellement) si $\hat{\varphi} \in \hat{\mathcal{E}}_\sigma$

Borel-Laplace dans les cellules lentes. Les opérateurs gamma et cogamma.

Borel-Laplace formel échange les cellules lentes $\mathcal{E}_\sigma$ et $\hat{\mathcal{E}}_\sigma$:

(IV.1.24) $\varphi(z) = z^{-\sigma}.A(z) \qquad \hat{\varphi}(\varsigma) = \varsigma^{\sigma-1}.B(\varsigma)$

selon les formules réciproques :

(IV.1.25) $B(\varsigma) = \text{cogam}_\sigma.A(1/\varsigma)\ ;\ A(z) = \text{gam}_\sigma.B(1/z) \quad (z \sim +\infty, \varsigma \sim +0)$

avec des opérateurs linéaires, gamma et cogamma, ainsi définis :

(IV.1.26) $\text{gam}_\sigma = \Gamma(\sigma - z.\frac{\partial}{\partial z}) = \sum_{n \geq 0} \frac{1}{n!} \Gamma^{(n)}(\sigma).(-z\frac{\partial}{\partial z})^n$ (Γ = fonction gamma)

(IV.1.27) $\text{cogam}_\sigma = \gamma(\sigma + \varsigma.\frac{\partial}{\partial \varsigma}) = \sum_{n \geq 0} \frac{1}{n!} \gamma^{(n)}(\sigma)(\varsigma\frac{\partial}{\partial \varsigma})^n$ (avec $\gamma(\sigma) \equiv 1/\Gamma(\sigma)$)

Pour $\sigma = 0$, la série de gam_σ commence par une intégration :

(IV.1.26 bis) $$\text{gam}_0 = (-z\frac{\partial}{\partial z})^{-1} + a_0 + \sum_{n \geq 1} a_n(-z\frac{\partial}{\partial z})^n$$

(*) pour des infiniment petits $C_1(z)$ et $C_2(\varsigma)$ *fixes* (indépendants de m) et pour $z \to +\infty, \varsigma \to +0$.

avec $\Gamma(\sigma) = \sigma^{-1} + \sum_{n\geq 0} a_n \sigma^n$

Borel-Laplace dans les cellules rapides. Formules cryptolinéaires "pratiques".

Soient deux classes réciproques $[f_0] \in [\mathcal{F}]$ et $[g_0] \in [\mathcal{G}]$. Borel-Laplace formel échange les cellules $\mathcal{E}_M = \mathcal{E}_{[f_0]}$ et $\hat{\mathcal{E}}_N = \hat{\mathcal{E}}_{[g_0]}$ correspondantes :

(IV.1.28) $\quad \varphi(z) = \Sigma\, a_{\underline{n}} \sqcup_{\underline{n}}(z) = a_0.e^{-F(z)} \qquad \hat{\varphi}(\varsigma) = \Sigma\, b_{\underline{n}} \sqcup_{\underline{n}}(\varsigma^{-1}) = b_0.e^{-G(\varsigma)}$

selon les formules suivantes :

(IV.1.29) $\quad \hat{\varphi} o f(z) = \frac{1}{\sqrt{2\pi}}.\varphi(z).e^{+zf(z)}.(-f'(z))^{-1/2}.F^{\#}(z)$ avec :

(IV.1.29 bis) $\quad f = F' > 0 \;;\; F_m = F^{(m)}/m!$

(IV.1.29 ter) $\quad F^{\#} = 1 + \sum_{p\geq 1} \frac{(-1)^p}{p!} \sum_{m\geq 1} \frac{(2m)!}{2^m\, m!} \sum_{m_i \geq 3\,.\,\Sigma m_i = 2m} F_{m_1}...F_{m_p}(F'')^{-m}$

(IV.1.30) $\quad \varphi\; o\; g(\varsigma) = \sqrt{2\pi}.\hat{\varphi}(\varsigma).e^{-\varsigma\, g(\varsigma)}.(-g'(\varsigma))^{-1/2}.G^{\#}(\varsigma)$ avec :

(IV.1.30 bis) $\quad g = -G' > 0 \;;\; G_m = G^{(m)}/m!$

(IV.1.30 ter) $\quad G^{\#} = 1 + \sum_{p\geq 1} \frac{(-1)^p}{p!} \sum_{m\geq 1} \frac{(2m)!}{2^m\, m!} \sum_{m_i\geq 3.\Sigma\, m_i = 2m} G_{m_1}...G_{m_p}(G'')^{-m}$

Borel-Laplace dans les cellules rapides. Formules cryptolinéaires théoriques.

Soit comme précédemment une transsérie $\varphi(z) = a_0\, e^{-F(z)} \in \mathcal{E}_M$ et sa transformée de Borel $\hat{\varphi}(\varsigma) = (a_0/2\pi).e^{-G(\varsigma)} \in \hat{\mathcal{E}}_N$. Posons encore $f = F', g = -G'$ et introduisons les réciproques formelles h et k de f et g :

$$f\; o\; h(\varsigma) \equiv \varsigma \;;\; f \in \mathcal{F},\; h \in \mathcal{G} \;;\; f \text{ et } k : +\infty \to +0 \text{ (formellement)}$$

$$g\; o\; k(z) \equiv z \;;\; g \in \mathcal{G} \;;\; k \in \mathcal{F} \;;\; g \text{ et } h : +0 \to +\infty \text{ (formellement)}$$

Avec les dérivées homogènes introduites en (IV.1.9) et (IV.1.10), la transformation de Borel-Laplace $\varphi(z) \rightleftarrows \hat{\varphi}(\varsigma)$ se traduit par les correspondances suivantes :

(IV.1.31) $\quad f \rightleftarrows k = g^{0(-1)} \qquad g \rightleftarrows h = f^{0(-1)}$

données par les formules (pour $n \geq -2$) :

(IV.1.32) $\quad f^{((n))} = k^{((n))} + \Sigma(-1)^n\, \beta^n_{n_1,\dots,n_m}\, k^{((n_1))}...\,k^{((n_m))}$

(IV.1.33) $\quad g^{((n))} = h^{((n))} + \Sigma\, \beta^n_{n_1,\dots,n_m}\, h^{((n_1))}...\,h^{((n_m))}$

(IV.1.34) $$k^{((n))} = f^{((n))} + \Sigma\ \gamma^{n}_{n_1 \ldots n_m} f^{((n_1))} \ldots f^{((n_m))}$$

(IV.1.35) $$h^{((n))} = g^{((n))} + \Sigma(-1)^n\ \gamma^{n}_{n_1 \ldots n_m} g^{((n_1))} \ldots g^{((n_m))}$$

où toutes les sommes Σ sont étendues aux m et n_i tels que :

(IV.1.36) $$\{m \geq 2\ ;\ n_i \geq 1\ ;\ n_1 + \ldots n_m = n + 2m - 2\}$$

et où les coéfficients β et γ sont des rationnels explicitement calculables à partir des formules cryptolinéaires "pratiques" (ci-avant).

On constate donc que ces transformations essentiellement linéaires que sont Borel-Laplace se décomposent en deux transformations non-linéaires à savoir, essentiellement, la prise d'un *inverse de composition* et la correspondance $f \leftrightarrow k$ ou $g \leftrightarrow h$ qui est explicitée par les formules ci-dessus et qu'on appelle le *quartage* (*).

Borel-Laplace dans les cellules rapides. Formules ouvertement linéaires.

Soit deux cellules rapides conjuguées $\mathcal{E}_M$ et $\hat{\mathcal{E}}_N$ et, dans ces cellules, une paire particulière, mais fixe, de transséries conjuguées φ et $\hat{\varphi}$. Alors les éléments génériques de $\mathcal{E}_M$ et $\hat{\mathcal{E}}_N$ se factorisent respectivement :

(IV.1.37) $$\varphi(z).A(z) \in \mathcal{E}_M \quad \text{et} \quad \hat{\varphi}(\varsigma).B(\varsigma) \in \hat{\mathcal{E}}_N \quad (^{**})$$

et se correspondent selon les formules ouvertement linéaires :

(IV.1.38) $$B\ o\ f(z) = A(z) + \sum_{m \geq 2} F^{\#}_m(z).(\frac{\partial}{\partial z})^m.A(z)$$

(IV.1.39) $$A\ o\ g(\varsigma) = B(\varsigma) + \sum_{m \geq 2} G^{\#}_m(\varsigma).(\frac{\partial}{\partial \varsigma})^m.B(\varsigma)$$

avec $f, g, F^{\#}, G^{\#}$ comme ci-avant et avec :

(IV.1.40) $$F^{\#}_m = \frac{1}{m!}.\frac{\partial}{\partial F_m}.\log F^{\#}$$

(*) parce que son itérée quatrième est l'identité ($(\text{quart})^4 = \text{id}$) ainsi d'ailleurs que l'itérée quatrième de la transformation de Fourier ($(\mathcal{F})^4 = \text{id}$).

(**) $\hat{\varphi}(\varsigma).B(\varsigma)$ désigne le produit ordinaire (des transséries formelles) et non la convolution.

(IV.1.41) $$G_m^{\#} = \frac{1}{m!}.\frac{\partial}{\partial G_m}.\log G^{\#}$$

Les correspondances ouvertement linéaires $A \rightleftarrows B$ des formules précédentes peuvent paraître beaucoup plus simples que les correspondances $\varphi = e^{-F} \rightleftarrows \hat{\varphi} = e^{-G}$ des formules (IV.1.29-30) ou des formules (IV.1.32-35), qui sont elles aussi linéaires (heureusement !), mais d'une linéarité cachée, et que pour cette raison on qualifie de cryptolinéaires. Toutefois, ces formules ouvertement linéaires nécessitent la connaissance préalable d'une paire conjuguée $(\varphi, \hat{\varphi})$ dans $(\mathcal{E}_M, \hat{\mathcal{E}}_N)$ et, pour des cellules générales, une telle paire ne peut s'obtenir qu'en recourant aux formules cryptolinéaires, qui s'avèrent donc incontournables.

IV-2. Fonctions résurgentes cohésives. Identité entre les fonctions cohésives et les accélérées faibles. Constructibilité du prolongement quasianalytique et du contournement des singularités quasianalytiques.

Nous allons voir qu'il y a identité entre les *fonctions cohésives* (les plus "régulières" des fonctions quasianalytiques de Denjoy-Carleman) et les *fonctions accélérées* (relativement aux accélérations faibles). La quasianalycité des accélérées et *providentielle*, car toute accélération $\hat{\varphi}_1(\zeta_1) \to \hat{\varphi}_2(\zeta_2)$ comporte en fait deux étapes distinctes : d'abord le *calcul du germe* $\hat{\varphi}_2(\zeta_2)$, grâce à l'intégrale d'accélération qui donne $\hat{\varphi}_2(\zeta_2)$ jusqu'à sa première singularité $\omega_1 > 0$; puis le *calcul de la fonction* $\hat{\varphi}_2(\zeta_2)$ (définie multiforme au-dessus de tout $\mathbf{R}^+$), calcul qui n'est possible, précisément, que par prolongement quasianalytique et contournement des singularités quasianalytiques.

Classes de Carleman. Critères de quasianalycité.

Soit $M. = \{M_n\}$ une suite de réels > 0 définie modulo la relation d'équivalence :

(IV.2.1) $$\{M_n\} \sim \{M'_n\} \quad \text{ssi} \quad 0 < \text{cste} < (M_n/M'_n)^{1/n} < \text{cste} < +\infty$$

Pour tout fermé fini $I = [ab] \subset \mathbf{R}$, on note $\mathcal{C}(M., I)$ la "classe de Carleman" des fonctions $\mathcal{C}^\infty$ sur I et telles que :

(IV.2.2) $$|\varphi^{(n)}(x)| < C_0 (C_1)^n M_n \quad \forall x \in I, \forall n \in \mathbf{N} \ (C_0, C_1 \text{ constantes})$$

Il est classique (Denjoy, Carleman, Bang, Ostrowski, etc...) que les trois conditions suivantes :

(IV.2.3) $$\Sigma\, 1/\beta_n = \infty \quad \text{avec} \quad \beta_n = \inf_{m \geq n} (M_m)^{1/m}$$

(IV.2.4) $$\int^{\infty} \log T(r)\,.\,r^{-2}.dr = +\infty \quad \text{avec} \quad T(r) = \sup_{m \geq 1} (r^m / M_m)$$

(IV.2.5) $$\Sigma\, M^{*}_{n-1}/M^{*}_{n} = \infty \quad \text{avec} \quad M^{*}_{n} = \sup_{r \geq 1} \inf_{m \geq 1} (r^{n-m} M_m)$$

$$= \text{régularisé convexe de } M_n.$$

sont équivalentes et que chacune d'elles est nécessaire et suffisante pour que la classe de Carleman $C(M.,I)$ soit *quasianalytique*, c'est-à-dire pour que chacune de ses fonctions φ soit entièrement déterminée par la suite $\varphi^{(n)}(x_0)$ de ses dérivées en un point $x_0 \in I$ quelconque (D–quasianalycité) ou encore par sa restriction à un sous-intervalle $I_0 \subset I$ quelconque (I–quasianalycité)(*).

Classes de Denjoy finies et transfinies. Fonctions cohésives

Pour tout ordinal transfini $\alpha < \omega^{\omega}$, notons L_α *un* itéré transfini d'ordre α de $L = \log$, c'est-à-dire un élément *du TCD* (type de croissance différentiable) $(\underline{L})^{0\alpha}$ itéré d'ordre α du TCD $\underline{L}$ (cf § IV.4).
La suite :

(IV.2.6) $${}^{\alpha}M_n = (L_\alpha(n)/L'_\alpha(n))^n \qquad (\alpha < \omega^{\omega}, n \in \mathbf{N}, L'_\alpha(x) = dL_\alpha(x)/dx)$$

dépend du représentant $L_\alpha \in \underline{L}_\alpha$, mais sa classe ${}^{\alpha}M.$ modulo (IV.2.1) ne dépend que de α et vérifie les critères de quasianalycité (IV.2.3,4,5). La classe de Carleman correspondante, soit ${}^{\alpha}D(I) = C({}^{\alpha}M.,I)$, est dite *classe de Denjoy* d'ordre transfini α. Contrairement aux classes générales de Carleman, les classes de Denjoy sont emboîtées (${}^{\alpha}D \subset {}^{\beta}D$ ssi $\alpha \leq \beta$) et, pour α fini, on retrouve les classes de Denjoy familières :

(IV.2.7) $${}^{\alpha}M_n = (n\ L_1 n\ L_2 n \ldots L_\alpha n)^n \qquad (\alpha \in \mathbf{N}, n \in \mathbf{N})$$

Une fonction φ sera dite cohésive sur $I = [a,b]$ si elle appartient à la réunion $\bigcup_\alpha {}^{\alpha}D(I)$ prise sur l'intervalle $\alpha < \omega^{\omega}$ (cela suffit à *tous* les besoins pratiques) ou sur des intervalles transfinis $\alpha < \tau$ beaucoup plus grands (cf § IV.4)(**).

(*) Pour les classes de Carleman, ces deux quasianalycités s'équivalent.
(**) Pour une définition "duale" des fonctions cohésives, en termes d'opérateurs réguliers de Nevanlinna, cf [E.5] et [E.8].

Echelle universelle de Dynkin. Prolongements pseudoanalytiques et développements pseudoanalytiques.

Une classe de Carleman est dite régulière si elle est définissable à partir d'une suite M_n telle que les nombres $m_n = M_n/n!$ vérifient :

(IV.2.8) $$(m_n)^2 \leq m_{n-1}\, m_{n+1} \quad \text{(convexité logarithmique)}$$

(IV.2.9) $$\sup(m_{n+1}/m_n)^{1/n} < \infty \quad \text{(stabilité de } C(M.) \text{ par dérivation)}$$

(IV.2.10) $$(m_n)^{1/n} \to \infty \text{ pour } n \to \infty \quad (C(M.) \text{ contient les fonctions analytiques})$$

Dynkin [Dyn] associe à chaque classe régulière un module $\mu(.)$, c'est-à-dire une application de $\mathbf{R}^+$ dans $\mathbf{R}^+$ $(0 \to +\infty)$ définie par les relations :

(IV.2.10 bis) $$\mu(r) = \sup\,(n!/r^n M_n) \text{ ou } \sup\,(n^n/r^n M_n)$$

modulo l'équivalence :

(IV.2.11) $$\mu_1 \sim \mu_2 \quad \text{ssi} \quad a\,\mu_2(br) < \mu_1(r) < c\,\mu_2(dr) \quad (a, b, c, d \text{ constantes})$$

Une classe de Carleman régulière est quasianalytique si et seulement si son module $\mu(.)$ vérifie :

(IV.2.12) $$\int_0^1 \log\log\mu(r)\,.\,dr = +\infty$$

et on a les équivalences suivantes (Dynkin, [Dyn]) :

(C_1) $f \in C(M., I)$

(C_2) f admet un *prolongement pseudoanalytique* F de module $\mu(.)$, c'est-à-dire un prolongement défini C^1 sur $\mathbf{C}$ et vérifiant :

(IV.2.13) $$|\,\bar\partial F(z)\,| < c_1/\mu(c_2 r) \quad (r = \text{dist}\,(z, I)\ ;\ r \text{ petit}\ ;\ \bar\partial = \tfrac{1}{2}\,(\partial_x + i\,\partial_y))$$

(C_3) f admet une représentation intégrale

(IV.2.14) $$f(z) = \int_{\mathbf{C}} \varphi(z_1)\ (z_1 - z)^{-1}\ dx_1\, dy_1 \quad (z_1 = x_1 + iy_1)$$

avec φ continue, à support compact et vérifiant :

(IV.2.15) $\quad | \varphi(z) | < c_3/\mu\,(c_4 r) \quad (r = \text{dist}\,(z, I)\ ;\ c_3, c_4 \text{ constantes})$

(C_4) f admet un *développement pseudoanalytique* :

(IV.2.16) $\quad f(z) = \sum_n f_n(z)$ avec f_n holomorphe sur $I_{r_n} = \{z\ ;\ \text{dist}(z, I) \leq r_n\}$

pour une suite $r_n \downarrow 0$ et des fonctions f_n holomorphes et petites sur I_{r_n} :

(IV.2.17) $\quad \| f_n \|_{I_{r_n}} \leq c_5/\mu(c_6.r_n) \quad (c_5, c_6 \text{ constantes})$

La preuve est dans Dynkin [Dyn], où les prolongements pseudoanalytiques et leurs modules $\mu(.)$ servent à construire une "échelle universelle" mesurant la "régularité" des fonctions sur $\mathbf{R}$, depuis les fonctions hölderiennes jusqu'aux classes de Carleman.

Notons que nos classes de Denjoy finies ou transfinies ${}^{\alpha}D = \mathcal{C}({}^{\alpha}M.)$ sont régulières et que leur module $\mu_\alpha(.)$ vaut (modulo (IV.2.11)) :

(IV.2.18) $\quad \log\log\ \mu_\alpha(r) = L'_\alpha\,(1/r)\,.\,r^{-2}$

Algèbres de germes cohésifs :

Fixons une classe de Carleman régulière, de suite $M.$ et de module $\mu(.)$ et convenons de dire que φ est $\mathcal{C}(M.)$ sur un intervalle ouvert $J \subset \mathbf{R}$ si elle est $\mathcal{C}(M.)$ sur tout sous-intervalle fermé $I \subset J$. Conservons aux expressions "mineur $\hat{\varphi}$ " "majeur $\check{\varphi}$ ", "classe $\overset{\triangledown}{\varphi}$ " leur sens habituel (cf § I.3).

Alors, pour la convolution (I.3.4) des mineurs, l'espace des mineurs $\hat{\varphi}$ qui sont définis et $\mathcal{C}(M.)$ en 0^+ (i.e. à la racine de $\mathbf{R}^+$, autrement dit sur un ouvert $]0, a[$) et intégrables en 0^+, constitue une algèbre, notée $\mathcal{S}^{\text{int}}(M., 0^+)$.

Pareillement, pour la convolution (I.3.3) des majeurs, l'espace des classes $\overset{\triangledown}{\varphi}$ de majeurs $\check{\varphi}$ qui sont :

(i) définis holomorphes à la racine du secteur $-2\pi < \arg \zeta < 0$

(ii) qui pour tout $\varepsilon > 0$ vérifient une majoration :

(IV.2.19) $\quad | \check{\varphi}(\zeta) | < c_3.\mu(\varepsilon^{-1}.\,|\,\text{Im}\,\zeta\,|) \quad (\varepsilon > 0, \zeta \text{ petit})$

limitant leur croissance en 0 sur toute direction autre que l'horizontale

(iii) qui possèdent sur les rayons-limites $\arg \zeta = 0$ et $\arg \zeta = -2\pi$ des valeurs limites $\check{\varphi}^+(\zeta)$ et $\check{\varphi}^-(\zeta)$ définies et $\mathcal{C}(M.)$ en 0^+ (i.e. sur un ouvert $]0, \omega_1[$),

cet espace, disons-nous, est une algèbre, que nous noterons $\mathcal{S}(M., 0^+)$.

Preuve : Dans le premier cas, la nature $\mathcal{C}(M.)$ du produit $\hat{\varphi}_3 = \hat{\varphi}_1 * \hat{\varphi}_2$ est élémentaire (cf [E.5]). Dans le second cas, l'existence, pour le produit $\check{\varphi}_3 = \check{\varphi}_1 * \check{\varphi}_2$, de valeurs-limites

φ_3^+ et φ_3^- et leur nature $C(M.)$ s'établissent en utilisant pour φ_1 et φ_2 des développements pseudoanalytiques (cf supra) qui permettent au chemin d'intégration de (I.3.3) de sortir légèrement du secteur $-2\pi <$ arg ς < 0 (cf [E.5] et [E.6]). A noter que la condition (IV.2.19) ne peut être affaibllie, mais qu'elle disparait pour la classe analytique $C(n!)$, car alors $\mu(r) \equiv +\infty$.

Tout ceci s'applique en particulier aux classes de Denjoy finies ou transfinies ${}^{\alpha}D$ et donne deux algèbres de germes cohésifs notées $S^{\text{int}}\ ({}^{\alpha}D, 0^+)$ et $S\ ({}^{\alpha}D, 0^+)$.

Algèbres de fonctions résurgentes cohésives au dessus de $\mathbf{R}^+$. Singularités quasi-analytiques.

Disons qu'une fonction φ définie et q.a. de classe régulière $C(M.)$ sur $]...,0[$ admet une *singularité q.a.* en 0, si elle n'admet pas de prolongement q.a. $C(M.)$ au delà de 0. La singularité en 0 est dite *contournable* si φ admet sur $]...,0[$ une décomposition :

(IV.2.20) $$\varphi(\varsigma) = \psi(\varsigma) + \Phi(\varsigma) \quad \text{avec :}$$

(IV.2.20.i) ψ est $q.a.C(M.)$ sur un certain intervalle $[-r, +r]$

(IV.2.20.ii) Φ est analytique sur le disque $D_r = \{|\ \varsigma\ | < r\}$ privé du rayon $[0, r]$.

(IV.2.20.iii) Φ possède sur le rayon $]0, r]$ des valeurs limites $\Phi^{\pm}\ q.a.C(M.)$.

$$\Phi^+(\varsigma) = \lim_{t \to 0} \Phi(\varsigma - \text{it}) \ ; \ \Phi^-(\varsigma) = \lim_{t \to 0} \Phi(\varsigma + \text{it}) \quad (t > 0)$$

(IV.2.20.iiii) la croissance de Φ en 0 est soumise aux contraintes :

$$\lim_{\varsigma \to 0} \Phi(\varsigma)/\mu(\varepsilon^{-1}.\ |\ \text{Im}\ \varsigma\ |) = 0 \quad (\forall\ \varepsilon > 0 \text{ fixe} \ ; \ \varsigma \in D_r - [0, r])$$

où $\mu(.)$ désigne le module de pseudoanalycité de la classe $C(M.)$.

Quand elle existe, la décomposition (IV.2.20) n'est pas unique, mais les fonctions $C(M.)$ définies sur $]0, ...[$ par :

(IV.2.21) $$\varphi^+(\varsigma) = \psi(\varsigma) + \Phi^+(\varsigma), \quad \varphi^-(\varsigma) = \psi(\varsigma) + \Phi^-(\varsigma)$$

sont, elles, uniques. On dira que φ^+ (resp φ^-) est le prolongement q.a. $C(M.)$ de φ au-delà de sa singularité 0 à droite (resp. à gauche).

Ce fait est établi en [E.5] et d'ailleurs on signalera ci-après un procédé constructif de contournement des singularités q.a.

Moyennant ces notions, on obtient aussitôt, calquées sur les algèbres de résurgences $\mathcal{R}^{\text{int}}(\mathbf{R}^+)$ et $\mathcal{R}\ (\mathbf{R}^+)$ introduites au §.I.3, des algèbres $\mathcal{R}^{\text{int}}(M., \mathbf{R}^+)$ et $\mathcal{R}(M., \mathbf{R}^+)$ de fonctions résurgentes q.a. au-dessus de $\mathbf{R}^+$ et en particulier des algèbres $\mathcal{R}^{\text{int}}({}^{\alpha}D, \mathbf{R}^+)$ et $\mathcal{R}({}^{\alpha}D, \mathbf{R}^+)$ de fonctions résurgentes cohésives au-dessus de $\mathbf{R}^+$.

Ici encore, la condition (IV.2.20.iiii) ne peut pas être affaiblie.

Semi-groupe des accélératrices et semi-groupe des coaccélératrices.

Considérons les accélérations $z_1 = F(z_2)$ qui sont faibles(*):

(IV.2.22) $$z_2/z_1 \to +\infty \text{ mais } \log z_2/\log z_1 \to 1 \ (z_1, z_2 \to +\infty;\ \arg z_1,\ \arg z_2 = \theta \in \mathbf{R})$$

et limitons-nous au cas où l'accélératrice F est *analysable*(**). Ces accélératrices engendrent un semi-groupe pour la composition 0. Pour tout F, posons $f = F', g =$ réciproque de f (g est définie analytique à la racine de $\mathbf{R}^+$) et notons G *la* primitive de $-$g qui vérifie :

(IV.2.23) $$F o\, g(\varsigma) = \varsigma g(\varsigma) + G(\varsigma) \text{ ou encore } F(z) = zf(z) + Gof(z)$$

G est dite *coaccélératrice.* Si $F_3 = F_2\ o\ F_1$ on aura $G_3 = G_2\ o\ G_1$ avec :

(IV.2.24) $$G_3(\varsigma) = \inf_{\varsigma_1 > 0} \{G_2(\varsigma/\varsigma_1).\varsigma_1 + G_1(\varsigma_1)\}$$

Ainsi, au semi-groupe des accélératrices F avec sa loi o, répond un semi-groupe de coaccélératrices G avec sa loi o et on a le schéma :

$$\begin{array}{ccc} F & \leftrightarrow & G \\ \partial \downarrow & & \downarrow -\partial \\ f & \leftrightarrow & g \end{array} \qquad \left\{ \begin{array}{l} F : +\infty \to +\infty \ ;\ f : +\infty \to +0 \\ G : +0 \to +\infty \ ;\ g : +0 \to +\infty \\ \qquad f\ o\ g = \text{ identité} \end{array} \right.$$

L'importance de la coaccélératrice vient de ce qu'elle règle les comportements asymptotiques du noyau d'accélération (cf §.I.2).

En particulier :

(IV.2.25) $$\log\ | C_F(\varsigma_2, \varsigma_1) | \sim -\varsigma_1\ G(\varsigma_2/\varsigma_1)\ (0 < \varsigma_2 \text{ fixe };\ 0 < \varsigma_1 \to +\infty)$$

(*) les autres ne font pas problème, puisque leurs accélérées sont automatiquement analytique

(**) pour un semi-groupe d'accélération plus vaste, voir [E.5] et surtout [E.6].

(IV.2.26) $\log \mid C_F(\varsigma_2,\varsigma_1) \mid \sim -\varsigma_1 \ G(\varsigma_2/\varsigma_1) \ (0 < \varsigma_1 \text{ fixe} \ ; \ 0 < \varsigma_2 \to +0)$

(IV.2.27) $\log \mid C_F(\varsigma_2,\varsigma_1) \mid < C(\theta) \ \mid \varsigma_1 \mid G(\mid \varsigma_2 \mid / \mid \varsigma_1 \mid)$

$$(0 \neq \varsigma_1 \text{ fixe} \ ; \ 0 \neq \varsigma_2 \to 0 \ ; \ \arg \varsigma_2 = \theta)$$

Pour plus de détails, voir [E.5] et surtout [E.6].

Les fonctions accélérées sont cohésives et les fonctions cohésives sont des accélérées.

Bien que le noyau $C_F(\varsigma_2,\varsigma_1)$ d'une accélération faible soit *entier* en ς_1 et *analytique* en $\varsigma_2 \neq 0$, ses dérivées successives en ς_2 ne vérifient pas de majorations analytiques *uniformes en* ς_1. On a seulement les inégalités suivantes, essentiellement optimales :

(IV.2.28) $\mid \partial^n_{\varsigma_2} C_F(\varsigma_2,\varsigma_1) \mid \leq \text{Cste } M_n . \mid a.\varsigma_2 \mid^{-n} .C_F(b\,\varsigma_2 \ ; \ b\,\varsigma_1) \quad (\text{ou } (C_F(\varsigma_2,\varsigma_1))^b)$

pour n'importe quels nombres complémentaires $a, b > 0 (a+b=1)$ et avec des constantes M_n qui vérifient le critère (IV.2.3) de quasianalycité, car $(M_n)^{1/n}$ croit et vaut :

(IV.2.29) $(M_n)^{1/n} = -F'(n)/F''(n) \sim H(n)/H'(n) \text{ où } H(z) = z/F(z)$

Par suite, dès qu'une fonction $\hat{\varphi}_1(\varsigma_1)$ (ou une mesure, ou une distribution) possède une abscisse d'accélération $\omega > 0$ (cf §.II.2), *l'intégrale d'accélération livre une accélérée* $\hat{\varphi}_2(\varsigma_2)$ *qui est toujours q.a. de classe* $C(M.)$ sur l'ouvert $]0,\omega[$ (et ceci quelle que soit la nature de l'accélérande $\hat{\varphi}_1(\varsigma_1)$) *mais en général non-analytique*(*).

En particulier, si l'accélération faible F est *analysable*, alors l'accélérée $\hat{\varphi}_2(\varsigma_2)$ appartient *toujours* à une classe de Denjoy finie ${}^nD(n \in \mathbb{N})$. Si F est encore plus faible, par exemple du type $F(z) = z/L_\alpha(z)$, pour une itérée transfinie L_α de L qui soit *analytique* (sur un voisinage complet ramifié de l'infini ; de tels L_α existent toujours pour $\alpha < \omega^\omega$ et même au-delà), alors $\hat{\varphi}_2(\varsigma_2)$ appartient à la classe de Denjoy transfinie ${}^\alpha D$.

Dans tous ces cas, on voit que l'accélérée est définie cohésive sur un ouvert $]0,\omega[$. Réciproquement, si on se donne $\hat{\varphi}_2(\varsigma_2)$ définie cohésive sur un semi-ouvert $[0,\omega[$, alors $\hat{\varphi}_2(\varsigma_2)$ est *toujours* l'accélérée d'une certaine $\hat{\varphi}_1(\varsigma_1)$, pour une accélération F suffisamment faible et qu'on peut toujours choisir du type $F(z) = z/L_\alpha(z)$ avec α fini ou transfini.

La preuve de la réciproque (voir [E.5] et surtout [E.6]) utilise l'intégrale de *décélération des majeurs* et les *développements pseudoanalytiques* introduits ci-dessus.

(*) sauf circonstance particulière, comme par exemple quand l'accélérande est définie - holomorphe - accélérable sur tout un secteur $-\theta < \arg \varsigma_1 < +\theta$, auquel cas l'accélérée est automatiquement définie-holomorphe *à la racine* du même secteur $-\theta < \arg \varsigma_2 < +\theta$.

Caractère constructif du prolongement q.a. et du contournement des singularités q.a.

Il existe déjà plusieurs méthodes constructives de prolongement quasianalytique - celles notamment de Carleman, Bang, Badalian -(*) mais les résultats ci-dessus conduisent à un procédé nouveau, particulièrement élémentaire et explicite :

$$\text{(IV.2.30)}\qquad \hat\varphi_2(\zeta_2) \overset{1}{\longrightarrow} \check\varphi_2(\zeta_2) \overset{2}{\longrightarrow} \check\varphi_1(\zeta_1) \overset{3}{\longrightarrow} \hat\varphi_1(\zeta_1) \overset{4}{\longrightarrow} \hat\varphi_2(\zeta_2)$$

qui se décompose en une suite d'intégrations élémentaires et qui permet de prolonger toute fonction cohésive (ou quasianalytique plus générale) sur $[0,\omega']$ à son intervalle maximal de cohésivité $[0,\omega]$. La flèche 1 figure la *prise d'un majeur* $\check\varphi_2(\zeta_2)$ à partir de l'intégrale :

$$\text{(IV.2.31)}\qquad \check\varphi_2(\zeta_2) = \frac{1}{2\pi i}\int_0^u \hat\varphi_1(\zeta_1)\frac{d\zeta_1}{\zeta-\zeta_1} \qquad (u\in]0\omega], -2\pi < \arg\zeta_2 < 0)(^{**}).$$

La flèche 2 figure la *décélération des majeurs*, pour une décélération $z_2 \to z_1 = F(z_2)$ suffisamment faible, au moyen de l'intégrale :

$$\text{(IV.2.32)}\qquad \zeta_1\check\varphi_1(\zeta_1) = -\int_0^v \zeta_2\,\check\varphi_2(\zeta_2)\; C_F(\zeta_2 e^{i\pi},\zeta_1 e^{i\pi})d\zeta_2 \qquad \begin{cases} -2\pi < \arg v < 0 \\ -2\pi < \arg\zeta_1 < 0 \end{cases}$$

qui transpose l'accélération (I.2.8) des mineurs et qui doit être prise le long d'un chemin approchant 0 selon la direction $\theta_1 = \arg\zeta_1$(***). La flèche 3 représente le prolongement analytique de $\check\varphi_1(\zeta_1)$ dans la totalité du secteur $-2\pi < \arg\zeta_1 < 0$ puis la prise de ses valeurs-limites au bord (*elles existent*) et de leur différence (qui donne $\hat\varphi_1(\zeta_1)$). Enfin, la flèche 4 représente le calcul de $\hat\varphi_2(\zeta_2)$ par l'intégrale (I.2.8) d'accélération des mineurs, qui converge automatiquement sur la totalité du segment $[0\omega[$ (****).

Le contournement, à droite et à gauche, des singularités quasianalytiques, est lui aussi constructif et démarque de très près le procédé usuel de contournement des singularités analytiques. Ainsi, le contournement à droite d'une singularité analytique ω comporte trois étapes :

(*) Pour une vue d'ensemble, voir [Man] et surtout [Vol].

(**) La classe $\overset{\triangledown}{\varphi}_2$ de $\check\varphi_2$ ne dépend pas de u.

(***) La classe $\overset{\triangledown}{\varphi}_1$ ne dépend *ni* de v *ni* du choix de $\check\varphi_2$ dans $\overset{\triangledown}{\varphi}_2$.

(****) Cela tient à ce que, pour une accélération faible ($\log z_1 \sim \log z_2$), l'abscisse d'accélérabilité coïncide avec la première singularité q.a. de l'accélérée $\hat\varphi_2(\zeta_2)$ sur $\mathbf{R}^+$. Pour les accélérations moyennes ($\log z_1 \sim \alpha\log z_2, 0<\alpha<1$), l'accélérée est ***analytique*** à la racine du secteur $\{-\frac{\pi}{2}\frac{\beta}{\alpha} < \arg\zeta_1 < +\frac{\pi}{2}\frac{\beta}{\alpha}\}$ $(\beta = 1-\alpha)$ et il faut prendre en compte les diverses singularités de ce secteur. Voir [E.6].

$$\hat{\varphi}\,(\varsigma) \xrightarrow{1} \hat{\varphi}\,(\varsigma - i\varepsilon) \xrightarrow{2} \hat{\varphi}\,(\varsigma' - i\varepsilon) \xrightarrow{3} \hat{\varphi}\,(\varsigma') \quad (0 < \varepsilon\ ;\ 0 < \varsigma < \omega < \varsigma') \tag{IV.2.33}$$

où la flèche 2 représente la prolongation analytique ordinaire de ς à ς' et où les flèches 1 et 3 figurent deux translations élémentaires, T^{-1} et T, de pas $-i\varepsilon$ et $+i\varepsilon$ ($\varepsilon > 0$ petit) :

$$T^{-1} = \exp\,(-i\varepsilon\,\partial_\varsigma)\ ;\ T = \exp\,(+i\varepsilon\,\partial_\varsigma) \tag{IV.2.34}$$

Le contournement quasianalytique s'effectue de la même manière, mais avec une flèche 2 qui représente le prolongement quasianalytique (constructif ; voir ci-avant) de ς à ς' et avec des flèches 1 et 3 qui figurent des *pseudotranslations* T^{-1} et T mutuellement inverses et ajustées (uniformément pour toute une classe $\mathcal{C}(M.)$) de manière à régulariser la singularité en ω du bon côté (ici, par exemple, en bas) sans détruire la quasianalycité ailleurs. Autrement dit, T^{-1} doit changer la fonction $\hat{\varphi}$, qui est q.a. sur $]0\omega]$, en une fonction $T^{-1}\ \hat{\varphi}$, qui est q.a. sur un intervalle $]0\omega'[$ plus grand (qui, dans l'accéléro-sommation, sera $\mathbf{R}^+$ tout entier). Techniquement, il est commode de prendre T de la forme.

$$T = A(e^{-i\theta}.\partial_\varsigma) \quad \text{avec} \quad 0 < \theta < \pi/2\ ;\ \partial_\varsigma = \partial/\partial\varsigma \tag{IV.2.35}$$

et avec pour $A(w)$ une fonction entière de croissance légèrement subexponentielle en w, par exemple :

$$A(w) = \prod_{n=1}^{\infty} (1 + \frac{w}{a_n}) \tag{IV.2.36}$$

avec des $0 < a_n \to +\infty$ donnés par :

$$a_n = -\partial\delta H(n) \quad \text{où} \quad H(z) = z/F(z)\ ;\ \delta H(z) = z\ H'(z)/H(z)\ ;\ \partial = \partial/\partial z \tag{IV.2.37}$$

IV-3. Notion de douceur pour les fonctions modérées ou très lentes. Equivalences universelles.

Les fonctions $\mathcal{L}$ à croissance très lente ont des dérivées "posthomogènes" qui ou bien ne présentent aucune régularité, ou bien possèdent des séries asymptotiques *universelles*, faites de logarithmes itérés. Dans ce cas, $\mathcal{L}$ est dite *douce*. Les fonctions très lentes se rencontrent rarement, mais leur douceur induit une notion analogue chez les fonctions modérées, elles très communes. Commençons par introduire deux algèbres qui serviront de "réservoirs" à series asymptotiques universelles.

Les algèbres graduées S et S*.

Soit comme d'habitude L_n = log...log (n fois) et E_n = exp...exp(n fois). Posons :

$$\lambda_n = L'_{n+1} = (x\ L_1 x\ L_2 x ... L_n x)^{-1} \tag{IV.3.1}$$

et introduisons les séries formelles :

$$\Lambda = L_1 + L_2 + L_3 + ... \tag{IV.3.2}$$

$$S^{(n_0, \dots, n_r)} = \sum\nolimits_{0 \leq i_0 < i_1 < \quad i_r < +\infty} (\lambda_{i_0})^{n_0} (\lambda_{i_1})^{n_1} ... (\lambda_{i_r})^{n_r} \tag{IV.3.3}$$

Notons $\mathbf{S}_n$ l'espace tendu par les $S^{(n_0, \dots n_r)}$ avec $n_0 + ... n_r = n$. La somme $\mathbf{S} = \mathbf{S}_1 \oplus \mathbf{S}_2 \oplus \mathbf{S}_3 ...$ est une algèbre différentielle graduée, avec pour dérivation $\partial = \partial/\partial x$ $(\mathbf{S}_m \to \mathbf{S}_{m+1})$:

$$\partial\, S^{(n_0, \dots n_r)} = -\sum\nolimits_{0 \leq i \leq r} (n_i + n_{i+1} + ... n_r).\{S^{(n_0, \quad .1+n_i, \dots, n_r)} + S^{(n_0, \quad .n_i - 1.1.n_i, \dots n_r)} \tag{IV.3.4}$$

et pour multiplication $(\mathbf{S}_m, \mathbf{S}_n \to \mathbf{S}_{m+n})$:

$$S^{(n^1)} S^{(n^2)} = \sum\nolimits_{n \subset n^1, n^2} S^{(n)} \quad (n = (n_i),\ n^1 = (n^1_i)\ ;\ n^2 = (n^2_i)) \tag{IV.3.5}$$

où $\subset$ désigne le "battage symétral" des multiindices n^1 et n^2, c'est-à-dire leur imbrication mutuelle avec préservation de l'ordre interne de n^1 et n^2 et fusion éventuelle de deux termes contigüs. Ainsi :

$$S^{(a)}.S^{(b.c)} = S^{(a.b.c)} + S^{(a+b.c)} + S^{(b.a.c)} + S^{(b.a+c)} + S^{(b.c.a)}. \tag{IV.3.5 bis}$$

Les séries formelles $\partial^n \Lambda$ dérivées de la série (IV.3.2) engendrent *librement* une sous-algèbre graduée $\mathbf{S}^* = \mathbf{S}^*_1 \oplus \mathbf{S}^*_2 ...$ de $\mathbf{S}$ et l'on a :

$$\dim(\mathbf{S}_m) = 2^{m-1} > \dim(\mathbf{S}^*_m) = p(m) = \#\{m_i, \Sigma\, m_i = m\ ;\ 0 < m_1 \leq m_2 ... \leq m\} \tag{IV.3.6}$$

Tout élément S de $\mathbf{S}$ ou $\mathbf{S}^*$ se présente comme somme transfinie :

$$S(x) = \sum_\sigma c_\sigma\, U_\sigma(x) \quad (0 \leq \sigma < \sigma(S) < \omega^\omega) \tag{IV.3.7}$$

de termes $U_\sigma = x^{-n_0(\sigma)} (L_1 x)^{-n_1(\sigma)} ... (L_r x)^{-n_r(\sigma)}$ comparables deux à deux et indexés par ordre décroissant. On note :

$$\text{deb}\,(S) = c_0\, U_0 = \text{début de } S \tag{IV.3.8}$$

(IV.3.9) $$\operatorname{as}(S)=\sum_{\sigma<\omega} c_\sigma\, U_\sigma = \text{partie asymptotique de } S$$

On aura ainsi :

(IV.3.8 bis) $$\begin{aligned}\operatorname{deb}(S^{(n_0,\dots,n_r)}) &= (\lambda_0)^{n_0}(\lambda_1)^{n_1}\dots(\lambda_r)^{n_r}\\ &= (x)^{-(n_0+\dots n_r)}(L_1x)^{-(n_1+\dots n_r)}\dots(L_rx)^{-(n_r)}\end{aligned}$$

(IV.3.9 bis) $$\operatorname{as}(S^{(n_0,\dots,n_r)}) = (\lambda_0)^{n_0}(\lambda_1)^{n_1}\dots(\lambda_{r-1})^{n_{r-1}}\sum\nolimits_{s\geq r}(\lambda_s)^{n_r}$$

Contrairement aux transséries associées aux fonctions *analysables* (cf.III.1), les séries formelles $S \in \mathbf{S}$ ont une partie asymptotique as (S) qui ne peut *jamais* être sommée. Il est donc impossible d'attribuer un sens "effectif" à ce qui vient après, c'est-à-dire aux U_σ d'indice $\sigma \geq \omega$ (ω = premier ordinal transfini).

Notion de douceur pour les fonctions très lentes. Séries asymptotiques universelles.

Une fonction $\mathcal{L} : [\dots,+\infty] \to [\dots,+\infty]$ est dite *très lente* si pour tout n, $\mathcal{L}(x)/L_n(x) \to 0$ quand $x \to +\infty$. Un opérateur différentiel D est dit *posthomogène* de degré n s'il est de la forme :

(IV.3.10) $$\begin{cases} D\,\mathcal{L} \equiv D^*(\log(1/\mathcal{L}')) \text{ avec } D^*H \equiv \Sigma\, a_{n_1,\dots,n_r}\, H^{(n_1)}\dots H^{(n_r)} \text{ (somme finie)}\\ \text{avec } H^{(n_i)} = \partial^{n_i}H\ ;\ n_i>0\ ;\ n_1+\dots n_r = n\ ;\ a_{n_1,\dots,n_r}\in\mathbf{R}\end{cases}$$

Un opérateur $\mathcal{D}$ est dit *logarithmico-différentiel* s'il est de la forme :

(IV.3.11) $$\mathcal{D}\mathcal{L} \equiv \sum\nolimits_D b^D(x).D\mathcal{L} \equiv \sum\nolimits_D b^D(x)\, D^*\log(1/\mathcal{L}') \quad \text{(somme finie)}$$

avec des D posthomogènes (de degré quelconques) et des coefficients $b^D(x)$ polynomiaux (ou rationnels) en les $(L_n x)^{-1}$.

Une fonction *très lente* $\mathcal{L}$ est dite *douce* si elle est C^∞ et vérifie l'une ou l'autre des conditions équivalentes (C_1) ou (C_2) :

(C_1) Pour tout opérateur *logarithmico-différentiel* $\mathcal{D}$, la fonction $\mathcal{D}\mathcal{L}$ possède la série asymptotique "universelle" suivante :

(IV.3.12) $$\operatorname{as}(\mathcal{D}\mathcal{L}) = S^{\mathcal{D}} = \operatorname{as}\{\Sigma\, b^D(x).D^*\Lambda(x)\} \quad (\mathcal{D} \text{ comme en (IV.3.11)})$$

qui ne dépend que de $\mathcal{D}$ et qui s'exprime à partir des dérivées de la série Λ introduite en (IV.3.2).

(C_2) Pour tout opérateur *posthomogène* D, les fonctions $D\mathcal{L}, D(\mathcal{L}oE), D(\mathcal{L}oE_2)$ etc... possèdent une même série asymptotique universelle S^D, fonction de D seul :

(IV.3.13) $\quad$ as $(D\mathcal{L}) =$ as $(D(\mathcal{L}oE_n)) = S^D =$ as $(D^{*}\Lambda)$ (D et D^{*} comme en (IV.3.10))

Considérons par exemple les trois opérateurs ∇_i, posthomogènes pour $i = 1$ et 2 :

(IV.3.14) $\nabla_0 \mathcal{L} = \log \mathcal{L}'$; $\nabla_1 \mathcal{L} = \mathcal{L}''/\mathcal{L}'$; $\nabla_2 \mathcal{L} = (\mathcal{L}''/\mathcal{L}')^2 - 2(\mathcal{L}''/\mathcal{L}')' =$ schwarzien de $\mathcal{L}$

qui incidemment sont les seuls à vérifier les identités :

(IV.3.15) $\quad \nabla_i(FoG) = (\nabla_i F) \; o \; G \; (G')^i + \nabla_i G \quad (i = 0, 1, 2)$

Considérons aussi les opérateurs posthomogènes ∂_n de dégré n définis par :

(IV.3.16) $\quad \partial_1 \mathcal{L} = -\nabla_1 \mathcal{L} = -\mathcal{L}''/\mathcal{L}'$ et $\partial_{n+1}\mathcal{L} = \partial(\partial_n \mathcal{L}) + n(\partial_1 \mathcal{L}).(\partial_n \mathcal{L}) \quad (n \geq 1)$

Pour toute fonction $\mathcal{L}$ très lente et douce, ces opérateurs donnent lieu aux développements universels :

(IV.3.17) $\quad$ as $(\nabla_o \mathcal{L}) = -\Lambda = -\sum_{n \geq 1} L_n x \quad (*)$.

(IV.3.18) $\quad$ as $(\nabla_i \mathcal{L}) = S^{\nabla_i} = -\sum_{n \geq 0} (L'_{n+1})^i = -\sum_{n \geq 0} (\lambda_n)^i \quad (i = 1, 2)$

(IV.3.19) $\quad$ as $(\partial_n \mathcal{L}) = S^{\partial_n} =$ as $(S^{(1, n-2\; 1)}) = (1/x)^{n-1}(1/Lx)^{n-2} \sum_{n \geq 2} \lambda_n$

Notons que pour tout D posthomogène, la série asymptotique universelle $S^D =$ as $(D\mathcal{L}) =$ as $(D^{*}\Lambda)$ revêt toujours la forme :

(IV.3.20) $\quad S^D =$ somme finie $+ (x)^{-n_0}(L_1 x)^{-n_1} \ldots (L_r x)^{-n_r} \sum_{m > r} (\lambda_m)^r$

et que le paramètre crucial $r = r(D)$ peut prendre n'importe quelle valeur entière positive.

Les $\mathcal{L}$ très lentes et douces forment *un semi-groupe* (pour la composition) qui contient en particulier toutes les itérées L_α d'ordre transfini $\alpha < \omega^\omega$ de la fonction $L = \log$ (cf IV.4).

(*) ∇_0 n'est pas posthomogène au sens de (IV.3.10), mais (IV.3.17) est une conséquence de (IV.3.18).

Notion de douceur pour les fonctions modérées. Equivalences universelles

Une fonction $F : [..., +\infty] \to [..., +\infty]$ est dite à croissance modérée si elle reste dans l'échelle logarithmique-exponentielle, autrement dit s'il existe n tel que :

(IV.3.21) $$L_n(x) \leq F(x) \leq E_n(x)$$

Par exemple, les fonctions algébrico-différentielles *analytiques* sur une demi-droite $[..., +\infty]$ sont automatiquement modérées(*).

Une fonction modérée F est dite *douce* si, pour toute $\mathcal{L}$ très lente et douce, les fonctions très lentes $Fo\mathcal{L}$ et $\mathcal{L}oF$ sont également douces. Comme tout opérateur posthomogène D vérifie une identité du type Faa di Bruno à coefficients $a_{i,j}$ scalaires :

(IV.3.22) $$D(FoG) = \Sigma\, a_{i,j}\, (D_i F)\; o\; G\; (G')^{\deg D_i} (D_j G) \text{ avec } \deg D = \deg D_i + \deg D_j$$

les conditions de douceur pour une F modérée :

(IV.3.23) $$\mathrm{as}(D\mathcal{L}) = \mathrm{as}\,(D(Fo\mathcal{L})) = \mathrm{as}\,(D(\mathcal{L}oF))$$

impliquent des équivalences universelles(**) du type :

(IV.3.24) $$S^D = \mathrm{as}\,\{\Sigma\, a_{i,j}\; (D_i F)o\mathcal{L}\;\; (\mathcal{L}')^{\deg D_i}.S^{D_j}\}$$

(IV.3.25) $$S^D = \mathrm{as}\,\{\Sigma\, a_{i,j}\;\; (S^{D_i}oF).(F')^{\deg D_i}.D_j F\}$$

La présentation d'un système "minimal" de conditions de douceur (IV.3.24) + (IV.3.25) nécessite le calcul d'une infinité d'opérateurs posthomogènes D spéciaux (les "hyper-schwarziens") qui sont "minimaux" en ce sens que les séries $D^{*}\Lambda$ correspondantes sont minimales pour l'ordre suivant sur $\mathbf{S}$:

(IV.3.26) $$\{A > B\} \text{ si } \{\text{le terme général de as}(A) >> \text{le terme général de as }(B)\}$$

Le schwarzien ∇_2 et (pour n assez grand) les opérateurs ∂_n de (IV.3.16) font partie de cette famille "spéciale", *mais ne l'épuisent pas.*

(*) Ceci n'est pas contredit par l'existence d'équations algébrico-différentielles (linéaires homogènes) aux solutions partout denses dans $C(\mathbf{R})$, car ces solutions ne sont analytiques que par *morceaux*. Voir [E.5].

(**) Les équivalences (IV.3.25) sont manifestement universelles (indépendantes de $\mathcal{L}$) mais c'est vrai *aussi* pour les équivalences (IV.3.24). Signalons que (IV.3.24) est "moins contraignant" que (IV.3.25).

Les fonctions modérées *et* douces forment un *groupe* pour la composition. Toutes les fonctions *analysables* (cf III) sont automatiquement *modérées et douces*, mais l'inverse n'est pas vrai. Ainsi, toute fonction croissante $\mathcal{C}^\infty$ qui vérifient $FoF = L$ (racine carrée d'itération de L) est automatiquement *modérée et douce*, mais *non analysable*. Comme exemple de F modérée et $\mathcal{C}^\infty$ mais *non douce*, citons les difféomorphismes périodiques non triviaux de $\mathbf{R}$, i.e. les F tels que $F(x+1) \equiv 1 + F(x)$ et $F'(x)$ non constant. Ces F préservent la douceur des $\mathcal{L}$ très lentes par précomposition $Fo\mathcal{L}$ mais pas par postcomposition $\mathcal{L}oF$.

IV-4. Types de croissance (TC) et types de croissance différentiables (TCD). Itération d'ordre transfini et échelle naturelle de croissance. Le Grand Cantor.

La mesure de la "régularité" ou de la "complexité" de tant d'objets mathématiques se ramène à des "types de croissance à l'infini" (*) qu'on ressent le besoin d'une *échelle naturelle de croissance* couvrant toutes les situations usuelles. Cependant, dès qu'on sort du domaine des logarithmes ou exponentielles itérés, on se heurte à une certaine *indétermination*, due à l'absence de fonctions-repères indiscutablement canoniques. On commence par circonscrire cette indétermination le plus possible en imposant aux fonctions envisagées un maximum de régularité - la *douceur* - puis on en prend son parti en regroupant dans des *"zones de croissance"* (ZC et ZCD) toutes les fonctions *"indiscernables"*. Les plus fines de ces zônes sont les *"types de croissance"* (TC et TCD), qui peuvent être soumis à une *itération transfinie* et reflètent la structure discontinue des ordinaux ou *types d'ordre* (TO). Localement, au contraire, l'échelle naturelle de croissance est faite de zônes (ZC et ZCD), de largeurs très diverses, et présente une remarquable structure de type *fractal*.

Types de croissance (TC) et types de croissance différentiables (TCD)

Une fonction $\mathcal{C}^\infty$: $[..., +\infty] \to [..., +\infty]$ sera dite *lente* si :

$$\delta F(x) = x\, F'(x)/F(x) \to 0 \quad \text{quand} \quad x \to \infty \tag{IV.4.1}$$

Cela implique $\log F(x) << \log x$. Les "moins lentes" des fonctions lentes sont de la forme :

$F(x) = \exp\left((\log x)/\mathcal{L}(x)\right)$ où $\mathcal{L}$ désigne une fonction *très lente* (cf § IV.3)

(*) Voir par exemple le *principe de Dynkin* au § IV.2 et aussi le classement des fonctions cohésives.

Quand une fonction lente (ou a fortiori très lente) est *douce* (cf § IV.3), elle vérifie des équivalences universelles (cf § IV.3) qui impliquent en particulier :

(IV.4.2) $$F^{(1+n)}(x) \sim (-1)^n.n!\ x^{-n} F'(x) \qquad (n \text{ fixe} ;\ x \to +\infty)$$

Bien entendu, les équivalences (IV.4.2) ne suffisent pas, de très loin, à assurer la douceur de F, mais elles sont quand même très fortes, puisqu'elles impliquent, pour la fonction *rapide* G réciproque de F, les équivalences :

(IV.4.3) $$G^{(n+1)}(x)/G^{(n)}(x) \sim G'(x)/G(x) \quad \text{et} \quad G^{(n)}(x)/G(x) \sim (G'(x)/G(x))^n$$

$(n$ fixe ; $x \to +\infty)$

qui entrainent en particulier l'infinie convexité de G(*).

Sur le semi-groupe LD des fonctions *lentes et douces* introduisons les relations d'équivalence :

(IV.4.4) $$F_1 \approx F_2 \quad \text{ssi} \quad F_2 = h_{21} o\, F_1 \quad \text{avec} \quad 0 < \text{cste} < h_{21}(x)/x < \text{cste} < +\infty$$

(IV.4.5) $$F_1 \sim F_2 \quad \text{ssi} \quad F_2 = h_{21} o\, F_1 \quad \text{avec} \quad 0 < \text{cste} < h'_{21}(x) < \text{cste} < +\infty$$

où h_{21} désigne n'importe quelle fonction croissante et C^∞. Les classes d'équivalence de F modulo $\approx$ et $\sim$ sont notées respectivement $\underline{\underline{F}}$ et $\underline{F}$ (un trait pour l'oubli d'un peu de structure ; deux traits pour l'oubli de beaucoup de structure). Les $\underline{\underline{F}}$ et $\underline{F}$ sont respectivement appelés *types de* croissance (TC) et *types de croissance différentiables* (TCD). On vérifie que sur le semi-groupe LD des fonctions lentes et douces, la composition o "passe aux classes" :

(IV.4.6) $$\underline{\underline{FoK}} = \underline{\underline{F}}\ o\ \underline{\underline{K}} \quad \text{et} \quad \underline{FoK} = \underline{F}\ o\ \underline{K} \qquad (\forall F, K \in LD)$$

Le demi-groupe LD donne donc naissance à deux semi-groupes $\underline{\underline{LD}}$ et $\underline{LD}$, faits de TC et de TCD. Bien entendu, tout TCD induit un TC, mais l'inverse n'est pas vrai : à tout TC sont subordonnés une infinité de TCD.

Justification de ces définitions : Les équivalences (IV.4.2) montrent que si deux fonctions F_1 et F_2 de LD (lentes + douces) sont proches ainsi que leurs dérivées premières F_1' et F_2', toutes leurs dérivées successives $F_1^{(n)}$ et $F_2^{(n)}$ seront proches. On est donc dispensé d'envisager une nouvelle relation d'équivalence pour chaque n. Il suffit d'en envisager

(*) Infinie convexité entendue au sens des germes : autrement dit il existe x_n tel que $G^{(n)}(x) > 0$ pour $x > x_n$. On peut avoir $x_n \to +\infty$, si bien que G peut ne pas être analytique, en dépit du théorème de Bernstein.

deux, pour $n = 0$ et $n = 1$. Ce sont les relations $\approx$ et $\sim$, qui en outre permettront d'identifier toutes les solutions "indiscernables" de l'équation (IV.4.7) ci-après et conduiront aux identités (IV.4.12) et (IV.4.13) raccordant les TC et TCD aux classiques TO. On pourrait d'ailleurs modifier légèrement les définitions de $\approx$ et $\sim$ mais l'admirable est que toutes ces variantes conduisent essentiellement à la même construction.

Les itérareurs F^* et les applications $\underline{F} \to \underline{F}^*$ et $\underline{\underline{F}} \to \underline{\underline{F}}^*$

Pour toute F lente et douce, on appelle *itérateur* F^* de F toute solution lente et douce de l'équation :

(IV.4.7) $$F^* oF(x) \equiv -1 + F^*(x) \quad (F \text{ donnée dans } LD \ ; \ F^* \text{ inconnue dans } LD)$$

De telles solutions existent toujours et $\underline{F}^*$ ou $\underline{\underline{F}}^*$ ne dépendent que de $\underline{F}$ ou $\underline{\underline{F}}$. Les F^* sont dits *itérateurs* de F parce qu'ils conjugent F à la translation T_{-1} de pas -1 et que le choix d'un F^* détermine un groupe d'itérées fractionnaires de F.

(IV.4.8) $$F^{ow} = {}^*FoT_{-w}oF^* \quad \text{avec} \quad {}^*FoF^* = \text{id} \ ; \ w \in \mathbf{Q} \text{ ou } \mathbf{R}; T_w(x) \equiv x + w$$

Quand F est d'exponentialité $-n < 0$(*), toutes les solutions F^* de (IV.4.7) qui sont C^∞ sont automatiquement *douces*. Il y a une infinité de tels F^*, liés les uns aux autres par précomposition par un difféomorphisme périodique de $\mathbf{R}$.

(IV.4.9) $$F^* \to hoF^* \quad \text{avec} \quad h(x+1) \equiv 1 + h(x) \quad \text{et} \quad h \in C^\infty, h' > 0$$

Mais les types $\underline{F}^*$ et $\underline{\underline{F}}^*$ ne dépendent que de F. Mieux encore, ils ne dépendent que des types $\underline{F}$ ou $\underline{\underline{F}}$ respectivement. Comme on va le voir, les applications $\underline{F} \to \underline{F}^*$ et $\underline{\underline{F}} \to \underline{\underline{F}}^*$ ne sont absolument pas injectives.

Itération transfinie des types de croissance.

Soit un ordinal transfini $\alpha < \omega^\omega$ d'expression :

(IV.4.10) $$\alpha = \omega^r.n_r + \omega^{r-1}.n_{r-1} + ...\omega^2.n_2 + \omega.n_1 + n_0 \qquad (n_i \in \mathbf{N})$$

Si t désigne un TC ou TCD et si $t^*, t^{**}, t^{***}...$ désignent les types successifs qui s'en déduisent par prise répétée de l'itérateur, on pose :

(IV.4.11) $$t^{0\alpha} = (t)^{0^{n_0}} o(t^*)^{0^{n_1}} o(t^{**})^{0^{n_2}} o(t^{***})^{0^{n_3}} o...(t^{*\cdots*})^{0^{n_r}}$$

(*) i.e. quand $L_r oF(x) \sim L_{r+n}(x)$ pour r assez grand (mais fixe) et $x \to +\infty$.

En particulier, $t^{0\omega} = t^{*}, t^{0\omega^2} = t^{**}$ *etc.* Le type $t^{0\alpha}$ est dit ***itéré transfini d'ordre*** α du type t. Cette terminologie apparaît tout à fait légitime car pour tous $\alpha, \beta < \omega^{\omega}$ on a les indentités cruciales :

(IV.4.12) $$(t^{0\alpha})o(t^{0\beta}) = t^{0(\beta+\alpha)}$$ (inversion!)(*).

(IV.4.13) $$(t^{0\alpha})^{0\beta} = t^{0(\alpha\beta)}$$ (pas d'inversion)

Par exemple :

(IV.4.14) $$t^{0\omega} \, o \, t = t^{0\omega} \text{ car } 1+\omega = \omega \text{ mais } t \, o \, t^{0\omega} \neq t^{0\omega} \text{ car } \omega + 1 \neq \omega$$

(IV.4.15) $$(t^{02})^{0\omega} = t^{0\omega} \text{ car } 2\omega = \omega \text{ mais } (t^{0\omega})^{02} \neq t^{0\omega} \text{ car } \omega.2 \neq \omega$$

En particulier, si $\underline{L}$ désigne le TCD de $L = \log$ et si $\alpha < \beta < \omega^{\omega}$, on aura $(\underline{L})^{0\alpha} >> (\underline{L})^{0\beta}$ en ce sens que cette "inéquivalence" vaut pour toute paire de représentants $L_\alpha \in (\underline{L})^{0\alpha}$ et $L_\beta \in (\underline{L})^{0\beta}$. Comme on l'a vu au § IV.2, ces L_α (itérés transfinis du logarithme L) permettent de définir les classes (quasianalytique) de Denjoy *généralisées*, qui couvrent la plupart des *fonctions cohésives* (et la totalité de celles qu'on rencontre en pratique).

Ainsi se trouve concrètement *réalisée*, en termes d'itération de types de croissance, la double structure additive et multiplicative (mais fortement non-commutative et associative d'un côté seulement) des ordinaux transfinis ou *types d'ordre* (TO), tout au moins jusqu'à l'ordinal ω^{ω} (exclu) qui, après ω, est le premier ordinal clos par addition *et* multiplication. Si l'on veut dépasser ω^{ω} et définir l'itération(**) pour des ordres α dénombrables et "représentables", il est plus commode de raisonner sur les types *rapides* (réciproques des types *lents*) et d'utiliser la "bonne suite naturelle" α_n qui tend vers α (ordinal sans antécédent) pour poser :

(IV.4.16) $$G^{0\alpha}(n) = G^{0\alpha_n}(n) \quad (\alpha_n \uparrow \alpha \,;\, G \text{ rapide } : \mathbf{N} \to \mathbf{N})$$

puis interpoler correctement. La construction est assez facile, tout au moins pour les TC et pour des intervalles transfinis raisonnables. Pour les TCD, il faut prendre plus de précautions et utiliser non directement la relation (IV.4.16) mais ses dérivées successives.

(*) Si l'on considérait les types de croissance *rapides* (réciproques des types lents), l'inversion disparaîtrait de (IV.4.12) sans pour autant apparaître dans (IV.4.13).

(**) avec préservation de (IV.4.12) et (IV.4.13).

Unicité à l'infini de l'échelle des TC et multiplicité des échelles de TCD.

Plus un ordinal α est "isolé", moins les itérés $t^{0\alpha}$ d'un *type de croissance t* dépendent de α. C'est déjà vrai (de façon spectaculaire) pour $\alpha = \omega$ et pour *toutes* les fonctions modérées, en particulier analysables, d'exponentialité < 0. Ainsi, malgré la prodigieuse diversité des fonctions F lentes et analysables (grosso modo : tout ce qu'on peut fabriquer avec des coefficients réels et avec des échafaudages de symboles $+, \times, 0$, log, exp) les itérateurs F^* de ces F (d'exponentialité < 0) sont tous extraordinairement "proches" : ils sont équivalents pour $\approx$ et définissent un unique type de croissance $\underline{\underline{F^*}} = \underline{\underline{L^*}}$ $(L = \log)$. On peut donc parler du "type itéré d'ordre ω des fonctions modérées" (d'exponentialité$<$ 0) sans préciser de quelles fonctions modérées on parle ! C'est encore plus vrai pour les α plus isolés. Dans le même ordre d'idées, toute paire t_1 et t_2 de TC, aussi différents soient-ils (mais constructivement définis), auront des itérés $t_1^{0\alpha}$ et $t_2^{0\alpha}$ qui *coïncideront* pour tous les α "suffisamment isolés" (et donc assez grands). En ce sens précis, on peut parler de l'unicité à l'infini de l'échelle de TC.

Il n'en va pas de même des TCD. Si t_1 et t_2 sont deux TCD différents, même très voisins, même subordonnés à un commun TC, mais non comparables(*) et non majorables par un troisième TCD, alors tous les TCD itérés $t_1^{0\alpha}$ et $t_2^{0\beta}$ sont deux à deux non comparables et non majorables par un même troisième. Il existe toutefois une échelle canonique des TCD, à savoir celle qui correspond aux itérés transfinis du TCD de $L = \log$ et qui permet en particulier de classer les *fonctions cohésives* (cf § IV.2).

Récapitulons dans un même tableau les quatre principaux domaines de croissance :

Domaine couvrant	balisé par	et stable pour les opérations
les fonctions contractantes : $F(x) << x$	$P_\sigma(x) = x^\sigma (0 < \sigma < 1)$	algébriques (1)
les fonctions lentes : $\log F(x) << \log x$	$L_n(x) \quad (n \in \mathbf{N}^*)$	algébrico-différentielles (2)
les fonctions très lentes : $F(x) << L_n(x) \quad (\forall n)$	$L_\alpha(x) \quad (\omega \leq \alpha < \omega^\omega)$	algébrico-fonctionnelles (3)
les fonctions très très lentes : $F(x) << L_\omega n(x) \quad (\forall n)$	$L_\alpha(x) \quad (\omega^\omega \leq \alpha)$	algébrico-logiques (4)

(*) Pour l'ordre naturel sur les TCD, qui porte évidemment sur les dérivées F' de leurs représentants F.

Ici, les qualificatifs "contractant", "lent", "très lent", sont entendus au sens strict, chacun excluant le suivant. (1) couvre la résolution d'équations algébriques. (2) couvre la prise de solutions *analytiques sur* $[..., +\infty]$ (*) des équations algébrico-différentielles.
(3) couvre la prise de solutions *analytiques ou cohésives sur* $[..., +\infty]$ des équations algébrico-fonctionnelles, dont (IV.4.7) est le prototype. (4) désigne des constructions du genre (IV.4.16), qui ne se rencontrent jamais en analyse, ne serait-ce que pour cette simple raison qu'elles comportent la réinjection de la *variable* (essentiellement continue) dans *l'ordre d'itération* (essentiellement discret).

Notion d'indiscernabilité. Le "Grand Cantor".

Voilà, très brièvement, pour *l'échelle à l'infini* des TC et des TCD canoniques : *elles se correspondent et reflètent la structure archi-discrète des* T0. Mais qu'en est-il de *l'echelle locale ?* Y-a-t'il, entre ces échelons naturels que sont les *types,* des mailles et sous-échelons naturels ? Il s'avère que la subdivision de l'échelle en mailles de plus en plus fines n'est pas possible *partout,* mais seulement par *endroits*, car on se heurte au principe *d'indiscernabilité* dont voici, très informellement, le sens et la portée : pour tout F d'exponentialité < 0, l'équation $F^{*} o F(x) \equiv -1 + F^{*}(x)$ (ainsi que la plupart des équations algébrico-fonctionnelles où F figure comme coefficient) ne possède pas de solution naturelle unique, qui se distinguerait de toutes les autres. Plus précisément, *aucun critère* fondé sur le comportement à l'infini(**) d'une famille finie de fonctions formées à partir de coefficients algébrico-différentiels et d'une suite finie d'opérations $(+, \times, 0, \partial)$ et de leurs inverses, ne permet de distinguer entre les F^{*} (si l'on n'utilise pas la composition 0, c'est essentiellement une conséquence de la *douceur* des F^{*} et des *séries universelles* ; cf § IV.3). Cela n'empêche pas tous ces F^{*} indiscernables d'appartenir à un même TC $\underline{F^{*}}$ et à un même TCD $\underline{\underline{F^{*}}}$, mais cela devient faux pour la plupart des fonctions engendrées à partir de F^{*}, à commencer par les itérées fractionnaires F^{0w} de F définies comme (IV.4.8) pour $w \in \mathbf{R}$. Certes, deux F^{0w_1} et F^{0w_2} définies à partir *d'un même* F^{*} seront comparables, maisce n'est plus vrai en général si F^{0w_1} et F^{0w_2} sont définis à partir de deux itérateurs F^{*} et hoF^{*} différents et si w_1 et w_2 ont même partie entière n. On est donc

(*) restriction essentielle, à cause de l'existence d'équations algébrico-différentielles à solutions analytiques par morceaux et denses dans $C(\mathbf{R})$.
(**) C'est-à-dire l'existence ou non d'un *signe fixe* au voisinage de $+\infty$ et l'existence ou non d'une limite, finie ou infinie, pour $x \to +\infty$. Notons d'ailleurs que le *passage au complexe* ne permettrait aucunement de lever ni même de réduire l'indiscernabilité. Pour $L = \log$, par exemple, tout itérateur L^{*} analytique sur $[..., +\infty]$ l'est automatiquement sur un voisinage complet (ramifié) de ∞ et présente à l'infini, sur tout axe $\arg z = \theta (\theta \in \mathbf{R})$ la totalité des comportements asymptotiques universels des fonctions très lentes et douces.

conduit à regrouper tous ces F^{0^w} (pour $n < w < n+1$) dans une même *zone de croissance* (ZC) ou *zone de croissance différentielle* (ZCD), beaucoup plus large qu'un TC ou qu'un TCD.

Ce principe général - regrouper dans une même zone tout ce qui est indiscernable - appliqué aux solutions des équations algébrico-fonctionnelles les plus générales, conduit à une espèce de complétion fractale de l'intervalle $[0, \omega^{\omega}[$ - "fractale", parce qu'elle ne contient ni **R** ni **Q**, se refuse à reproduire le continu et présente, plus peut-être que tout autre fractal au monde, une extraordinaire tendance à l'autosimilitude. L'échelle des ZC et ZCD canoniques ainsi obtenue possède une richissime structure et en particulier *trois ordres totaux,* largement indépendants, à savoir :

i) *l'ordre naturel*

ii) l'ordre défini par leur *largeur* et qui consiste à comparer (grâce à l'ordre (i)) la taille de *leur groupe de stabilité* (pour la composition à gauche)

iii) l'ordre mesurant la *complexité* de leur définition, en termes de longueurs de mots.

Les ZC et ZCD canoniques jouent ici le même rôle que les ouverts que Cantor ôtait successivement de [01] pour construire son premier "cantor" historique, mais la structure obtenue ici - le Grand Cantor - est beaucoup plus complexe, puisqu'elle reflête les T0 (jusqu'à ω^{ω} ou bien au-delà si on le souhaite) non seulement à *l'infini* mais aussi *localement.* D'ailleurs, tout comme la suite des T0 contient et code tous les *bons ordres,* il semble que le Grand Cantor contienne et code (mais dans un sens assez différent) tous les *"cantors linéaires"*. Pour une description moins vague du Grand Cantor avec ses ZC et ZCD, voir [E.5].

V - REFERENCES

[Ba] R. BAMON - Solution of Dulac's problem for quadratic vector fields, An. Acad. Bras. 57 (1985). 3.

[Br] A.D. BRJUNO - Analytic form of differential equations. Trans. Moscow. Math. Soc. Vol. 25 (1971).

[D] H. DULAC - Sur les cycles limites, Bull. Soc. Math. France, 51 (1923), p 45-188.

[Dyn] E.M. DYN'KIN - Pseudoanalytic extension of smooth functions. The uniform scale. Ann. Math. Soc. Transl., (2) Vol.115 (1980).

[E.1] J. ECALLE - Les fonctions résurgentes. T.1. Algèbres de fonctions résurgentes, Pub. Math. Orsay (1981).

[E.2] J.E. - Les fonctions résurgentes T.2. Les fonctions résurgentes appliquées à l'itération, Pub. Math. Orsay (1981).

[E.3] J.E. - Les fonctions résurgentes T.3. L'équation du pont et la classification analytique des objets locaux. Pub. Math. Orsay (1985).

[E.4] J.E. - Cinq applications des fonctions résurgentes. Prépub. Math. Orsay (1984).

[E.5] J.E. - Preuve de la conjecture de Dulac et accéléro-sommation de l'application de retour (soumis à Travaux en Cours).

[E.6] J.E. - Calcul accélératoire et applications (à paraître à Travaux en Cours).

[E.7] J.E. - Calcul compensatoire et linéarisation quasianalytique des champs de vecteurs locaux (à paraître à Travaux en Cours).

[E.8] J.E. - Fonctions résurgentes, calcul étranger, calcul accélératoire et applications (cours général - en préparation).

[E.9] J.E. - Memento sur les fonctions résurgentes, l'accélération et leurs applications (En préparation ; sur demande chez l'auteur à partir de septembre 1990 - Recueil de définitions, énoncés, formules et figures).

[E.M.M.R] J. ECALLE, J. MARTINET, R. MOUSSU, J.P. RAMIS - Non-accumulation des cycles-limites. C.R.A.S., t.304, série I, n° 14 (1987). (I) p. 375-378 (II) p. 431-434.

[F.P] J.P. FRANCOISE and C.C. PUGH - Keeping track of limit cycles, Journal of Diff. Eq., 65 (1986) p. 139-157).

[Il.1] Yu. S. IL'YASHENKO - Limit cycles of polynomial vector fields with non degenerate singular points on the real plane, Funk. anal. i ego prim., 18, 3 (1984) p. 32-34 ; Func. Ana. and Appl., 18,3 (1985) p. 199-209.

[Il.2] Yu. S.I. - Uspehi Mat. Nauk, 42, 3 (1987), p. 223.

[Il.3] Yu. S.I. - Finiteness theorems for limit cycles (En russe ; à paraître).

[P.L] I.G. PETROVSKI et E.M. LANDIS - On the number of limit cycles of the equation $dy/dx = P(x,y)/Q(x,y)$ where P and Q are polynomials, Ann. Math. Soc. Trans., 2, 14 (1960) p. 181-200.

[Man] S. MANDELBROJT - Séries adhérentes, régularisation des suites et applications. Gauthier-Villars, Paris, 1952.

[M.R.1] J. MARTINET et J.P. RAMIS - Problèmes de modules pour des équations différentielles non-linéaires du premier ordre, Pub. Math. IHES, 55 (1982) p.63-164.

[M.R.2] J.M. et J-P. R - Classification analytique des équations différentielles non-linéaires résonnantes du premier ordre, Ann. Sc. Ec. Norm. Sup., 4ème série, t.16 (1983) p. 571-625.

[Mou] R. MOUSSU - Le problème de finitude du nombre de cycles-limites (d'après R. Bamon et Yu. S. Il'yashenko). Séminaires Bourbaki, 37ème année, 1984-85, n° 655 (Nov. 1985).

[Rou] R. ROUSSARIE - A note on finite cyclicity property and Hilbert's 16 th problem ; Preprint de l'Univ. de Bourgogne, Déc. 1987.

[S] A. SEIDENBERG - Reduction of singularities of the differentiable equation $A.dy = B.dx$, Amer. J. Math (1968) p. 248-269.

[Vol] A. VOLBERG - Quasianalytic functions - a survey (à paraître).

[Y] J-Ch. YOCCOZ - Non-accumulation des cycles-limites - Séminaires Bourbaki, 40ème année, 1987-88, n°690 (Nov. 1987).

LIMIT CYCLES AND ZEROES OF ABELIAN INTEGRALS SATISFYING THIRD ORDER PICARD - FUCHS EQUATIONS

Ljubomir Gavrilov[*,**]
Institute of Mathematics
Bulgarian Academy of Sciences
1090 Sofia

Emil Horozov[*]
Faculty of Mathematics
and Informatics
Sofia University
5 A. Ivanov
Sofia

1. INTRODUCTION

The second part of Hilbert 16th problem raises the question about an upper bound c(N) for the maximal number of the limit cycles (i.e. isolated periodic solutions) of a planar polynomial vector field of degree N :

$$P(x,y)dx + Q(x,y)dy = 0 \tag{1.1}$$

This problem is still open, and it is not known even whether $c(N) < \infty$. The above question splits into a few subquestions, the first of which is: *is the number of limit cycles of a fixed vector field finite*? It was believed almost 60 years that the answer is yes, as claimed Dulac in his memoir "Sur les cycles limites"[8].However, in 1985 Yu. Il'yashenko[12] found a trivial technical gap in the final step of Dulac's proof. The correct proof was announced in the articles of Il'yashenko[21] and Ecalle, Martinet, Moussu, Ramis[9] . An important part of this proof containing the crucial idea appeared recently in [24]

A possible approach to the Hilbert 16th problem is, starting from a known system, to study the possible subsequent bifurcations

[*] *Partially supported by the Committee of Science to the Bulgarian Council of Ministers, contract* No 911 .

[**] *Part of this work was done while visiting (L.P.A) Universite de Pau et des Pays de l'Adour from December* 1989 *until March* 1990 .

of the vector field [18]. Consider the perturbed integrable system of ordinary differential equations

$$dH(x,y) + \varepsilon.(R_1(x,y)dx + R_2(x,y)dy) = 0 \tag{1.2}$$

where H, R_1, R_2 are polynomials. If the level sets $\{H = h\} \subset \mathbb{R}^2$ of the integrable system contain a family of ovals then, after perturbing this system some of the ovals sustain the perturbation, i.e. become limit cycles. To determine whether it happens we consider the Poincaré map p(h), corresponding to a given oval $o(h) \subset \{H = h\}$. The first approximation of p(h) - h with respect to ε is given by the Abelian integral

$$I(h) = \oint_{o(h)} R_2.dx - R_1.dy \quad . \tag{1.3}$$

The maximal number $\tilde{c}(N)$ (deg H $\le$ N + 1, deg $R_1 \le$ N, deg $R_1 \le$ N) of the zeroes of I(h) (including the multiplicities) is an upper bound of the number of the ovals, sustaining the bifurcation.

The next question is: *Determine the number* $\tilde{c}(N)$. It is also known as the "weakened 16th Hilbert problem"[23]. Recently Khovansky[15] and Varchenko[20] proved independently that $\tilde{c}(N) < \infty$.

Consider the simplest case N = 2. The number of zeroes of (1.3) is known for certain fixed Hamiltonian functions H, and certain perturbations [7,13,16,17]. In [7] the authors consider a small one-parameter cubic perturbation of a fixed cubic polynomial H. As the moduli space of all cubic Hamiltonians, whose level sets contain ovals, is a two-dimensional one (see section 2) then these perturbed Hamiltonians represent a co-dimension one "local" family (i.e. defined for "small" values of the parameter) of cubic polynomials.

In the present paper we consider another co-dimension one family of Hamiltonians, namely Hamiltonians with an invariant line. This family is "global", in the sense that it depends upon a parameter taking all real values. We prove the monotonicity of the period functions (which are given by special Abelian integrals) - Theorem 1 of section 2. Our second result - Theorem 2 of section 2, gives an upper bound for the limit cycles of any quadratic perturbation of these Hamiltonians. The method exploited here applies to general cubic Hamiltonians. The case considered in this paper has the only advantage that the computations needed for the proofs are simpler. We hope to return to the general case in another publication.

We use well known methods of algebraic geometry : Picard-Fuchs equations, Picard-Lefschetz theory. The Abelian integral I(h) satisfies a third order Picard-Fuchs equation, and $J(h) = \frac{d^2}{dh^2}I(h)$ satisfies a second order Picard-Fuchs equation. As the coefficients of this equation are rational functions in h, then the standard technique [2,3,5,6,7,10,16,17] implies a bound for the number of zeroes of $J(h)$, and hence of $I(h)$. As we want to find an exact bound for the number of zeroes then we face two problems :

- by differentiating $I(h)$ twice we obviously loose information about the behaviour of $I(h)$
- we need an estimate of the number of the zeroes of $J(h)$, which is a priori lower than this number for an arbitrary solution of the Picard-Fuchs equation satisfied by $J(h)$.

To overcome the second problem we have to know what is the distinction between $J(h)$ and the remaining solutions of the Picard-Fuchs equation. From a geometrical point of view the answer is surprisingly simple. Namely, consider the corresponding Riccati equation as a two-dimensional autonomous system of ordinary differential equations. *Then the solution corresponding to* $J(h)$ *is a separatrix solution of this system.* To make use of this fact we study the global phase portrait, and apply the "Rolle's theorem for dynamical systems"[14].

To compensate the loss of information after differentiating $I(h)$ we use the "local" information about the values of $I(h)$, $\frac{d}{dh}I(h)$, $\frac{d^2}{dh^2}I(h)$, at the ends of the interval $\Delta \ni h$ on which the ovals are defined. Probably, however, that is not enough to find the exact value of the number of zeroes of $I(h)$.

The paper is organized as follows. In section 2 we formulate our results and give some definitions. Section 3 and Section 4 are more or less routine. There we derive the Picard-Fuchs equations satisfied by our Abelian integrals, and describe their asymptotic behaviour. The reader may skip these two sections, and then to come back when the occasion arises. In section 5 we prove the monotonicity of the period functions corresponding to the Hamiltonian functions under consideration. At last, all these results are used in section 6, where we prove Theorem 2.

Remark. When this text was already prepared we learned that W.A.Coppel [23] has proved that any quadratic system with an invariant line has at most one limit cycle. This implies immediately our Theorem 2. Nevertheless we keep it as the proof contains the main ideas needed to find an upper bound for the limit cycles of quadratic perturbations of the general quadratic Hamiltonian system - namely the second derivative of the Abelian integral (1.3) satisfies a second order Picard-Fuchs equation.

We are grateful to A. Zegeling for pointing out to us the paper [23]. We are also grateful to H. Zoladek for the extremely important suggestions which helped us to improve the text.

2. STATEMENT OF THE RESULTS

Let $H(x,y)$ be a polynomial and $\omega = P_1(x,y)dx + P_2(x,y)dy$ be a polynomial one-form, where $\deg(H) = 3$, $\deg(P_1), \deg(P_2) \le 2$, and H, P_1, P_2 have real coefficients. Further we shall suppose that for some $h \in \mathbb{R}$ the set $\{H = h\} \subset \mathbb{R}^2$ contains an oval (i.e. a compact real curve Γ, such that $(H_x, H_y) \neq (0,0)$ on Γ). Let $\Delta \subset \mathbb{R}$ be the set of those $h \in \mathbb{R}$ for which $\{H = h\} \subset \mathbb{R}^2$ contains an oval. Δ is either an open interval or a union of two open intervals.

Let $H = x.(y^2 + (x - g)^2 - 1)$. Define the period function $T(h)$ to be the period of the unique periodic solution lying on the level set $\{H= h\}$, $h \in \Delta$ (see [6]). The central results of the present paper are the following theorems.

<u>Theorem 1.</u> If $h \in \Delta$, then $T'(h) \neq 0$.

<u>Theorem 2.</u> The Abelian integral $I(h) = \int_{\delta(h)} \omega$ either vanishes identically in Δ, or it has no more than three zeroes (including the multiplicities) in Δ. If $|g| \ge 1$, then $I(h)$ has no more than one simple zero in Δ.

If Δ is not empty, the Hamiltonian system corresponding to H has a continuous family of periodic solutions. It is known that in this case the system also has a center [23]. Hence, after suitable $\mathbb{R}$ - linear change of the variables H takes the form

$$H = \frac{x^2 + y^2}{2} + Ax^3 + Bx^2y + Cxy^2 + Dy^3 \tag{2.1}$$

The reader may check that there always exists a rotation which brings H into the form (2.1) with $D = 0$. If $B \neq 0$ we shall also suppose (without loss of generality) that $B = 1$, or if $B = 0$, we shall also suppose that $A = 1$.Thus we get the following two normal forms

$$H = \frac{x^2 + y^2}{2} + Ax^3 + x^2y + Cxy^2 \tag{2.2}$$

$$H = \frac{x^2 + y^2}{2} + x^3 \quad + Cxy^2 \tag{2.3}$$

Definition. We shall say that the two polynomials $H_1(x,y)$, $H_2(x,y)$ with real coefficients are equivalent, provided that there exists a $\mathbb{R}$ - linear change of the variables which brings $H_1(x,y)$ into the form $c_1 + c_2.H_2(x,y)$, for some c_1, $c_2 \in \mathbb{R}$, $c_2 \neq 0$.

The constants A, C defined by (2.2), (2.3) provide moduli for the space of all non-equivalent cubic polynomials, whose level sets contain ovals. To obtain a one to one correspondance between the parameters A, C and this space we have to reduce $\mathbb{R}^2\{A,C\}$ and $\mathbb{R}\{C\}$ by identifying the points lying in one and the same orbit of the free action of a finite group, generated by the rotations and the reflections preserving the normal form (2.2) or (2.3) , and the rank two group $(x,y) \to (x_0-x, y_0-y)$, where (x_0, y_0) is the other center (if it exists).The upshot is that the moduli space of all non-equivalent cubic polynomials is a two dimensional one.

Definition. The Hamiltonian H (with real coefficients) is reducible, provided that for some $h \in \mathbb{R}$ the set $\{H = h\} \subset \mathbb{C}^2$ is reducible.

In the case when deg(H) is an odd number, one easely checks that the above definition is equivalent to

Definition. The Hamiltonian H (with real coefficients) is reducible, provided that for some $h \in \mathbb{R}$ holds $H - h = H_1.H_2$, where H_1 and H_2 are polynomials with real coefficients.

Example. The Hamiltonian (2.3) is reducible. Indeed

$$\frac{x^2 + y^2}{2} + x^3 + Cxy^2 - \frac{3}{8}C = (x + \frac{1}{2}C).(y^2 + \frac{3}{2}x - \frac{3}{4}C) \quad .$$

The present paper deals only with a codimension one subspace of reducible cubic polynomials, given by $H = x.(y^2 + (x-g)^2 - 1)$.

Let H be a reducible cubic polynomial with real coefficients. Then without loss of generality $H = H_1.H_2$, where $H_1 = x$ and $\deg(H_2) = 2$, and let

$$H = x.(ax^2 + bxy + cy^2 + dx + ey + f).$$

If $c = 0$ then $\{H = h\} \cap \{x = \text{const.}\}$ is the empty set, or it consists of one point, and hence $\{H = h\}$ does not contain ovals. Thus, without loss of generality we may put $c = 1$ and also $b = e = 0$. Further $\mathbb{R}$ - linear changes show that H is equivalent to one of the following Hamiltonians

$$x.(y^2 \mp (x-g)^2 \mp 1)$$
$$x.(y^2 \mp (x-g)^2)$$
$$x.(y^2 + x \mp 1) \tag{2.4}$$
$$x.(y^2 \mp 1)$$
$$x.y^2$$

The only reducible Hamiltonians (2.4) contained in the family (2.2) or (2.3) are

$$x.(y^2 + (x-g)^2 \pm 1) \text{ (elliptic case)}$$
$$x.(y^2 - (x-g)^2 \pm 1) \quad \text{(hyperbolic case)}$$
$$x.(y^2 + x - 1) \quad \text{(parabolic case)}$$
$$x.(y^2 \pm (x-1)^2) \quad \text{(linear case)} .$$

In this paper we shall study only one of the above cases.

3. THE PICARD - FUCHS EQUATIONS

Consider the Hamiltonian

$$H = x.(y^2 + (x-g)^2 - 1) \quad . \tag{3.1}$$

The level sets $\{H = h\} \subset \mathbb{R}$ are shown on fig.1. For any fixed $g \in \mathbb{C}$ there are three critical values of h. They are $h = 0$, $h = h^{\pm}(g)$, where

$$h^{\pm} = 2.(g.(g^2 - 9) \pm (g^2 + 3)^{3/2})/27 \quad .$$

For all non-critical values of h the affine algebraic curve $\Gamma_{g,h} = \{H = h\} \subset \mathbb{C}^2$ is a smooth complex manifold. The union of the curves defined by the equations $h = 0$ and $h = h^{\pm}(g)$ in $\mathbb{C}\{h,g\}$ form the bifurcation diagram. It is given on fig.2. For each fixed g let us define on $\Gamma_{g,h}$ the following differential one-forms

$$\alpha = ydx \ , \ a = \frac{d}{dh}\alpha = \frac{dx}{2xy} \ ,$$
$$\beta = x(x-g).y.dx \ , \ b = \frac{d}{dh}\beta = \frac{(x-g)}{2y}dx \ , \tag{3.2}$$
$$\gamma = xy.dx, \ c = \frac{d}{dh}\gamma = \frac{dx}{2y} \ .$$

Here $\frac{d}{dh}$ is a covariant derivative in the Gauss-Manin connection of

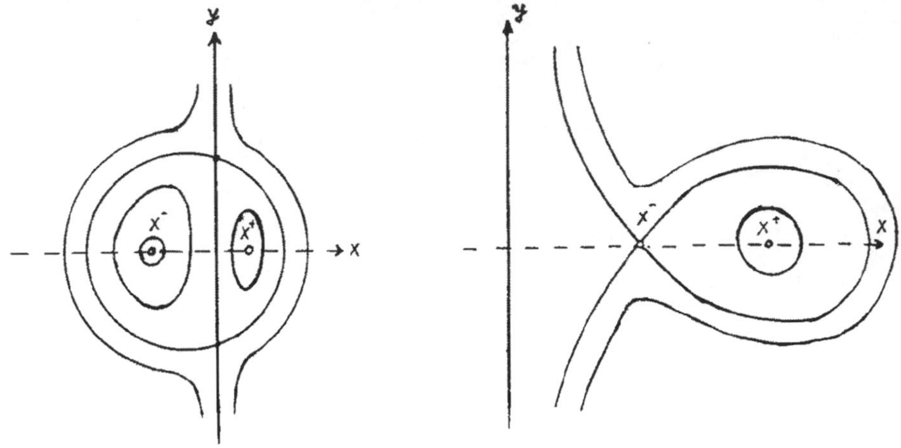

a) $|g| < 1$ b) $|g|>1$

$$x^{\mp} = \frac{2g \mp \sqrt{g^2+3}}{3}$$

Fig.1.Level sets of Hamiltonian (3.1).

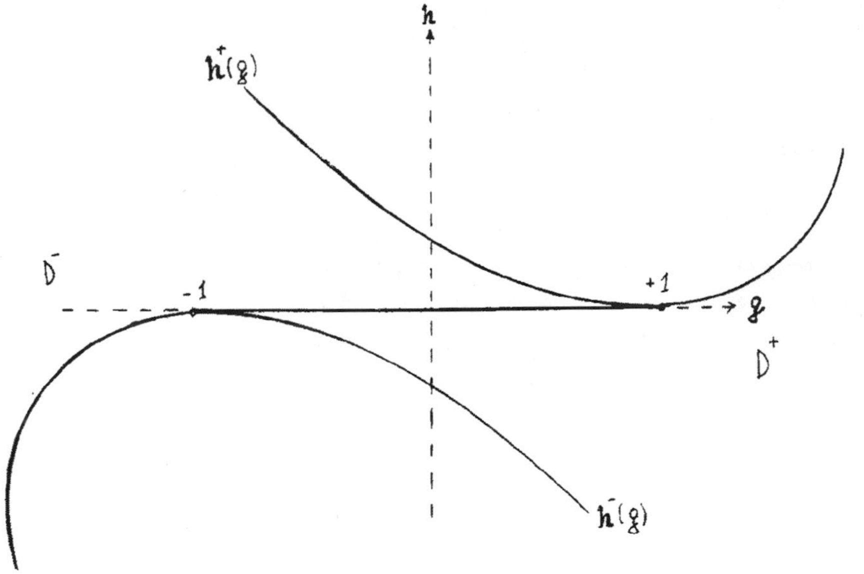

Fig 2. Bifurcation diagram of (3.1).

the bundle $H^1(\Gamma_{g,h},\mathbb{C}) \to h \in \mathbb{C}\setminus\{h^+(g),h^-(g),0\}$, associated with the local trivial bundle $\mathbb{C}^2 \supset \Gamma_{g,h} \to h \in \mathbb{C}\setminus\{h^+(g),h^-(g),0\}$ [1].

Proposition 3.1. For any fixed point (g,h) which does not lie on the bifurcation diagram, the one-forms a, b, c defined by (3.2) form a basis of the three dimensional linear space $H^1(\Gamma_{g,h},\mathbb{R})$.

Proof. It is more convenient to make the computations in the proof of this Proposition using the following new coordinates

$$x \to x, \quad x.y \to z \quad .$$

The elliptic curve $\Gamma_{g,h} = \{H = h\} \subset \mathbb{C}^2$ takes the form

$$\{ z^2 = x.(h - x.(x - g)^2 + x)\} \ . \qquad (3.3)$$

One may easely check that the above change of variables provides a biholomorphic mapping between the affine curve (3.3) and $\Gamma_{g,h}$. In the new coordinates we have $a = \frac{dx}{2z}$, $b = \frac{x.(x-g)}{2z}\,dx$, $c = \frac{x}{2z}\,dx$. According to Grothendieck's theorem [11] the polynomial one-forms generate a basis of the first de Rham cohomology group of the affine algebraic curve (3.3). Hence it is enough to prove that each polynomial one-form restricted to (3.3) is equal, up to an addition of an exact form, to a linear combination of the forms a, b, c. Namely, each polynomial one-form equals on (3.3) to a sum of one forms $\frac{x^s}{z}\,dx$, s = 0, 1, 2. The degrees of the one forms $\frac{x^s}{z}\,dx$ is reduced with the help of the identity

$$d(x^{s-3}.z) = \frac{2(s-3)x^{s-4}.P(x) + x^{s-3}.P'(x)}{2z}\,dx$$

where $P(x) = x.(h - x(x - g)^2 + x)$, $P'(x) = \frac{d}{dx}P(x)$. The leading term in the coefficient of the above one-form is $x^s.(2(s-3)+4)$, and hence $\frac{x^s}{z}\,dx$, $s \geq 3$, is equivalent on (3.3) to a linear combination of forms $\frac{x^r}{z}\,dx$, $r = 0, 1, \ldots, s-1$. ■

Proposition 3.2 For each fixed (g,h) the following identities hold on $\Gamma_{g,h}$

$$\alpha = \frac{3}{2}.h.a + g.b + c + \frac{1}{2}.d(xy)$$

$$\beta = \frac{(3-2g^2).h}{24}.a + \frac{(9h + 5g - 2g^3)}{12}.b + \frac{1}{4}.c + d\left(\frac{xy(6x^2-10gx + 2g^2-3)}{24}\right)$$

$$\gamma = \frac{gh}{6}.a + \frac{2+g^2}{3}.b + (h+g).c + d\left(\frac{xy(2x-g)}{6}\right) \quad . \qquad (3.4)$$

The proof is a straightforward computation : we use (3.2) and the identity $y = ((h - x.(x - g)^2 + x)/x)^{1/2}$.

Let $\delta(h) \to h$ be a locally constant section of the bundle $H_1(\Gamma_{g,h}, \mathbb{C}) \to h \in \mathbb{C}\setminus\{h^+(g), h^-(g), 0\}$. Suppose that $\delta(h)$ are represented by (homological) ovals on $\Gamma_{g,h}$. Thus we can associate the above section with a continuous family of ovals on $\Gamma_{g,h}$. We shall denote these ovals again by $\delta(h)$.

Proposition 3.3. The Abelian integrals $A = \int_{\delta(h)} a$, $B = \int_{\delta(h)} b$, $C = \int_{\delta(h)} c$, satisfy the following Picard-Fuchs system

$$d(h).\frac{d}{dh}\begin{pmatrix} A \\ B \\ C \end{pmatrix} = \begin{pmatrix} h(8g+9h) & ,2(2g^2+3gh-2) & ,0 \\ h(2g-2g^3+3h-2g^2h) & ,-h(8g+9h) & ,0 \\ h(2g^2-2+3gh) & ,\ 2h(g^2+3) & ,0 \end{pmatrix}.\begin{pmatrix} A \\ B \\ C \end{pmatrix} \tag{3.5}$$

where

$$d(h) = -27h(h - h^+(g))(h - h^-(g)) = h(4g^4+4-8g^2+4hg^3-36hg-27h^2).$$

Proof. We integrate (3.4) along $\delta(h)$ and then differentiate with respect to h. As $\frac{d}{dh}\alpha = a$, $\frac{d}{dh}\beta = b$, $\frac{d}{dh}\gamma = c$, then we obtain a linear system for A, B, C, $\frac{d}{dh}A$, $\frac{d}{dh}B$, $\frac{d}{dh}C$. Solving this system in $\frac{d}{dh}A$, $\frac{d}{dh}B$, $\frac{d}{dh}C$ (which in view of Proposition 3.1. is always possible), we obtain (3.5).■

It is easy to check that the one forms a, b have no residues on $\Gamma_{g,h}$. Hence they represent elements of $H^1(\overline{\Gamma}_{g,h}, \mathbb{R})$, where $\overline{\Gamma}_{g,h}$ is the compactification of $\Gamma_{g,h}$. As $\dim(H^1(\overline{\Gamma}_{g,h}, \mathbb{Z})) = \dim(H_1(\overline{\Gamma}_{g,h}, \mathbb{Z})) = 2$ ($\overline{\Gamma}_{g,h}$ is an elliptic curve), then A, B, satisfy a second order Picard-Fuchs equation (see (3.5)).

Let $\delta(h)$, $\theta(h)$ form a basis of $H_1(\overline{\Gamma}_{g,h}, \mathbb{Z})$. For arbitrary rational one-forms ω_1, ω_2, which do not possess residues, consider the Wronskian

$$W(\omega_1, \omega_2) = \det\begin{pmatrix} \int_{\delta(h)}\omega_1 & , & \int_{\delta(h)}\omega_2 \\ \int_{\theta(h)}\omega_1 & , & \int_{\theta(h)}\omega_2 \end{pmatrix}.$$

It is a rational function in g, h [1]. Now the Liouville theorem, applied to the Picard-Fuchs system satisfied by A, B, implies that W(a,b) does not depend upon h, as the trace of this (linear) system is equal to zero. On the other hand Proposition 3.1. implies that $W(a,b) \neq 0$, if $d(h) \neq 0$, and hence W(a,b) is equal to a onstant $\rho \neq 0$.

$A' = \frac{d}{dh}A$, $C' = \frac{d}{dh}C$ satisfy the following Picard-Fuchs system

$$\frac{d}{dh}\begin{pmatrix} C' \\ A' \end{pmatrix} = \begin{pmatrix} a_{11} a_{12} \\ a_{21} a_{22} \end{pmatrix}.\begin{pmatrix} C' \\ A' \end{pmatrix} \tag{3.6}$$

where $a_{11} = W(c'',a')/W(c',a')$, $a_{12} = -W(c'',c')/W(c',a')$, $a_{21} = W(a'',a')/W(c',a')$, $a_{22} = -W(a'',c')/W(c',a')$. $A = A(h)$ satisfies the following Picard-Fuchs equation

$$p.A'' + q.A' + r.A = 0 \tag{3.7}$$

where $p = W(a',a).d^2$, $q = - W(a'',a).d^2$, $r = W(a'',a').d^2$, $d = d(h)$. Below we shall compute the above Wronskians explicitely. For that purpose we shall use (3.5). For example to compute $W(c',a')$ we note that

$$d(h).\frac{d}{dh}\begin{pmatrix} C \\ A \end{pmatrix} = M(h).\begin{pmatrix} A \\ B \end{pmatrix}$$

where the matrix $M(h)$ is given by (3.5). Then $W(c',a') = W(a,b).\det(M(h)/d^2 = \rho.\det(M(h)/d^2$. However, one can prove along the same lines as in [13] that $W(c',a')$ has only simple poles. It means that $d = d(h)$ divides the polynomial $\det(M(h))$. The same cancelations occur for the remaining Wronskians, and the explicite expressions are rather simple. The direct computation gives

Proposition 3.4. Let $d(h)$ be as in Proposition 3.3. Then

$$W(a',a)/\rho = 2.(2 - 2g^2 - 3gh)/d(h) \quad ,$$

$$W(c',a')/\rho = 2/d(h) \quad ,$$

$$W(c'',a')/\rho = 6h.(2g^2h + 15h + 8g)/d(h)^2 \quad ,$$

$$W(c'',c')/\rho = 2h.(2g^3h + 3gh + 2 - 2g^2)/d(h)^2 \quad ,$$

$$W(a'',c')/\rho = 4.\{h^2.(3g^2 - 18) + h.(4g^3 - 24g) + 2.(g^2 - 1)^2\}/d(h)^2,$$

$$W(a'',a')/\rho = 12.(3gh^2 + 4g^2h - 5h + 2g.(g^2 - 1))/d(h)^2 \quad ,$$

$$W(a'',a)/\rho = 4.\{-81gh^3 + h^2.(6g^4 - 135g^2 + 81) + h.(8g^5 - 80g^3 + 72g)$$
$$+ 4.(g^2 - 1)^3\}/d(h)^2 \quad , \rho = \text{const.} \neq 0.$$

4. ASYMPTOTIC BEHAVIOUR OF ABELIAN INTEGRALS

Consider the open subsets of $\mathbb{R}^2$

$$D^+ = \{(g,h) \in \mathbb{R}^2 : h^-(g) < h < h^+(g),\ g \geq 1\}$$
$$\cup\{ (g,h) \in \mathbb{R}^2 : h^-(g) < h < 0,\ |g| < 1 \}$$

and

$$D^- = \{ (g,h) \in \mathbb{R}^2 : h^-(g) < h < h^+(g),\ g \leq -1 \}$$
$$\cup\{ (g,h) \in \mathbb{R}^2 : 0 < h < h^+(g),\ |g| < 1 \} ,$$

shown on fig.2. For each $(g,h) \in D^+ \cup D^-$ the set $\{H = h\} \subset \mathbb{R}^2$

contains an oval $\delta(h)$ (see fig.1). Note that for $|g| > 1$, $h = 0$ the complex curve $\Gamma_{g,h}$ is singular (i.e. (g,h) is a bifurcation point). Nevertheless it still contains an oval. For all the remaining bifurcation points $\Gamma_{g,h}$ does not contain an oval.

Consider now the open set $\Delta_g = \{h \in \mathbb{R} : (g,h) \in D^+ \cup D^-\}$. For each fixed $g \in \mathbb{R}$ it coincides with the set Δ of Theorem 1. Let us define on Δ_g the following two meromorphic functions

$$\zeta_g(h) = \int_{\delta(h)} c \Big/ \int_{\delta(h)} a, \quad \xi_g(h) = \int_{\delta(h)} \frac{d}{dh} c \Big/ \int_{\delta(h)} \frac{d}{dh} a ,$$

where the one forms a, b, c are defined by (3.2). The importance of these functions were suggested to us by [5,7]. In this section we shall study the asymptotic behaviour of $\zeta_g(h)$, $\xi_g(h)$, when h tends to a point on the boundary $\overline{\Delta_g} \setminus \Delta_g$ of Δ_g. Notice the obvious relations :

$$(g,h) \in D^+ \Leftrightarrow (-g,-h) \in D^-, \ \zeta_g(h) = -\zeta_{-g}(-h), \ \xi_g(h) = -\xi_{-g}(-h).$$

<u>Proposition 4.1.</u> $\zeta_g(h)$ has the following asymptotic behaviour

i) $\lim\limits_{h \downarrow h^-(g)} \zeta_g(h) > \lim\limits_{h \uparrow h^+(g)} \zeta_g(h)$, for $g \geq 1$

ii) $\lim\limits_{h \downarrow h^-(g)} \zeta_g(h) > \lim\limits_{h \uparrow 0} \zeta_g(h) = 0 = \lim\limits_{h \downarrow 0} \zeta_g(h) > \lim\limits_{h \uparrow h^+(g)} \zeta_g(h)$

for $-1 < g < 1$

iii) $\lim\limits_{h \downarrow h^-(g)} \frac{d}{dh} \zeta_g(h) < 0$, for $g > -1$

iv) $\lim\limits_{h \uparrow 0} \frac{d}{dh} \zeta_g(h) = \lim\limits_{h \downarrow 0} \frac{d}{dh} \zeta_g(h) = -\infty$, for $-1 < g < 1$

v) $\lim\limits_{h \uparrow h^+(g)} \frac{d}{dh} \zeta_g(h) = -\infty$, for $g \geq 1$.

<u>Proposition 4.2.</u> $\xi_g(h)$ has the following asymptotic behaviour

i) $\lim\limits_{h \uparrow h^+(g)} \xi_g(h) = \lim\limits_{h \uparrow h^+(g)} \zeta_g(h)$, for $g \geq 1$

ii) $\lim\limits_{h \uparrow 0} \xi_g(h) = \lim\limits_{h \uparrow 0} \zeta_g(h) = 0$, for $-1 \leq g \leq 1$

iii) $\lim\limits_{h \uparrow h^+(g)} \frac{d}{dh} \xi_g(h) = \infty$, for $g \geq 1$

iv) $\lim\limits_{h \uparrow 0} \frac{d}{dh} \xi_g(h) = \infty$, for $-1 \leq g \leq 1$.

v) $\lim\limits_{h \downarrow h^-(g)} \frac{d}{dh} \xi_g(h) < 0$, for $g > 1$.

Like in section 3 we can associate the oval $\delta(h)$ with the corresponding homological cycle in $H_1(\Gamma_{g,h},\mathbb{Z})$, which we shall denote again by $\delta(h)$. Now $\delta(h)$ can be defined also for complex values $h \in \mathbb{C}\setminus\{h^+(g),h^-(g),0\}$, in such a way that $\delta(h) \to h$ is a locally constant section of the bundle $H_1(\Gamma_{g,h},\mathbb{Z}) \to h \in \mathbb{C}\setminus\{h^+(g),h^-(g),0\}$. Thus each Abelian integral $\int_{\delta(h)} \omega$ (and hence $\zeta_g(h)$, $\xi_g(h)$) becomes a multivalued meromorphic function on $\mathbb{C}\setminus\{h^+(g),h^-(g),0\}$, with branch points at $h^+(g)$, $h^-(g)$, and 0. To prove Proposition 4.1. and Proposition 4.2. we shall need a formula for the branching of the Abelian integrals at these points. It can be derived from the Picard-Lefschetz formula [1].

For any fixed g the affine curve $\Gamma_{g,h}$ is singular only if $h = 0$, or $h = h^{\pm}(g)$. Suppose that $g \neq \pm 1$ is a fixed real number. If $h = 0$, the level set $\Gamma_{g,h}$ orresponding to the critical value $h = 0$ contains two Morse critical points ($x = 0$, $y = \pm\sqrt{1 - g^2}$) of H. Let us denote the corresponding vanishing cycles by $\theta^{\pm}$. Without loss of generality we may suppose that θ^+ and θ^- are homological and let us denote them by $\theta \equiv \theta^+ \sim \theta^-$. If $h = h^{\pm}(g)$, $\Gamma_{g,h}$ has one double point at $x = x^{\mp} = \dfrac{2g \mp \sqrt{g^2+3}}{3}$, $y = 0$, which is a Morse critical point of H with critical value $H = h^{\pm}(g)$. Let us denote the corresponding vanishing cycle by $\delta^{\pm}$. Denote $D_\rho = \{|z| < \rho\}\setminus\{h^-(g),\ 0,\ h^+(g)\} \subset \mathbb{C}$, where ρ is a sufficiently big fixed real number. Let z_0 be a point on the boundary $|z| = \rho$ of D_ρ. Any loop $\ell \in \pi_1(D_\rho,z_0)$ induces an isomorphism ℓ_* (monodromy) in the first homology group

$$\ell_* : H_1(\Gamma_{g,h},\mathbb{Z}) \to H_1(\Gamma_{g,h},\mathbb{Z}) .$$

Let ℓ^0, $\ell^{\pm} \in \pi_1(D_\rho,z_0)$ be loops around 0 and $h^{\pm}(g)$ respectively.The (generalized) Picard-Lefschetz formula [4,1] reads

$$\begin{aligned} \ell^0_*(\delta) &= \delta + (\theta^+\circ\delta)\theta^+ + (\theta^-\circ\delta)\theta^- = \delta + 2(\theta\circ\delta)\theta \\ \ell^{\pm}_*(\delta) &= \delta + (\delta^{\pm}\circ\delta)\delta^{\pm} \end{aligned} \tag{4.1}$$

where $(\delta^{\pm}\circ\delta)$, $(\theta^{\pm}\circ\delta)$, $(\theta\circ\delta)$ are the intersection indexes of the corresponding cycles.

Consider an arbitrary Abelian integral $I(h) = \int_{\delta(h)} \omega$, where ω is a meromorphic one-form without residues on $\Gamma_{g,h}$. The Picard-Lefschetz formula (4.1) implies [1] that in a neighbourhood of $h = 0$ holds

$$I(h) = \frac{\log(h)}{\pi i} \cdot \int_{\theta(h)} \omega + P(h) \quad , \tag{4.2}$$

and in a neighbourhood of $h = h^{\pm}(g)$ holds

$$I(h) = \frac{\log(h - h^{\pm}(g))}{2\pi i} \cdot \int_{\delta^{\pm}(h)} \omega + Q(h - h^{\pm}(g)) \quad , \tag{4.3}$$

where P, Q are meromorphic functions.

Consider now the case $g = \pm 1$. The Milnor number [1] of the critical point $x = 0$, $y = 0$ of H is two, and hence the Picard-Lefschetz formula (4.1) can not be directly applied. Denote

$$D_\rho = \{|z| < \rho\} \setminus \{h^-(g), h^+(g)\} \subset \mathbb{C}$$

where ρ is a sufficiently big fixed real number, and let z_0 be a fixed point on the boundary $|z| = \rho$ of D_ρ. Let ℓ be a loop around (1,0) lying in the complex plane $\{g = 1\} \times \mathbb{C}\{h\}$, and let $\ell_1, \ell_2 \in \pi_1(D_\rho, z_0)$ be loops around $(1+\varepsilon, h^+(1+\varepsilon))$ and $(1+\varepsilon, 0)$ in the complex plane $\{g = 1+\varepsilon\} \times \mathbb{C}\{h\}$, where $\varepsilon > 0$ is a sufficiently small number (fig. 3).

Denote by M_ℓ, M_{ℓ_1}, M_{ℓ_2}, the corresponding monodromy matrices acting upon $H_1(\Gamma_{g,h}, \mathbb{Z})$. It is well known that $M_\ell = M_{\ell_1} \circ M_{\ell_2}$. Hence to ompute the monodromymatrix M_ℓ acting upon $H_1(\Gamma_{g,h}, \mathbb{Z})$ it is enough to compute M_{ℓ_1} and M_{ℓ_2}. The Picard-Lefschetz formula implies that in θ, δ^+, δ^- coordinates (having suitable orientations) holds

$$M_{1_1} = \begin{pmatrix} 1 & 1 & 0 \\ 0 & 1 & 0 \\ 0 & 1 & 1 \end{pmatrix}, \quad M_{1_2} = \begin{pmatrix} 1 & 0 & 0 \\ -2 & 1 & 0 \\ -2 & 0 & 1 \end{pmatrix}, \quad \Rightarrow \quad M_{1_1} \circ M_{1_2} = \begin{pmatrix} -1 & 1 & 0 \\ -2 & 1 & 0 \\ -4 & 1 & 1 \end{pmatrix}.$$

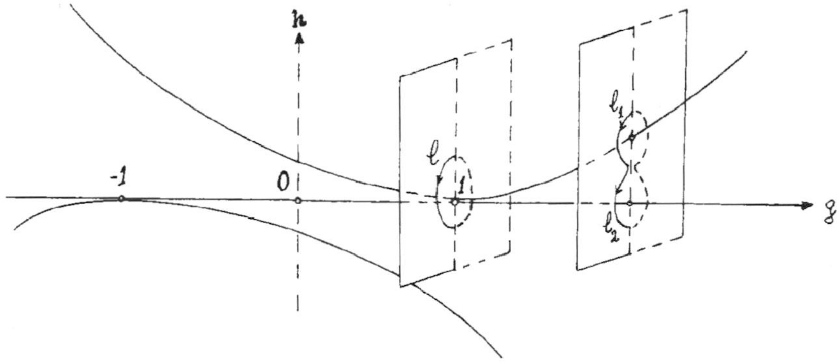

Fig.3. Deformation of the loop l into the loop $l_1 \circ l_2$.

As the eigenvalues of the last matrix are 1, $\pm$ i, then it is concluded that the Abelian integral I(h) is a meromorphic function in a neighbourhood of $h = 0$, with respect to $h^{1/4}$ [1].

<u>Proof of Proposition 4.1.</u>

i) As $\delta(h)$ vanishes at $h = h^-(g)$, $g \neq \mp 1$, into a Morse critical point of H, then $\int_{\delta(h)} \gamma$, $\int_{\delta(h)} \alpha$ are holomorphic functions in h, in a sufficiently small neighbourhood of $h = h^+(g)$. As $\lim_{h\downarrow h^-(g)} \int_{\delta(h)} \gamma = \lim_{h\downarrow h^-(g)} \int_{\delta(h)} \alpha = 0$, then

$$\lim_{h\downarrow h^-(g)} \int_{\delta(h)} \gamma \Big/ \int_{\delta(h)} \alpha = \lim_{h\downarrow h^-(g)} \int_{\delta(h)} c \Big/ \int_{\delta(h)} a = \lim_{h\downarrow h^-(g)} \zeta_g(h) =$$

$$= \operatorname{Res}\Big|_{x=x^+}\left(\frac{x.dx}{\sqrt{x.[h^-+x.(1-(x-g)^2)]}}\right)\Big/\operatorname{Res}\Big|_{x=x^+}\left(\frac{dx}{\sqrt{x.[h^-+x.(1-(x-g)^2)]}}\right)=$$

$= x^+$, where $x^+= \dfrac{2g + \sqrt{g^2 + 3}}{3}$ (see fig.1). In quite a similar way one computes $\lim_{h\uparrow h^+(g)} \int_{\delta(h)} \gamma \Big/ \int_{\delta(h)} \alpha = \lim_{h\uparrow h^+(g)} \zeta_g(h) = x^-$.Obviously $x^- < x^+$ (see fig.1) and hence to prove i) it remains to consider the case $g = 1$. The two cycles θ_1, θ_2 vanish simultaneously when h 0. For $h < 0$, $h \in \mathbb{R}$, the projections of these two cycles, and the cycle $\delta(h)$ on the $\mathbb{C}\{x\}$-plane are shown in fig.4.

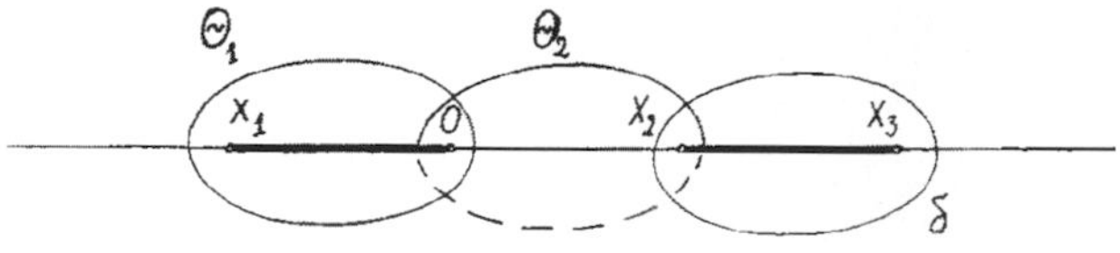

Fig.4. The cycles θ_1, θ_2, δ in the complex x - plane.

$x_1(h), x_2(h), x_3(h)$, are roots of the polynomial $h-x.(x-g)^2+x$, and $x_1(0) = x_2(0) = 0$. Fig.4 implies that $\delta(h)$ is homological in $H_1(\Gamma_{g,h}, \mathbb{Z})$ to $\theta_1 \pm \theta_3$ where θ_3 is a cycle represented by a loop around one of the "infinite" points on $\Gamma_{g,h}$.Hence

$$\int_{\delta(h)} c = \int_{\theta_1(h)} c \pm 2\pi i.\operatorname{Res}\Big|_{x=\infty}\left(\frac{dx}{2.\sqrt{(1-(x-g)^2}}\right) = \pi + \int_{\theta_1(h)} c \ .$$

(the sign of π is fixed by the condition $\int_{\delta(h)} c > 0$), and also

$$\int_{\delta(h)} a = \int_{\theta_1(h)} a \pm 2\pi i.\mathrm{Res}\Big|_{x=\infty}\left(\frac{dx}{2x.\sqrt{(1-(x-g)^2}}\right) = \int_{\theta_1(h)} a \quad .$$

The Picard-Lefschetz formula implies that the integrals $\int_{\delta(h)} c$ and $\int_{\delta(h)} a$ are holomorphic functions in $h^{1/4}$ (in a neighbourhood of $h = 0$). Changing the variable x as $x \to z.\sqrt{-h}$, where $\sqrt{-h} > 0$ for $h < 0$ we obtain

$$2.\int_{\delta(h)} c = (-h)^{1/4}. \int_{\theta_1(h)} \frac{zdz}{\sqrt{z(-1 + 2z^2 - z^3.\sqrt{-h}\ )}} + \pi =$$

$$= \pi + (-h)^{1/4}.(2.\int_{-1/\sqrt{2}}^{0} \sqrt{\frac{z}{(-1 + 2z^2)}}\ dz + O((-h)^{1/4})), \text{ and}$$

$$2.\int_{\delta(h)} a = (-h)^{-1/4}. \int_{\theta_1(h)} \frac{dz}{\sqrt{z(-1 + 2z^2 - z^3.\sqrt{-h}\ )}} =$$

$$= (-h)^{-1/4}.(2.\int_{-1/\sqrt{2}}^{0} \frac{dz}{\sqrt{z(-1 + 2z^2)}} + O((-h)^{1/4}))\ .$$

Hence in a neighbourhood of $h = 0$ we have

$$\zeta_g(h) = c_1.(-h)^{1/4} + O((-h)^{1/2}), \quad c_1 > 0 \qquad (4.4)$$

which implies $\lim_{h\downarrow h^-(1)} \zeta_1(h) > \lim_{h\uparrow h^+(1)} \zeta_1(h)$.

ii) Suppose that $-1 < g < 1$, let $\theta(h)$ be a vanishing cycle at $h = 0$, and $(\theta(h)\circ\delta(h)) = 1$.The Picard - Lefshetz formula implies

$$\zeta_g(h) = (-\frac{\log(h)}{\pi i}\int_{\theta(h)} c + P(h))/(-\frac{\log(h)}{\pi i}\int_{\theta(h)} a + Q(h)) \qquad (4.5)$$

where P(h) and Q(h) are holomorphic functions in a neighbourhood of $h = 0$.

$$\text{As} \quad \int_{\theta(0)} c = \mathrm{Res}\Big|_{x=0}\left(\frac{dx}{2.\sqrt{(1-(x-g)^2}}\right) = 0,$$

$$\int_{\theta(0)} a = \mathrm{Res}\Big|_{x=0}\left(\frac{dx}{2x.\sqrt{(1-(x-g)^2}}\right) \neq 0,$$

we conclude that $\lim_{h\uparrow 0} \zeta_g(h) = \lim_{h\downarrow 0} \zeta_g(h) = 0$. As $x^+ > 0$, then ii) is proved.

iii) $\delta(h)$ vanishes at $h = h^-(g)$ and hence

$$\lim_{h\downarrow h^-(g)} \frac{d}{dh} \zeta_g(h) = \lim_{h\downarrow h^-(g)} \frac{d}{dh} \Big(\int_{\delta(h)} c \Big/ \int_{\delta(h)} a \Big) =$$

$$\lim_{h\downarrow h^-(g)} \Big\{\Big(\frac{d}{dh} \int_{\delta(h)} c\Big) . \int_{\delta(h)} a - \Big(\frac{d}{dh} \int_{\delta(h)} a\Big) . \int_{\delta(h)} c\Big\}\Big/\Big(\int_{\delta(h)} a\Big)^2 .$$

All integrals above are holomorphic for h in a sufficiently small neighbourhood of $h^-(g)$ and their limits are equal to the corresponding residues at $x = x^+$. After computing these residues we obtain

$$\lim_{h\downarrow h^-(g)} \frac{d}{dh} \zeta_g(h) = \frac{3x^- - 5x^+}{6.x^+.(x^+ - x^-)} < 0 .$$

iv) Let $\theta(h)$ is defined as in ii). We note that in (4.5) $\int_{\theta(0)} c = 0$, $\int_{\theta(0)} a \neq 0$. Also as $\int_{\delta(h)} c = -\frac{\log(h)}{\pi i} \int_{\theta(h)} c + P(h)$, then taking the limit $h\uparrow 0$ we conclude that

$$P(0) = \lim_{h\uparrow 0} \Big(\int_{\delta(h)} c + \frac{\log(h)}{\pi i} \int_{\theta(h)} c\Big) = \lim_{h\uparrow 0} \int_{\delta(h)} c =$$

$$= \int_0^{1+g} \frac{dx}{\sqrt{1 - (x - g)^2}} = \arccos(-g) \neq 0.$$

Differentiating (4.5) and using the above relations we obtain $\lim_{h\uparrow 0} \frac{d}{dh} \zeta_g(h) = -\infty$.

v) If $g = 1$ then (4.4) implies $\lim_{h\uparrow 0} \frac{d}{dh} \zeta_g(h) = -\infty$. Suppose that $g > 1$ and let $\theta(h)$ be a vanishing cycle at $h = h^+(g)$. The Picard - Lefshetz formula implies that in a neighbourhood of $h = h^+(g)$ holds $\zeta_g(h) - x^- =$

$$\Big(- \frac{\log(h - h^+(g))}{2\pi i} P_1(h) + P_2(h)\Big)\Big/\Big(- \frac{\log(h - h^+(g))}{2\pi i} Q_1(h) + Q_2(h)\Big)$$

where $P_1(h)$, $P_2(h)$, $Q_1(h)$, $Q_2(h)$ are holomorphic functions, and $P_1(h) = \int_{\theta(h)} (c - a.x^-)$, $Q_1(h) = \int_{\theta(h)} a$ .Differentiating the above identity we obtain

$$\frac{d}{dh} \zeta_g(h) = \frac{2\pi i}{\{h - h^+(g)\}.\{\log(h - h^+(g))^2\}} . \frac{P_1.Q_2 - P_2.Q_1}{Q_1^2} .(1+O(h)) .$$

As $P_1(0) = 0$ (see i)) then it is enough to prove that

$\lim_{h\uparrow h^+(g)} \dfrac{2\pi i . P_2(h)}{Q_1(h)} < 0$. The formula

$$\int_{\delta(h)} c - a.x^- = - \frac{\log(h - h^+(g))}{2\pi i} P_1(h) + P_2(h)$$

implies $P_2(h^+(g)) = \lim_{h\uparrow h^+(g)} \int_{\delta(h)} c - a.x^- = \lim_{h\uparrow h^+(g)} \int_{\delta(h)} \dfrac{x - x^-}{2xy} dx =$

$= \displaystyle\int_{x^-}^{2g-2x^-} \frac{dx}{\sqrt{x.(-x - 2x^- + 2g)}} \neq 0$. Also $Q_1(h^+(g)) = \lim_{h\uparrow h^+(g)} \int_{\theta(h)} a =$

$= 2\pi i . \operatorname{Res}\Big|_{x=x^-} \left(\dfrac{dx}{2.\sqrt{x.[h^+ + x.(1-(x-g)^2)]}} \right) \neq 0$. Thus we have proved that $\lim_{h\uparrow h^+(g)} \frac{d}{dh} \zeta_g(h) = \pm \infty$. At last we note that for $h < h^+(g)$

$\zeta_g(h) - x^- = \displaystyle\int_{\delta(h)} \frac{x - x^-}{y} dx \Big/ \int_{\delta(h)} \frac{dx}{y} > 0$, and if h is in a sufficiently small neighbourhood of $h^+(g)$, then

$\zeta_g(h) - x^- = \dfrac{2\pi i . P_2(h)}{Q_1(h) . \log(h-h^+(g))} . (1+O(h))$. It implies that

$\lim_{h\uparrow h^+(g)} \dfrac{2\pi i . P_2(h)}{Q_1(h)} < 0$, and hence Proposition 4.1. is proved.■

<u>Proof of Proposition 4.2.</u>

The parts i) and ii) of Proposition 4.2. are proved directly, after applying the Picard-Lefschetz formula. For example if $g > 1$ then $\xi_g(h) =$

$$\frac{d}{dh} \left(- \frac{\log(h - h^+(g))}{2\pi i} \int_{\theta(h)} c + P(h) \right) \Big/ \frac{d}{dh} \left(- \frac{\log(h - h^+(g))}{2\pi i} \int_{\theta(h)} a + Q(h) \right)$$

which , after differentiating and taking the limits, implies $\lim_{h\uparrow h^+(g)} \xi_g(h) = \lim_{h\uparrow h^+(g)} \zeta_g(h) = x^-$. If $g = 1$ then $\lim_{h\uparrow h^+(g)} \xi_g(h) = \lim_{h\uparrow h^+(g)} \zeta_g(h) = 0$, as may be seen from the asymptotics of the Abelian integrals , derived in the proof of Proposition 4.1. i).

iii) Differentiating the above formula for $\xi_g(h)$ we compute that for h sufficiently close to $h^+(g)$ holds

$$\frac{d}{dh} \xi_g(h) = \log(h - h^+(g)) . \frac{d}{dh} \left(\int_{\theta(h)} c \Big/ \int_{\theta(h)} a \right) . (1+O(h)) \qquad (4.6)$$

As in the proof of Proposition 4.1. iii) we compute that

$$\lim_{h\uparrow h^+(g)} \frac{d}{dh}\left(\int_{\theta(h)} c \Big/ \int_{\theta(h)} a\right) = \frac{3x^+ - 5x^-}{6.x^-.(x^- - x^+)} < 0 ,$$

and hence $\lim_{h\uparrow h^+(g)} \frac{d}{dh}\xi_g(h) = \infty$.

iv) If $g = 1$, then the formulae derived in the proof of Propsition 4.1. imply that in a neighbourhood of $h = 0$ in $\mathbb{C}$ holds $\frac{d}{dh}\xi_g(h) = c.(-h)^{-1/2} + \ldots$, where $c < 0$, and hence $\lim_{h\uparrow 0} \frac{d}{dh}\xi_g(h) = \infty$. At last consider the case $-1 < g < 1$. The Picard - Lefschetz formula implies that in a neighbourhood of $h = 0$ holds

$$\frac{d}{dh}\xi_g(h) = \log(h).\frac{d}{dh}\left(\int_{\theta(h)} c \Big/ \int_{\theta(h)} a\right).(1+O(h)) \quad . \tag{4.7}$$

As $\int_{\theta(0)} c = 0$ then we obtain

$$\frac{d}{dh}\left(\int_{\theta(h)} c \Big/ \int_{\theta(h)} a\right)\Big|_{h=0} =$$

$$= \left[\left(\frac{d}{dh}\int_{\theta(h)} c\right)\Big/\left(\int_{\theta(h)} a\right)\right]\Big|_{h=0} =$$

$$\{\operatorname{Res}\big|_{x=0}\left[\frac{d}{dh}\left(\frac{x.dx}{\sqrt{x.[h + x.(1-(x-g)^2)]}}\right)\right]\big|_{h=0}\}\Big/\{\operatorname{Res}\big|_{x=0}\left(\frac{dx}{x.\sqrt{1-(x-g)^2}}\right)\}$$

$= -\dfrac{1}{2.(1-g^2)} < 0$, and hence $\lim_{h\uparrow 0} \frac{d}{dh}\xi_g(h) = \infty$.

v) As $\delta(h)$ vanishes at $h = h^-(g)$ then we have

$$\lim_{h\downarrow h^-(g)} \frac{d}{dh}\xi_g(h) = \lim_{h\downarrow h^-(g)} \frac{d}{dh}\left(\frac{d}{dh}\int_{\delta(h)} c \Big/ \frac{d}{dh}\int_{\delta(h)} a\right) =$$

$$\lim_{h\downarrow h^-(g)} \{(\frac{d^2}{dh^2}\int_{\delta(h)} c).\frac{d}{dh}\int_{\delta(h)} a - (\frac{d^2}{dh^2}\int_{\delta(h)} a).\frac{d}{dh}\int_{\delta(h)} c\}/(\frac{d}{dh}\int_{\delta(h)} a)^2.$$

All integrals above are holomorphic functions for h in a sufficiently small neighbourhood of $h^-(g)$ and their limits are equal to the corresponding residues at $x = x^+$. After computing these residues we obtain

$$\lim_{h\downarrow h^-(g)} \frac{d}{dh}\xi_g(h) = -\frac{80.(g.(16g^2 + 27) + (20g^3 - 3).\sqrt{g^2+3}\,)}{9.R(x^+)^2.x^+.\{R(x^+)^2 - 2.R(x^+).x^+ + 5(x^+)^2\}^2} < 0$$

for $g > 1$, and hence Proposition 4.2. is proved.■

5. MONOTONICITY OF THE PERIOD

In this section we prove Theorem 1.

As $\frac{dx}{dt} = \partial H/\partial y$ then the period $T(h)$ of the only periodic solution contained in the level set $\{H = h\}$ is equal to $\oint dt = \oint \frac{dx}{\partial H/\partial y} = \int_{\delta(h)} \frac{dx}{2xy} = \int_{\delta(h)} a$. Introduce the following notation (see fig. 2)

$$\Delta_g^+ = \{h \in \mathbb{R} : (g,h) \in D^+\},\ \Delta_g^- = \{h \in \mathbb{R} : (g,h) \in D^-\} .$$

Δ_g^+ and Δ_g^- are open intervals and $\Delta = \Delta_g = \Delta_g^+ \cup \Delta_g^-$. As $T(h)$ takes the same values at (g,h) and at $(-g,-h)$, it is enough to prove the theorem for $h \in \Delta_g^+$. In section 3 we derived the Picard-Fuchs equation (3.7) satisfied by $A(h) = \int_{\delta(h)} a \quad (= T(h))$

$$p.A'' + q.A' + r.A = 0 \qquad ' = \frac{d}{dh}$$

where p, q, r are polynomials. As a is the holomorphic one form on the compact elliptic curve $\overline{\Gamma}_{g,h}$, and $A(h) \in \mathbb{R}$, then (without loss of generality) $A(h) > 0$ for $h \in \Delta_g^+$, and hence the function A'/A takes only finite values in Δ_g^+. We shall prove that the equation $A'/A = 0$ does not possess solutions in Δ_g^+. The function A'/A satisfies the following Riccati equation

$$p.x' + q.x + p.x^2 + r = 0 \quad .$$

Consider the autonomous system

$$\begin{cases} \overset{\circ}{x} = -q.x - p.x^2 - r & , \quad x = x(t), \quad {}^{\circ} = \frac{d}{dt} , \\ \overset{\circ}{h} = p & , \quad h = h(t) \quad . \end{cases} \tag{5.1}$$

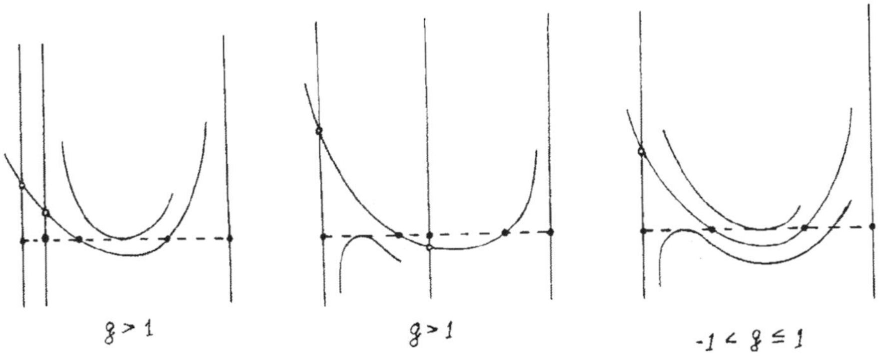

Fig.5. Phase portrait of system (5.1) .

Suppose that $-1 < g < 1$. Then $\Delta_g^+ = (h^-(g),0)$ and $\{h = h^-(g)\}$, $\{h = 0\}$, $\{x = A'(h)/A(h)\}$ are invariant sets of (5.1). Straightforward computation gives

$$\lim_{h\downarrow h^-(g)} A'(h)/A(h) = -\frac{3((R(x^+))^2 + 4(x^+)^2)}{16(x^+)^2.R^3(x^+)} \tag{5.2}$$

where $x.(h-x.(x-g)^2+x)\Big|_{h=h^-(g)} =_- x.(x - x^+)^2.R(x)$. One easely computes $R(x) = - x - 2x^+ + 2g$, $R(x^+) = -\sqrt{g^2+3} < 0$, and hence

$$\lim_{h\downarrow h^-(g)} A'(h)/A(h) > 0 .$$

For h in a sufficiently small neighbourhood of h = 0 in the complex domain, the Picard-Lefschetz formula implies $A(h) = \frac{\log(h)}{\pi i}. \int_{\theta(h)} a + P(h)$, where $\theta(h)$ is a vanishing cycle at h = 0 and P(h) is a holomorphic function (see section 4).This implies $A'(h)/A(h) = (1+O(h))/\{h.\log(h)\}$, and hence $\lim_{h\uparrow 0} A'(h)/A(h) = +\infty$ .Suppose now that the phase curve $x = A'(h)/A(h)$ intersects the line x = 0 . Then it intersects x = 0 at least twice (fig.5).Denote these points by P_2, P_3, and put $P_1 = (0,h^-(g))$. It is easely seen that on the line $h = h^-(g)$ there is only one equilibrium point, which is a saddle. As $p = p(h) \neq 0$ in Δ_g^+, then the direction of the vector field at the points P_1,P_2,P_3, implies that there exist at least two points on the interval $x = 0$, $h^-(g) < h < 0$, and the vector field (5.1) is tangent to the line x = 0 at these points. In other words the polynomial $r(h) = r_g(h) = 12.(3gh^2+h.(4g^2-5)+2g.(g^2-1))$ has two zeroes on the interval $(h^-(g),0)$. This is, however, impossible as it can be seen after some tedious but straightforward computations (see fig.6, where the level set $\{(g,h) \in \mathbb{R}^2: r_g(h) = 0\}$ is pictured).

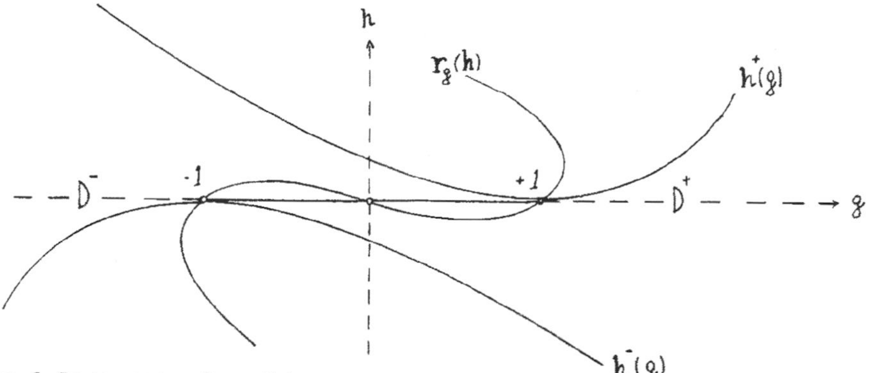

Fig.6.Zero set of $r_g(h)$.

The case $g > 1$ can be studied along the same lines : formula (5.2) holds for any $g > -1$, and $\lim_{h \uparrow h^-(g)} A'(h)/A(h) = +\infty$. The polynomial $p = p(h)$ vanishes exactly once in the interval Δ_g^+, and hence $r_g(h)$ has at least one zero in Δ_g^+ (fig.5). This is ,however , a contradiction (fig.6) .At last if $g = 1$ then the asymptotic expansion for $\int_{\delta(h)} a$ around $h = 0$, derived in the proof of Proposition 4.2. i), implies that $A'(h)/A(h) = -(1+O(h))/4h$, and hence $\lim_{h\uparrow 0} A'(h)/A(h) = +\infty$. This completes the proof of Theorem 1.■

<u>Remark.</u> In the above proof we used that there is only one equilibrium point of (5.1) on the line $h = h^-(g)$, and that it is a saddle. Of course it can be checked directly in (5.1). However, we do not need to make the precise computations. Generically this equilibrium is a saddle, or a node. It is a saddle iff $\delta(h)$ vanishes at $h = h^-(g)$, and a node if it is not so (for example the "infinite"point on the line $h = h^-(g)$ is a node). In the non-generic cases the equilibrium point is a standard saddle-node [12]. Indeed, all phase curves of the system (5.1) are given by

$$x_{\alpha,\beta} = (\alpha . \int_{\delta(h)} \mathring{a} + \beta . \int_{\theta(h)} \mathring{a}) / (\alpha . \int_{\delta(h)} a + \beta . \int_{\theta(h)} a) \ , \ "\circ" = \frac{d}{dh} \ ,$$

where $\delta(h), \theta(h)$ form a basis of $H_1(\overline{\Gamma}_{g,h}, \mathbb{Z})$, and $\overline{\Gamma}_{g,h}$ is the compactification of $\Gamma_{g,h}$. Let $\delta(h)$ vanishes at $h = h^-(g)$. Applying the Picard-Lefshetz formula we obtain

$$\lim_{h \downarrow h^-(g)} x_{\alpha,\beta} = \begin{cases} -\dfrac{3((R(x^+)^2 + 4(x^+)^2)}{16(x^+)^2 . R^3(x^+)} & , \ \beta = 0 \\ -\infty & , \ \beta \neq 0 \end{cases}$$

which implies that the "finite" equilibrium point is a saddle, and the "infinite" one is a node.

6. ZEROES OF THE ABELIAN INTEGRAL

In this section we prove Theorem 2.

<u>Definition</u> (Petrov[16]). We say that the polynomial one-forms ω_1 and ω_2 are equivalent, provided that $\omega_1 - \omega_2 = P_1 dH + dP_2$ where P_1 and P_2 are polynomials.

Let Ω_2 be the factor space of all polynomial one forms with real coefficients $\omega = P_1 dx + P_2 dy$ where $\deg(P_1) \leq 2$ modulo the above equivalency. Then $\dim(\Omega_2) = 3$ and

$$\Omega_2 = \mathbb{R}^3\{ydx,\ xydx,\ y^2dx\}\quad .$$

It is enough to prove Theorem 2 for $I(h) = \int_{\delta(h)} \omega$, where $\omega \in \Omega_2$, and $h \in \Delta_g$.

<u>Remark</u> If $g = 0$, it is known ([7] Lemma 3.9.) that $I(h)$ has no more than one zero in Δ_g. One can also prove that for all $g \in (-1,1)$ $\frac{d}{dh}\zeta_g(h)$ has no zeroes in a neighbourhood of $h = h^-(g)$, $h^+(g)$, 0 $(h \in \Delta_g)$. That is why we conjecture that $I(h)$ has no more than one zero in Δ_g for all values of g.

As $I(h)$ vanishes at $h = h^-(g)$, $g > -1$, and at $h = h^+(g)$, $g < 1$, then the number of the zeroes of $I(h)$ does not exceed the number of the zeroes of $\frac{d}{dh} I(h)$ in Δ_g. For an arbitrary one form $\omega \in \Omega_2$ consider the Abelian integral

$$I(h) = \int_{\delta(h)} \omega = p.(\int_{\delta(h)} xydx) + q.(\int_{\delta(h)} ydx) + r.(\int_{\delta(h)} y^2dx)\ ,\ p,q,r \in \mathbb{R}$$

and

$$\frac{d}{dh} I(h) = p.(\int_{\delta(h)} \frac{dx}{2y}) + q.(\int_{\delta(h)} \frac{dx}{2xy}) + r.(\int_{\delta(h)} \frac{dx}{x}) =$$

$$= p.(\int_{\delta(h)} \frac{dx}{2y}) + q.(\int_{\delta(h)} \frac{dx}{2xy}) = p.\int_{\delta(h)} c + q.\int_{\delta(h)} a\quad .$$

As $\int_{\delta(h)} \frac{dx}{2xy} \neq 0$ in Δ_g we shall prove that the equation $\zeta_g(h) = (\int_{\delta(h)} c)/(\int_{\delta(h)} a) = \text{const.}$ has no more than three solutions in Δ_g, for $|g| < 1$, and no more than one solution in Δ_g, for $|g| \geq 1$.

Proposition 4.1. implies certain restrictions on the possible graphics of $\zeta_g(h)$ (fig.7). Namely, if there exists $c_1 = \text{const.}$ and $\zeta_g(h) = c_1$ has more than three (one) solutions in Δ_g, then there exists a constant c_2, and $\zeta_g(h) = c_2$ has at least five (three) solutions in Δ_g.

As $\int_{\delta(h)} a \neq 0$, $\frac{d}{dh}(\int_{\delta(h)} a) \neq 0$ for $h \in \Delta_g$ (see section 5), then $\frac{d}{dh}\zeta_g(h) = 0 \Leftrightarrow \xi_g(h) = \zeta_g(h)$. If the equation $\zeta_g(h) = c_2$, $|g| < 1$, has at least five solutions in Δ_g, then Proposition 4.2. implies that there exists a constant c_3 such that the equation $\xi_g(h) = c_3$ has at least three solutions in Δ_g^+ or in Δ_g^-. As $\xi_g(h) = -\xi_{-g}(-h)$, then we may suppose that the latter equation has *at least three solutions in* Δ_g^+.

Let us suppose that the equation $\zeta_g(h) = c_2$, $|g| \geq 1$, has at least three solutions in Δ_g. Now the same reasonings, together with Proposition 4.2. v) imply that there exists a constant c_4, and the equation $\xi_g(h) = c_4$ has *at least three solutions in* $\Delta_g = \Delta_g^+$.

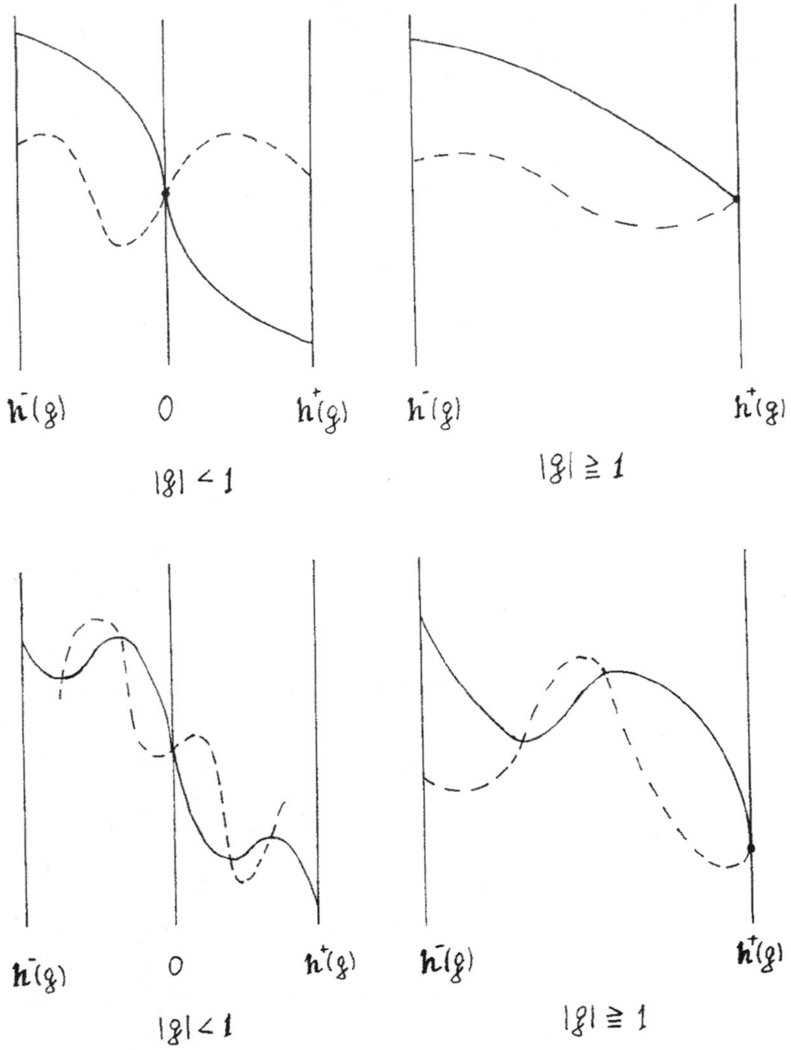

Fig.7. Possible graphics of $\zeta_g(h)$ (bold line) and $\xi_g(h)$ (dottedline) in the interval Δ_g .

The next lemma shows that it is impossible

<u>Lemma 6.1.</u> The equation $\xi_g(h)$ = const. has no more than two solutions in Δ_g^+.

We arrived at the desirable contradiction. To the end of this section we shall prove Lemma 6.1.

The one-forms $\frac{d}{dh}$ c, $\frac{d}{dh}$ a, satisfy a second order Picard-Fuchs system (3.6), derived in section 3. The Riccati equation, satisfied by $\xi_g(h)$ has the form

$$p.x' + q.x + r.x^2 + s = 0$$

where p is a cubic, and q, r, s, are quadratic polynomials in h. Now Rolle's theorem for dynamical systems [14,15] implies that, if x(h) is a solution of the Riccati equation defined for all $h \in \Delta_g^+$, then for any $x_o \in \mathbb{R}$, the function $x(h) - x_o$ has no more than three zeroes in Δ_g^+. We need, however, a stronger statement. To prove Lemma 6.1. we shall study (as in the proof of Theorem 1) the global phase portrait of the system

$$\begin{cases} \mathring{x} = -q.x - r.x^2 - s & , \quad x = x(t), \quad \mathring{} = \frac{d}{dt} , \\ \mathring{h} = p & , \quad h = h(t) . \end{cases} \tag{6.1}$$

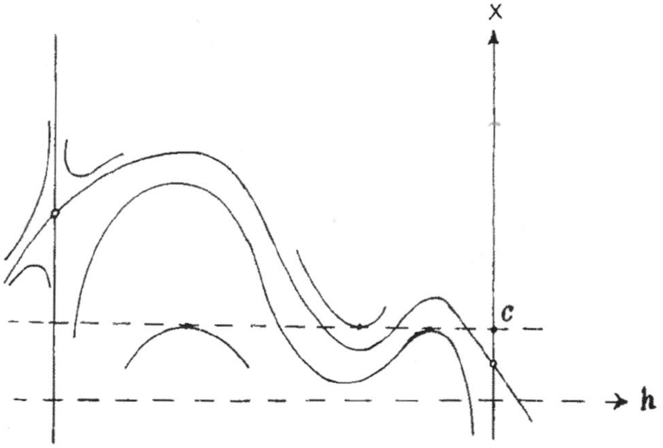

Fig.8. Phase portrait of system (6.1).

All phase curves of the system (6.1) are given by

$$x_{\alpha,\beta}(h)=(\alpha.\int_{\delta(h)}\mathring{c}+\beta.\int_{\theta(h)}\mathring{c})/(\alpha.\int_{\delta(h)}\mathring{a}+\beta.\int_{\theta(h)}\mathring{a}),\ "\circ"=\frac{d}{dh},\ \alpha,\ \beta\in\mathbb{R},$$

where $\delta(h),\theta(h)$ form a basis of $H_1(\bar{\Gamma}_{g,h},\mathbb{Z})$, and $\bar{\Gamma}_{g,h}$ is the compactification of $\Gamma_{g,h}$. Let $\delta(h)$ vanishes at $h = h^-(g)$. Applying the Picard-Lefshetz formula we obtain

$$\lim_{h\downarrow h^-(g)} x_{\alpha,\beta} = \begin{cases} 4.x^+.R(x^+).(3.x^+-R(x^+))/\{3.((R(x^+)-x^+)^2+4(x^+)^2)\} + \\ + x^+ , \quad \beta = 0 \\ x^+ , \quad \beta \neq 0 \end{cases}$$

where $R(x)$ is defined in section 5. As

$$R(x^+).(3.x^+-R(x^+)) = - 2.(g + \sqrt{g^2+3}).\sqrt{g^2+3} < 0,$$

then there are exactly two (different) equilibrium points on the line $h = h^-(g)$, which are a node (with coordinates $(h^-(g),x^+)$) and a saddle (see fig.8). Notice that $\xi_g(h) \equiv x_{1,0}(h)$. Also, as we have proved in Proposition 4.2.,

$$\lim_{h\uparrow h^+(g)} \xi_g(h) = x^- \text{ for } g > 1, \text{ and } \lim_{h\uparrow 0} \xi_g(h) = 0 \text{ for } |g| < 1.$$

As $\lim_{h\downarrow h^-(g)} \xi_g(h) = x^+ > \max(x^-,0)$, then there always exists a constant c, and the vector field (6.1) is tangent to the line $x = c$ at three points (at least) - see fig.7. In other words the quadratic polynomial $P(h) = q.c + r.c^2 + s$ has three roots in Δ_g^+ which is a contradiction. Thus Lemma 6.1., and hence Theorem 2, is proved.∎

ACKNOLEDGEMENT

The last part of this manuscript was prepared while the first author (L.G.) was visiting Laboratoire de Physique Applique, Universite de Pau et des Pays de l'Adour. He is grateful for its hospitality.

REFERENCES

1. V.I. Arnol'd, A.N. Varchenko, S.M. Gusein-Zade, Singularities of differentiable maps, Vol.2, Basel, Birkhauser Verlag, 1988 .

2. J. Carr, S.-N. Chow, J.K. Hale, Abelian integrals and bifurcation theory, Journal Diff. Eqns. 59 (1985), 413-437.

3. J. Carr, S.A. van Gils, J.A. Sanders, Nonresonant bifurcations with symmetry , SIAM J. Math. Anal., 18 (1987), No 3, 579-591.

4.C.H. Jr. Clemens, Picard-Lefschetz theorem for families of nonsingular algebraic varieties acquiring ordinary singularities, Trans. Am. Math. Soc. 136 (1969) 93-108.

5. R.Cushman, J.A.Sanders, A codimension two bifurcation with a third order Picard-Fuchs equation, Journal Diff. Eqns, 59 (1985),243-256.

6. S.N. Chow, J.A. Sanders, On the number of the critical points of the period, Journal. Diff. Eqns, 64 (1986) 51-66.

7. B. Drachman, S.A. van Gils, Zhang Zhi-fen, Abelian integrals for quadratic vector fields, J. reine angew. Math.382 (1987), 165-180.

8. H. Dulac, Sur les cycles limites, Bull. Soc. Math. France, 51 (1923), 45-188.

9. J. Ecalle, J. Martinet, R. Moussu, J.P. Ramis, Compt. Rendu de l'Acad. sci.(Paris), 304, Série 1, (1987), 375-378, 431 - 434.

10. S.A. van Gils, E. Horozov, Uniqueness of limit cycles in planar vector fields which leave the axes invariant, Contemporary Mathematics 56 (1986), 117-129.

11 A. Grothendieck, On the de Rham cohomology of algebraic varieties, Publ. Math. Inst. HES, 29 (1966), 351-359.

12. Yu. Il'yashenko, Dulac's memoir "Sur les cycles limites" and related problems of the local theory of differential equations, Russian Math. Surveys, 40 (1985), 1-49.

13. Yu. Il'yashenko, Zeroes of special Abelian integrals in a real domain, Funct. Anal. Appl. 11 (1977), 309-311.

14. A.G. Khovansky, Cycles of dynamical systems on a plane and Rolle's theorem , Siberia Math. Journal, 25 (1984), n°3, 198-203.

15. A.G. Khovansky, Real analytic manifolds with finiteness properties and complex abelian integrals, Funct. Anal. Appl. 18 (1984), 119-128.

16. G.S. Petrov, Number of zeroes of complete elliptic integrals Funct. Anal. Appl.18 (1984), 148-150.

17. G.S. Petrov, Elliptic integrals and their nonoscillations, Funct. Anal. Appl.20 (1986), 37-40.

18. C. Rousseau, Bifurcation methods in quadratic systems, Canadian Math. Soc., Conference Proc., 8 (1987), 637-653.

19. D. Schlomiuk, J. Guckenheimer, R. Rand, Integrability of plane quadratic vector fields, to appear.

20. A.N. Varchenko, Estimate of the number of zeroes of abelian

integrals depending on parameters and limit cycles, Funct. Anal. Appl., 18 (1984), 98-108.

21. Yu. Il'yashenko, Finiteness theorems for limit cycles, Russian Math. Surveys, 42 (1987), No. 3, p. 223.

22. W. A. Coppel, A survey of quadratic systems, J. Differential Equations, 2 (1966), 293-304.

23. W. A. Coppel, Some quadratic systems with at most one limit cycle, Dynamics Reported, Vol. 2 (1989), 61-88. Ed.U.Kirchgraber & H.O.Walther, John Wiley & Sons Ltd. and B.G. Teubner.

24. Yu. Il'yashenko, Finiteness theorems for limit cycles, Russian Math. Surveys, 45 (1990), No. 2, p. 143-200.

ON THE BASIN OF ATTRACTION OF DISSIPATIVE PLANAR VECTOR FIELDS

A. GASULL[1] AND J. SOTOMAYOR[2]

Abstract. An estimate for the size of the basin of attraction of an equilibrium point for a class of planar dissipative vector fields is given here. Our main result, which generalizes a theorem of Krasowskii, is applied to give several sufficient conditions for global asymptotic stability.

1. INTRODUCTION.

Consider the (x, y)-plane $\mathbf{R}^2$, endowed with the Euclidean norm $|p| = (x^2 + y^2)^{1/2}$, $p = (x, y)$.

A C^1 planar vector field $V = X\frac{\partial}{\partial x} + Y\frac{\partial}{\partial y}$, is called *dissipative* on a region D if $divV = \frac{\partial X}{\partial x} + \frac{\partial Y}{\partial y} \leq 0$ on D and the equality only holds on a set with Lebesgue measure zero. If the equality never holds on D we will say that V is *strictly dissipative* on D.

This paper is concerned with the estimate of the size of the *basin of attraction* A_V of an equilibrium point p of a C^1 vector field V. The aim is to determine a disk, as large as posible, centered at p, contained in A_V, i.e. such that the orbit through any of its points tends to p as the time goes to infinity. Conditions on V for $A_V = \mathbf{R}^2$ (global asymptotic stability) have been studied in [**GLS**], where references for the bakground of this problem can be found. The present paper focuses on situations not necessarily covered in [**GLS**], as discussed in Section 4.

Define the following functions associated to a planar vector field W

$$m(r) = \min\{|W(p)|;\ |p| = r\},$$
$$M(r) = \max\{|W(p)|;\ |p| = r\}.$$

For $0 < R \leq S \leq \infty$ define

$$L_S(R) = \frac{1}{2\pi}\int_R^S m(r)\,dr.$$

This is a non increasing function when $S < \infty$ or when $\int_R^\infty m(r)\,dr < \infty$, otherwise it is constant equal to infinity.

If B is a positive C^1 function such that $B \cdot W$ is (strictly) dissipative on D we will say that B is a *(strict) Dulac function* for W on D.

The appropiate choice of Dulac functions has been shown to actually enlarge the domain where a vector field is dissipative.

Call $T = \sup\{r > 0;\ L_S(r) - rM(r) \geq 0\}$ where the functions L_S and M are associated to $B \cdot V$; $T = \infty$ if this supremum is infinity. Note that always $0 < T \leq S$.

The main result of this paper is the following Theorem.

---

[1] Partially supported by CIGYT grant number PB86-0351.

[2] Current address : IHES, 35 Route de Chartres, Bures sur Yvette, 91440 FRANCE

THEOREM A. *For a C^1 planar vector field V with a singularity at 0, assume that the following conditions hold:*

(i) *V has a Dulac function B on $D_S = \{p;\ |p| < S\}$ $(D_\infty = \mathbb{R}^2)$.*

(ii) *Every equilibrium point of V in D_S is local asymptotically stable.*

Then, the orbit of V through p, $|p| < T$, tends to 0 as the time goes to infinity.

This theorem gives an explicit estimate for the size of the basin of attraction of 0 for V. It can be computed by studying the inequality

$$\frac{1}{2\pi}\int_R^S m(r)\,dr - RM(R) \geq 0,$$

in which the only information on V that we need is the norm of $B \cdot V$ on concentric circumferences.

This theorem extends (and is inspired, in proof and contents) by a result of Krasowskii [**K1**, Th 2.2], see also [**K2**, p. 114]. The proof is given Section 2.

In Krasowskii's formulation instead of conditions (i) and (ii), appears the following one:

iii) All the solutions $\lambda(x, y) \in \mathbb{C}$ of the equation

$$\begin{vmatrix} \frac{\partial X}{\partial x} - \lambda & \frac{\partial X}{\partial y} \\ \frac{\partial Y}{\partial x} & \frac{\partial Y}{\partial y} - \lambda \end{vmatrix} = 0,$$

on D_S have negative real parts.

Obviously (iii) $\Rightarrow$ (i) and (ii), taking as B the constant function 1.

Other results and related Corollaries to Theorem A will be discussed in Section 3.

2. PROOF OF THEOREM A

Assume that there exists a point p, $|p| = R < T$ such that $p \notin A_V$. Here A_V denotes the basin of attraction of 0 for V. Consider first the case $S < \infty$.

Take now the new vector field $B \cdot V$. Obviously $A_{B \cdot V} = A_V = A$. For the sake of simplicity we will call V the vector field $B \cdot V$. Then we have that V is dissipative on D_S. Therefore, from Bendixson-Dulac Criterion, V has no periodic orbits in D_S. Furthermore we know from hypothesis (ii) that all equilibrium points of V in D_S are local asymptotically stable.

Denote by $\overline{rs}$ the segment joining r, $s \in \mathbb{R}^2$ and by o the point $(0, 0)$.

From the above considerations we know that there exists $q \in \partial A \cap \overline{op}$ such that the ω-limit set of its orbit γ_q is not a rest point of V in D_S.

The orbit γ_q has the point q in D_R and leaves D_S. Of course there is another point $r \in [o, q]$ such that the positive semiorbit through it, γ_r^+, is totally contained in D_S.

So we can take $s \in \overline{rq}$ with $\gamma_s^+ \subset D_S$ and the properties: (i) for $s' \in \overline{sq}$ $\gamma_{s'}^+ \not\subset D_S$; (ii) for $s' \in \overline{os}$ $\gamma_{s'}^+ \subset D_S$. Furthermore we have that γ_s^+ is tangent to ∂D_S in a point u. See Figure 1.

Consider the vector field W orthogonal to V in such a way that W points towards the origin in u. Denote by Ψ_u^+ the positive semiorbit of W through u.

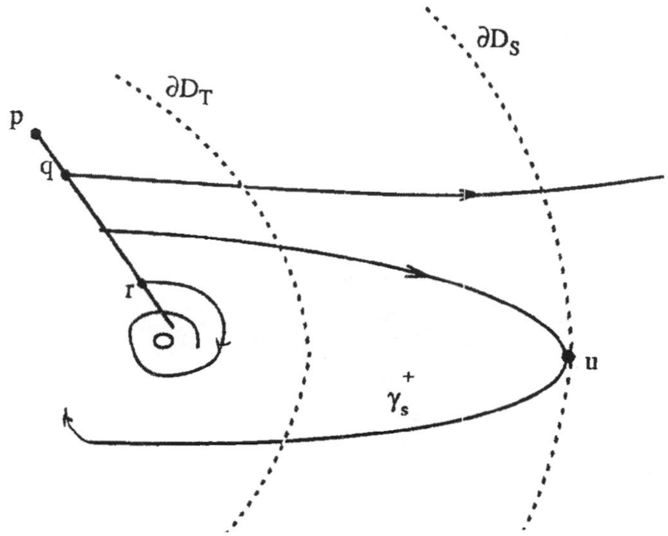

Figure 1. Choice of the orbit γ_s^+.

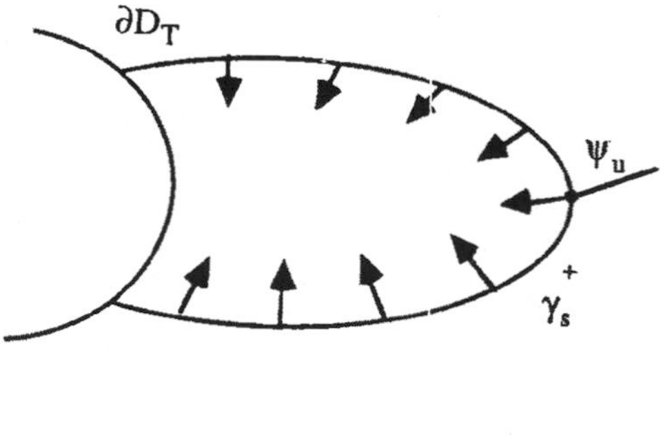

Figure 2. Vector field W on γ_s^+.

Now we prove that $\Psi_u^+ \cap \partial D_T \neq \emptyset$. Assume that this intersection is empty. So Ψ_u^+ is completely contained in the region K bounded by ∂D_T and γ_s^+, because W has transversal contact with Ψ_u^+. See Figure 2.

The ω-limit set of Ψ_u^+ will be either a periodic orbit or a set with critical points. Since in the interior of a periodic orbit there are always critical points we can ensure that W has a critical point in K. Then there is also a critical point of V in K.

Hence we can use similar arguments to those in the beginning of the proof to show that there exists a $p' \in K$ such that $\gamma_{p'}^+$ leaves D_S. So this orbit that, of course can not cut γ_s^+, must cut $\overline{os}$ in contradiction with the choice of s.

Denote by v the first intersection of Ψ_u^+ with ∂D_T. We can consider the region $\overline{K}$ bounded by ∂D_T, Ψ_u^+ and γ_s^+. See Figure 3.

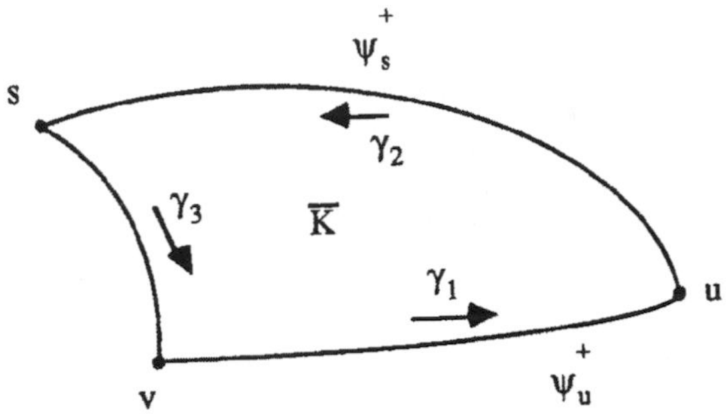

Figure 3. Region $\overline{K} \subset K \subset D_S$.

We call γ to $\partial\overline{K}$ and γ_1, γ_2, γ_3 the partition of $\partial\overline{K}$ given by Ψ_u^+, γ_s^+ and ∂D_T respectively. We take in γ_1 and γ_2 the parameterization given by the time of the integral curves of W and V respectively, with minus sign, and in ∂D_T the angle of polar coordinates, also with sign reversed. This orientation defined on $\partial\overline{K}$ is the one induced by $\overline{K}$, positively oriented.

From the Divergence Theorem applied to $\overline{K}$, we have

$$\int_\gamma X\,dy - Y\,dx = \iint_{\overline{K}} div(X,Y)\,dx\,dy < 0, \tag{1}$$

since $\overline{K} \subset D_S$ and V is dissipative in D_S. On the other hand

$$\int_\gamma X\,dy - Y\,dx = \sum_{i=1}^{3} \int_{\gamma_i} X\,dy - Y\,dx.$$

Now we compute these three terms.

Since V is orthogonal to γ_1,

$$\begin{aligned}\int_{\gamma_1} X\,dy - Y\,dx &= \int_{w_0}^{w_1} (X^2(w) + Y^2(w))\,dw = \\ &= \int_{s_0}^{s_1} \sqrt{X^2(s) + Y^2(s)}\,ds \geq \\ &\geq \int_T^S m(s)\,ds = 2\pi\, L_S(T) \geq 2\pi\, T \cdot M(T).\end{aligned}$$

In the above expression s is the arc lenght parameter for γ_1. In last inequality, the choice of T has been used.

$\int_{\gamma_2} X\,dy - Y\,dx = 0$ since V is tangent to γ_2.

Finally,

$$\begin{aligned}\left|\int_{\gamma_3} X\,dy - Y\,dx\right| &= \left|\int_{\theta_0}^{\theta_1} \langle (X,Y), (-T\sin\theta, -T\cos\theta)\rangle\,d\theta\right| \leq \\ &\leq T \cdot M(T) \int_{\theta_0}^{\theta_1} d\theta \leq 2\pi\, T \cdot M(T)\end{aligned}$$

Then from the above calculations

$$\int_\gamma X\,dy - Y\,dx \geq 0$$

in obvious contradiction with (1). So for any p, $|p| < T$, $p \in A$ and the Theorem follows for $S < \infty$.

If $S = \infty$ we can make the above arguments for any $S > 0$ and conclude that the theorem also follows in this case. ∎

3. Consequences of Theorem A.

Assume in what follows that V is a C^1 planar vector field with an equilibrium point at $0 = (0,0)$.

It is clear that in Theorem A we can replace conditions (i) and (ii) by a set of stronger conditions. In this way, as we have already mentioned, replacing (i) and (ii) of Theorem A, by (iii) we obtain Theorem 2.1 of Krasowskii. Other possible replacement gives the following result.

THEOREM 1. *Assume that the following conditions hold:*

(iv) *V has a strict Dulac function B on D_S.*
(v) *$\det D(BV)(p) > 0$ at all singular points on D_S.*

Then we have the same conclusions for V as in Theorem A.

From Theorem A we can also obtain the following corollaries, on global asymptotic stability of 0. In all them the functions M and L that appear are associated to $B \cdot V$.

COROLLARY 2. *Assume that on $D_\infty = \mathbb{R}^2$, V satisfies conditions (i) and (ii) of Theorem A . Then if there exists $\{r_n\}_{n\in\mathbb{N}}$ with $r_n \uparrow \infty$ such that $L_\infty(r_n) - r_n M(r_n) \geq 0$ for all $n \in \mathbb{N}$ then 0 is global asymptotically stable.*

COROLLARY 3. *Assume that V satisfies conditions (i) and (ii) of Theorem A on D_∞. Then if $L_\infty(r) - rM(r) \geq 0$ for all $r > 0$ then 0 is global asymptotically stable.*

COROLLARY 4. *Assume that V satisfies conditions (i) and (ii) of Theorem A on D_∞. Then if $L_\infty(r) \equiv \infty$ then 0 is global asymptotically stable.*

We could have similar results to the Corollaries above using Theorem 1 instead of Theorem A. For instance, assuming that $m(r) \geq kr$ on $\mathbb{R}^2$ it is clear that 0 is the unique equilibrium point of V, that $\det DV(0) \neq 0$ and that $L_\infty(r) = \int_r^\infty m(s)\,ds \geq \int_r^\infty s\,ds = \infty$. Therefore we have the following corollary.

COROLLARY 5. *Assume that V satisfies conditions (iv), that $m(r) \geq kr$ on $\mathbb{R}^2$ and that $\det DV(0) > 0$. Then 0 is global asymptotically stable.*

This corollary generalizes Theorem 25.2 of [**K2**].

4. FINAL REMARKS.

From Corollary 3 we have that a sufficient condition to ensure that 0 is global asymptotically stable for a dissipative vector field on $\mathbb{R}^2$ is that $L_\infty(r) - r\,M(r) \geq 0$ for all r or, in other words, that

$$\int_r^\infty m(s)\,ds \geq 2\pi r \cdot M(r).$$

We study this inequality in the following Proposition.

PROPOSITION 6. *Assume that $m(x)$ and $M(x)$ are continuous functions such that*

$$a) \quad 0 < m(x) \leq M(x),$$

$$b) \quad \int_0^\infty m(s)\,ds < \infty,$$

$$c) \quad \int_x^\infty m(s)\,ds \geq 2\pi x \cdot M(x), \quad \text{for all } x > 0.$$

Then, given $y > 0$, for $x \in [0, y)$ it holds that

$$m(x) \leq \frac{w(y)\, y^{\frac{1}{2\pi}}}{2\pi} \frac{1}{x^{\frac{1+2\pi}{2\pi}}},$$

where $w(y) = \int_y^\infty m(s)\, ds$.

PROOF: Note that

$$\int_x^\infty m(s)\, ds \geq 2\pi x\, M(x) \geq 2\pi x\, m(x).$$

Introducing the function w, $w'(x) = -m(x)$ and this last inequality is equal to $-2\pi x\, w'(x) \leq w(x)$. Since $w(x) > 0$ on $[0, y)$ $w'(x)/w(x) \geq -1/2\pi x$. Integrating between x and y

$$\int_x^y \frac{w'(s)}{w(s)}\, ds \geq \int_x^y \frac{-1}{2\pi s}\, ds.$$

Hence $\frac{w(y)}{w(x)} \geq \left(\frac{y}{x}\right)^{\frac{-1}{2\pi}}$. Using that $w(x) \geq 2\pi x\, m(x)$ we have that

$$\frac{w(y)}{2\pi x\, m(x)} \geq \left(\frac{y}{x}\right)^{-\frac{1}{2\pi}},$$

and so

$$m(x) \leq \frac{w(y)\, y^{\frac{1}{2\pi}}}{2\pi} \frac{1}{x^{\frac{1+2\pi}{2\pi}}}. \quad \blacksquare$$

This proof is reminiscent of that of Gronwall's Lemma.

REMARK 7. *Consider the integral inequality*

$$(2) \qquad \int_x^\infty m(s)\, ds \geq 2\pi x \cdot M(x),$$

under the hypotheses of Proposition 6. Consider also the space E_k of continuous integrable functions m on $\mathbf{R}^+ = [0, \infty)$ such that

$$m(0) = 0,\ m(x) > 0, \text{ and } \liminf_{x \to \infty} x^{\frac{1}{2\pi}} \int_x^\infty m(s)\, ds = k.$$

Then a solution m of (1) in E_k satisfies

$$m(x) \leq \frac{k}{2\pi x^{\frac{1+2\pi}{2\pi}}}.$$

This shows that the range of applicability of Theorems A and 1 is restricted to vector fields whose associated function m satisfies this last inequality.

Note that by l'Hôpital's rule a solution of the inequality (2) in the subset of E_k for which $\lim_{x\to\infty} x^{\frac{1}{2\pi}} \int_x^\infty m(s)\,ds$ exists, satisfies

$$\lim_{x\to\infty} x^{\frac{1}{2\pi}} \int_s^\infty m(s)\,ds = 2\pi \lim_{x\to\infty} x^{1+\frac{1}{2\pi}} m(x) = k.$$

On the other hand all functions in E_∞ satisfy that $\int_x^\infty m(s)\,ds \geq 2\pi x\, m(x)$; some examples are given by $m(x) = x^{-1-\varepsilon}$ with $0 < \varepsilon < \frac{1}{2\pi}$.

In **[O]**, **[HO]** and **[GLS]** a related problem on global asymptotic stability was considered. The problem yet unsolved, in the notation the present paper, is the following:

P1. Do hypotheses (iv) and (v) of Theorem 1 on D_∞ for $B \equiv 1$, imply that V has 0 as global asymptotically stable? It was proved in **[GLS]** that **P1** is equivalent to the following problem :

P2. Do hypotheses (iv) and (v) of Theorem 1 on D_∞ for $B \equiv 1$, imply for V that $\int_0^\infty m(r) = L_\infty(0) = \infty$?

From the equivalence between **P1** and **P2** and our present results, the following considerations arise:

C1) Assume that **P1** has a positive answer. Then if V satisfies hypotheses (iv) and (v) of Theorem 1 on D_∞ for $B \equiv 1$, then for any $r > 0$, $L_\infty(r) = \infty$. Therefore , for $S = \infty$, Theorem 1 gives no additional information.

C2) Assume that **P2** has a negative answer. Then there are strictly dissipative vector fields on D_∞ also satisfying (v) of Theorem 1 (with $B \equiv 1$), for which $L_\infty(0) < \infty$. For these vector fields may or may not be 0 is global asymptotically stable. Then in these cases Theorem A with $B \equiv 1$ and its corollaries give information about the basin of attraction of 0. Results of Proposition 6 and Remark 7 give sufficient conditions on $m(x)$ to ensure that 0 is global asymptotically stable for a vector field for which $L_\infty(0) < \infty$.

C3) The existence of a strictly dissipative vector field on D_∞ satisfying condition (v) of Theorem 1 (with $B \equiv 1$) and such that $L_\infty(0) < \infty$, would answer in the negative the open problem**P1**.

REMARK 8. *The actual computation of the estimate T in Theorem A, must appeal in most cases to numerical methods. To this end, two main steps must be achieved: the evaluation of extremal values of a positive real function and the computation of a definite integral on an interval.*

References

[GLS] Gasull, A., Llibre, J. and Sotomayor, J., *Global asymptotic stability of differential equations in the plane*, to appear in J. of Diff. Equations, 1990.

[HO] Hartman, P. and Olech, C., *On global asymptotic stability of differential equations*, Trans. Amer. Math. Soc. **104** (1962), 159–178.

[K1] Krasowskii, N.N., *On the behaviour in the large of the integral curves of a system of two differential equations*, Prikl. Mat. Mek. **18** (1954), 149–154. (in russian).

[K2] Krasowskii, N.N., Stability of Motion, Stanford University Press, Stanford, California 1963.

[O] Olech, C., *On the global stability of an autonomous system on the plane*, Cont. to Diff. Eq. **1** (1963), 389–400.

Departament de Matemàtiques
Universitat Autònoma de Barcelona
08193 - Bellaterra (Barcelona), SPAIN

Instituto de Matemática Pura e Aplicada
Est. Dona Castorina, 110, CEP 22460
Rio de Janeiro, R.J., BRAZIL

Periodic lines of curvature bifurcating from Darbouxian umbilical connections

C.Gutierrez and J.Sotomayor

Instituto de Matemática Pura e Aplicada
Estrada Dona Castorina 110
Rio de Janeiro R.J. 22460. Brasil

Abstract

The configurations of lines of curvature around separatrices joining Darbouxian umbilical points as well as their patterns of bifurcation to closed principal lines are studied in this work.

1. Introduction

Let M be a compact connected, oriented, two dimensional smooth manifold. An immersion α of M into $\mathbf{R}^3$ is a map such that $D\alpha_p : TM_p \to \mathbf{R}^3$ is one to one, for every $p \in M$. Denote by $\mathcal{I}^r = \mathcal{I}^r(M, \mathbf{R}^3)$ the set of C^r-immersions of M into $\mathbf{R}^3$. When endowed with the C^s-topology, $s \leq r$, this set is denoted by $\mathcal{I}^{r,s} = \mathcal{I}^{r,s}(M, \mathbf{R}^3)$.

Associated to every $\alpha \in \mathcal{I}^r$ is defined the C^{r-1} normal map $N_\alpha : M \to S^2$:

$$N_\alpha(p) = \frac{\alpha_u \wedge \alpha_v}{\| \alpha_u \wedge \alpha_v \|}$$

where $(u, v) : (M, p) \to (\mathbf{R}^2, 0)$ is a positive chart of M around p, $\wedge$ denotes the exterior product of vectors in $\mathbf{R}^3$, determined by a once for all fixed orientation of $\mathbf{R}^3$, $\alpha_u = \frac{\partial \alpha}{\partial u}$, $\alpha_v = \frac{\partial \alpha}{\partial v}$ and $\| \, \| = \sqrt{< , >}$ is the Euclidean norm in $\mathbf{R}^3$.

Since $DN_\alpha(p)$ has its image contained in the image of $D\alpha(p)$ the endomorphism $\omega_\alpha : TM \to TM$ is well defined by

$$D\alpha \cdot \omega_\alpha = DN_\alpha$$

It is well known that ω_α is a self adjoint endomorphism, when TM is endowed with metric $< , >_\alpha$ induced by α from the metric $< , >$ in $\mathbf{R}^3$.

Let $\mathcal{K}_\alpha = det(\omega_\alpha)$ and $\mathcal{H}_\alpha = -1/2(trace(\omega_\alpha))$ be the *Gaussian* and *Mean Curvatures* of the immersion α.

A point $p \in M$ is called an *umbilical point* of α if $(\mathcal{H}_\alpha(p))^2 - \mathcal{K}_\alpha(p) = 0$. This means that the eigenvalues of ω_α are equal at p. The set of umbilical points of α will be denoted by $\mathcal{U}_\alpha$.

Outside $\mathcal{U}_\alpha$ the eigenvalues of ω_α are distinct. Their opposite values given by $K_\alpha = -\mathcal{H}_\alpha + \sqrt{(\mathcal{H}_\alpha)^2 - \mathcal{K}_\alpha}$ and $k_\alpha = -\mathcal{H}_\alpha - \sqrt{(\mathcal{H}_\alpha)^2 - \mathcal{K}_\alpha}$ are called respectively *maximal* and *minimal principal curvatures* of α. The eigenspaces associated to the principal curvatures define two C^{r-2} line fields $\mathcal{L}_\alpha$ and l_α mutually orthogonal in TM (with the metric $< , >_\alpha$), called *the principal line fields of* α. They are characterized by Rodrigues' equations ([Sp], [St])

$$\mathcal{L}_\alpha = \{v \in TM; \omega_\alpha v + K_\alpha v = 0\}$$

$$l_\alpha = \{v \in TM; \omega_\alpha v + k_\alpha v = 0\}$$

Elimination of k_α, K_α in these equations lead to a single quadratic differential equation, which in a chart (u, v) writes as

$$det \begin{pmatrix} dv^2 & -dudv & dv^2 \\ E_\alpha & F_\alpha & G_\alpha \\ e_\alpha & f_\alpha & g_\alpha \end{pmatrix} = 0,$$

where the second and third lines are the coefficients of the first I_α and second II_α fundamental forms of α in the chart (u, v) : $I_\alpha =< , >_\alpha$, $II_\alpha =< N, D^2\alpha >$ ([Sp], [St]).

The integral curves of $\mathcal{L}_\alpha$ (resp. l_α) are called *lines of maximal* (resp. *minimal*) principal curvature. The family of such curves i.e. the integral foliation of $\mathcal{L}_\alpha$ (resp. l_α) in $M - \mathcal{U}_\alpha$ will be denoted by $\mathcal{F}_\alpha$ (resp. f_α) and called the *maximal* (resp. *minimal*) *principal foliation of* α.

The triple $\mathcal{P}_\alpha = (\mathcal{U}_\alpha, \mathcal{F}_\alpha, f_\alpha)$ will be called the *principal configuration of* α. The local study of principal configurations around an umbilical point received considerable attention in the classical works of Cayley [Ca], Darboux [Da], and Gullstrand [Gu] among others. The analysis of the relation between the principal configurations and the focal set of a surface, which is the caustic set of Geometric Optics, was initiated by Gullstrand [Gu]. The study of the structure of the caustic set fits the context of Thom's Theory of Catastrophes [Th]. The structural stability of the focal set at umbilical points under small perturbations of the surface leads to the so called hyperbolic and elliptic umbilical points. See the works of Bruce and Giblin [B–G], Porteous [Po] and Montaldi [Mo].

Recently, physicists have investigated the statistical proportion in which the umbilical points considered by Darboux in [Da] as well as the hyperbolic and elliptic umbilical points appear in optical experiments [B–H].

The global structure of principal configurations however, happens to be known only for very rare classical examples: Surfaces of revolution and surfaces which belong to a triply orthogonal system of surfaces [St]. See also the book of Fischer [Fi] for a commented description, as well as pictorial representation, of the principal configuration of quadrics and other surfaces.

The general properties of principal configurations on surfaces immersed with constant mean curvature have been established by Gutierrez and Sotomayor [GS6].

In a more general setting, the geometry of foliations with singularities, not directly connected to principal lines, have been studied for instance by Rosenberg [Ro], Levitt [Le], Guiñez [Gui] and Gutierrez[Gut].

The study of the global features of principal configurations P_α which remain undisturbed under small perturbations of the immersion α, *structural stability*, was initiated by Gutierrez and Sotomayor in [G-S.1]. There were established sufficient conditions for immersions α of class C^r of a compact oriented surface M into $\mathbf{R}^3$ to have a *C^s-structurally stable configuration*, $r > s \geq 3$. This means that for any immersion β sufficiently C^s-close to α, there must be a homeomorphism h of M which maps $\mathcal{U}_\alpha$ onto $\mathcal{U}_\beta$ and maps lines of $\mathcal{F}_\alpha$ and f_α onto lines of $\mathcal{F}_\beta$ and f_β, respectively.

Some definitions will be introduced below to review the sufficient conditions for structural stability, proved in [G-S.1], and to state the main results of this paper.

Let $(u,v) : (M,p) \to (\mathbf{R}^2, 0)$ be a chart on M with $p \in \mathcal{U}_\alpha$ and Γ be an isometry of $\mathbf{R}^3$ with $\Gamma(\alpha(p)) = 0$, such that $\Gamma \circ \alpha(u,v) = (u, v, h(u,v))$, with 3-jet at 0 given by

$$J_0{}^3 h(u,v) = (k/2)(u^2+v^2) + (a/6)u^3 + (b/2)uv^2 + (c/6)v^3$$

Below are defined three different types of umbilical points and their local principal configurations are illustrated in Fig. 1.1. These three points, denoted D_1, D_2 and D_3, are called Darbouxian or of type D [Da], [G-S.1].

$D_1 : b(b-a) \neq 0,$
$a/b > (c/2b)^2 + 2$

$D_2 : b(b-a) \neq 0,$ $a \neq 2b$
$1 < a/b < (c/2b)^2 + 2$

$D_3 : b(b-a) \neq 0,$
$1 > a/b$

figure 1.1

The index $i = 1,2,3$ of D_i denotes the number of *umbilical separatrices* of p. These are principal lines which tend to the umbilical point p and separate regions of different patterns of approach to p. It can be proved that these are the only umbilical points at which $\alpha \in \mathcal{I}^r$ is locally C^s structurally stable, $r > s \geq 3$ [G-S.1].

A compact line c of $\mathcal{F}_\alpha$ (resp. f_α) is called *maximal* (resp. *minimal*) *principal cycle* of α.

Call $\pi = \pi_c$ the Poincaré first return map (holonomy) defined by the lines of the foliation to which c belongs, defined on a segment of a line of the orthogonal foliation through 0 in c.

A cycle is called of *type H* or *hyperbolic* if $\pi_0' \neq 1$. It was proved in [G-S.1] that hyperbolicity of c is equivalent to either one of the following conditions:

(1) $$\int_c \frac{dk_\alpha}{K_\alpha - k_\alpha} = \int_c \frac{dK_\alpha}{K_\alpha - k_\alpha} \neq 0,$$

(2) $$\int_c \frac{d\mathcal{H}_\alpha}{\sqrt{\mathcal{H}_\alpha^2 - K_\alpha}} \neq 0.$$

A principal line z of α which is a separatrix of two different umbilical points or twice a separatrix of the same umbilical point is called an *umbilical connection* of α; in the second case z is also called an *umbilical loop*.

Call $\mathcal{S}^r(j)$, $j = a,b,c,d$, respectively, the set of $\alpha \in \mathcal{I}^r$, $r \geq 4$ such that:

(a) all the umbilical points of α are of type D,

(b) all the principal cycles of α are of type H,

(c) α has no umbilical connection and

(d) the limit set of every principal line of α is the union of umbilical points, principal cycles and umbilical connections.

1.1 Theorem. *([G-S.1],[G-S.2])*

$\mathcal{S}^r = \cap \mathcal{S}^r(j)$, $j = a,b,c,d$, $r \geq 4$, is open in $\mathcal{I}^{r,3}$ and dense in $\mathcal{I}^{r,2}$. Every $\alpha \in \mathcal{I}^r$ is C^3–structurally stable.

This paper will be devoted to the description of the simplest patterns of topological change –bifurcations– in one parameter families of immersed surfaces α_t, depending smoothly on a real parameter t, which occur at values $t = t_0$ (bifurcation values) where α_t fails to satisfy condition c); that is, when $\alpha_{t_0} \in \mathcal{I}_1^r(c)$, where $\mathcal{I}_1^r(j) = \mathcal{I}^r - \mathcal{S}^r(j)$, $j = a,b,c,d$. The other bifurcations, which occur when $\alpha_{t_0} \in \mathcal{I}_1^r(j)$, $j = a,b$, have been described in [G-S.3], [G-S.4] and [G-S.5].

Following the approach of Andronov-Leontovich [A-L] and Sotomayor [So] for the study of bifurcations of ordinary differential equations, and also adopted by Gutierrez and Sotomayor in the study of bifurcations of umbilical points [G-S.3], [G-S.5], and principal cycles [G-S.4], in this paper will be precisely formulated certain conditions which violate c) in the mildest possible way while respecting a), b) and d). These conditions define a set $S_1^r(c)$ which will be shown here to be a one to one immersed submanifold of class C^{r-4} of I^r, open in $I_1^{r,3}(c)$ and dense in $I_1^{r,2}(c)$, where $I_1^{r,s}(c)$ denotes the set $I_1^r(c)$ endowed with the C^s-topology, $s \leq r$. Furthermore, the set $S_1^r(c)$ separates locally a connected component of S^r from other immersions with distinct principal configurations and every $\alpha \in S_1^r(c)$ is shown to be principally structurally stable along $I_1^r(c)$.

Let $i, j \in \{1,2,3\}$. A C_{ij} *umbilical connection* z is one which joins umbilical points p and q of types D_i and D_j, respectively. This C_{ij} umbilical connection is called *simple* if one of the following three conditions is satisfied:

- z is neither a C_{11} umbilical connection nor a C_{kk} umbilical loop, with $k \in \{2,3\}$.
- $i = j = 1$, $p \neq q$ and the Poincaré return map T on a segment transversal to z, having $0 \in z$ as an endpoint, is such that $\lim_{x \to 0}[T(x)/x] \neq 1$.
- $i = j \in \{2,3\}$, $p = q$ and the Poincaré return map T on a segment transversal to z, having $0 \in z$ as an endpoint, is such that $\lim_{x \to 0}[\log(T(x))/\log(x)] \neq 1$.

Here, T is induced by the foliation to which z belongs.

The limits involved in the definition of simple C_{ii} umbilical connections, $i = 1,2,3$, are proved to always exist and to only depend on the three jet of α at the umbilical points p and q. This study is carried out in sections 2, 3, 4.

The bifurcations of these connections are illustrated in Figs. 1.2 and 1.3.

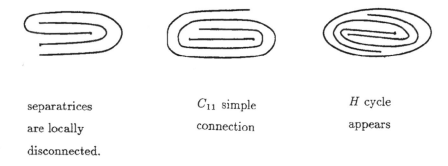

separatrices are locally disconnected. | C_{11} simple connection | H cycle appears

figure 1.2: C_{11} connection and bifurcation

Call $S_1^r(c)$ the set of $\alpha \in \cap S^r(j)$, $j = a, b, d$, such that α has exactly one umbilical connection which is simple. Call $\tilde{S}_1^r(c)$ the set of $\alpha \in S_1^r(c)$ such that its umbilical connection is not contained in the limit of other umbilical separatrix.

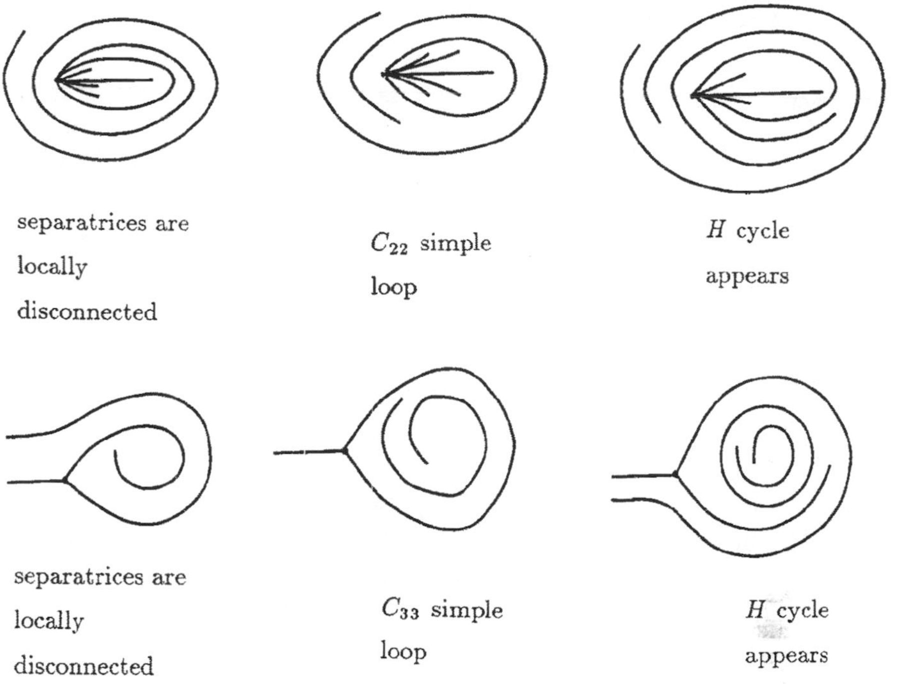

figure 1.3: Simple loops and bifurcations

An element α belonging to $\mathcal{C} \subset I^r$. is said to be C^s*-structurally stable relative to* $\mathcal{C}$, (resp. C^s*-structurally stable along* $\mathcal{C}$) if there is a neighborhood $\mathcal{V}$ of α in $I^{r,s}$ such that for every $\beta \in \mathcal{V} \cap \mathcal{C}$ (resp. β in the connected component of α in $\mathcal{V} \cap \mathcal{C}$), there is a homeomorphism $h = h_\beta$ of M which maps $\mathcal{U}_\alpha$ onto $\mathcal{U}_\beta$ and the lines of $\mathcal{F}_\alpha$ and f_α respectively onto those of $\mathcal{F}_\beta$ and f_β. When $\mathcal{C}$ is the whole I^r, α is called simply C^s-*structurally stable*

1.2 Theorem. *(Stability and Smooth Structure). Let* $r \geq 6$.

a) $S_1^r(c)$ *(resp.* $\tilde{S}_1^r(c)$*) is a one to one immersed (resp. embedded) Banach submanifold of* $I^{r,r}$ *of codimension 1 and of class* C^{r-4}.

b) $\tilde{S}_1^r(c)$ *is an open subset of* $I_1^{r,4}(c)$.

c) Every $\alpha \in S_1^r(c)$ *(resp.* $\alpha \in \tilde{S}_1^r(c)$*) is* C^4*-structurally stable along* $S_1^r(c)$ *(resp. relative to* $I_1^r(c)$*).*

1.3 Theorem. *(Density). Let $r \geq 6$. The set $\tilde{S}_1^r(c)$ is dense in the subspace $S^r(a) \cap I_1^r(c)$ of $I_1^{r,2}(c)$.*

These theorems will be proved in Sections 5 and 6 after a preparation consisting of the detailed study of C_{11}, C_{22} and C_{33}, carried out, respectively, in sections 2, 3 and 4.

2. C_{11} umbilical connections

In 2.3 will be shown that the return map T which characterizes the behaviour of the principal lines near a connection z of D_1 umbilical points p_1 and p_2 is of class C^1. The logarithm of T' at a point $p \in z$ is shown to be an algebraic function on the coefficients of the three jet of α at $\{p_1, p_2\}$.

In 5.2 is proved the unicity and hyperbolicity of the principal cycle bifurcating from a simple ($T'(p) \neq 1$) C_{11} umbilical connection, as illustrated in Fig. 1.2.

2.1 Lemma. *([G-S.1])*

Let $p \in \mathcal{U}_\alpha$ be of type D_1 . There exists a coordinate system $(u,v) : (M,p) \to (\mathbf{R}^2, 0)$ belonging to the positive orientation of M and an orientation preserving isometry Γ of $\mathbf{R}^3$ with $\Gamma(\alpha(p)) = 0$, such that $\Gamma \circ \alpha(u,v) = (u, v, h(u,v))$, with 3-jet at 0 given by

$$J_0{}^3 h(u,v) = (k/2)(u^2 + v^2) + (a/6)u^3 + (b/2)uv^2 + (c/6)v^3$$

where

i) $b \neq 0$ and the roots of $bt^2 - ct + a - 2b = 0$ are not real (i.e. $a/b > 2 + (c/2b)^2$), and
ii) the ray $\{(u,0)/u > 0\}$ is tangent at p to the separatrix of p belonging to $\mathcal{F}_\alpha$
These conditions determine uniquely (u,v), Γ, and (a,b,c).

In view of this lemma, the following definition is meaningful. Let $p \in \mathcal{U}_\alpha$ be of type D_1 and $a, b, c \in \mathbf{R}^3$ as in Lemma 2.1. The *asymmetry of α at p* is the number:

$$\chi(p) = \frac{c(3b - a)}{(b - a)\sqrt{(a - 2b)b - (c^2/4)}}$$

Let $\alpha \in I^r(M)$, $r \geq 5$ and z be a C_{11} umbilical connection joining umbilical points p_1 and p_2.

Let V be a neighborhood of z. Suppose V small and orient $\mathcal{F}_\alpha|_{V-z}$ so that its integral curves go around z following the negative orientation of M. Let

$$\lambda : (-\delta, \delta) \to V$$

be a regular curve transversal to $\mathcal{F}_\alpha$ meeting z exactly at $p_0 = \lambda(0)$. Assume that λ crosses z in such a way that if z is oriented from p_1 to p_2, then the ordered pair $\{z'(0), \lambda'(0)\}$, is a positive basis for $T_{p_0}(M)$. Denote by

$$T : [0, \delta) \to [0, \delta)$$

the coordinate expression of the forward Poincaré return map $\lambda([0,\delta)) \to \lambda([0,\delta))$ induced by $\mathcal{F}_\alpha|_{V-z}$. See Fig. 2.1.

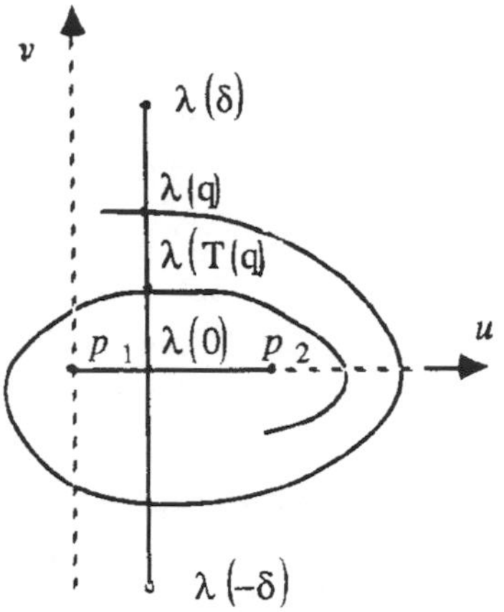

figure 2.1

2.2 Proposition.

1) *T is of class C^1,*

2) $\frac{2}{\pi}\log(T'(0)) = \chi(p_1) + \chi(p_2)$, *where $\chi(p)$ is the asymmetry of p.*
In particular, the umbilical connection z is simple if and only if $T'(0) \neq 1$,

The proof will be given below, in 2.6, after some preliminaries.

To show that T is differentiable at 0 and explicitly compute $T'(0)$, it will be convenient to decompose T into $T_1 \circ T_2$, where

$$T_1 : (0,\delta) \to (0,\delta) \qquad \text{and}$$
$$T_2 : (0,\delta) \to (0,\delta)$$

are the natural coordinate expressions of the forward maps $\lambda((-\delta,0)) \to \lambda((0,\delta))$ and $\lambda((0,\delta)) \to \lambda((-\delta,0))$, respectively, induced by the oriented foliation $\mathcal{F}_\alpha|_{V-z}$.

To study the maps T_i, consider the differential equation for the principal lines of α, around $p \in \{p_1, p_2\}$, when the coordinate system is that of Lemma 2.1. [Sp]:

(1) $(bv + M_1(u,v))dv^2 - ((b-a)u + cv + M_2(u,v))dudv - (bv + M_3(u,v))du^2 = 0$

Each M_i, $i = 1,2,3$, is of class C^{r-2} and $M_i(u,v) = 0(u^2 + v^2)$.

Consider the vector field $Y = P\frac{\partial}{\partial u} + Q\frac{\partial}{\partial v}$ with

$$(2) \qquad \begin{aligned} P &= 2\rho(bv + M_1) \\ Q &= [(b-a)u + cv + M_2] + \sqrt{[(b-a)u + cv + M_2]^2 + 4(bv + M_1)(bv + M_3)} \end{aligned}$$

where $\rho = \frac{b}{|b|} = \frac{a-b}{|b-a|}$. In what follows it will be assumed $b > 0$ and so $\rho = 1$. The case $b < 0$ ($\rho = -1$) is similar.

When $P \neq 0$, $\frac{dv}{du} = \frac{P}{Q}$ solves equation (1). By the analysis in [G–S.1, pag. 203], which is essentially reproduced in the proof of lemma 2.5, it is obtained that Y is tangent to $\mathcal{F}_\alpha$, except possibly when $P = 0$. From the fact that $P(0, v) > 0$, for small $v > 0$, follows that along the v–axis the orientations of $\mathcal{F}_\alpha$ and Y coincide. (See Fig. 2.1, when $p = p_1$).

Given $i, \tilde{i} \in \{-1, 1\}$ and $t, \tilde{t} \in (0, \infty)$, consider the following rays starting at the origin: $L(i,t) = \{(u, tu)/iu > 0\}$, and denote by $T(i, t, \tilde{i}, \tilde{t}) : L(i,t) \to L(\tilde{i}, \tilde{t})$ the forward map induced by the foliation $\mathcal{F}_\alpha|_{V-z}$. Notice that the domain of definition of $T(i, t, \tilde{i}, \tilde{t})$ must be a small open arc of $L(i,t)$ having (0,0) as an endpoint.

Let $\Omega_1 : \mathbf{R}^2 \to \mathbf{R}$ be given by $\Omega_1(u, v) = |u|$. This map restricted to any ray $L(i,t)$ defines a global coordinate. Let

$$f_\epsilon, \ g_1, \ g_2, \ g_3 \quad \text{and} \quad h_\epsilon$$

be the Ω_1–coordinate expression of

$$T(1, -\epsilon, 1, -1), \quad T(1, -1, -1, 1), \quad T(-1, 1, -1, -1) \quad T(-1, -1, 1, 1) \quad \text{and} \quad T(1, 1, 1, \epsilon),$$

respectively, where $\epsilon \in (0, 1)$. See fig. 2.2.

2.3 Lemma.

Let $\vartheta : \mathbf{R} \to \mathbf{R}$ be given by

$$\vartheta(t) = 2bc \int_t^\infty \frac{(1 + t^2)}{(bt^2 + ct - (2b - a))(bt^2 - ct - (2b - a))} dt$$

Then

a) $\vartheta(0) = \frac{\pi(b-a)}{2(2b-a)} \chi(p)$

b) *Both g_1 and g_3 are differentiable at 0 and* $\log(g_1'(0) \cdot g_3'(0)) = \vartheta(1)$

Proof:

$a)$ follows directly from the calculus of residues [Car].

To study g_1 and g_3 it will be convenient to analyze the phase portrait of Y in an angular sector containing the domain of definition and the image of both g_1 and g_3. To do so perform the blowing up

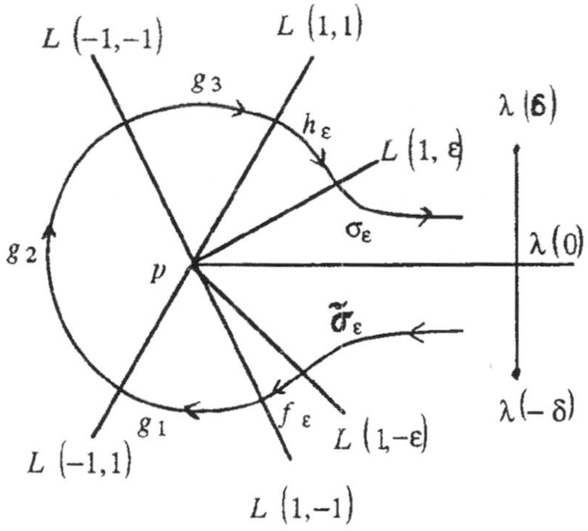

figure 2.2

$$H(t,s) = (ts,s) = (u,v)$$

The map H is a diffeomorphism of the complement of the line $\{ s = 0\}$ onto the complement of the line $\{ v = 0\}$. More precisely, given $\sigma \in \{-1,1\}$, H is a diffeomorphism of the plane $\sigma s > 0$ onto the plane $\sigma v > 0$; it maps rectangles onto distorted angular sectors preserving orientation when $\sigma = 1$ and reversing it otherwise.

Call

$$Z(\sigma) = H_*^{-1} Y = T_\sigma \frac{\partial}{\partial t} + S_\sigma \frac{\partial}{\partial s}$$

the induced vector field $H_*^{-1} Y(t,s) = DH^{-1}_{H(s,t)} Y H(t,s)$ defined on the plane $\sigma s > 0$. Therefore the function S_σ and T_σ are given in the plane $\sigma s > 0$ by.

(1)
$$T_\sigma = \frac{R_1}{s} - \frac{tR_2}{s} - \frac{t}{s}\sqrt{R_2{}^2 + R_1 R_3}$$
$$S_\sigma = R_2 + \sqrt{R_2{}^2 + R_1 R_3}$$

where

$$R_1 = 2(bs + M_1(ts, s))$$
$$R_2 = (b - a)ts + cs + M_2(t, s)$$
$$R_3 = 2(bs + M_3(ts, s))$$

By Hadamard's Lemma, for $i \in \{1, 2, 3\}$, the following expression holds:

(2) $$M_i(ts, s) = sU_i(t, s), \qquad U_i(t, 0) \equiv 0.$$

See in fig.2.3 the phase portraits of $Z(1)$ and $Z(-1)$.

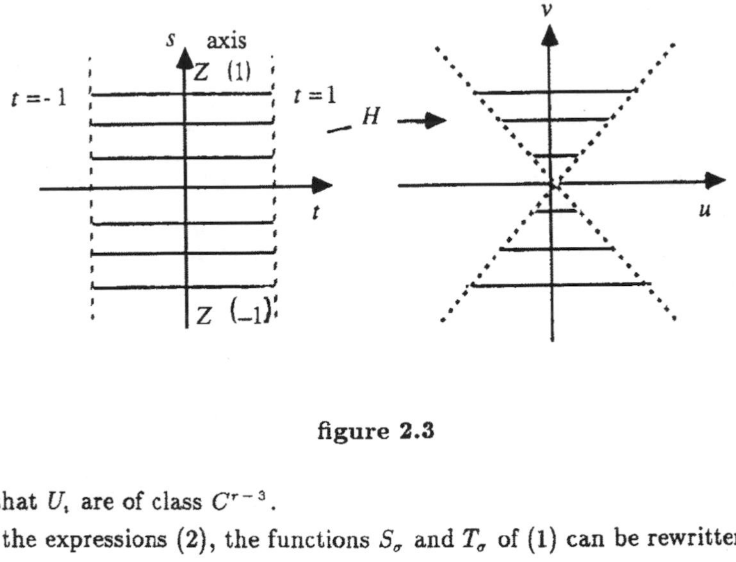

figure 2.3

Note that U_i are of class C^{r-3}.
Using the expressions (2), the functions S_σ and T_σ of (1) can be rewritten as follows:

(1') $$T_\sigma = 2b + U_1 - [(b - a)t^2 + ct + U_2] - t\sigma\sqrt{[(b - a)t + c + U_2]^2 + (2b + U_1)(2b + U_3)}$$

$$S_\sigma = s\left((b - a)t + c + U_2 + \sigma\sqrt{[(b - a)t + c + U_2]^2 + (2b + U_1)(2b + U_3)}\right)$$

Using $(1')$ and (2), it follows that $S_\sigma(t,s)$ and $T_\sigma(t,s)$, $\sigma \in \{-1,1\}$, are actually restrictions of functions of class C^{r-3} in a neighborhood of $\mathbf{R} \times \{0\}$, which are denoted by the same symbols. Moreover

(3) For all $t \in \mathbf{R}$, $T_\sigma(t,0) \neq 0$.

In fact,

$$T_\sigma(t,0) \cdot \left(2b - ((b-a)t^2 + ct) + t\sigma\sqrt{((b-a)t+c)^2 + 4b^2}\right)$$

is equal to $4b((a-2b)t^2 - ct + b)$ which is not zero because $c^2 < 4b(a-2b)$.

This implies that g_1 and g_3 extend differentiably to 0 and that $g_1'(0).g_3'(0). \neq 0$. Therefore:

$$\text{(4)} \qquad \log(g_1'(0)) = \int_{-1}^{1} \left(\frac{\partial}{\partial s}\left(\frac{S_{-1}}{T_{-1}}\right)(0,t)\right) dt \quad \text{and}$$

$$\log(g_3'(0)) = \int_{-1}^{1} \left(\frac{\partial}{\partial s}\left(\frac{S_1}{T_1}\right)(0,t)\right) dt$$

Calculations,using (1') and (2), show that

$$\theta(t,\sigma) = \frac{\partial}{\partial s}\left(\frac{S_\sigma}{T_\sigma}\right)(0,t) = \frac{\frac{\partial S_\sigma}{\partial s}}{T_\sigma}(0,t).$$

Therefore,

$$\text{(5)} \qquad \theta(t,\sigma) = \frac{[(b-a)t+c] + \sigma\sqrt{[(b-a)t+c]^2 + 4b^2}}{(2b + ct - (b-a)t^2 - \sigma t\sqrt{[(b-a)t+c]^2 + 4b^2}}$$

Follows from (4) and (5) that

$$\text{(6)} \quad \log(g_1'(0) \cdot g_3'(0)) = \int_{-1}^{0} \theta(t,1)\,dt + \int_{0}^{1} \theta(t,1)\,dt + \int_{-1}^{0} \theta(t,-1)\,dt + \int_{0}^{1} \theta(t,-1)\,dt$$

Since

$$\int_{-1}^{0} \theta(t,1)\,dt = \int_{0}^{1} \theta(-t,1)\,dt \quad \text{and} \quad \int_{-1}^{0} \theta(t,-1)\,dt = \int_{0}^{1} \theta(-t,-1)\,dt,$$

follows that

$$\log(g_1'(0)\cdot g_3'(0)) = \int_0^1 [\theta(t,-1)+\theta(t,1)]\,dt + \int_0^1 [\theta(-t,-1)+\theta(-t,1)]\,dt =$$

$$= \int_0^1 \left(\frac{c+(3b-a)t}{b-ct-(2b-a)t^2}\right) dt + \int_0^1 \left(\frac{c-(3b-a)t}{b+ct-(2b-a)t^2}\right) dt$$

$$= 2bc \int_0^1 \frac{1+t^2}{(b-ct-(2b-a)t^2)(b+ct-(2b-a)t^2)}\,dt$$

$$= 2bc \int_1^\infty \frac{(1+t^2)}{(bt^2+ct-(2b-a))(bt^2-ct-(2b-a))}\,dt$$

$$= \vartheta(1) \quad \blacklozenge$$

2.4 Lemma. *Given ϵ in $(0,1)$, the maps f_ϵ, g_2, h_ϵ are differentiable at 0 and*

$$log(f_\epsilon'(0)\cdot g_2'(0)\cdot h_\epsilon'(0)) = \vartheta(\epsilon) - \vartheta(1) + \Theta(\epsilon)$$

where $\Theta = \Theta(\epsilon)$ is continuous and $\Theta(0) = 0$.

Proof: To study $f_\epsilon, g_2, h_\epsilon$ it is convenient to analyze the phase portrait of Y in an angular sector containing the domain of definition and the image of the maps

$$T(1,-\epsilon,1,-1), \quad T(-1,1,-1,-1) \quad \text{and} \quad T(1,1,1,\epsilon).$$

To do so, perform the following blowing up:

$$H(s,t) = (s, ts + v(s)) = (u,v)$$

Here, $v = v(s)$ is the unique solution of $P(s,v(s)) = 0$, with $v(0) = 0$. Since $\frac{\partial P}{\partial v}(0,0) = 2b \neq 0$, v is well defined and of class C^{r-2}, by the Implicit Function Theorem; also $v'(0) = 0$.

Given $\sigma \in \{-1,1\}$, call $Z(\sigma) = H_*^{-1}Y = S_\sigma \frac{\partial}{\partial s} + T_\sigma \frac{\partial}{\partial t}$ the induced vector field defined in the plane $\sigma s > 0$ by

$$(1) \qquad \begin{aligned} S_\sigma &= 2[bts + bv(s) + M_1(s, st+v(s))] \\ T_\sigma &= 1/s\left(-tR - v'(s)R + R_1 + \sqrt{R_1{}^2 + RR_2}\right) \end{aligned}$$

where

$$\begin{aligned} R_1 &= (b-a)s + cst + cv(s) + M_2(s, st+v(s)) \\ R_2 &= 2[bts + bv(s) + M_3(s, st+v(s))] \\ R &= S_\sigma \end{aligned}$$

It follows from Hadamard 's Lemma (See [G–S.1, pag. 204]) that

(2)
$$R(s,t) = 2[bst + stU(s,t)], \quad \text{with} \quad U(0,t) \equiv 0$$
$$R_1(s,t) = (b-a)s + cst + sU_1(s,t), \quad \text{with} \quad U_1(0,t) \equiv 0$$
$$R_2(s,t) = 2[bst + sU_2(s,t)], \quad \text{with} \quad U_2(0,t) \equiv 0$$

where U, U_1, U_2 are functions of class C^{r-3}.

Using the expressions (2), the functions S_σ and T_σ of (1) can be rewritten as follows:

(1')
$$S_\sigma = 2st(b+U)$$
$$T_\sigma = -2t(t+v'(s))(b+U) + b - a + ct + U_1 + $$
$$+\sigma\sqrt{(b-a+ct+U_1)^2 + 4t^2(b+U)(b+U_2}$$

It follows from (1') and (2) that $S_\sigma(s,t)$ and $T_\sigma(s,t)$, $\sigma \in \{-1,1\}$, are actually restrictions of functions of class C^{r-3} in a neighborhood of $\mathbf{R} \times \{0\}$, which are denoted by the same symbols. Moreover,

(3)
$$\text{For all} \quad t \in \mathbf{R} - \{0\}, \quad T_\sigma(t,0) \neq 0 \quad \text{and} \quad T_{-1}(0,0) \neq 0.$$

In fact, $T_\sigma(t,0)$ is equal to $-2bt^2 + b - a + ct + \sigma\sqrt{(b-a+ct)^2 + 4b^2t^2}$. Thus, $T_{-1}(0,0) = 2(b-a) \neq 0$. Moreover, for all $t \in \mathbf{R} - \{0\}$,

$$T_\sigma(t,0) \cdot \left(-2bt^2 + b - a + ct - -\sigma\sqrt{(b-a+ct)^2 + 4b^2t^2}\right)$$

is equal to $4bt^2(bt^2 - ct + a - 2b)$ which is different from 0 because $bt^2 - ct + a - 2b$ has no real roots. This implies that:

(4) The maps $f_\epsilon, g_2, h_\epsilon$ extend differentiably to 0 and $f'_\epsilon(0) \cdot g'_2(0) \cdot h'_\epsilon(0) \neq 0$.

Therefore:

(5)
$$log(f'_\epsilon(0) \cdot g'_2(0) \cdot h'_\epsilon(0)) = \int_1^{-1} \left[\frac{\partial}{\partial s}\left(\frac{S_{-1}}{T_{-1}}\right)(0,t)\right] dt +$$
$$+\int_1^{\epsilon} \left[\frac{\partial}{\partial s}\left(\frac{S_1}{T_1}\right)(0,t)\right] dt + \int_{-\epsilon}^{-1} \left[\frac{\partial}{\partial s}\left(\frac{S_1}{T_1}\right)(0,t)\right] dt$$

Calculations, using (2), show that for all $t \in \mathbf{R} - \{0\}$

$$\frac{\partial}{\partial s}\left(\frac{S_\sigma}{T_\sigma}\right)(0,t) = \frac{\frac{\partial S_\sigma}{\partial s}(0,t)}{T_\sigma(0,t)} = \theta(t,\sigma),$$

where

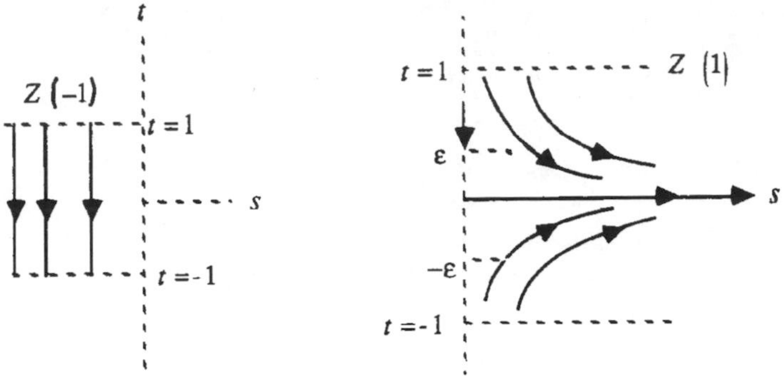

figure 2.4

(6) $$\theta(t,\sigma) = \frac{-2bt^2 + b - a + ct - \sigma\sqrt{(b-a+ct)^2 + 4b^2t^2}}{2t(bt^2 - ct + a - 2b)}$$

Observe that, since $b - a < 0$,

$$\sqrt{(b-a+ct)^2 + 4b^2t^2} = -(b-a) - ct + t^2 f(t)$$

where $f(t)$ is a smooth function on $\mathbf{R}$. This implies, by (6) that

(7) $\theta(t,-1)$ extends smoothly to $t = 0$ and $\theta(0,-1) = 0$.

It follows from (5), (6) and (7) that $log(f'_\epsilon(0) \cdot g'_2(0) \cdot h'_\epsilon(0))$ is equal to

$$\int_1^\epsilon \theta(t,1)\,dt + \int_{-\epsilon}^{-1} \theta(t,1)\,dt + \int_1^0 \theta(t,-1)\,dt + \int_0^{-1} \theta(t,-1)\,dt$$

Since

$$\int_1^{\epsilon} \theta(t,1)\,dt = -\int_{\epsilon}^{1} \theta(t,1)\,dt,$$

$$\int_{-\epsilon}^{-1} \theta(t,1)\,dt = -\int_{\epsilon}^{1} \theta(-t,1)\,dt,$$

$$\int_1^{0} \theta(t,-1)\,dt = -\int_0^{\epsilon} \theta(t,-1)\,dt - \int_{\epsilon}^{1} \theta(t,-1)\,dt \quad \text{and}$$

$$\int_0^{-1} \theta(t,-1)\,dt = -\int_{-\epsilon}^{0} \theta(t,-1)\,dt - \int_{-1}^{-\epsilon} \theta(t,-1)\,dt,$$

it results that $log(f'_{\epsilon}(0) \cdot g'_2(0) \cdot h'_{\epsilon}(0))$ is equal to

$$-\int_{\epsilon}^{1} [\theta(t,1)+\theta(t,-1)]\,dt - \int_{\epsilon}^{1} [\theta(-t,1)+\theta(-t,-1)]\,dt - \int_{-\epsilon}^{\epsilon} \theta(t,-1)\,dt =$$

$$= 2bc \int_{\epsilon}^{1} \frac{1+t^2}{(bt^2+ct-(2b-a))(bt^2-ct-(2b-a))}\,dt + \int_{\epsilon}^{-\epsilon} \theta(t,-1)\,dt$$

It follows from (7) that

$$\Theta(\epsilon) = \int_{\epsilon}^{-\epsilon} \theta(t,-1)\,dt$$

satisfies the conditions required in this lemma. ◆

Assuming the coordinates $\lambda(t) \to -t$ (resp. $\lambda(t) \to t$) for $\lambda((-\delta,0))$ (resp. $\lambda((0,\delta))$), let $\tilde{\sigma}_{\epsilon}$ and σ_{ϵ} be the Ω_1–coordinate expressions of the maps $\lambda[(-\delta,0)] \to L[(1,-\epsilon)]$ and $L[(1,\epsilon)] \to \lambda[(0,\delta)]$, respectively, which are forwardly induced by $\mathcal{F}_{\alpha}|_{V-z}$. See Fig. 2.5.

2.5 Lemma.

Both σ_{ϵ} and $\tilde{\sigma}_{\epsilon}$ extend to 0. Moreover, there exist orientation preserving real valued C^1 diffeomorphisms r_{ϵ}, s_{ϵ} and v defined on $(0,\delta)$, taking 0 to 0, varying continuously with ϵ in the C^1–topology, such that:

$$v \circ \sigma_{\epsilon} \circ r_{\epsilon}^{-1}(x) = A_{\epsilon} x^{\nu} \qquad \text{and}$$

$$s_{\epsilon}^{-1} \circ \tilde{\sigma}_{\epsilon} \circ v^{-1}(x) = \tilde{A}_{\epsilon} x^{1/\nu},$$

where $\lim_{\epsilon\to 0}(r_{\epsilon}) = \lim_{\epsilon\to 0}(s_{\epsilon})$ *is the identity diffeomorphism,* $\lim_{\epsilon\to 0}(A_{\epsilon}\cdot\tilde{A}_{\epsilon}^{\nu}) = 1$ and $\nu = \frac{2b-a}{b-a}$.

Proof: To study σ_{ϵ} and $\tilde{\sigma}_{\epsilon}$, the phase portrait of Y will be analyzed in an angular sector containing $L(1,\epsilon)$ and $L(1,-\epsilon)$. To do so, perform the blowing up

$$H(s,t) = (s, ts + v(s)) = (u,v),$$

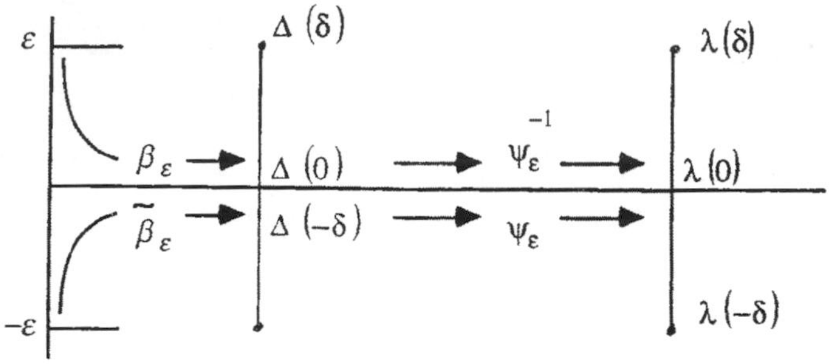

figure 2.5

where $v(s)$ is the unique solution of $P(s, v(s))$, with $v(0) = 0$. As in the proof of Lemma 2.4, it is obtained

$$Z = H_*^{-1} Y = S\frac{\partial}{\partial s} + T\frac{\partial}{\partial t}$$

defined on a neighborhood of 0, where

(1) $$S = 2[bts + stU(s,t)]$$

$$T = -t\left(\frac{S}{s}\right) - v'(s)\left(\frac{S}{s}\right) + \left(\frac{R_1}{s}\right) + \sqrt{\left(\frac{R_1}{s}\right)^2 + \left(\frac{S}{s}\right)\left(\frac{R_2}{s}\right)},$$

It holds that:

(2) $$R_1 = (b-a)s + cst + sU_1(s,t),$$
$$R_2 = 2bst + sU_2(s,t),$$

with $U(0,t) = U_1(0,t) = U_2(0,t) \equiv 0$.

Instead of Z, it will be convenient to consider the vector field:

$$\tilde{Z} = \omega(s,t) \cdot Z = \tilde{S}\frac{\partial}{\partial s} + \tilde{T}\frac{\partial}{\partial t},$$

where $\omega(s,t) = \frac{1}{ts}\left(-tS - v'(s)S + R_1 - \sqrt{R_1{}^2 + SR_2}\right)$. It follows that

$$\tilde{S} = 2(b+U(s,t))[-tS - v'(s)S + R_1 + (b-a)sR(s,t)], \tag{3}$$

$$\tilde{T} = t\left(\frac{S}{s}\right)^2 + v'(s)^2\left(\frac{S}{st}\right)\left(\frac{S}{s}\right) + 2v'(s)\left(\frac{S}{s}\right)^2 - 2\left(\frac{S}{s}\right)\left(\frac{R_1}{s}\right)$$
$$- 2v'(s)\left(\frac{S}{st}\right)\left(\frac{R_1}{s}\right) - \left(\frac{S}{st}\right)\left(\frac{R_2}{s}\right),$$

where

$$R(s,t) = \sqrt{\left(1 - \frac{c}{b-a}t + \frac{U_1}{b-a}\right)^2 + \frac{2t}{(b-a)^2}(b+U)(2bt+U_2)}.$$

Using (2) and (3) it results that $\tilde{Z}$ is of class C^1 around (0,0). Moreover, the Jacobian matrix of $\tilde{Z}$ at (0,0) is given by

$$D\tilde{Z}(0,0) = \begin{pmatrix} 4(b-a)b & 0 \\ \frac{\partial \tilde{T}}{\partial s}(0,0) & -4b(2b-a) \end{pmatrix}, \tag{4}$$

which is a hyperbolic saddle.

Let $E_1 = (u_1, v_1)$ be the unit eigenvector of $D\tilde{Z}(0,0)$ with $u_1 > 0$. Certainly $E_2 = (0,1)$ is other eigenvector of $D\tilde{Z}(0,0)$. For small $\tilde{\delta} > 0$, the image of the map $\Lambda_\epsilon(y) = \epsilon^{1/\nu}E_1 + yE_2$, $y \in (-\tilde{\delta}, \tilde{\delta})$, is a cross section to $\tilde{Z}$. Let ψ_ϵ be the Ω_1–coordinate expression of the map $\lambda((-\delta,\delta)) \to \Lambda_\epsilon((-\tilde{\delta},\tilde{\delta}))$ induced by $\mathcal{F}_\alpha|_{V-z}$. The domain of definition of ψ_ϵ is in general a small proper subinterval of $(-\delta,\delta)$ containing 0.

Let $\beta_\epsilon = \psi_\epsilon \circ \sigma_\epsilon$. See fig. 2.5. It follows from [Hr, Chap. 9] that, in a small neighborhood of (0,0), there is a C^1 change of coordinates tangent to the identity at (0,0), on which $\tilde{Z}$ is written as $D\tilde{Z}(0,0)$. Therefore, since ν is the absolute value of the quotient of eigenvalues of $D\tilde{Z}(0,0)$, in such linearizing coordinates,

$$y = \beta_\epsilon(x) = x^\nu. \tag{5}$$

In fact, integrating $D\tilde{Z}(0,0)$, for some $t \in \mathbf{R}$,

$$x(\exp(4(b-a)bt)E_1 + \epsilon(\exp(-4b(2b-a)t)E_2 = \epsilon^{1/\nu}E_1 + yE_2,$$

and so $y = x^\nu$.

Similarly, in the linearizing coordinates above, if $\tilde{\beta}_\epsilon = \tilde{\sigma}_\epsilon \circ \psi_\epsilon^{-1}$ writes as

(6) $$y = \tilde{\beta}_\epsilon(x) = x^{1/\nu}.$$

Under these conditions, The lemma follows easily from [Hr, Chap. 9], (5), (6) and the fact that $\sigma_\epsilon = (\psi_\epsilon^{-1}) \circ \beta_\epsilon$ and $\tilde{\sigma}_\epsilon = \tilde{\beta}_\epsilon \circ \psi_\epsilon|_{(-\delta,0]}$. In fact, since the change of coordinates is tangent to the identity at (0,0), it gives rise to the diffeomorphisms r_ϵ, s_ϵ and v which have the required asymptotic properties, as ϵ goes to 0.

2.6 Proof of Proposition 2.2.:

From the considerations preceding Lemma 2.3, given $\epsilon \in (0,1)$ small,

(1) $$T_1 = v^{-1} \circ (v \circ \sigma_\epsilon \circ r_\epsilon^{-1}) \circ \Gamma_\epsilon \circ (s_\epsilon^{-1} \circ \tilde{\sigma}_\epsilon v^{-1}) \circ v,$$

where

$$\Gamma_\epsilon = r_\epsilon \circ h_\epsilon \circ g_3 \circ g_2 \circ g_1 \circ f_\epsilon \circ s_\epsilon.$$

By Lemmas 2.3, 2.4 and 2.5, Γ_ϵ is a C^1-diffeomorphism and

(2) $$(h_\epsilon \circ g_3 \circ g_2 \circ g_1 \circ f_\epsilon)'(0) = \exp(\Theta(\epsilon) + \vartheta(\epsilon)).$$

The diffeomorphism Γ_ϵ can be written as

$$\Gamma_\epsilon(y) = y\Gamma_\epsilon^1(y),$$

where Γ_ϵ^1 is continuous and $\Gamma_\epsilon^1(0) \neq 0$. It follows from (2) and Lemma 2.5 that

(3) $$lim_{\epsilon \to 0}\Gamma_\epsilon^1(0) = exp(\Theta(\epsilon) + \vartheta(\epsilon)).$$

Differentiating (1) and simplifying the expression, using Lemma 2.5, obtain:

(4) $$T_1{}'(x) = \frac{dv^{-1}}{dx}\left(\tilde{A}_\epsilon\left\{\Gamma_\epsilon[A_\epsilon(v(x))^{1/\nu}]\right\}^\nu\right) \cdot \tilde{A}_\epsilon \cdot A_\epsilon^\nu \cdot$$
$$\cdot \left(\Gamma_\epsilon^1[A_\epsilon(v(x))^{1/\nu}]\right)^{\nu-1} \cdot \frac{d\Gamma_\epsilon}{dx}[A_\epsilon(v(x))^{1/\nu}] \cdot v'(x).$$

This shows that T_1 is of class C^1.

It follows from (3), (4) and Lemma 2.5 that

$$\begin{aligned} T_1'(0) &= \lim_{\epsilon \to 0} ((\Gamma_\epsilon^1(0))^\nu) \\ &= \lim_{\epsilon \to 0} (\cdot \exp(\nu\vartheta(\epsilon) + \Theta(\epsilon))) \\ &= \exp(\nu\vartheta(0)) \\ &= \exp(\frac{\pi}{2}\chi(p_1)) \end{aligned}$$

The proposition is proved. ♦

3. C_{22} umbilical loops

In 3.2 will be established the expression of the return map, which characterizes the principal configuration near an umbilical D_{22} loop. This expression shows that the generic behaviour depends only on the three jet of α at the umbilical point of the loop. In 5.2 is proved the unicity and hyperbolicity of the principal cycle bifurcating from a simple C_{22} umbilical loop, as illustrated in Fig 1.3.

Let $\alpha \in \mathcal{I}^r$, $r \geq 5$, and z be a C_{22} umbilical loop of $\mathcal{F}_\alpha$ at an umbilical point p. Denote by $L_1 = L_1(p)$ and $L_2 = L_2(p)$ the rays in TM_p which are tangent to the separatrices of $\mathcal{F}_\alpha$ at p. Denote by $L_0 = L_0(p)$ the ray which is the common tangent to all other lines of $\mathcal{F}_\alpha$ approaching p. Following the positive orientation of M, denote by $\theta_i = \theta_i(p)$, $i = 1, 2$, the oriented angle from L_0 to L_i. See Fig. 3.1.

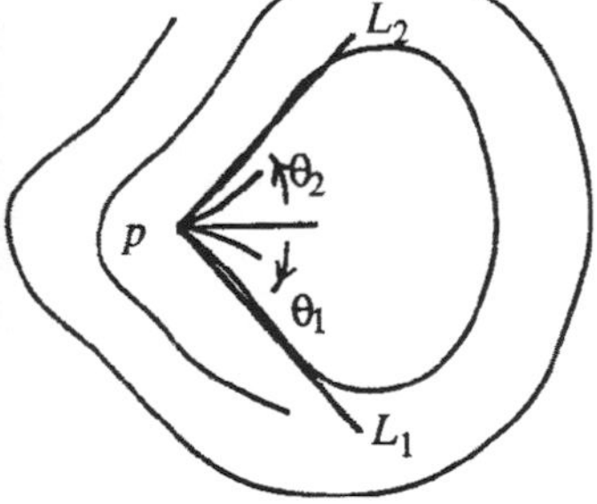

figure 3.1

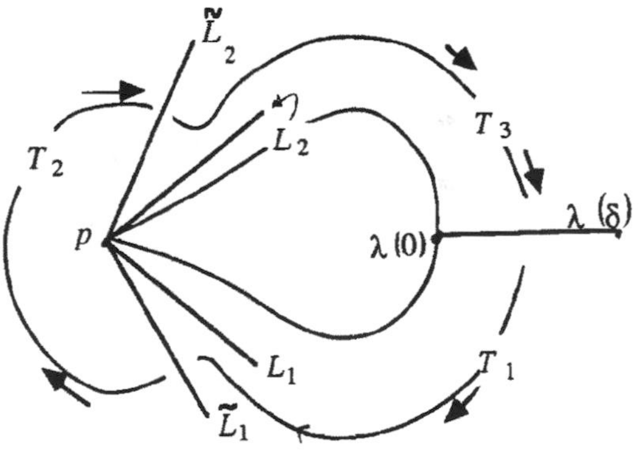

figure 3.2

The *asymmetry of α at the D_2 umbilical point p* is defined to be the number $\tau_1 + \tau_2$, where $\tau_1 = \tan(\theta_1)$ and $\tau_2 = \tan(\theta_2)$. Compare with the definition for the D_1 case.

Orient $\mathcal{F}_\alpha|_z$ starting at p in the L_1 direction. Let V be a small neighborhood of z and $\tilde{W}$ be the connected component of $V - z$ containing the principal lines of $\mathcal{F}_\alpha|_V$ which go around z. Extend the orientation of $\mathcal{F}_\alpha|_z$ to $\mathcal{F}_\alpha|_W$, where $W = \tilde{W} \cup z$.

Let

$$\lambda : [0, \delta) \to W$$

be a regular curve transversal to $\mathcal{F}_\alpha$ meeting z exactly at $\lambda(0)$. Denote by

$$T : [0, \delta) \to [0, \delta)$$

the λ–coordinate expression of the forward Poincaré return map $\lambda([0,\delta)) \to \lambda([0,\delta))$ induced by the oriented foliation $\mathcal{F}_\alpha|_W$. Suppose that

$$\tau_1 > 0 > \tau_2$$

3.1 Lemma [G-S.1].

There exists a coordinate system $(u, v) : (M, p) \to (\mathbf{R}^2, 0)$ belonging to the positive orientation of M and an orientation preserving isometry Γ of $\mathbf{R}^3$ with $\Gamma(\alpha(p)) = 0$, such that $\Gamma \circ \alpha(u,v) = (u, v, h(u,v))$, with 3–jet at 0 given by

$$J_0{}^3 h(u,v) = (k/2)(u^2 + v^2) + (a/6)u^3 + (b/2)uv^2 + (c/6)v^3$$

where

i) $b \neq 0$, the roots $\frac{c}{2b} + \sqrt{\left(\frac{c}{2b}\right)^2 - \frac{a}{b} + 2}$ and $\frac{c}{2b} - \sqrt{\left(\frac{c}{2b}\right)^2 - \frac{a}{b} + 2}$ of $bt^2 - ct + a - 2b = 0$ are precisely τ_1 and τ_2, respectively.

Moreover $0 > \tau_1 \cdot \tau_2 > -1$ and

ii) the derivative of (u,v) at p is the Identity and so, $L_0 = \{(u,0)/u > 0\}$, $L_1 = \{(u, \tau_1 u)/u > 0\}$ and $L_2 = \{(u, \tau_2 u)/u > 0\}$.

These conditions determine uniquely $((u,v), \Gamma, (a,b,c))$.

3.2 Proposition.

Let

$$\varsigma = \left| \frac{({\tau_1}^2 + 2 - a/b)({\tau_2}^2 + 1)}{({\tau_2}^2 + 2 - a/b)({\tau_1}^2 + 1)} \right|,$$

then

$$T(x) = Ax^{\varsigma}(1 + r(x)),$$

where $r(x)$ is continuous, $r(0) = 0$ and A is a positive constant.

Proof:

Consider the differential equation for the principal lines of α, around p, when the coordinate system is that of Lemma 3.1 [Sp]:

(1) $$(bv + M_1(u,v))dv^2 - ((b-a)u + cv + M_2(u,v))dudv - (bv + M_3(u,v))du^2 = 0.$$

Each M_i, $i = 1,2,3$, is of class C^2 and $M_i(u,v) = 0(u^2+v^2)$.
Consider the vector field $Y = P\frac{\partial}{\partial u} + Q\frac{\partial}{\partial v}$, with

(2) $$\begin{aligned} P &= 2(bv + M_1) \\ Q &= (b-a)u + cv + M_2 + \\ &\quad - \rho\sqrt{[(b-a)u + cv + M_2]^2 + 4(bv + M_1)(bv + M_3)}, \end{aligned}$$

where $\rho = \frac{b-a}{|b-a|}$. When $P \neq 0$, $\frac{dv}{du} = \frac{P}{Q}$ solves equation (1); moreover for small $v > 0$, $bP(0,v) > 0$. Therefore Y is tangent to $\mathcal{F}_\alpha$ except possibly when $P = 0$ and its orientation when $b > 0$, along the v–axis, is the same as that of $\mathcal{F}_\alpha|_{V-z}$. See Fig. 3.1.

In what follows, it will be assumed that $b > a$ and so $\rho = 1$. the case $b < a$ is similar.

Let $\tilde{L}_1 = \{(u, 2\tau_1 u)/u > 0\}$ and $\tilde{L}_2 = \{(u, 2\tau_2 u)/u > 0\}$. Decompose T into $T_3 \circ T_2 \circ T_1$, where $T_1 : [0.\delta) \to \tilde{L}_2$, $T_2 : \tilde{L}_2 \to \tilde{L}_1$, $T_3 : \tilde{L}_1 :\to [0,\delta)$ are the forward maps induced by $\mathcal{F}_\alpha|_W$. See Fig. 3.2. It holds that:

(3) Both T_1 and T_3 extend to 0, $T_1(x) = A_1 x^B(1 + r_1(x))$ and $T_3(x) = A_3 x^C(1 + r_3(x))$, where r_1 and r_3 are continuous, $r_1(0) = r_3(0) = 0$,

$$B = \left|\frac{4b^2(\tau_2{}^2 + 1)}{4b[b(\tau_2{}^2 + 1) + b - a]}\right| \quad \text{and} \quad C = \left|\frac{4b[b(\tau_1{}^2 + 1) + b - a]}{4b^2(\tau_1{}^2 + 1)}\right|.$$

(4) T_2 is of class C^1 and $T_2'(0) \neq 0$.

To prove (3), it will be convenient to perform the blowing up

$$H(s,t) = (s, ts + w(s)) = (u,v),$$

defined on the half-plane $s > 0$, where $v = w(s)$ is the unique solution of $P(s, w(s)) \equiv 0$ satisfying $w(0) = w'(0) = 0$. As in the proof of Lemma 2.4, it is obtained

$$Z = H_*^{-1} Y = S\frac{\partial}{\partial s} + T\frac{\partial}{\partial t}$$

defined on a neighborhood of 0, where

(5) $$S = 2[bts + stU(s,t)]$$

$$T = -t\left(\frac{S}{s}\right) - w'(s)\left(\frac{S}{s}\right) + \left(\frac{R_1}{s}\right) - \sqrt{\left(\frac{R_1}{s}\right)^2 + \left(\frac{S}{s}\right)\left(\frac{R_2}{s}\right)},$$

It holds that:

(6) $$R_1 = (b-a)s + cst + sU_1(s,t),$$
$$R_2 = 2bst + sU_2(s,t),$$

where $U = U(s,t)$, $U_1 = U_1(s,t)$, and $U_2 = U_2(s,t)$ are of class C^1 and, for all $t \in \mathbf{R}$, $U(0,t) \equiv U_1(0,t) \equiv U_2(0,t) \equiv \frac{\partial}{\partial t}(0,t) \equiv 0$.

Instead of Z, it will be convenient to consider the vector field:

$$\tilde{Z} = \omega(s,t) \cdot Z = \tilde{S}\frac{\partial}{\partial s} + \tilde{T}\frac{\partial}{\partial t},$$

where $\omega(s,t) = \frac{1}{ts}\left(-tS - w'(s)S + R_1 + \sqrt{R_1{}^2 + SR_2}\right)$.

Proceeding as in the proof of Lemma 2.5, it follows that if τ is a root of $bt^2 - ct + a - 2b$, then

(7) $$\frac{\partial}{\partial t}\tilde{T}(0,\tau) = 4b[b(\tau^2+1) + b - a], \quad \frac{\partial}{\partial t}\tilde{S}(0,\tau) = 0, \quad \frac{\partial}{\partial s}\tilde{S}(0,\tau) = -4b^2(\tau^2+1).$$

Under these conditions, similarly to the proof of Lemma 2.5, (3) follows from [Hr]. The proof of (4) will be omitted because it is analogous to that of (3).

Since $\varsigma = BC$, the proposition follows from (3) and (4). ♦

3.3 Corollary.

The following conditions are equivalent

1) *z is simple (see Section 1),*

2) $c \neq 0$,

3) *The asymmetry, $\tau_1 + \tau_2$, of α at p is different from 0, and*

4) $(\log(\varsigma))\,(\tau_1 + \tau_2) < 0$.

Therefore, if $\tau_1 + \tau_2 > 0$ (resp. $\tau_1 + \tau_2 < 0$), then z is an attracting (resp. repelling) cycle of the oriented foliation $\mathcal{F}_\alpha|_W$.

Proof:

Calculations show that

$$\varsigma = \left| \frac{(\tau_2{}^2+1)(\tau_1{}^2+1)+(1-a/b)(\tau_2{}^2+1)}{(\tau_2{}^2+1)(\tau_1{}^2+1)+(1-a/b)(\tau_1{}^2+1)} \right|$$

Since $1 < a/b < 2$, it follows that

(8) $$\begin{aligned} &\varsigma < 1 \Leftrightarrow (1-a/b)(\tau_2{}^2+1) < (1-a/b)(\tau_1{}^2+1) < 0 \Leftrightarrow \tau_2{}^2 > \tau_1{}^2. \\ &\varsigma > 1 \Leftrightarrow (1-a/b)(\tau_1{}^2+1) < (1-a/b)(\tau_2{}^2+1) < 0 \Leftrightarrow \tau_2{}^2 < \tau_1{}^2. \end{aligned}$$

The corollary follows immediately from this and Proposition 3.2. ♦

4. C_{33} umbilical loops

In 4.2 will be established the expression of the return map, which is needed to describe the principal configuration near an umbilic D_{33} loop. As for umbilical D_{22} loops, the generic behavior depends only on the three jet of α at the umbilical point of the loop. In 5.2 is proved the unicity and hyperbolicity of the principal cycle bifurcating from a simple C_{33} umbilical loop, as illustrated in Fig. 1.3.

Let $\alpha \in \mathcal{I}^r$, $r \geq 5$, and z be a C_{33} umbilical loop of $\mathcal{F}_\alpha$ with involved umbilical point p. Denote by $L_0 = L_0(p)$, $L_1 = L_1(p)$ and $L_2 = L_2(p)$ the rays in TM_p which are tangent to the separatrices of $\mathcal{F}_\alpha$ at p. Following the positive orientation of M, denote by $\theta_i = \theta_i(L_0)$, $i = 1, 2$, the oriented angle from L_0 to L_i. See fig.4.1.

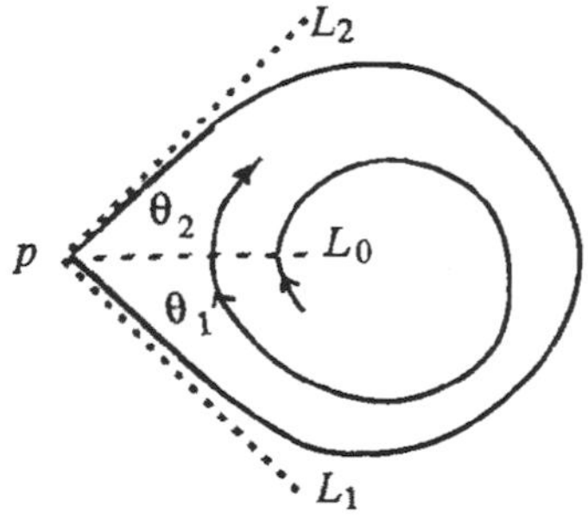

figure 4.1

The *asymmetry of α at the D_3 umbilical point p, with respect to L_0* is defined to be the number $\tau_1 + \tau_2$, where $\tau_1 = \tan(\theta_1)$ and $\tau_2 = \tan(\theta_2)$. Compare with the previous definitions for the D_1 and D_2 cases.

Among the rays L_0, L_1 and L_2, suppose that L_0 is precisely the one which is not tangent to z at p. Orient $\mathcal{F}_\alpha|_z$ starting at p in the L_1 direction. Let V be a small neighborhood of z and $\tilde{W}$ be the connected component of $V - z$ containing the principal lines of $\mathcal{F}_\alpha|_V$ which go around z. Extend the orientation of $\mathcal{F}_\alpha|_z$ to $\mathcal{F}_\alpha|_W$, where $W = \tilde{W} \cup z$.

Let

$$\lambda : [0, \delta) \to W$$

be a regular curve transversal to $\mathcal{F}_\alpha$ meeting z exactly at $\lambda(0)$. Denote by

$$T : [0, \delta) \to [0, \delta)$$

the λ–coordinate expression of the forward Poincaré return map $\lambda([0,\delta)) \to \lambda([0,\delta))$ induced by the oriented foliation $\mathcal{F}_\alpha|_W$. Suppose that

$$\tau_1 > 0 > \tau_2$$

4.1 Lemma [G-S.1].

There exists a coordinate system $(u, v) : (M, p) \to (\mathbf{R}^2, 0)$ belonging to the positive orientation of M and an orientation preserving isometry Γ of $\mathbf{R}^3$ with $\Gamma(\alpha(p)) = 0$, such that $\Gamma \circ \alpha(u, v) = (u, v, h(u, v))$, with 3–jet at 0 given by

$$J_0{}^3 h(u,v) = (k/2)(u^2 + v^2) + (a/6)u^3 + (b/2)uv^2 + (c/6)v^3$$

where

i) $b \neq 0$, the roots $\frac{c}{2b} + \sqrt{\left(\frac{c}{2b}\right)^2 - \frac{a}{b} + 2}$ and $\frac{c}{2b} - \sqrt{\left(\frac{c}{2b}\right)^2 - \frac{a}{b} + 2}$ of $bt^2 - ct + a - 2b = 0$ are precisely τ_1 and τ_2, respectively, moreover $\tau_1 \cdot \tau_2 < -1$, and

ii) The derivative of (u, v) at p is the identity and so, $L_0 = \{(u, 0)/u > 0\}$, $L_1 = \{(u, \tau_1 u)/u > 0\}$ and $L_2 = \{(u, \tau_2 u)/u > 0\}$.

These conditions determine uniquely $((u, v), \Gamma, (a, b, c))$.

4.2 Proposition.

Let

$$\varsigma = \left| \frac{(\tau_1{}^2 + 2 - a/b)(\tau_2{}^2 + 1)}{(\tau_2{}^2 + 2 - a/b)(\tau_1{}^2 + 1)} \right|,$$

then

$$T(x) = Ax^{\varsigma}(1 + r(x)),$$

where $r(x)$ is continuous, $r(0) = 0$ and A is a positive constant.

Proof:

Similar to Proposition 3.2. In fact, let

$$\tilde{L}_1 = \{(u, (1/2)\tau_1 u)/u > 0\} \quad \text{and} \quad \tilde{L}_2 = \{(u, (1/2)\tau_2 u)/u > 0\}.$$

Decompose T into $T_3 \circ T_2 \circ T_1$, where $T_1 : [0.\delta) \to \tilde{L}_2$, $T_2 : \tilde{L}_2 \to \tilde{L}_1$, $T_3 : \tilde{L}_1 :\to [0, \delta)$ are the forward maps induced by $\mathcal{F}_\alpha|_W$. See Fig. 4.1. It holds that:
(2) Both T_1 and T_3 extend to 0, $T_1(x) = A_1 x^B (1 + r_1(x))$ and $T_3(x) = A_3 x^C (1 + r_3(x))$, where r_1 and r_3 are continuous, $r_1(0) = r_3(0) = 0$,

$$B = \left| \frac{4b^2(\tau_2{}^2 + 1)}{4b[b(\tau_2{}^2 + 1) + b - a]} \right| \quad \text{and} \quad C = \left| \frac{4b[b(\tau_1{}^2 + 1) + b - a]}{4b^2(\tau_1{}^2 + 1)} \right|.$$

(2) T_2 is of class C^1 and $T_2'(0) \neq 0$.

The proof of (1) and (2) are identical to their counterparts in Proposition 3.2. The proposition follows from (1) and (2). ♦

4.3 Corollary.

The following conditions are equivalent

1) *z is simple (see Section 1),*
2) *$c \neq 0$,*
3) *The asymmetry, $\tau_1 + \tau_2$, of α at p, with respect to L_0, is different from 0, and*
4) *$(\log(\varsigma))\,(\tau_1 + \tau_2) > 0$.*

Therefore, if $\tau_1 + \tau_2 < 0$ (resp. $\tau_1 + \tau_2 > 0$), then z is an attracting (resp. repelling) cycle of the oriented foliation $\mathcal{F}_\alpha|_W$.

Proof:

Similar to that of Corollary 3.3. Take into account, in (8), that now $a/b < 1$. ♦

5. Proof of Theorem 1.2

Proceed to define $S_1^r(c)$ implicitly near $\alpha_0 \in S_1^r(c)$. Let z be the umbilical connection of α_0 joining the Darbouxian umbilical points p and q. Suppose that z belongs to the minimal principal foliation f_{α_0}. Let $\tilde{v} : \Sigma \to [-2, 2]$ be a coordinate system for a small arc $\Sigma \subset M$ which belongs to $\mathcal{F}_{\alpha_0}$ and meets z exactly at $\tilde{v}^{-1}(0)$. A small perturbation α of α_0 uniquely determines umbilical points $p[\alpha]$ and $q[\alpha]$ and real numbers $s[\alpha]$ and $t[\alpha]$, belonging to (-2,2), characterized by the following two conditions:

- $p = p[\alpha_0]$ (resp. $q = q[\alpha_0]$) and $p[\alpha]$ (resp. $q[\alpha]$) depends differentiably on α.
- $s[\alpha]$ (resp. $t[\alpha]$) depends differentiably on α, $\tilde{v}^{-1}(s[\alpha])$ (resp. $\tilde{v}^{-1}(t[\alpha])$) belongs to the separatrix of $p[\alpha]$ (resp. $q[\alpha]$) originated from z and so $s[\alpha_0] = t[\alpha_0] = 0$.

In fact, observe that the condition $b(b - a) \neq 0$ appearing in the definition of umbilical points of type D_i, $i = 1, 2, 3$, (see Section 1) amounts to the transversality of $J^2\alpha$ to the submanifold (of codimension 2) of umbilical 2–jets. It follows that $p(\alpha)$ and $q(\alpha)$ depend C^{r-2}–differentiably on α. Also, the umbilical separatrices, and so $s[\alpha]$ and $t[\alpha]$, depend differentiably on α; this follows from the fact that they can be interpreted –in terms of blowing up– as separatrices of hyperbolic singularities of vector fields. See sections 2, 3 and 4. The lost of differentiability, from $r-2$ to $r-4$, results from the divisions involved in these operations.

Locally $S_1^r(c) = B^{-1}(0)$, where B is the C^{r-4}–differentiable function

(1) $$B(\alpha) = s[\alpha] - t[\alpha]$$

See Fig. 5.1 for an illustration for a connection of type C_{13}

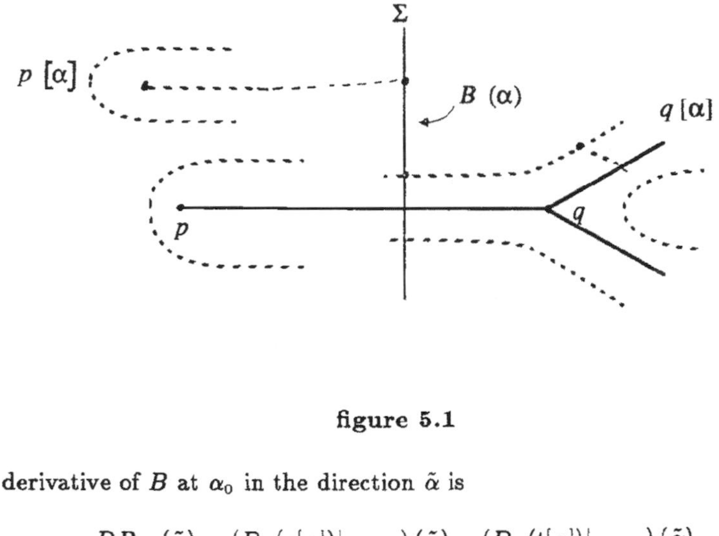

figure 5.1

The derivative of B at α_0 in the direction $\tilde{\alpha}$ is

(2) $$DB_{\alpha_0}(\tilde{\alpha}) = (D_\alpha(s[\alpha])|_{\alpha=\alpha_0})(\tilde{\alpha}) - (D_\alpha(t[\alpha])|_{\alpha=\alpha_0})(\tilde{\alpha})$$

To show that $DB_{\alpha_0} \neq 0$, the following lemma will be needed.

5.1 Lemma.. *[G-S.2,Lemma 4.3]*

Let $\alpha \in I^s$, $\infty \geq s \geq 3$, and $p \in M$ be such that $dk|_{\ell_\alpha(p)} \neq 0$. Let $(u,v) : M \to 2I \times 2I$ be coordinates of α such that $(u(p), v(p)) = (0,0)$, $\ell_\alpha = \mathbf{R}(\partial/\partial u)$ and $\mathcal{L}_\alpha = \mathbf{R}(\partial/\partial v)$, where $I = [-1,1]$. Then given any $\epsilon > 0$ and any sequence of C^r-norms $\| \; \|_r$, $r = 2,3,...$, on I^s, $s \geq r+1$, there are numbers $\delta = \delta(\epsilon, \| \; \|_2) > 0$ and $\tau = \tau(\epsilon, \| \; \|_2) > 0$ such that for any $\rho \in (0,\delta]$ and any $p_0 \in u^{-1}(-1) \cap v^{-1}((1-\delta)I)$ it is possible to construct a C^{s-1} family $\{\alpha_\mu\}$, $\mu \in [-1,1]$, of C^{s-1} immersions which satisfy the following conditions:

i) The support of $\alpha_\mu - \alpha$ *is contained in* $D = (u,v)^{-1}([-1,1] \times \{v(p_0) + 2\rho I\})$ *and* $\alpha_0 = \alpha$.

ii) For all $\mu \in [-1,1]$, $\| \alpha_\mu - \alpha \|_2 < \epsilon$

iii) The minimal principal arc of $\alpha_\mu|_D$ *which passes through* p_0 *meets the segment* $u^{-1}(1)$ *in a point denoted by* $\xi_\mu(p_0)$. *The range of the map* $\mu \to v(\xi_\mu(p_0))$, $\mu \in [0,1]$, *contains the interval* $[v(p_0), v(p_0) + \rho\tau]$. *See fig 4.1. Moreover* $(\partial/\partial\mu)(\xi_\mu(p_0)) \neq 0$.

iv) There exists $\mu_0 = \mu_0(r) > 0$ *such that* $\| \alpha_{\mu 0} - \alpha \|_r < \epsilon$ *and* $v(\xi_{\mu 0}(p_0)) > v(p_0)$.

Since k_α is not constant along z, Lemma 5.1 can be applied to α_0 and a point of z appropriately chosen. Under the conditions of Lemma 5.1 assume that $\Sigma = u^{-1}(1)$, $\tilde{v} = v|_\Sigma$, and that the arc of separatrix joining $q[\alpha]$ and $v^{-1}(t[\alpha])$ does not meet

$$(u,v)^{-1}((-1,1) \times (-1,1)).$$

Certainly $(\partial/\partial\mu)(t[\alpha_\mu])|_{\mu=0} = 0$. However, by Lemma 5.1 $(\partial/\partial\mu)(s[\alpha_\mu])|_{\mu=0} \neq 0$. Therefore $DB_{\alpha_0}(\beta) \neq 0$, where $\beta = (\partial/\partial\mu)([\alpha_\mu])|_{\mu=0}$.

Now take a C^r family $\{\tilde{\alpha}_\mu\}$ of C^r immersions which is C^3 close to $\{\alpha_\mu\}$ and satisfies $\tilde{\alpha}_0 = \alpha_0$. It follows that $DB_{\alpha_0}(\tilde{\beta}) \neq 0$, where $\tilde{\beta} = (\partial/\partial\mu)([\tilde{\alpha}_\mu])|_{\mu=0}$.

This proves the differentiability of the manifold $S_1^r(c)$.

5.2 Proposition. *Assume the notations above. When* z *is either a simple* C_{11} *umbilical connection or a simple* C_{kk}, $k = 2,3$, *umbilical loop, by crossing* $S_1^r(c)$ *the umbilical connection disappears generating exactly two possibilities: the resulting separatrices are either trapped by the unique H cycle which appears close to* z *or there must exit a neighborhood of* z *which does not contain any principal cycle. (See Figs. 1.2 and 1.3 for an illustration).*

Proof: In fact $S_1^r(c)$ (i. e. $B = 0$) locally separates I^r into two connected components which correspond to $B > 0$ and $B < 0$, where the separatrix is broken. The existence of the principal cycle and neighborhood in the proposition follows from Poincaré–Bendixon Theorem. The unicity and hyperbolicity of the principal cycle can be argued as follows.
Case C_{11}*:* The mapping $T_1 = T_1(\alpha)$, defined after Proposition 2.2 depend continuously on α, in the C^1-sense. This can be verified by checking the steps involved in the results used in the proof of 2.2, particularly the results of [Hr]. The simplicity hypothesis on z amounts to the transversality of the graphs of T_1 and T_2^{-1}. Therefore, from the C^1–continuity (on α) of these mappings, follows the unicity and hyperbolicity (which amounts to the transversality) of the three alternatives illustrated in Fig. 5.2.
Cases C_{22} *and* C_{33}*:* By means of blowing up in Sections 3 and 4. These cases can be interpreted as the simple singular cycles with saddle corners, illustrated in Fig. 5.3.

The perturbation of the immersion have as effect the corresponding breaking of the connection z, but never of the connection w. These perturbations lead to a unique hyperbolic cycle as was already established in a similar situation in [P–S], following the methods in [Re]. ♦

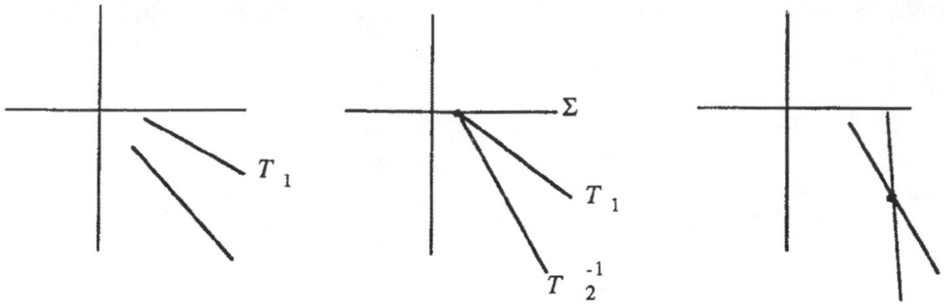

no connection, no cycle | C_{11}–connection, $B = 0$ | no connection, H cycle

figure 5.2

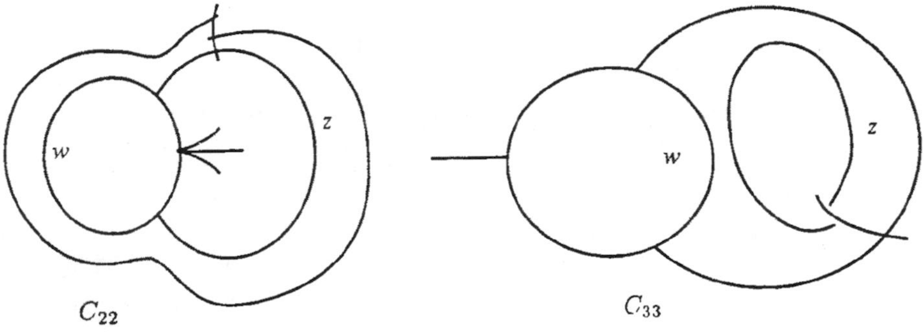

figure 5.3

The conclusion that $S_1^r(c)$ (resp. $\tilde{S}_1^r(c)$) is a one–to–one immersed (resp. embedded) submanifold of I^r can be obtained by expressing it as the projection of the embedded submanifold S_1^r of $\mathbf{R}_+ \times I^r$ which consists of pairs (s, α) such that $\alpha \in S_1^r(c)$ and s is the length of the umbilical connection of α. The proof that $\tilde{S}_1^r(c)$ is an embedded submanifold follows from b). This proves part a) of 1.2.

The proof of part b) is similar to that of the case of vector fields ([So],[A-L]). In fact, when the simple umbilical connection is destroyed by a small perturbation of $\alpha \in \tilde{S}_1^r(c)$, the resulting immersion belongs to S^r. Actually only for α in $S_1^r(c) - \tilde{S}_1^r(c)$ an arbitrary small perturbation, on the side where no cycle bifurcates from the connection, produces immersions in $I_1^r(c)$ which have umbilical connections with arbitrarily large length. See Fig. 5.4 for an illustration.

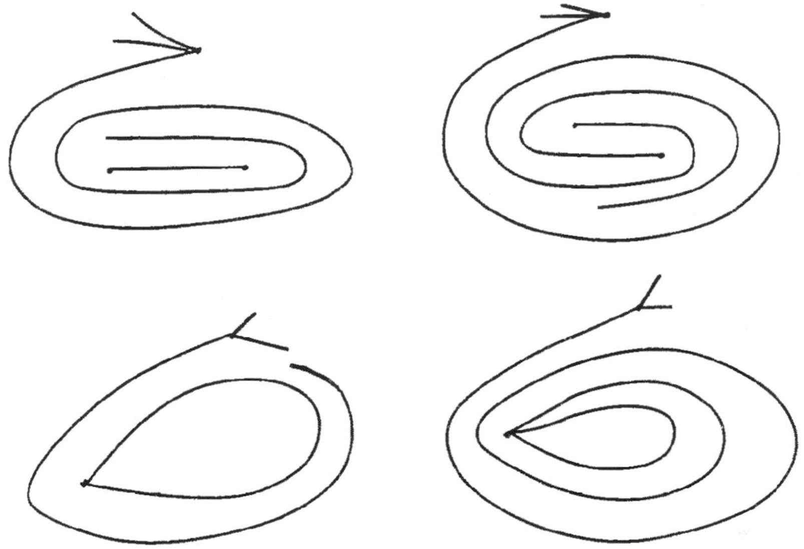

$\alpha \in S_1^r(c) - \tilde{S}_1^r(c)$ perturbation of $\alpha \in I_1^r(c)$

figure 5.4

As in [G-S.1, Section 5], the method of canonical regions applies to construct the homeomorphisms required to prove c.

6. Proof of Theorem 1.3

Let $\alpha \in I_1^r(c) \cap S^r(a)$ and let z be a non-simple $C_{i,}$, $i = 1, 2, 3$, umbilical connection of α joining umbilical points (whether equal or not) p and q. In what follows, all perturbations and objects will be referred to the foliation to which z belongs.

Firstly, it will be shown that

(1) α can be arbitrarily C^2 approximated by an $\alpha_1 \in I^{r-1}$ which has a simple C_{11} umbilical connection nearby z.

In fact, by a small perturbation of α it can be obtained that the umbilical points p and q have asymmetries appropriate to produce a simple umbilical connection. If this perturbation splitted apart the separatrices of p and q that formed z, by using another perturbation -localized around some point of z and away from umbilics- as provided by Lemma 5.1, a simple umbilical connection joining p and q can be obtained.

Secondly, it is claimed that

(2) Any $\beta \in I_1^r(c) \cap S^r(a)$ having a simple umbilical connection z can be arbitrarily C^2 approximated by a $\beta_1 \in S_1^\infty(c)$ whose principal configuration, around the union of z and the set of umbilical points, is equivalent to that of β.

In fact, The same argument used in section 5 to prove that $S_1^r(c)$ is a codimension one submanifold of $I^{r,r}$, shows that if V and $\mathcal{V} \subset I^{r,r}$ are small open neighborhoods of z and β, respectively; the set of immersions of $\mathcal{V}$ having a simple umbilical connection contained in V is a codimension one submanifold of $I^{r,r}$. Therefore, an immersion $\beta_1 \in I^\infty$, C^2 close to β, and having a simple umbilical connection in V can be found as the transversal intersection of this submanifold with an appropriate curve of immersions of I^∞. Under these conditions, the proof that β_1 can be arbitrarily C^2-approximated by an element $\beta_2 \in S_1^\infty(c)$ which has a simple umbilical connection is similar to that of [G-S.2, Theorem 3.1].

From (1) and (2) follows that

(3) α can be arbitrarily C^2 approximated by an $\alpha_2 \in S_1^r(c)$

If $\alpha_2 \in S_1^r(c) - \tilde{S}_1^r(c)$ and so its simple umbilical connection z_2 is accumulated by an umbilical separatrix y, then by a small perturbation, as provided by Lemma 5.1, it can be produced an umbilical connection formed by y and an umbilical separatrix that was previously forming z_2.

(4) The only way that all the possible resulting immersions obtained by combining this method with (2) and (3) belong to $S_1^r(c) - \tilde{S}_1^r(c)$ is when: z_2 is a C_{11} umbilical connection and all separatrix that accumulates on z_2 belongs to an umbilical point of type D_1.

In fact, otherwise z_2 must be a C_{33} umbilical loop such that the only separatrix which accumulates on it is precisely the extra separatrix of the umbilical point, not involved in the loop. See fig. 6.1.

This implies that the set U_2 formed by the union of principal lines whose limit set is exactly z_2 is open, connected and non-empty. Because of the structure of the limit set of the principal lines of α_2, the boundary of U_2 is made up of umbilical connections and their corresponding umbilical points. Since foliations on compact two-manifolds cannot have exactly one umbilical point of type D_3, there must be a connected component $\tilde{z}$, different

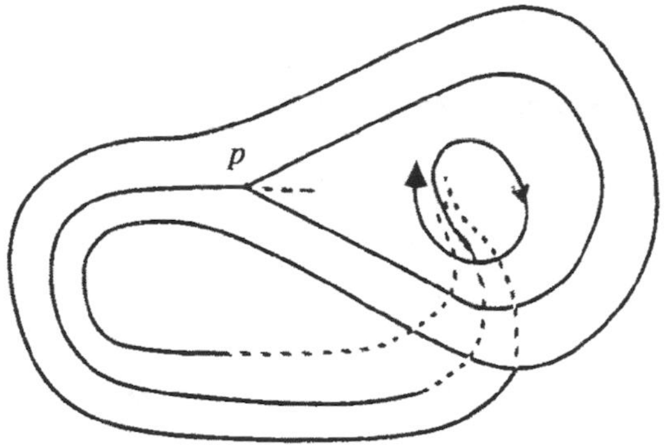

figure 6.1

from z, of the boundary of U_2. This implies that there exists an open cylinder on M having $\tilde{z}$ as one of its boundary components. However, by Poincaré–Bendixson Theorem [Hr, Chap. 7], there cannot exist principal lines whose limit set is z and which approach $\tilde{z}$ through this open cylinder. This contradiction proves (4).

The last argument also implies that when (4) is true then necessarily

(5) M is the sphere and the foliation has four umbilical points all of type D_1, whose separatrices either accumulate on or belong to z_2.

Let p_1 denote one of the umbilical points such that its separatrix accumulates on z_2. By a small C^2 perturbation of α_2, p_1 can be changed into an umbilical point of type D_2. Under these conditions, using (2), the resulting immersion can be arbitrarily approximated by an element of $\tilde{S}_1^r(c)$. ◆

7. Final remarks

(1) Examples of simple C_{11}–connections can be easily found by small local perturbations breaking the symmetry of the ellipsoid

$$\frac{x^2}{a^2} + \frac{y^2}{b^2} + \frac{z^2}{c^2} = 1, \qquad a > b > c,$$

which exhibits C_{11}–connections.

Examples of the other connections can be obtained from local models of Darbouxian umbilical points by cutting and gluing at separatrices. It would be interesting to exhibit more natural examples as in the C_{11} case.

(2) The general structure of transition maps on cross sections, induced by orbits on hyperbolic sectors, as those appearing in the analysis of the return map associated to

umbilical connections, were originally studied by Dulac [Du] for analytic vector fields. The work of [Iy1] and the paper of [Mou] showed the revival of interest on the problem left open in [Du]. Recently, J. Ecalle [Ec] and Yu Il'yasenko [Iy2] have announced to have independent solutions of this problem.

References

[AL] Andronov A., Leontovich E. et al., *Theory of Bifurcations of Dynamical Systems on the plane.* John Wiley, New York (1973)

[B–G] Bruce J. W., Giblin P. J., *Generic curves and Surfaces.* J. London Math. Soc. (2), **24** (1981).

[Ca] Cayley A., *On differential equations and umbilici.* Philos. Mag., **26**, (1963), 373–379, 441–452, (Collected works, Vol V).

[Car] Cartan H., *Théorie èlèmentaire des fonctions analytiques d'une ou plusieurs variables complexes.* Paris, Hermann, (1961).

[Da] Darboux G., *Sur la forme des lignes de courbure dans le voisinage d'un ombilic.* Note VII, Leçons sur la théorie générale des surfaces. **IV**. Gauthier–Villars. (1986).

[Du] Dulac H., *On limit cycles.* Bull. Soc. Math. Franc. (1923).

[Ec] Ecalle J., *Finitude des cycles limites et accelero sommations de l'application de retour.* To appear. (1990).

[Fi] Fischer G., *Mathematical Models.* Friedr Vieweg and Sohn, (1986).

[Gu] Gullstrand A.,*Zur Kenntniss der Kreispunkte.* Acta Math. **29**, (1905).

[Gui] Guiñez V., *Positive quadratic differential forms and foliations with singularities on surfaces.* Trans. Amer. Math. Soc. **309**, No. 2, (1988).

[Gut] Gutierrez C., *Smoothing continuous flows on two-manifolds and recurrences.* Ergod. Th. and Dynam. Sys. **6**, 17–44, (1986).

[GS1] Gutierrez C.,Sotomayor J. *Structurally stable configurations of lines of principal curvature.* Asterisque **98–99** (1982).

[GS2] Gutierrez C.,Sotomayor J. *An approximation theorem for immersions with stable configurations of lines of principal curvature.* Springer Lectures Notes in Math. **1007** (1983).

[GS3] Gutierrez C.,Sotomayor J. *Stability and bifurcations of configurations of principal lines.* Aport. Mat. **1** (1985), Soc. Mat. Mex.

[GS4] Gutierrez C.,Sotomayor J. *Closed principal lines and bifurcations.* Bol. Soc. Bras. Mat. **17** (1986).

[GS5] Gutierrez C.,Sotomayor J. *Bifurcations of umbilical points and related principal cycles.* In preparation.

[GS6] Gutierrez C.,Sotomayor J. *Principal lines on surfaces immersed with constant mean curvature.* Trans. Amer. Math. Soc. **293**, No. 2 (1986).

[Hr] Hartman P. *Ordinary differential equations.* J. Wiley (1964).

[Iy1] Il'yasenko Yu. S. *Limit cycles of polynomial vector fields with nondegenerate singular points on the real plane.* Funk. Anal. Ego. Pri., **18**, 3, (1984), 32–34. (Func. Anal. and Appl., **18**, 3, (1985), 199–209).

[Iy2] Il'yasenko Yu. S. *Fineteness theorems for limit cycles.* (In russian, to appear).

[Le] Levitt G. *La decomposition dynamique et la différentiabilité des feuilletages des surfaces.* Ann. Inst. Fourier, **37**, 3, (1987), 85–116.

[Mo] Montaldi J. *Contact with applications to submanifolds of* $\mathbf{R}^n$. Thesis. Liverpool (1983).

[Mou] Moussu R. *Le problème de la finitude du nombre de cycles limites.* Séminaire Bourbaki. 38'eme anée, **655** (1985–86).

[Po] Porteous I. R., *The normal singularities of a submanifold.* J. Diff. Geom. **5** (1971).

[P–S] Paterlini R., Sotomayor J. *Bifurcations of planar polinomial vector fields.* Can. Math. Soc. Conference Proceedings **8**, (1987).

[Re] Reyn J. W., *Generation of limit cycles from separatrix polygons in the phase plane.* Springer Lecture Notes in Math. **810**. (1980).

[Ro] Rosenberg H. *Labyrinths in discs and surfaces.* Ann. Math. **117** (1983), 1–33.

[So] Sotomayor J. *Generic one-parameter families of vector fields on two-dimensional manifolds.* Publ.Math. IHES **43** (1974).

[Sp] Spivak M. *A comprehensive Introduction to differential geometry.* Berkeley, Publish or Perish Inc. (1979).

[St] Struik D. *Lectures on classical differential geometry.* Addisson Wesley. (1950).

[Th] Thom R. *Stabilité structurelle et morphogénèse.* Benjamin, (1972).

Conditions for a centre and the bifurcation of limit cycles in a class of cubic systems

N G Lloyd and J M Pearson
Department of Mathematics, The University College of Wales, Aberystwyth, UK

1. Introduction

For a class of cubic systems

$$\dot{x} = P(x,y), \quad \dot{y} = Q(x,y) \tag{1.1}$$

in which P is linear, we consider the closely related problems of the number of limit cycles which can bifurcate out of the origin and the conditions under which the origin is a centre.

In [8] the Russian mathematician Kukles gave conditions which were said to be necessary and sufficient for the origin to be centre for systems of the form

$$\dot{x} = \lambda x+y, \quad \dot{y} = -x+\lambda y+a_1x^2+a_2xy+a_3y^2+a_4x^3+a_5x^2y+a_6xy^2+a_7y^3. \tag{1.2}$$

Our interest in these particular systems was stimulated by the work of Jin and Wang as reported in [7]. They describe computations suggesting that the conditions proposed by Kukles are incomplete (we give details in Section 3). In [3] we proved that this is indeed so, and we also gave a full description of the bifurcation of limit cycles from the origin in the case when $a_7 = 0$. The computations for the full system (1.2) present quite severe technical problems, and we shall describe this work elsewhere. There the possibility $a_2 = 0$ is excluded, and our purpose in this paper is to deal with this special case. We give necessary and sufficient conditions for the origin to be a centre, and prove that up to six limit cycles can bifurcate from the origin. The conditions for a centre are not covered by those given by Kukles, and are obtained by means of the technique recently developed by Colin Christopher, described in [4], exploiting the consequences of the existence of invariant algebraic curves. This approach also yields other conditions for a centre when $a_2 \neq 0$; these are different from the Kukles conditions and those given here, and are described in [4].

This investigation is part of our group's continuing programme of research on the limit cycles of polynomial systems. We have concentrated on so–called *small amplitude* limit cycles, that is, limit cycles which bifurcate out of a critical point under perturbation of the coefficients arising in the equations themselves, and much of our recent work has been on cubic systems.

It has been known for some time that if P and Q are symmetric cubics (i.e. there are no quadratic terms) then (1.1) has at most five small–amplitude limit cycles [2]. Various classes of cubic systems with several limit cycles are described in [12]. In particular, an example is given in which six limit cycles bifurcate from the origin; Wang gives another such example in [18]. More recently, instances of cubic systems with seven small-amplitude limit cycles have been given [1,9]. In [5] we describe such an example in which there is only one quadratic term, so that the introduction of this one term increases the number of possible bifurcating limit cycles from five to seven. The main part of [5] is concerned with the description of a class of cubic systems with eight small–amplitude limit cycles. Systems of the form

$$\dot{x} = \lambda x+y+p_2(x,y)+xs(x,y), \quad \dot{y} = -x+\lambda y+q_2(x,y)+ys(x,y),$$

where p_2,q_2 and s are homogeneous quadratic forms, have also been investigated recently. James and Yasmin [6] give necessary and sufficient conditions for a centre and have shown that the maximum number of bifurcating limit cycles is again five (note that there are nine parameters in the nonlinear part of the equations). For a description of other recent developments on the number of limit cycles of polynomial systems, we refer to the survey papers [10,11] and the references contained therein.

The structure of this paper is that in Section 2 we give a brief description of the technique which we use to investigate bifurcating limit cycles, while Section 3 consists of an account of our investigaticn of systems of the form (1.2) with $a_2 = 0$.

2. Small–amplitude limit cycles

We consider polynomial systems in which the origin is a critical point of focus type. In canonical coordinates such systems are of the form

$$\dot{x} = \lambda x+y+p(x,y), \quad \dot{y} = -x+\lambda y+q(x,y), \tag{2.1}$$

and we write

$$p(x,y) = p_2(x,y)+p_3(x,y)+\ldots+p_n(x,y),$$
$$q(x,y) = q_2(x,y)+q_3(x,y)+\ldots+q_n(x,y),$$

where p_k and q_k are homogeneous polynomials of degree k. Recall that the origin is a *fine focus* if $\lambda=0$. It is well known that there is a function V defined in a neighbourhood of the origin such that $\dot{V}$, its rate of change along orbits, is of the form

$$\dot{V} = \eta_2 r^2 + \eta_2 r^4 + \ldots,$$

where $r^2 = x^2 + y^2$. The coefficients η_{2k} are polynomials in the coefficients arising in p and q, and are the *focal values*. It is easily verified that $\eta_2 = \lambda$. The origin is a fine focus of *order k* if

$$\eta_{2j} = 0 \text{ for } 1 \leq j \leq k \text{ but } \eta_{2k+2} \neq 0.$$

Clearly, the stability of the origin is determined by the first non–vanishing focal value. Consequently the significant quantities are the *reduced focal values* or *Liapunov quantities* L(0),L(1),... . These are the non–trivial expressions obtained by computing each η_{2k} subject to the conditions $\eta_{2j} = 0$ for $j < k$; positive multiplicative numerical factors can be ignored. The origin is a centre if all the focal values are zero; moreover, for a given class $\mathcal{C}$ of systems, there is $\kappa(\mathcal{C})$ such that the origin is a centre if $\eta_{2j} = 0$ for $j \leq \kappa$. Let K be the smallest such κ; then the origin is a fine focus of order at most K–1 for systems in $\mathcal{C}$. The idea is to start with a fine focus of as high order as possible, and then to introduce perturbations into p and q each of which reverses the stability of the origin and reduces its order as a fine focus by one. It is easily seen that if the origin is a fine focus of order k, then at most k small–amplitude limit cycles can bifurcate from it [2]; however there is no guarantee that it is possible to bifurcate this maximum number -- an example where it is not possible is given in [11] for instance.

The details of this technique are described in several other papers; we refer the reader to [10,13,14], for instance. As explained in these papers, the procedure consists of four elements:

(1) The calculation of the focal values.

(2) The reduction of the focal values.

(3) Verification of the maximum possible order of the fine focus.

(4) Introduction of appropriate perturbations.

In phase (1) it is necessary to use an appropriate Computer Algebra system. The algorithm and the computational issues involved are described in [14]. The basic idea is to write

$$V = \sum_{i+j\geq 2} V_{ij}x^i y^j$$

and use the equations (2.1) to obtain an expression for $\dot{V}$; comparing coefficients, sets of linear equations for the focal values and the V_{ij} are obtained and these are solved symbolically The implementation of the algorithm, called FINDETA, uses REDUCE, and the computations described in this paper have been performed on the Amdahl 5890 at the Manchester Computing Centre which we access via the JANET network. FINDETA is described in detail in [14] and has been designed to be very 'user friendly'.

Phase (2) of the procedure also requires the use of REDUCE. As we shall see in Section 3, rational substitutions are taken from the relations $\eta_2 = \eta_4 = \ldots = \eta_{2k} = 0$ to 'reduce' η_{2k+2}. This process is continued until it appears that all subsequent focal values are zero. A proposed value of K is thus obtained.

In phase (3) it is confirmed that this proposed value of K is indeed the correct one. This is done by proving that the origin is a centre if $\eta_{2k} = 0$ for $k \leq 1 + K$. Deriving conditions for a critical point to be a centre is often a difficult problem, and necessary and sufficient conditions are known for only a few classes of systems: conditions for quadratic systems are known and for symmetric cubic systems [16], but in very few other instances. The approach using invariant algebraic curves has enabled us to derive necessary and sufficient conditions in a number of cases.

The final phase of the procedure involves an appropriate selection of perturbations each of which reduces the order of the fine focus by one and reverses its stability. Thus, if we start with a fine focus of order k, the first step is to arrange for a perturbation such that $\eta_2 = \ldots = \eta_{2k-2} = 0$ and $\eta_{2k}\eta_{2k+2} < 0$; we then continue with a sequence of perturbations until k limit cycles bifurcate.

The computations described in this paper were done using REDUCE 3.2. We reverted to version 3.2 after finding that version 3.3 was considerably slower for our purposes (some timings are given in [14]). The version of FINDETA used here utilises the REDUCE functions COEFFN and SOLVE, and much use is made of the FACTORIZE facility in the second phase of the programme described above. Since completing the work described in this paper a new version of FINDETA has been written in which the use of the function SOLVE is avoided. This and other improvements have made the procedure significantly more efficient, and have reduced the required cpu time by more than a half in cases such as those considered in this paper.

3. The Kukles conditions

We consider systems of the form

$$\dot{x} = \lambda x+y, \quad \dot{y} = -x+\lambda y+a_1x^2+a_2xy+a_3y^2+a_4x^3+a_5x^2y+a_6xy^2+a_7y^3. \qquad (3.1)$$

The conditions proposed by Kukles, and given on page 124 of Nemytskii and Stepanov [15], are that the origin is a centre if $\lambda = 0$ and one of the following four conditions are satisfied.

(K1) $M_1 = M_2 = M_3 = M_4 = 0$, where

$$M_1 = a_4a_2^2 + a_5\mu,$$
$$M_2 = (3a_7\mu+\mu^2+a_6a_2^2)a_5 - 3a_7\mu^2 - a_6a_2^2\mu,$$
$$M_3 = \mu + a_1a_2 + a_5,$$
$$M_4 = 9a_6a_2^2 + 2a_2^4 + 9\mu^2 + 27a_7\mu,$$

and $\mu = 3a_7 + a_2a_3$;

(K2) $a_7 = M_1 = M_2 = M_3 = 0$;

(K3) $a_7 = a_5 = a_2 = 0$;

(K4) $a_7 = a_5 = a_3 = a_1 = 0$.

In [7] Jin and Wang report their computation of focal values for certain systems of the form (3.1). For the system

$$\dot{x} = y, \quad \dot{y} = -x + a_1x^2 - 2a_1y^2 - a_1x^2y/3 - 3a_7x^2y + a_7y^3 \qquad (3.2)$$

with $18a_7^2 = a_1^4$ and $a_7 \neq 0$, they found that $\eta_{2k} = 0$ for $k \leq 9$, and rightly regarded this as evidence that the origin is a centre even though the system is not covered by any of the conditions (K1) – (K4). In [3] we proved that the origin is indeed a centre for (3.2) when $18a_7^2 = a_1^4$. Doubt having been cast on the conditions given by Kukles, it was clearly necessary to investigate the whole question afresh. Jin and Wang considered the subclass of systems of the form (3.1) with $a_7 = 0$, and found that $\eta_{2k} = 0$ for $k \leq 6$ if and only if one of (K2), (K3) or (K4) holds. In [3] we also considered the case $a_7 = 0$ and derived the following result, which implies that the Kukles conditions are in fact complete for this particular subclass.

Theorem 3.1 [3] *Let $a_7 = 0$ in system (3.1). The origin is a centre if and only if $\lambda = 0$ and one of the following conditions holds.*

(i) $a_2 = a_5 = 0$,
(ii) $a_1 = a_3 = a_5 = 0$,
(iii) $a_4 = a_5 = a_6 = 0,\ a_1 + a_3 = 0$,
(iv) $a_4 = (a_1+a_3)a_3,\ a_5 = -(a_1+a_3)a_2,\ a_6 = -(a_1+a_3)a_3^2(a_1+2a_3)^{-1}$.

Furthermore, at most five limit cycles can bifurcate from the origin and this maximum is attained.

We now consider another subclass of (3.1), namely that in which $a_2 = 0$, noting that this contains the example given by Jin and Wang. We shall give necessary and sufficient conditions for a centre and show that up to six limit cycles can bifurcate from the origin. The interesting feature of this result is that there are as many small–amplitude limit cycles as there are parameters in the system. In most cases which have been studied before there are many fewer bifurcating limit cycles than parameters. The possibility $a_2 = 0$ was excluded in the investigation of the full system (3.1) because the required substitutions for the reduction of the focal values are defined only for $a_2 \neq 0$. Significantly, if $a_2 = 0$ then the system (3.1) without the cubic terms is symmetric about the x-axis, and so the origin is a centre. As noted in Section 1, the centre conditions are not covered by (K1) - (K4), and we emphasise that other such conditions are given in [4] when $a_2 \neq 0$.

We therefore consider systems of the form

$$\dot{x}=y,\quad \dot{y}=-x+a_1x^2+a_3y^2+a_4x^3+a_5x^2y+a_6xy^2+a_7y^3 \tag{3.3}$$

and suppose throughout that $a_7 \neq 0$. It is convenient to replace a_1 and a_3 by b_1 and b_2, where $b_1 = a_1+a_3$ and $b_2 = 5a_1+3a_3$.

We use FINDETA to compute the focal values. First we find that $L(1) = a_5+3a_7$. For a fine focus of order greater than one we must have $L(1) = 0$; we take

$$a_5 = -3a_7. \tag{3.4}$$

Further computation then gives $L(2) = a_7(3a_4+b_1b_2+a_6)$. For a fine focus of order greater than two we need $L(2) = 0$; since $a_7 \neq 0$, we take

$$a_4 = -(b_1b_2+a_6)/3. \tag{3.5}$$

Continuing with FINDETA we find that $L(3) = a_7(-36a_7^2+f(a_6,b_1,b_2))$, where

$$f(a_6,b_1,b_2) = -4a_6^2+4a_6b_2^2+8a_6b_1b_2-104a_6b_1^2-2b_2^3b_1+9b_2^2b_1^2+130b_2b_1^3-135b_1^4.$$

Since $a_7 \neq 0$, for a fine focus of order greater than three we take

$$a_7^2 = f(a_6,b_1,b_2)/36, \tag{3.6}$$

noting that a_6,b_1 and b_2 must be such that $f > 0$; in particular, b_1 and b_2 cannot both be zero. Further computation now gives $L(4) = a_7(Aa_6^2+Ba_6+C)$, where

$$A = -80(b_2-4b_1)^2,$$
$$B = 2(b_2-4b_1)(6b_2^3+58b_2^2b_1-1102b_2b_1^2-1935b_1^3),$$
$$C = -b_2(b_2-b_1)(6b_2^4-16b_2^3b_1-1325b_2^2b_1^2+6165b_2b_1^3+26730b_1^4).$$

For a fine focus of order greater than four, we must, of course, have $L(4) = 0$. We need $B^2-4AC \geq 0$, and suppose that

$$a_6 = (-B+D)/2A, \tag{3.7}$$

where $D^2 = B^2-4AC$, noting for future reference that D has two possible values. Let $b_3 = b_2-4b_1$. We compute that $B^2-4AC = b_3{}^2\phi(b_1,b_2)$, where

$$\phi(b_1,b_2) = 653625b_1{}^6+291060b_1{}^5b_2+43416b_1{}^4b_2{}^2-5148b_1{}^3b_2{}^3 -900b_1{}^2b_2{}^4+24b_1b_2{}^5+4b_2{}^6.$$

We exclude for the present the possibility that $b_3 = 0$. Continuing the computation of focal values using FINDETA, we find that $L(5) = -a_7(GD+H)b_3{}^{-1}$, where

$$G = -(184b_2{}^6-2892b_2{}^5b_1-31596b_2{}^4b_1{}^2+356142b_2{}^3b_1{}^3+1551708b_2{}^2b_1{}^4-4210434b_2b_1{}^5 -25431345b_1{}^6)$$

and

$$H = 368b_2{}^9-4680b_2{}^8b_1-123600b_2{}^7b_1{}^2+1091796b_2{}^6b_1{}^3+14126832b_2{}^5b_1{}^4 -59449766b_2{}^4b_1{}^5-648352836b_2{}^3b_1{}^6-355378374b_2{}^2b_1{}^7 +8039416860b_2b_1{}^8+20526992325b_1{}^9.$$

For a fine focus of order greater than five we need $L(5) = 0$, and we therefore have $D = -H/G$ if $G \neq 0$, or $G=H=0$ otherwise. Using the RESULTANT function in REDUCE, the resultant of G and H is non–zero unless $b_1 = b_2 = 0$, a possibility which we have already excluded. It follows that G and H cannot be zero simultaneously, and so we suppose that $G \neq 0$. There are now two expressions for D: for consistency we require that

$$B^2 - 4AC = (H/G)^2; \qquad (3.8)$$

moreover, D has the sign of $-H/G$. A straightforward calculation tells us that (3.8) is satisfied if and only if

$$\psi(b_1,b_2) = b_1{}^6(b_2-b_1)F(b_1,b_2) = 0,$$

where

$$F(b_1,b_2) = -3186845749080b_1^{11} + 1048502846790b_1^{10}b_2 + 193976300259b_1^9b_2^2 - 54272788740b_1^8b_2^3 - 22059722583b_1^7b_2^4 + 7321569165b_1^6b_2^5 - 570046311b_1^5b_2^6 - 34456359b_1^4b_2^7 + 8127591b_1^3b_2^8 - 495966b_1^2b_2^9 + 12302b_1b_2^{10} - 92b_2^{11}.$$

Hence $b_1 = 0$ (and $b_2 \neq 0$) or $b_1 = b_2 \neq 0$ or $F(b_1,b_2) = 0$. We suppose for the moment that $b_1 \neq 0$ and $b_1-b_2 \neq 0$. With $F(b_1,b_2) = 0$, we continue the calculation of focal values using FINDETA, and find that

$$L(6) = -a_7b_1^6(b_2-b_1)\chi(b_1,b_2)/b_3^2G,$$

where $$\chi(b_1,b_2) = 7124b_2^{11} - 951494b_2^{10}b_1 + 39115942b_2^9b_1^2 - 737410267b_2^8b_1^3 + 6661763643b_2^7b_1^4 - 18667396833b_2^6b_1^5 - 130220619285b_2^5b_1^6 + 1047738480831b_2^4b_1^7 - 2171066759280b_2^3b_1^8 + 4793413413537b_2^2b_1^9 - 3663280994270b_2b_1^{10} + 80202151292760b_1^{11}.$$

We again use the RESULTANT function of REDUCE and find that the resultant of F and χ is non–zero unless $b_1 = b_2 = 0$. Thus $\chi \neq 0$ when $F = 0$.

We summarise the discussion so far in the following lemma.

Lemma 3.2 *Suppose that the following conditions hold.*

(1) $a_7 \neq 0$, $b_1 \neq 0$, $b_1-b_2 \neq 0$, $4b_1-b_2 \neq 0$, $f(a_6,b_1,b_2) > 0$, $\phi(b_1,b_2) \geq 0$,
(2) $a_5 = -3a_7$,
(3) $a_4 = -(b_1b_2+a_6)/3$,
(4) $a_7^2 = f(a_6,b_1,b_2)/36$,
(5) $a_6 = (-B+D)/2A$,
(6) $F(b_1,b_2) = 0$.

Then the origin is a fine focus of order six. If any of the conditions (2) – (6) are violated, then the order of the fine focus at the origin is less than six.

We shall see later that these conditions can be satisfied simultaneously. First, however, we consider the cases excluded in Lemma 3.2. Suppose that $b_1 = 0$, but $b_2 \neq 0$. Referring to the expressions given above for the Liapunov quantities and recalling that $a_7 \neq 0$, for a fine focus of order at least four we have $a_5 = -3a_7$, $a_4 = -a_6/3$, $a_7^2 = 4a_6(b_2^2-a_6)/36$. Then $L(4) = a_7(12b_2^2-80a_6)b_2^2a_6$. Now $a_6 = 0$ implies that $a_7 = 0$, which is contrary to hypothesis; hence we take $a_6 = 3b_2^2/20$. We compute $L(5) = a_7b_2^8$. Since both a_7 and b_2 are non-zero by hypothesis, the origin is of order at most five.

In the second excluded case, $b_1 = b_2 \neq 0$. In terms of a_1 and a_3 this means that $a_1 = -a_3/2 \neq 0$. For a fine focus of order at least four,

$$a_5 = -3a_7,\ a_4 = -(4a_6+a_3^2)/12,\ 288a_7^2 = -(32a_6^2+184a_6a_3^2-a_3^4).$$

Computing η_{10}, we have

$$L(4) = a_3^2a_6a_7(160a_6-991a_3^2).$$

If $a_3 = 0$ then $a_1 = 0$, which is excluded. If $a_6 = 991a_3^2/160$ we compute that $L(5) = a_3^8a_7$, and since neither a_3 nor a_7 is zero, for a fine focus of order greater than five we must have $a_6 = 0$; it follows that

$$a_1 = -a_3/2,\ a_4 = -a_3^2/12,\ a_5 = -3a_7,\ a_7^2 = a_1^4/18.$$

This is the situation covered by the example of Jin and Wang, which was proved in [3] to imply that the origin is a centre.

It remains to consider the case in which $b_3=0$; this means that $a_1=a_3$. If $L(1) = L(2) = 0$, we have $a_5 = -3a_7$ and $a_4 = -(a_6+16a_3^2)/3$. From $L(3) = 0$ we have $9a_7^2 = -(a_6^2+8a_6a_3-160a_3^4)$, and further computation gives $L(4) = a_3^6a_7$. Now $a_3 = 0$ implies that $a_1 = 0$ and $9a_7^2 = -a_6^2$, which is impossible. Thus the origin is a fine focus of order at most four.

We therefore have the following result.

Theorem 3.3 *Suppose that $a_7 \neq 0$. The origin is a fine focus of order at most six. It is of order six if $f(a_6,b_1,b_2) > 0$, $\phi(b_1,b_2) \geq 0$ and conditions (2) to (6) of Lemma 3.2 are satisfied. The origin is a centre if and only if*

$$a_3 = -2a_1,\ a_4 = -a_1^2/3,\ a_5 = -3a_7,\ a_6 = 0,\ a_7^2 = a_1^4/18.$$

We proceed to show that there are indeed systems of the form (3.3) for which all the conditions of Lemma 3.2 are satisfied, and then show how six limit cycles can bifurcate. The first task is to locate the zeros of $F(b_1,b_2)$, which is a homogeneous polynomial of degree 11. Real zeros of polynomials in one variable can be located by means of the classical theorem of Sturm (see [17], page 220, for example). The associated computations are impossible to complete by hand in the case of polynomials of high degree, and a REDUCE procedure, called STURM, was written. This takes as parameters a polynomial p, the independent variable and an interval I, and returns the number of zeros of p in I.

Let $f_1(u) = F(1,u)$ and $f_2(v) = F(v,1)$. Using STURM, f_1 has seven roots in the interval (−100,100), all of which are simple, and no roots in (−1,1); furthermore, there are only seven zeros of f_2 in (−1,1). Hence F has exactly seven distinct real factors, all of multiplicity one. More precisely, one root of f_1 is located in each of the following intervals: (−10,−9), (3,4), (4,5), (10,12), (12,15), (15,20) and (78,79). Since D has the sign of $-G/H$, we consider G and H in each of these intervals. Using STURM, it transpires that GH is of one sign in three of them: $GH>0$ in (3,4), while $GH<0$ in (12,15) and (78,79). We select the interval (12,15), and after some experimentation using REDUCE find that the zero of f_1 in (12,15) lies between 13.7 and 13.9. Since we simply seek one instance where the conditions for a fine focus of order six are satisfied, we do not need to enter into more precise calculations in the other intervals.

We fix an arbitrary non-zero b_1 and let $b_2 = ub_1$; we then regard functions of b_1 and b_2 as functions of u alone. For a fine focus of order six, $u = u_*$, where u_* is the unique zero of f_1 in $I = (13.7,13.9)$. Since G and H are of opposite sign for $u \in I$, we have that $D > 0$. In addition, $A < 0$, so a_6 is given by the smaller root of Aw^2+Bw+C. Furthermore, again using STURM, we find that ϕ is non-zero for $u \in I$, and it is easily checked that $\phi > 0$. Next we consider $f(a_6,b_1,b_2)$; with a_6 as given by (3.7) we compute that $f = (\alpha D+\beta)/\gamma$, where $\gamma = 9600b_3^2$ and

$$\alpha = -6095b_1^3 + 258b_1^2b_2 + 138b_1b_2^2 - 34b_2^3,$$
$$\beta = 4863675b_1^6 - 955620b_1^5b_2 + 126552b_1^4b_2^2 + 25444b_1^3b_1^3 - 9300b_1^2b_2^4 -72b_1b_2^5+68b_2^6.$$

We first confirm by means of STURM that neither α nor β is zero for $u \in I$, and we check that $\alpha < 0$ and $\beta > 0$ in this interval; therefore $\alpha D - \beta < 0$. To determine the sign of $\alpha D+\beta$, we consider $\alpha^2D^2-B^2$ and find that this is negative for $u \in I$. Hence $f(a_6,b_1,b_2) > 0$. With b_2 as chosen above we therefore find that the origin is a fine focus of order six when a_5,a_4,a_7 and a_6 are given by (3.4), (3.5), (3.6) and (3.7), respectively.

It is now possible to describe how six limit cycles can bifurcate out of the origin. We start, of course, with a fine focus of order six, and suppose that the coefficients a_i are as chosen above. We have already seen that F and χ do not have a common zero; however, at this stage it is necessary to be more precise, for we need to know the sign of χ when $u = u_*$. It turns out that χ has a zero with $u \in I$, and so we locate u_* more precisely than hitherto. We find that $u_* \in I_1 = (13.7, 13.81)$ and note that $\chi < 0$ in I_1. Since $G > 0$, initially L(6) has the sign of a_7. The first perturbation is of b_2. We require $L(5)L(6) < 0$, and so $(GD+H) > 0$. Now $G^2D^2-H^2 = b_1{}^6(b_2-b_1)F(b_1,b_2)$, and it is readily verified that F changes sign from positive to negative as u increases through u_*; hence $G^2D^2-H^2$ is decreasing at $u = u_*$. But, since $G > 0$ and $H < 0$ for $u \in I_1$, we have $D > 0$, and so $GD-H > 0$ for $u \in I_1$. Consequently, GD+H decreases as u increases through u_*, and a limit cycle bifurcates if b_2 is decreased. To ensure that the origin is a fine focus of order five after perturbation, a_5,a_4,a_7 and a_6 are adjusted in accordance with the relations (3.4) – (3.7).

The second perturbation is of a_6: we require $L(4)L(5) < 0$, that is, $Aa_6{}^2+Ba_6+C > 0$. Now $A < 0$ and initially a_6 is the smaller root of Aw^2+Bw+C; therefore we increase a_6, and at the same time adjust a_5,a_4 and a_7 so that (3.4) – (3.6) continue to hold. Thus the stability of the origin is again reversed, and its order as a fine focus is reduced by one. A limit cycle bifurcates and provided that the perturbation of a_6 is small enough the first limit cycle persists. The third limit cycle is generated by perturbing a_7; L(3)L(4) has the sign of

$$(Aa_6{}^2+Ba_6+C)(-36a_7{}^2+f(a_6,b_1,b_2)).$$

We therefore increase a_7 and adjust a_5 to maintain (3.4). The remaining three limit cycles appear by perturbing a_4,a_5 and introducing a non-zero λ. It may be checked that a_4 is increased, a_5 is increased or decreased according to whether $a_7 < 0$ or $a_7 > 0$, and λ is chosen so that $\lambda a_7 > 0$. We arrive at the following result.

Theorem 3.4 *There are systems of the form (3.3) with six small-amplitude limit cycles, and this is the maximum possible number.*

References

1. J M Abdulrahman, *Bifurcation of limit cycles of some polynomial systems* (PhD thesis, The University College of Wales, Aberystwyth, 1989).
2. T R Blows and N G Lloyd, 'The number of limit cycles of certain polynomial differential equations', *Proc. Roy. Soc. Edinburgh Sect. A* 98 (1984) 215-239.
3. C J Christopher and N G Lloyd, 'On the paper of Jin and Wang concerning the conditions for a centre in certain cubic systems', *Bull. London Math. Soc.*, 94 (1990) 5-12.
4. C J Christopher and N G Lloyd, 'Invariant algebraic curves and conditions for a centre', pre print, The University College of Wales, Aberystwyth, 1989.
5. E M James and N G Lloyd, 'A cubic system with eight small–amplitude limit cycles', preprint The University College of Wales, Aberystwyth, 1990.
6. E M James and N Yasmin, 'Limit cycles of a cubic system', preprint, The University College of Wales, Aberystwyth, 1989.
7. Jin Xiaofan and Wang Dongmin, 'On Kukles' conditions for the existence of a centre', *Bull London Math. Soc.* , 94 (1990) 1-4.
8. I S Kukles, 'Sur quelques cas de distinction entre un foyer et un centre', *Dokl. Akad. Nauk. SSSR* 42 (1944) 208-211.
9. Li Jibin and Bai Jinxin, 'The cyclicity of multiple Hopf bifurcation in planar cubic differential systems: $M(3) \geq 7$', preprint, Kunming Institute of Technology, 1989.
10. N G Lloyd, 'Limit cycles of polynomial systems', *New directions in dynamical systems,* London Math. Soc. Lecture Notes Series No.127 (ed. T Bedford and J Swift, Cambridge University Press, 1988), pp.192-234.
11. N G Lloyd, 'The number of limit cycles of polynomial systems in the plane', *Bull. Inst. Math. Appl. 24* (1988), 161-165.
12. N G Lloyd, T R Blows and M C Kalenge, 'Some cubic systems with several limit cycles', *Nonlinearity* 1 (1988) 653–669.
13. N G Lloyd and S Lynch, 'Small amplitude limit cycles of certain Lienard systems', *Proc. Royal Soc. London Ser A* 418 (1988) 199–208.
14. N G Lloyd and J M Pearson, 'REDUCE and the bifurcation of limit cycles', *J. Symbolic Comput.,* to appear.
15. V V Nemytskii and V V Stepanov, *Qualitative theory of differential equations* (Princeton University Press, 1960).
16. K S Sibirskii and V A Lunkevich, 'On the conditions for a centre', *Differencial'nye Uravneniya* 1 (1965) 53–66.
17. B L van der Waerden, *Modern Algebra, Volume 1* (English translation, Frederick Ungar, New York, 1949).
18. Wang Dongmin, 'A class of cubic differential systems with 6–tuple focus', Technical report 88–47.0, RISC, University of Linz, 1988.

On first integrals of linear systems, Frobenius integrability theorem and linear representations of Lie algebras

Jean MOULIN OLLAGNIER*
Jean-Marie STRELCYN†

Abstract

A necessary condition to be satisfied by $n-1$ vector fields in $\mathbb{R}^n$ in order to have a common first integral is supplied by the compatibility condition of Frobenius integrability theorem. This condition is also generically sufficient for the local existence of such a common first integral. We study here the question of the existence of a global common first integral for compatible linear vector fields in $\mathbb{R}^n$.

For the dimension 3, we prove that any two compatible linear vector fields have a common global first integral.

On the contrary, we give an example for the dimension 4, in which three compatible linear vector fields cannot have a common global first integral.

This leads us to ask many simple and natural questions, some of them about representations of Lie algebras by Lie algebras of linear vector fields.

Some historical comments and abundant references are also provided.

1 Introduction

Let us consider two systems of homogeneous linear differential equations with constant coefficients in $\mathbb{R}^3$:

$$\frac{du}{dt} = L_1(u) = A_1 u, \quad \frac{du}{dt} = L_2(u) = A_2 u \tag{1}$$

where u belongs to $\mathbb{R}^3$ and where A_1 and A_2 are real 3×3 matrices.

The point of departure of the present paper is the problem of the existence of a common non-trivial first integral for both systems (1). This problem seems to

*Département de Mathématiques et Informatique, UA CNRS 742, C. S. P., Université Paris-Nord, Avenue J. B. Clément 93430 VILLETANEUSE, FRANCE

†Département de Mathématiques, Université de Rouen, B.P. 118, 76134 MONT-SAINT-AIGNAN CEDEX, FRANCE, UA CNRS 742 & 1378

have never been studied before. If F is such a common integral, the level surfaces $\{F = const.\}$ are tangent to both vector fields L_1 and L_2. The compatibility condition from the Frobenius integrability theorem is thus necessarily satisfied, i. e. at any point u of $\mathbb{R}^3$, the three vectors $L_1(u)$, $L_2(u)$ and $[L_1, L_2](u)$ are linearly dependent. As usual, $[L_1, L_2]$ denotes the Lie bracket of the two vector fields; here, as L_1 and L_2 are linear, one has $[L_1, L_2](u) = -[A_1, A_2](u)$, where $[A_1, A_2] = A_1A_2 - A_2A_1$ is the matrix commutator.

The compatibility condition is equivalent to the following one:

$$\det(L_1(u), L_2(u), [L_1, L_2](u)) = 0 \tag{2}$$

for every u in $\mathbb{R}^3$.

It is worth noting that this property does not imply that the three vector fields L_1, L_2 and $[L_1, L_2]$ are linearly dependent over $\mathbb{R}$.

In the following, any two, not necessarily linear, smooth vector fields satisfying condition (2) will be called *compatible.*

Although the Frobenius integrability theorem guarantees that two compatible vector fields have a common first integral around any point at which these vector fields are linearly independent, nothing can be said on the existence of a global first integral without a further study of the concrete framework.

Our first result asserts that two compatible linear vector fields defined on $\mathbb{R}^3$ always have a common first integral, typically with some singularities.

Let us note that a similar result was also obtained by P. Basarab-Horwath and S. Wojciechowski [4].

Let us give an example. Consider the two matrices A_1 and A_2

$$A_1 = \begin{pmatrix} 0 & 1 & 0 \\ 0 & 0 & 1 \\ 0 & 0 & 0 \end{pmatrix} \quad A_2 = \begin{pmatrix} 1 & 0 & 1 \\ 0 & 1 & 0 \\ 0 & 0 & 1 \end{pmatrix}$$

The following function $F(x, y, z)$ is easily shown (cf. [77, 32]) to be a common global first integral for both systems (1) corresponding to the matrices A_1 and A_2

$$F(x, y, z) = \frac{y^2 - 2xz}{2z^2} + \log |\, z \,| \, .$$

Having succeeded in proving the existence of a common global first integral for systems (1), we learnt that the solution of this problem was in fact almost entirely, but implicitly, contained in the classical works of C. G. Jacobi [36] and D. Poisson (cf. [24]). Their result are clearly stated in the classical textbooks of E. Goursat [31] and E. L. Ince [35].

Our point of view is somewhat different so that our solution has some special features. In particular, we consider very carefully the question of the uniformity of our integrals; we also use the intrinsic, coordinate free, very economical approach with the differential forms.

After having sought a common first integral of two linear compatible vector fields in $\mathbb{R}^3$, we were naturally led to consider a similar problem in higher dimensions.

Let $L_1, \ldots, L_k$ be some smooth vector fields defined on $\mathbb{R}^n$, $2 \leq k < n$, satisfying for all indices i and j and every point u in $\mathbb{R}^n$ the compatibility condition of the Frobenius integrability theorem:

$$L_1(u), L_2(u), \ldots, L_k(u), [L_i, L_j](u) \text{ are linearly dependent.} \tag{3}$$

Recall that the Frobenius integrability theorem asserts that the compatibility condition (3) is sufficient to find, around any point u at which the k vectors $L_1(u), \ldots, L_k(u)$ are linearly independent, $n-k$ functionally independent common first integrals $\Phi_1, \ldots, \Phi_{n-k}$.

The level manifolds $\{\Phi_1 = c_1, \ldots, \Phi_{n-k} = c_{n-k}\}$ thus define a smooth k-dimensional foliation of some neighborhood of u. This local foliation can be extended to a global one [76].

The simplest higher dimensional problem is the following one. Is is true that three linear compatible vector fields in $\mathbb{R}^4$ always have a common global first integral, perhaps with some singularities ? The answer is surprisingly no and we give an example; this is our main result and the true "raison d'être" of this paper.

Anticipating a little, let us say that, in this example, all but two global leaves of the associated foliation are everywhere dense in $\mathbb{R}^4$.

Let us stress that such an example is of direct interest in control theory in relation with the notion of set of accessibility (cf. [28, 78]). Indeed, we have here an approximate controlability despite of the compatibilty (integrability) of our linear vector fields.

All this leads us to ask many simple and natural questions, some of them about representations of Lie algebras by Lie algebras of linear vector fields, that do not seem to have been formulated before.

Although completely independent and self-contained, the present paper is a sequel of [77] and of [32] where the compatibility condition (2) is used as an effective tool for the search of first integrals of some non-linear systems of three autonomous ordinary differential equations.

The paper is organized as follows. In section 2, we recall some facts from exterior calculus; in section 3, we describe, in the exterior form framework, the integrability results in $\mathbb{R}^3$ and give the outline of the proof. The complete proof of the result is presented in section 4 while section 5 consists of the description of our example of non-integrability in $\mathbb{R}^4$ and its easy extension to higher dimensional cases. In section 6, we formulate some questions and give additionnal remarks, mainly of historical nature.

The problems related to those studied in the present paper were intensively investigated by many people. See, for example, [1, 2, 3, 4, 8, 9, 10, 11, 13, 21, 22, 23, 24, 26, 27, 30, 31, 32, 35, 36, 37, 38, 39, 41, 42, 43, 44, 47, 50, 51, 54, 55, 56, 60, 61, 62, 63, 64, 65, 66, 68, 69, 70, 71, 77, 81], and this list is very far from being complete.

Let us stress that the important books by D. Cerveau and F. Mattei [11] and by J.-P. Jouanolou [37] are devoted to problems directly related to ours.

Acknowledgements We are very indebted to R. Moussu (Université de Dijon) who told us that our topic had in fact a long and rich history about which we were completely unaware. In particular we owe him the knowledge of the basic work of G. Darboux.

We are also grateful to B. Bru (Université Paris 5), C. Houzel (Université Paris 13) and M. Loday-Richaud (Université Paris 11) for their help in bibliographical queries.

We want to thank P. Basarab-Horwath (University of Linköping), P. Cartier (Ecole Normale Supérieure, Paris), D. Cerveau (Université Rennes 1), M. Chaperon (Université Paris 7), A. Chenciner (Université Paris 7), J. P. Françoise (CNRS, Université Paris 11), B. Grammaticos (Université Paris 7), W. Hebisch (University of Wrocław), B. Jakubczyk (Polish Acad. of Sciences, Warsaw), T. Józefiak (University of Toruń), J. Sam Lazaro (Université de Rouen), A. Ramani (Ecole Polytechnique, Palaiseau), R. Roussarie (Université de Dijon) and A. Tyc (Polish Acad. of Sciences, Toruń) for very helpful discussions.

Last but not least we thank the anonymous referee for very interesting comments.

2 Some facts from exterior calculus

We recall now some well-known and useful definitions and results about exterior differentiation, inner products and volume form (cf. [7, 18, 53, 58, 76] for more details) as well as Euler's theorem on homogeneous functions.

All differential forms and vector fields are supposed to be defined and sufficiently differentiable on a non-empty open subset U of $\mathbb{R}^n$. We denote by w^r an exterior r-form.

The exterior differentiation. The exterior derivative is a linear map d from the set of differential forms into itself that increases the degree by 1 (so that $dw = 0$ for n-forms), whose square $d \circ d$ is the null map; moreover, d is an antiderivation with respect to the exterior product of differential forms, i. e.

$$d\,(w^p \wedge w^q) = (dw^p) \wedge w^q + (-1)^p\, w^p \wedge (dw^q). \tag{4}$$

The inner product. The inner product $i(X).(.)$ by a vector field X is a linear map from the set of differential forms into itself that decreases the degree by 1; thus $i(X).(f) = 0$ for 0-forms, i. e. functions. If w^p is a p-form, $i(X).(w^p)$ is the $(p-1)$-form given by

$$i(X).(w^p)\,(v_1,\ldots,v_{p-1}) = w^p(X, v_1,\ldots,v_{p-1})$$

where $v_1,\cdots,v_{p-1}$ are vector fields.

With respect to the differential forms, the inner product is an antiderivation:

$$i(X).(w^p \wedge w^q) = (i(X).(w^p)) \wedge w^q + (-1)^p\, w^p \wedge (i(X).(w^q)). \tag{5}$$

On the other hand, the inner product is obviously anticommutative with respect to the vector fields and, in particular, two successive inner products by the same vector field yield 0.

In coordinate form, the inner product $i(X).(w)$ of a 1-form $w = \sum_{i=1}^{n} w_i dx_i$ by a vector field $X = \sum_{i=1}^{n} X_i \partial/\partial x_i$ is equal to $\sum_{i=1}^{n} X_i w_i$.

In particular, a smooth function F is a first integral of a vector field X if it satisfies:

$$i(X).(dF) = X(dF) = \sum_{i=1}^{n} X_i\, \partial F/\partial x_i = 0.$$

Volume form. Denote by $x_1,\cdots,x_n$ the cartesian coordinates in $\mathbb{R}^n$. The volume form Ω is the exterior n-form $\Omega = dx_1 \cdots dx_n$.

Given n vectors $X_1,\cdots,X_n$ in $\mathbb{R}^n$, $\Omega(X_1,\cdots,X_n)$ is equal to the determinant $\det((X_1,\cdots,X_n))$, where $(X_1,\cdots,X_n)$ is the $n \times n$ matrix, whose columns are the vectors $X_1,\cdots,X_n$.

Euler's theorem on homogeneous functions. A function f defined on $\mathbb{R}^n$ is said to be homogeneous of degree k if, for every point x in $\mathbb{R}^n$ and every positive real number t, $f(tx) = t^k f(x)$.

The famous Euler's theorem on homogeneous functions in $\mathbb{R}^n$ asserts that a smooth function f defined on $\mathbb{R}^n$ is homogeneous of degree k if and only if the following identity holds:

$$\sum_{i=1}^{n} x_i\, \partial f/\partial x_i = k\, f. \tag{6}$$

From the previous identity, a generalized Euler's formula can be deduced; if w is a p-form in $\mathbb{R}^n$, all of whose components are homogeneous functions of degree k, and if I is the so-called radial vector field $I = \sum_{i=1}^{n} x_i\, \partial/\partial x_i$, then the following identity holds:

$$i(I).(dw) + d\,(i(I).(w)) = (p+k)\, w. \tag{7}$$

Euler's original identity (6) is a special case, when p is equal to 0, of the generalized one and can then be written as:

$$i(I).(df) = k\,f.$$

3 Integrability in $\mathbb{R}^3$: outline of the proof

In this section, we give the outline of the proof of the following theorem:

> Any two compatible linear vector fields in $\mathbb{R}^3$ have a global common first integral.

We must first carefully define the notion of a global first integral; let us give it in a general context, not only for linear fields.

Let us consider a smooth vector field X defined on $\mathbb{R}^n$ or on some open subset U of it. A *global first integral* F of X is a smooth function defined on a dense open X-invariant subset V of U, which satisfies the identity $XF = i(X).(dF) = 0$ at every point of V, and which is not constant on any open subset of V.

A subset E of U is said to be X-invariant if it consists of complete trajectories of the field X; this means that no segment of an X-trajectory can join a point of E to a point of $U \setminus E$. Equivalently, E and $U \setminus E$ are *locally* X*-invariant*, i. e. invariant under the local flow induced by X.

Let us remark that the escape to infinity in finite time cannot be generally excluded, so that the complete X-trajectories are not necessarily described by a time parameter going from $-\infty$ to $+\infty$. Nevertheless, this phenomenon never occurs with linear vector fields.

A *linear vector field* in $\mathbb{R}^n$ is a vector field whose components are homogeneous linear polynomials with respect to the space variables $x_1, \cdots, x_n$. In the case $n = 3$, we write naturally the variables x_1, x_2, x_3 as x, y and z.

Due to the algebraic aspect of the problem, the invariant subset V, on which we define the common first integral of two linear vector fields on $\mathbb{R}^3$ will be the complement of the set of zeros of a finite number of real polynomials, i. e. a dense Zariski open subset of $\mathbb{R}^3$.

Given two linear vector fields L_1 and L_2, denote by w the 1-form defined by $w = i(L_1).(i(L_2).(\Omega))$ where Ω is the volume 3-form $\Omega = dx\,dy\,dz$. These two vector fields are compatible if and only if w is *integrable* i. e. satisfies the integrability condition:

$$w \wedge dw = 0. \tag{8}$$

If w vanishes everywhere, L_1 and L_2 are either linearly dependent vector fields or multiples of the same constant vector field. In this case our theorem relies on the easily proven fact that a linear or constant vector field has always a global first integral.

We shall therefore only consider pairs (L_1, L_2) of compatible linear vector fields such that the 1-form w, whose coefficients are homogeneous quadratic polynomials in the space variables, does not identically vanish; this 1-form w is then different from 0 on a dense open subset U of $\mathbb{R}^3$.

The derivative dF of a common first integral of the two fields is everywhere colinear to w; indeed, consider the obvious identity $\Omega \wedge dF = 0$, take its inner product by L_2, then by L_1 (cf. (5)) to get $w \wedge dF = 0$, which means that the two 1-forms w and dF are colinear.

The first step then consists in finding an integrating factor for w, i. e. a function ϕ such that $d(\phi w) = 0$. A primitive F of this closed 1-form will then be the desired first integral, provided that F is uniform, i. e. univalued on its domain of definition.

The Frobenius integrability theorem yields the local existence of an integrating factor for a 1-form w satisfying the integrability condition (8). But we are interested in a global solution to the problem; we show that there exists a non-zero homogeneous cubic polynomial Q such that $1/Q$ is the desired integrating factor of w. Our proof is then in fact independent of the Frobenius theorem.

Let us first suppose that w is *irreducible* (and this is typically the case), which means that the components of w have no non-trivial polynomial common factor.

Consider the homogeneous cubic polynomial $P = i(I).(w)$, where I is the radial linear vector field $I = x\,\partial/\partial x + y\,\partial/\partial y + z\,\partial/\partial z$. If P does not vanish identically, let V be the dense open subset of U where P is different from 0. In this case, $1/P$ can be choosen as an integrating factor of w on V. Moreover P satisfies $dP \wedge r = 0$ where $r = dw$ is the exterior differential of w.

When $P = i(I).(w)$ vanishes identically, the inverse $1/Q$ of a cubic homogeneous non-zero polynomial Q is an integrating factor for w if and only if $dQ \wedge r = 0$; and such polynomials do exist; and, in this case, we call V the dense open subset of U where Q is different from 0.

We are then faced with two global problems. Knowing that w/Q is a closed 1-form defined on the dense open subset V of $\mathbb{R}^3$, on which w and Q do not vanish, we have to integrate it, i. e. to study the topology of the connected components of V; in order to show that V is natural with respect to our problem, we must also prove that it consists of complete trajectories of the two original linear vector fields L_1 and L_2.

To solve the first geometrical question, we apply the classification of the closed non-zero 2-forms r in $\mathbb{R}^3$, whose coefficients are homogeneous linear polynomials, under the action of $SL(3, \mathbb{R})$; and we give, in each case, a description of the vector space of all homogeneous cubic polynomials Q such that $dQ \wedge r = 0$.

In this way, besides the fact that such non-zero cubic polynomials always exist, it also appears that the connected components of the complement of the set of zeros of any such polynomial are either simply connected, or of degree of connectivity two, i. e. their fundamental group is isomorphic to the group $\mathbb{Z}$ of all relative integers.

On a simply connected component V_0 of V, a closed 1-form is exact and we get our first integral. Otherwise the integration of w/Q can lead us to consider a first integral with values in the circle $S^1 = \mathbb{R}/k\mathbb{Z}$ instead of the real line $\mathbb{R}$, where k is the smallest strictly positive jump of the integral on a closed non contractible curve in V_0.

This kind of multivalued first integral can nevertheles be considered as a good parametrization of the set of leaves of the foliation given by the 1-form. Let us note that whenever such a first integral F is known, $G(u) = \sin(\frac{2\pi}{k}u)$ defines a first integral in the usual sense.

On the other hand, to prove that the involved open set $V = \{Q \neq 0, w \neq 0\}$ consists of complete trajectories, we prove the local invariance of its complement under the two linear vectors fields. More precisely, we prove that the set $\{Q = 0, w \neq 0\}$ is locally invariant for any polynomial vector fields X such that $i(X).(w) = 0$. To prove the local X-invariance of $\{w = 0\}$, we use essentially the fact that w is irreducible; if w is not irreducible, the result follows from the consideration of some irreducible 1-forms of lower degree.

4 Integrability in $\mathbb{R}^3$: the proof

We begin this section with a classification of the closed 2-forms in $\mathbb{R}^3$, whose coefficients are real homogeneous linear polynomials, with respect to a linear change of variables in $\mathbb{R}^3$.

This classification relies on the corresponding classification of linear vector fields in $\mathbb{R}^3$, which in turn is nothing else but the well-known classification of linear mappings from $\mathbb{R}^3$ to itself. In what follows, we do not distinguish between linear vector fields and linear mappings.

In $\mathbb{R}^3$, it can indeed be easily verified that the mapping ϕ

$$l_x\,\partial/\partial x + l_y\,\partial/\partial y + l_z\,\partial/\partial z \xrightarrow{\phi} i(l_x\,\partial/\partial x + l_y\,\partial/\partial y + l_z\,\partial/\partial z).(\Omega)$$

establishes an isomorphism between the vector space of all linear vector fields $L = l_x\,\partial/\partial x + l_y\,\partial/\partial y + l_z\,\partial/\partial z$ in $\mathbb{R}^3$ and the vector space of all 2-forms whose coefficients are homogeneous linear polynomials in $\mathbb{R}^3$.

Moreover, this mapping ϕ commutes with a linear change of variables, provided that this change belongs to the special linear group $SL(3, \mathbb{R})$, which preserves the volume form Ω.

Indeed, let ρ_L be the image $\rho_L = \phi(L) = i(L).(\Omega)$ of a linear vector field L

under ϕ. Recall that, for any point u of $I\!R^3$ and any two vectors A and B of $I\!R^3$

$$\rho_L(u)(A,B) = \det(Lu, A, B).$$

Now, if T is an invertible linear mapping from $I\!R^3$ to itself, it can easily be shown that

$$\rho_L(Tu)(TA,TB) = \det(T)\, \rho_{T^{-1}LT}(u)(A,B).$$

Thus, up to a non-zero multiplicative constant, the classification of our 2-forms under the action of $GL(3, I\!R)$ is the same as the real linear classification of 3×3 real matrices; and, similarly, the classification under the action of $SL(3, I\!R)$ of the 2-forms, whose coefficients are linear polynomials, agrees with the corresponding classification of matrices.

Moreover, the 2-forms r, that we are interested in, are closed; this corresponds to the vanishing trace of the linear mappings $L = \phi^{-1}(r)$. Let us now state a general remark: a smooth function Q is a first integral of a smooth vector field X defined on $I\!R^3$ if and only if $dQ \wedge (i(X).(\Omega)) = 0$.

As we are interested in the description of the vector space of all cubic homogeneous polynomials Q which are first integrals of L, i. e. such that $dQ \wedge r = 0$, our classification is more detailled than the linear classification of vanishing trace linear mappings.

Proposition 1 *Consider the following nine canonical forms $L_1, \cdots, L_9$ of vanishing trace linear mappings of $I\!R^3$*

$$L_1 = \begin{pmatrix} \lambda & 0 & 0 \\ 0 & \mu & 0 \\ 0 & 0 & \nu \end{pmatrix} \quad \begin{matrix} \lambda \neq 0, \mu \neq 0, \nu \neq 0, \lambda \neq \mu \neq \nu \neq \lambda \\ \lambda + \mu + \nu = 0 \end{matrix}$$

$$L_2 = \begin{pmatrix} \lambda & 0 & 0 \\ 0 & -\lambda & 0 \\ 0 & 0 & 0 \end{pmatrix} \quad \lambda \neq 0$$

$$L_3 = \begin{pmatrix} a & -b & 0 \\ b & a & 0 \\ 0 & 0 & -2a \end{pmatrix} \begin{matrix} a \neq 0 \\ b \neq 0 \end{matrix}, \qquad L_4 = \begin{pmatrix} 0 & -b & 0 \\ b & 0 & 0 \\ 0 & 0 & 0 \end{pmatrix} \quad b \neq 0$$

$$L_5 = \begin{pmatrix} a & 0 & 0 \\ 0 & a & 0 \\ 0 & 0 & -2a \end{pmatrix} \quad a \neq 0, \qquad L_6 = \begin{pmatrix} a & 1 & 0 \\ 0 & a & 0 \\ 0 & 0 & -2a \end{pmatrix} \quad a \neq 0$$

$$L_7 = \begin{pmatrix} 0 & 1 & 0 \\ 0 & 0 & 1 \\ 0 & 0 & 0 \end{pmatrix}, \quad L_8 = \begin{pmatrix} 0 & 0 & 1 \\ 0 & 0 & 0 \\ 0 & 0 & 0 \end{pmatrix}, \quad L_9 = \begin{pmatrix} 0 & 0 & 0 \\ 0 & 0 & 0 \\ 0 & 0 & 0 \end{pmatrix}$$

Let L be a vanishing trace linear mapping of $I\!R^3$. Then L is conjugate in $GL(3, I\!R)$ with exactly one of the canonical forms. Moreover, if $L = TL_iT^{-1}$, where L_i is one of the canonical form, mapping T can be choosen in $SL(3, I\!R)$.

Proof. We built this classification according to the multiplicity and non-nullity of eigenvalues.

The nine cases then correspond to:

1. three different real non-zero eigenvalues,
2. three different real eigenvalues, one of them is 0,
3. three different non-zero eigenvalues, with only one of them real,
4. three different eigenvalues, one of them 0 and the two other conjugate,
5. a double (real, non-zero) eigenvalue, diagonalizable case,
6. a double (real, non-zero) eigenvalue, non-diagonalizable case,
7. a triple 0 eigenvalue, rank 2,
8. a triple 0 eigenvalue, rank 1,
9. a triple 0 eigenvalue, rank 0, i. e. the 0 matrix.

It is not difficult to verify, in each case, that matrices with an arbitrary real determinant exist in the commutant of a canonical matrix; then, the element T of $GL(3,\mathbb{R})$ such that $L = TL_iT^{-1}$ can be choosen in $SL(3,\mathbb{R})$. ∎

The above proposition implies immediately the following one.

Proposition 2 *Let r be a non-zero closed 2-form in $\mathbb{R}^3$, the coefficients of which are homogeneous linear polynomials. Under a linear change of variables belonging to $SL(3,\mathbb{R})$, r is conjugate to one and only one of the following canonical forms:*

1. $r = \lambda\, x\, dy \wedge dz + \mu\, y\, dz \wedge dx + \nu\, z\, dx \wedge dy$
   with $\lambda \neq 0, \mu \neq 0, \nu \neq 0, \lambda \neq \mu \neq \nu \neq \lambda$ *and* $\lambda + \mu + \nu = 0$
2. $r = \lambda\, x\, dy \wedge dz - \lambda\, y\, dz \wedge dx$, *with* $\lambda \neq 0$
3. $r = (a\, x - b\, y)\, dy \wedge dz + (b\, x + a\, y)\, dz \wedge dx - 2a\, z\, dx \wedge dy$
   with $a \neq 0, b \neq 0$
4. $r = -b\, y\, dy \wedge dz + b\, x\, dz \wedge dx$, *with* $b \neq 0$
5. $r = a\, x\, dy \wedge dz + a\, y\, dz \wedge dx - 2\, a\, z\, dx \wedge dy$, *with* $a \neq 0$
6. $r = (a\, x + y)\, dy \wedge dz + a\, y\, dz \wedge dx - 2\, a\, z\, dx \wedge dy$, *with* $a \neq 0$
7. $r = y\, dy \wedge dz + z\, dz \wedge dx$
8. $r = z dy \wedge dz$

It is now easy, although slightly cumbersome, to compute the general form of a cubic homogeneous polynomial Q such that $dQ \wedge r = 0$ in each of the previous cases. These computations are summarized in the following proposition.

Proposition 3 *In each case of the previous classification, the corresponding vector space of all third degree homogeneous real polynomials Q such that $dQ \wedge r = 0$ is generated by the following polynomials:*

1. $Q_1 = xyz$,
2. $Q_1 = xyz$ *and* $Q_2 = z^3$,
3. $Q_1 = (x^2 + y^2)z$,
4. $Q_1 = (x^2 + y^2)z$ *and* $Q_2 = z^3$,
5. $Q_1 = x^2 z$, $Q_2 = xyz$ *and* $Q_3 = y^2 z$,
6. $Q_1 = y^2 z$,
7. $Q_1 = z^3$ *and* $Q_2 = z(y^2 - 2xz)$,
8. $Q_1 = y^3$, $Q_2 = y^2 z$, $Q_3 = yz^2$ *and* $Q_4 = z^3$.

After these algebraic preliminaries, we pass on to the heart of the matter.

Let us denote by w an arbitary exterior 1-form defined in $\mathbb{R}^3$, whose coefficients are homogeneous quadratic polynomials in the space variables x, y and z; r stands for the closed 2-form dw and the integrability condition $w \wedge r = 0$ holds.

Let us now describe how inverses of cubic polynomials can be used as integrating factors for such exterior 1-forms.

Proposition 4 *If the inverse of a non-zero homogeneous polynomial Q is a integrating factor for w, then Q satisfies $dQ \wedge r = 0$.*

Proof. The hypothesis means that the 1-form w/Q is closed on the open set $W = \{u \in \mathbb{R}^3, Q(u) \neq 0\}$, i. e. that the following identity holds on W (cf. (4))

$$d\,(w/Q) = (1/Q^2)(Q dw - dQ \wedge w) = 0.$$

Differentiating the numerator yields the result

$$0 = d\,(Q dw - dQ \wedge w) = 2dQ \wedge r.$$

That completes the proof. ∎

Proposition 5 *Let P be the cubic homogeneous polynomial $P = i(I).(w)$. If P does not vanish identically, then $1/P$ is a integrating factor for w; if $P = 0$, then for every non-zero cubic homogeneous polynomial Q such that $dQ \wedge r = 0$, $1/Q$ is a integrating factor for w.*

Proof. When P is different from 0, it suffices, in order to show that $1/P$ is an integrating factor w, to prove that the numerator of $d(w/P)$ is equal to 0, i. e. that $Pdw = dP \wedge w$. Thanks to the generalized Euler's formula (7), applied to the homogeneous 1-form w, the following identity holds

$$3w = i(I).(r) + dP. \tag{9}$$

Exterior multiplication by w yields

$$0 = i(I).(r) \wedge w + dP \wedge w.$$

The desired equality then follows from the inner product by I of the identity $w \wedge r = 0$ (cf. (5))

$$0 = i(I).(w \wedge r) = i(I).(w)\, r + i(I).(r) \wedge w.$$

Comparing the last two equalities, one obtains $P\,dw = dP \wedge w$ as needed.

When P is equal to 0, formula (9) allows us to define w from its exterior differential $dw = r$ by $3w = i(I).(r)$. To prove that the inverse $1/Q$ is an integrating factor for w it suffices to show that

$$Qr = dQ \wedge w.$$

To prove this identity, we apply the inner product by I to equality $dQ \wedge r = 0$ (cf. (5) and (6)):

$$0 = i(I).(dQ \wedge r) = i(I).(dQ)\, r - dQ \wedge i(I).(r) = 3Q\, r - 3dQ \wedge w.$$

And the proof is now complete. ∎

The following proposition is the key result to show that an exterior integrable 1-form, whose coefficients are homogeneous quadratic polynomials, has a global first integral. To formulate this proposition in a concise manner, as explained at the end of section 3, by functions we will not only understand real-valued functions, but also circle-valued ones.

Proposition 6 *Let r be a non-zero closed 2-form defined in $\mathbb{R}^3$, whose coefficients are linear homogeneous polynomials, and let Q be a non-zero cubic homogeneous polynomial such that $dQ \wedge r = 0$.*

Then the closed form $(1/Q)i(I).(r)$ is exact on every connected component of the open dense set $W = \{u \in \mathbb{R}^3, Q(u) \neq 0\}$.

Proof. The above mentioned property does not depend on a linear change of variables; it then suffices to show it for any canonical form of r listed in Proposition 2.

In all cases, the connected components of W are easily shown either to be simply connected, or to have a fundamental group isomorphic to the group $\mathbb{Z}$ of all relative integers.

In the first case, a closed 1-form is exact, which means that a real-valued function F such that $dF = (1/Q)i(I).(r)$ does exist. The same is true in the second case, if the value k of the integral of $(1/Q)i(I).(r)$ on a closed curve corresponding to a generator of the fundamental group is equal to 0. If this value k is not zero, the 1-form is the derivative of a function from W to the circle $\mathbb{R}/k\mathbb{Z}$. This situation cannot be avoided in some cases of the classification (cases 3 and 5). ∎

We need now a proposition to ensure that, if the cubic polynomial Q yields an integrating factor for a 1-form w, then the subset $\{Q \neq 0, w \neq 0\}$ consists of global leaves of the foliation defined by w.

Proposition 7 *Let w be a smooth integrable 1-form defined on some open subset U of $\mathbb{R}^3$ on which it does not vanish. Let Q be a smooth function defined on U such that $Q\,dw = dQ \wedge w$, i. e. such that $1/Q$ is an integrating factor of w on the set $\{u \in U, Q(u) \neq 0\}$. Let m and m' be the beginning and the end of a smooth path lying in some leaf of the foliation of U defined by w and suppose that $Q(m) = 0$. Then, $Q(m')$ also vanishes.*

Proof. Because w does not vanish on U, a smooth vector field X such that $i(X).(w) = 1$ can be defined on U. The inner product by X of the identity $Q\,dw = dQ \wedge w$ yields (cf. (5))

$$Q\,i(X).(dw) \;=\; i(X).(dQ)\,w \;-\; dQ. \tag{10}$$

Consider now a smooth path lying in the leaf of m from m to m'. This path is a smooth mapping ϕ from some real interval $[0, a]$ to U. Denote by ψ the function $Q \circ \phi$.

Because the tangent vector $d\phi/dt$, $0 \leq t \leq a$, is everywhere tangent to the leaf, the previous equality (10) yields by an inner product by $d\phi/dt$

$$\psi(t)\,i(d\phi/dt).(i(X[\phi(t)]).(dw[\phi(t)])) \;=\; -d\psi/dt.$$

The continuous function $i(d\phi/dt).(i(X[\phi(t)]).(dw[\phi(t)]))$ is bounded on the compact interval $[0, a]$ and we deduce an a priori estimate

$$|\,d\psi/dt\,| \leq C\,|\,\psi\,| \tag{11}$$

everywhere on the interval $[0, a]$ with some positive constant C.

But as $Q(m) = 0$, $\psi(0) = 0$ and, thanks to a Gronwall lemma, $\psi(a)$ is also 0, i. e. $Q(m') = 0$, which completes the proof. ∎

Given an integrable 1-form w, a *first integral* of w is any smooth function Φ such that $d\Phi = \alpha w$ for some function α and such that Φ is not constant on any open set.

We can now conclude with two theorems.

Theorem 1 *Let w be an integrable non-zero 1-form defined in $\mathbb{R}^3$, whose coefficients are homogeneous quadratic polynomials. This form has a global first integral defined on a open dense subset V of $\mathbb{R}^3$ consisting of global leaves of the foliation given by w.*

Proof. It follows from the identity (9) and from Propositions 3–7. More precisely, let as usual P be the inner product $P = i(I).(w)$.

If $P \neq 0$, $1/P$ is an integrating factor for w and w/P can be written

$$w/P = (1/3)(dP/P + (i(I).(r))/P)$$

where $r = dw$.

The closed form dP/P is exact and the 1-form $(i(I).(r))/P$ is exact on the connected components of the open set $V = \{u \in \mathbb{R}^3, P(u) \neq 0, w(u) \neq 0\}$.

If $P = 0$, then w is equal to $(1/3)i(I).(r)$ (cf. (9)), where $r = dw$. Let then Q be a non-zero homogeneous cubic polynomial such that $dQ \wedge r = 0$; r does not vanish identically and such polynomials do exist according to the previously described classification given in Proposition 3.

Function $1/Q$ is then an integrating factor of w on the open set V defined by $V = \{u \in \mathbb{R}^3, Q(u) \neq 0, w(u) \neq 0\}$ and the 1-form $i(I).(r)/Q$ is exact on the connected components of V.

In both cases, a closed form can be integrated up to a real or circle valued function on the connected components of the set V.

Finally Proposition 7 shows that the open set V consists of global leaves of the foliation defined by w on the open set $U = \{u \in \mathbb{R}^3, w(u) \neq 0\}$. ∎

Let us now recall that a non-zero 1-form $w = w_x dx + w_y dy + w_z dz$ defined in $\mathbb{R}^3$, whose coefficients are homogeneous real polynomials, is irreducible if w_x, w_y and w_z have no non-trivial polynomial common factor.

If a polynomial S is a common factor of the coefficients of w, then the reduced form w/S defines the same foliation as w, but perhaps on a larger open subset of $\mathbb{R}^3$; it is therefore natural to consider irreducible 1-forms.

The next proposition shows the interest of this assumption for the trajectories of vector fields that are orthogonal to such a form.

Proposition 8 *Let w be an integrable 1-form defined on $\mathbb{R}^3$, whose coefficients are homogeneous polynomials of the same degree. Let X be a non-zero vector field defined on $\mathbb{R}^3$, whose coefficients are homogeneous polynomials of the same* *degree; suppose that the identity $i(X).(w) = 0$ holds on $\mathbb{R}^3$ and that w is irreducible.*

Then, the subset $\{w = 0\}$ is locally X-invariant, so that the open subset $\{w \neq 0\}$ consists of complete trajectories of the field X.

Proof. Consider the integrability relation $w \wedge dw = 0$ and take its inner product by X to get $w \wedge (i(X).(dw)) = 0$.

The polynomial 1-forms w and $i(X).(dw)$ are then colinear on some non-empty open subset of $\mathbb{R}^3$, and there exists an irreducible rational function N/D such that $i(X).(dw) = (N/D)\,w$, i. e. such that $N\,w = D\,i(X).(dw)$.

Because w is irreducible, the polynomial D is a constant and we thus get

$$i(X).(dw) = N_1\,w \tag{12}$$

where N_1 is some polynomial.

Recall now the well known formula for the Lie derivative $\mathcal{L}_X w$ of an exterior form with respect to a vector field X (cf. [53, 58])

$$\mathcal{L}_X w = i(X).(dw) + d[i(X).(w)].$$

As $i(X).(w) = 0$, $i(X).(dw)$ is the Lie derivative $\mathcal{L}_X w$. Taking into account (12), an a priori estimate, like (11), can then be established showing that, if $w(m) = 0$, then w remains equal to 0 along the trajectory of the field X passing through m, which proves the result.

As proven by easy examples, the irreducibility assumption is essential here. ∎

Theorem 2 *Any two compatible linear vector fields L_1 and L_2 defined in $\mathbb{R}^3$ have a common global first integral defined on a open dense subset of $\mathbb{R}^3$ consisting of complete trajectories of both fields.*

Proof. Consider the 1-form $w = i(L_1).(i(L_2).(\Omega))$. If $w = 0$, the fields are proportional and it is a well known fact that a global first integral exists for a linear field.

If $w \neq 0$, Theorem 1 shows that an integrating factor $1/Q$ exists for w. As $i(L_i).(w) = 0$ for $i = 1, 2$, the function F, defined on $V = \{w \neq 0, Q \neq 0\}$ and such that dF is proportional to w, is a first integral for both fields.

To prove that this integral is a global one, it remains to be shown that the open dense subset V consists of complete trajectories of L_1 and L_2. It therefore suffices to show that the subsets $\{Q = 0, w \neq 0\}$ and $\{w = 0\}$ are locally L-invariant, L being one of the two linear vector fields L_1, L_2, or more generally some vector field with polynomial coefficients such that $i(L).(w) = 0$.

As far as the first subset is concerned, take the inner product by L of the usual equality $Q\,dw = dQ \wedge w$ to get

$$Q\,i(L).(dw) = i(L).(dQ)\,w.$$

As w is supposed to be different from 0 in some neighborhood of a point m at which $Q(m) = 0$, this relation yields one more time an a priori estimate like (11) and Q vanishes on the L-trajectory around m.

As far as the second subset is concerned, Proposition 8 gives the proof if w is irreducible.

If w is not irreducible, we replace it by a simpler irreducible 1-form $w' = w/S$, whose coefficients are of the same degree 0 or 1. To complete the proof in this case, we must find an integrating factor $1/Q$ for w such that S is a factor of Q. Then the open subset on which the first integral is defined is $\{Q' \neq 0, w' \neq 0\}$ where $Q' = Q/S$.

In the case where $Q = i(I).(w) \neq 0$, S is of course a factor of Q. Otherwise, $(1/3)\,i(I).(r)$ is irreducible in each of the eight cases but two (5 and 8) of the classification given by Proposition 2. These are precisely the two cases in which one of the eigenvalues has a two-dimensional eigenspace. It is nevertheless possible to choose, in the vector space of all cubic polynomials Q such that $dQ \wedge r = 0$, a non-zero polynomial which is a multiple of the greatest common divisor of the coefficients of $i(I).(r)$. ∎

Let us finally note that the present proof of the existence of a common first integral also gives an algorithm to find it. For each of the canonical forms $L_1, \cdots, L_8$ of $r = dw$, it is easy to see that the corresponding first integral is expressed in finite terms. In what concerns case L_9, it follows from identity (9) that the cubic homogeneous polynomial $P = i(I).(w)$ is a common global first integral.

Let us underline that the appearence of an *arctan* means that the integral is in fact circle-valued.

5 An example of non-integrability in $\mathbb{R}^4$

We shall now find three compatible linear vector fields L_1, L_2 and L_3 in $\mathbb{R}^4$ which generate together a three dimensional foliation with dense leaves. This foliation is also defined by the 1-form $w = i(L_1).(i(L_2).(i(L_3).(\Omega)))$, where Ω is the volume 4-form $\Omega = dx\,dy\,dz\,dt$. The coefficients of w are cubic homogeneous polynomials and w satisfies the integrability condition $w \wedge dw = 0$.

In our example, a local integrating factor for w is easily found: it is equal to $1/P$, where $P = i(I).(w)$, I being the radial vector field $I = x\,\partial/\partial x + y\,\partial/\partial y + z\,\partial/\partial z + t\,\partial/\partial t$.

The local situation is thus very similar to the three dimensional one.

On the other side, the global non-integrability relies on the density of all non-singular leaves of the foliation, which impedes the existence of a continuous common first integral for the three fields.

The example. Let α and β be two rationally independent real numbers. Consider now the three linear vector fields L_1, L_2 and L_3:

$$\begin{aligned} L_1 &= (x-y)\,\partial/\partial x + (x+y)\,\partial/\partial y \\ L_2 &= (z-t)\,\partial/\partial z + (z+t)\,\partial/\partial t \\ L_3 &= -\alpha y\,\partial/\partial x + \alpha x\,\partial/\partial y - \beta t\,\partial/\partial z + \beta z\,\partial/\partial t. \end{aligned}$$

These three vector fields commute with one another, i. e. the three pairewise Lie brackets vanish, and they are thus compatible.

The 1-form $w = i(L_1).(i(L_2).(i(L_3).(\Omega)))$ is equal to

$$\begin{aligned} w &= \beta\,(z^2+t^2)((x+y)\,dx + (y-x)\,dy) \\ &- \alpha\,(x^2+y^2)((z+t)\,dz + (t-z)\,dt). \end{aligned}$$

Let then P be the fourth degree homogeneous polynomial

$$P = i(I).(w) = (\beta-\alpha)\,(x^2+y^2)(z^2+t^2).$$

This polynomial is different from 0 outside of the two 2-dimensional planes $x^2+y^2=0$ and $z^2+t^2=0$. Function $(\beta-\alpha)/P$ is an integrating factor of w on the complement of these two planes in $\mathbb{R}^4$.

The closed form $(\beta-\alpha)\,w/P$ is equal to

$$\begin{aligned} (\beta-\alpha)\,w/P &= \beta\,((x+y)\,dx + (y-x)\,dy)/(x^2+y^2) \\ &- \alpha\,((z+t)\,dz + (t-z)\,dt)/(z^2+t^2) \end{aligned}$$

and it can be formally integrated up to

$$\frac{\beta}{2}\log(x^2+y^2) - \beta\arctan(y/x) - \frac{\alpha}{2}\log(z^2+t^2) + \alpha\arctan(t/z).$$

The rational independence of α and β then prevents us from giving any signification to the sum "$\beta\arctan(y/x) - \alpha\arctan(t/z)$".

We now give a more precise explanation of this fact by showing that every leaf of the foliation of the open invariant set $V = \{m \in \mathbb{R}^4,\ P(m) \neq 0\}$ defined by the 1-form w is dense in V.

To prove this, it suffices, given two points m and m' of V, to find a path starting from m, lying in the leaf of the foliation passing through m, and ending in a point arbitrarily close to m'. Denote by (x,y,z,t) the coordinates of m and by (x',y',z',t') those of m'. Such a path then consists of three pieces.

The first piece is a path from (x, y, z, t) to (x_1, y_1, z, t), where $x_1{}^2 + y_1{}^2 = x'^2 + y'^2$, the second one is a path from (x_1, y_1, z, t) to (x_1, y_1, z_2, t_2), where $z_2{}^2 + t_2{}^2 = z'^2 + t'^2$; and the third path goes from (x_1, y_1, z_2, t_2) to (x_3, y_3, z_3, t_3), where $x_3{}^2 + y_3{}^2 = x'^2 + y'^2$, $z_3{}^2 + t_3{}^2 = z'^2 + t'^2$, and where (x_3, y_3, z_3, t_3) is close to (x', y', z', t').

Each of the three pieces follows the trajectory of one of the three linear vector fields L_1, L_2 an L_3. Taking polar coordinates (ρ, θ) in th xy-plane and (ρ', θ') in the zt-plane, we get

$$\begin{aligned} L_1 &= \rho\, \partial/\partial\rho + \partial/\partial\theta \\ L_2 &= \rho'\, \partial/\partial\rho' + \partial/\partial\theta' \\ L_3 &= \alpha\, \partial/\partial\theta + \beta\, \partial/\partial\theta'. \end{aligned}$$

Trajectories of L_1 are the logarithmic spirals $\{\rho = Ce^{\theta},\ z = z_0,\ t = t_0\}$ and those of L_2 are the logarithmic spirals $\{\rho' = Ce^{\theta'},\ x = x_0,\ y = y_0\}$ while, due to the rational independence of α and β, every trajectory of L_3 is dense in the two dimensional torus $\{\rho = \rho_0,\ \rho' = \rho'_0\}$ in which it lies.

As suggested by P. Cartier, a similar construction can be done in higher dimensions. Indeed, there exist $n + 1$ compatible linear vector fields in $\mathbb{R}^{2n}$, which define together a foliation of an invariant dense open subset V of $\mathbb{R}^{2n}$, whose every leaf is dense.

Therefore choose cartesian coordinates x_i, y_i, for i between 1 and n and corresponding polar coordinates ρ_i, θ_i and consider n rationally independent real numbers $\alpha_1, \cdots, \alpha_n$. Define then n "spiral" linear vector fields $L_1, \cdots, L_n$ by $L_i = \rho_i\, \partial/\partial\rho_i + \partial/\partial\theta_i$ and another linear vector field L by $L = \sum \alpha_i\, \partial/\partial\theta_i$. The open set V is the set where all ρ_i are positive.

Following trajectories of the L_i, we can draw a path from an arbitrary point m of V to some point m' of an arbitrary fixed n-dimensional torus $\{\rho_1 = r_1, \cdots, \rho_n = r_n\}$ without leaving the leaf of m; thereafter, the L-trajectory passing through m' approaches arbitrarily any given point m'' of the torus.

This construction can be used to show that there exist $n+2$ compatible linear vector fields in $\mathbb{R}^{2n+1}$, which define together a foliation of a dense open subset V of $\mathbb{R}^{2n+1}$, whose every leaf is dense. For this aim, it suffices to consider the $(2n + 1)$-th coordinate t and to add the $(n + 2)$-th linear vector field $L' = t\, \partial/\partial t$ to the previous ones. In this case, V is the invariant dense open set where all ρ_i are positive and where $t \neq 0$ and the n-dimensional torus are defined by $\{\rho_1 = r_1, \cdots, \rho_n = r_n, t = t_0\}$.

6 Questions and final remarks

The positive result on the integrability of two arbitrary compatible linear vector fields in $I\!R^3$ and the negative result that non-integrability can occur for thre compatible (even commuting) linear vector fields in $I\!R^4$ lead in a natural way to many interesting questions. Let us formulate some of them.

Let us fix natural numbers k and n, $2 \le k < n$. Describe, or rather classify, the set of all k-tuples of compatible linear vector fields in $I\!R^n$, which are linearly independent at some point of $I\!R^n$. Such a k-tuple will be noted shortly CLVF.

It will also be interesting to study the set of all k-tuples of CLVF, viewed as an algebraic manifold.

In fact we are rather interested in the global foliation of $I\!R^n$ (of the $(n-1)$-dimensional real projective space) corresponding to such a k-tuple in virtue of Frobenius integrability theorem. For $k = 2$ and $n = 3$, such a classification can in principle be deduced from our results. But in general the problem seems to be quite intricate (cf. [37]).

Let us note, by the way, that the important Hermann-Nagano refinement of Frobenius integrability theorem ([33, 57], cf. also [28] and Sec. 3.1 of [78]) applies to our framework.

Given a k-tuple $\mathcal{A}$ of CLVF in $I\!R^n$, we define its index, $\mathrm{ind}(\mathcal{A})$, as the number of its global functionally independent first integrals $(0 \le \mathrm{ind}(\mathcal{A}) \le n-k)$. More precisely, the preceeding problem can be stated as a problem of the description of the level set of the function ind, when k and n are fixed.

From a more algebraic point of view, it seems that particular attention should be paid to the first integrals whose gradients consist of rational functions. As proved before, this is always the case when $k = 2$ and $n = 3$. It is rather doubtful that this is a general feature (cf. [13, 32]); nevertheless such integrals seem to appear quite frequently.

Given a k-tuple of CLVF in $I\!R^n$, one can ask for a maximal dimension of the closure of an individual leaf of the assiociated k-dimensional foliation and its relation to the number of functionally independent global first integrals.

For a given $n \ge 4$ one can also ask for a minimal number $k = k(n) \ge 2$ such that there exists a k-tuple of CLVF with dense leaves. As follows from Sec. 5,

$$k(n) \le \begin{cases} p+1 & \text{if } n = 2p \\ p+2 & \text{if } n = 2p+1 \end{cases}$$

where $p \ge 2$. Are $p+1$ and $p+2$ the true lower bounds ? If not, determine $k(n)$. In particular, is it true that two CLVF in $I\!R^4$ always have a global first integral ?

Although the description of all k-tuples of CLVF in $I\!R^n$ seems to be quite complicated, there is a natural subclass of them, which seems to be much easier to handle.

A k-tuple $(A_1, \cdots, A_k)$ of CLVF in $\mathbb{R}^n$ will be called $\mathbb{R}$-compatible ($\mathbb{R}$-CLVF) if it generates a k-dimensional real Lie algebra, i. e. if for every $1 \leq i < j \leq k$

$$[A_i, A_j] = \sum_{r=1}^{k} C_{i,j}^r A_r$$

for some real numbers $\{C_{i,j}^r\}$.

The foliation associated to k-tuples of $\mathbb{R}$-CLVF have a simple description in terms of linear representations of Lie algebras.

Let $\mathcal{A}$ be a k-dimensional real Lie algebra and let $\pi : \mathcal{A} \longrightarrow \mathcal{L}(\mathbb{R}^n)$ be an injective linear representation of $\mathcal{A}$ in the space $\mathcal{L}(\mathbb{R}^n)$ of all linear mappings of $\mathbb{R}^n$ into itself. To any basis $(a_1, \cdots, a_k)$ of $\mathcal{A}$ we assiociate the k-tuple $(A_1, \cdots, A_k)$ of $\mathbb{R}$-CLVF defined on $\mathbb{R}^n$ by

$$A_k(u) = \pi(a_k)(u)$$

for $u \in \mathbb{R}^n$.

Although $(A_1, \cdots, A_k)$ depends on the choice of the basis $(a_1, \cdots, a_k)$, the associated global k-dimensional foliation of $\mathbb{R}^n$ (at least of an open dense subset of it) does not depend on it, but only depends on the linear representation π.

When π in not injective, one considers the global foliation of $\mathbb{R}^n$ associated to the induced representation

$$\tilde{\pi} : \mathcal{A}/\mathrm{Ker}(\pi) \longrightarrow \mathcal{L}(\mathbb{R}^n)$$

which is injective.

Thus to any linear representation π of a real Lie algebra $\mathcal{A}$, one associates in a canonical way a foliation of the space in which the representation acts.

Consequently, all properties of these foliations can be described in terms of algebraic properties of $\mathcal{A}$ and π, in particular the density of leaves, the existence or non-existence of first integrals and so on.

Explanation of these points is a very important problem. As an example, let us note that the commuting linear vector fields on $\mathbb{R}^n$ with dense leaves described in Sec. 5 arise from a representation of the commutative Lie algebra $\mathbb{R}^n$, but some other linear representations of the same Lie algebra have many independent first integrals.

The study of $\mathbb{R}$-CLVF would only be a first step because, already in $\mathbb{R}^3$, there exist foliations corresponding to some pairs of CLVF that do not correspond to any pair (M, N) of $\mathbb{R}$-CLVF.

Consider indeed such an example, due to W. Hebisch: the foliation in spheres centered at the origin. This foliation corresponds, for instance, to the pair (L_x, L_y) of CLVF, where $L_x = y\,\partial/\partial z - z\,\partial/\partial y$ and $L_y = z\,\partial/\partial x - x\,\partial/\partial z$, which generate the rotations around the x-axis and the y-axis respectively.

Let us note that a linear vector field whose trajectories lie on spheres centered at 0 is always given by a matrix in so$(3, \mathbb{R})$. Consider now two non-proportional such linear vector fields M and N. Then, their Lie bracket $[M, N]$ is not a linear combination of them. This proves that the foliation in spheres cannot correspond to a pair of $\mathbb{R}$-CLVF.

It will be interesting to distinguish geometrically the global foliations corresponding to $\mathbb{R}$-CLVF among those corresponding to CLVF.

A careful examination proves that pairs (A, B) of non-commuting $\mathbb{R}$-CLVF in $\mathbb{R}^3$ are rather rare. Indeed, taking into account that there exists only one non-commutative 2-dimensionnal Lie algebra, one can assume, without any restriction of generality that $[A, B] = A$. Then, $\mathrm{tr}(A) = 0$ and B can only be found in cases L_7 and L_8 of the classification of Prop. 1.

It will be interesting to obtain a detailled analysis of the same phenomenon in higher dimension.

It is also natural to ask if there exist some classes of Lie algebras of dimension at least three for which the global foliation corresponding to an arbitrary injective linear representation of any algebra of the class always has a first integral.

As we were informed by W. Hebisch, an example of such a class is given by compact semisimple real Lie algebras, where a quadratic first integral always exists.

Indeed, according to H. Weyl's theorem, for an injective linear representation ϕ of such an algebra L, any connected Lie group corresponding to $\phi(L)$ is compact. The result now follows from a standard argument: integration of the translates of some positive definite quadratic form with respect to the Haar measure (see [82, 72]).

It is worth noting that the above remark admits a far reaching development in invariant theory, a topic intimately related to ours (see, for instance, Chap. 5 of [25] and also [67] for a development in another direction).

Although up to now the complete classification of all Lie algebras does not exist, it exists for small dimensions [59]. In particular, one has exactly nine types of three dimensional Lie algebras (see [59] and also [18]). The first stage in the realization of the above program will be the careful study of the possibilities occuring for their linear representations in low dimensional Euclidean spaces.

Three variations of our topics arize in a natural way.

First, instead of considering real linear vector fields, one can study affine vector fields, i. e. vector fields $A(u) = Bu + b$ where $A \in \mathcal{L}(\mathbb{R}^n)$ and $b \in \mathbb{R}^n$.

Secondly, the complex space C^n and the complex time can be considered instead of their real counterparts.

Finally, as the Frobenius integrability theorem remains valid in infinite dimensional setting (cf. [5, 7, 15, 19, 48]), it is natural to ask similar questions about it.

In fact, such problems can be considered in every specific framework where a counterpart of Frobenius integrability theorem is proved. See, for example, Sec. 9.4 of [5] for manifolds over valued fields with a non-zero characteristic.

Let us finish with some scattered remarks, mainly of historical nature.

We apologize for the complete lack of any explicit mention of contributions of differential algebra to the problem of the integration in finite terms of ordinary differential equations.

Our main tool was the search of an integrating factor written in finite terms, for integrable 1-forms. The method of the integrating factor goes back to L. Euler ([22, 23], cf. also [75]).

In particular, for $n = 2$, he knew the first part of our proposition 5. The second one is implicit in C. G. Jacobi's paper [36]. The search of integrating factors and the related problem of the search of a first integral for 1-forms, mainly in two dimensions, were very intensively studied during the nineteenth century. This is clearly shown in the treatises of A. R. Forsyth [26] and E. Goursat [30]. Let us stress the wealth of examples collected in the problems at the end of the corresponding chapters of these treatises.

This research culminated at the end of the nineteenth century in the fundamental works of G. Darboux [13], P. Painlevé [60, 61, 62, 63, 64, 65] and A. N. Korkine [41, 42, 43, 44]. Already in Darboux's paper [13] the close relations with algebraic geometry appears. This aspect was strongly emphazised by P. Painlevé [60, 61, 62, 63, 64, 65] and H. Poincaré [68, 69, 70].

After this fruitful period, partly as a consequence of the growing influence of the qualitative methods, this research was (almost) completely abandoned, although many open problems remain.

It is a pity that nowadays no published survey is available on the history and results obtained in this area up to the second world war.

One of the inherent difficulties in the preparation of such a survey is the fact that many interesting texts in this and related fields were published in Russian, some in journals and books which are now very difficult to find. It is worth noting that the strong activity of russian mathematicians in this area during the second half of the nineteenth century was the direct consequence of the great interest of P. L. Chebyshev in these problems [12, 29].

The lack of such a survey is only partly compensated by the historical notes at the end of B. M. Koialovich's book [38], which provides a very interesting annotated bibliography. The same remains true for D. Morduhai-Boltovskoi's book [55] and E. Vessiot's survey [80].

On the other hand, the unpublished thesis [52] of N. V. Lokot' is a very comprehensive study of the history of integration in finite terms of elementary functions.

Let us also quote [49] and the recent books [2, 16, 17] and [40]. In a forthcoming paper [6], we will publish an incomplete but nonetheless quite extensive bibliography of the subject up to the second world war.

Until recent times, integration in finite terms and related topics seemed to be marginal compared to the main stream of mathematics. But now, with a revival of the interest in the explicit integration of differential equations (cf. [58, 73, 34]), in the problem of non-integrability (cf. [46]) and above all with the development of the applications of computer algebra to the automatised study of differential equations (cf. [45, 20, 74, 79]), we hope that many of these, now almost completely forgotten works, will regain their importance and will find a contemporary understanding and development.

Let us stress that the Liouville theory of integration in finite terms resulted one hundred and fifty years later in the computer algebra programs for the integration of elementary functions (cf. [14] and bibliography therein).

The Liouville theory, together with the ideas of S. Lie, E. Picard and E. Vessiot on the Galois theory of differential equations finally resulted one century later in the computer algebra programs for the integration of second order linear differential equations with rational coefficients (cf. [20] and [45]).

As the algorithmic search for integrating factors written in finite terms is capital in the effective study of differential equations, one can look forward to an algorithm and then to a computer algebra program for an automatised search for them.

A. N. Korkine's papers and B. M. Koialovich's book will surely be very useful for this purpose. Recently the algorithm of B. M. Koialovich was substantial for the elaboration of a computer algebra program used for the discovery of new cases of integrability of Abel's differential equation of second kind $yy' - y = R(x)$ for hundreds of appropriate functions R [83].

References

[1] V. V. Amel'kin, Autonomous and linear multidimensional differential equations (in Russian), ed. Universitetskoe, Minsk (1985).

[2] V. V. Amel'kin, N. A. Lukashevich, A. P. Sadovskii, Nonlinear oscillations in two-dimensional systems (in Russian), ed. of Bielorussian State University, Minsk (1982).

[3] P. Basarab-Horwath, A classification of vector fields in involution with linear fields in $\mathbb{R}^3$, preprint, Linköping (1990).

[4] P. Basarab-Horwath, S. Wojciechowski, Classification of linear vector fields in $\mathbb{R}^3$ (to be published).

[5] N. Bourbaki, Eléments de mathématique, Fasc. XXXVI : Variétés différentiables et analytiques, Fascicule de résultats, Hermann, Paris (1971).

[6] B. Bru, J. Moulin Ollagnier, J.-M. Strelcyn, Integration in finite terms : selected bibliography up to the second world war (to be published).

[7] H. Cartan, Formes différentielles, Hermann, Paris (1967).

[8] C. Camacho, A. Lins Neto, The topology of integrable differentiable forms near a singularity, *Public. Math. IHES* **55** (1982), 5–36.

[9] D. Cerveau, Equations différentielles algébriques : remarques et problèmes, *J. Fac. Sci. Univ. Tokyo, Sect. I-A, Math.*, **36** (1989), 665–680.

[10] D. Cerveau, F. Maghous, Feuilletages algébriques de C^n, *C. R. Acad. Sci. Paris*, **303** (1986), 643-645.

[11] D. Cerveau, J. F. Mattei, Formes intégrables holomorphes singulières, *Astérisques*, **97** (1982).

[12] P. L. Chebyshev (P. L. Tchebychef), Œuvres, Vol. I, II, Chelsea Publ. Comp., New York.

[13] G. Darboux, Mémoire sur les équations différentielles algébriques du premier ordre et du premier degré, *Bull. Sc. Math. 2ème série* **t. 2** (1878), 60–96, 123–144, 151–200.

[14] J. H. Davenport, On the Integration of Algebraic Functions, *Lecture Notes in Computer Science* **102**, Springer-Verlag, Berlin (1981).

[15] J. Dieudonné, Eléments d'analyse, tome I, Gauthier-Villars, Paris (1971).

[16] V. A. Dobrovolskii, Outline of the development of analytical theory of differential equations (in Russian), ed. Vischa Shkola , Kiev (1974).

[17] V. A. Dobrovolskii, Vasilii Petrovich Jermakov (in Russian), Nauka, Moscow (1981).

[18] B. A. Dubrovin, A. T. Fomenko, S. P. Novikov, Modern Geometry - Methods and Applications, vol. I *Graduate Texts in Math.* **93**, Springer-Verlag, Berlin (1984).

[19] G. Duchamp, On Frobenius integrability theorem : the analytic case (to be published).

[20] A. Duval, M. Loday-Richaud, A propos de l'algorithme de Kovačic, preprint, Orsay (1989).

[21] F. Engel, K. Faber, Die Liesche Theorie der partiellen Differentialgleichungen erster Ordnung, B. G. Teubner, Leipzig (1932).

[22] L. Euler, De aequationis differentialibus secundi gradus, *Nov. Comm. Acad. Sci. Petrop.* **7** (1761), 163–202.

[23] L. Euler, Institutiones calculi integralis, vol. 3, Petropoli (1770); reprinted in Opera Mathematica, vol. 13, B. G. Teubner, Leipzig (1914).

[24] A. Fais, Intorno all'integrazione delle equazioni differenziali totali di 1° ordino e di 1° grado, *Giornale di Matematiche* **13** (1875), 344–351.

[25] J. Fogarty, Invariant Theory, W. A. Benjamin, New York (1969).

[26] A. R. Forsyth, Theory of differential equations, vol 1–6, Cambridge University Press (1890–1906), reprinted by Dover Public., New York (1959).

[27] I. V. Gaishun, Completely solvable multidimensional differentiable equations (in Russian), Ed. Nauka i Technika, Minsk (1983).

[28] J. P. Gauthier, Structure des systèmes non-linéaires, Ed. du CNRS, Paris (1984).

[29] V. V. Golubiev, The work of P. L. Chebychev on integration of algebraic functions (in Russian), in Scientific heritage of P. L. Chebychev, Part 1 : Mathematics, Ed. of Acad. of Sciences of the USSR, Moscow-Leningrad (1945), 88–121.

[30] E. Goursat, Leçons sur l'intégration des équations aux dérivées partielles du premier ordre, second edition, Hermann, Paris (1921).

[31] E. Goursat, Cours d'analyse mathématique, Vol. 2, Paris (4th. ed. 1924). English translation : A Course of Mathematical Analysis Vol II , Part Two : Differential Equations, Dover Public. , New York (1959).

[32] B. Grammaticos, J. Moulin Ollagnier, A. Ramani, J.-M. Strelcyn, S. Wojciechowski, Integrals of quadratic ordinary differential equations in $\mathbb{R}^3$: the Lotka-Volterra system, *Physica A* **163** (1990), 683–722.

[33] R. Hermann, Cartan connections and the equivalence problem for geometric structures, in Contributions to Differential Equations **3** (1964), 199–248.

[34] J. Hietarinta, Direct methods for the search of the second invariant, *Phys. Reports* **147** (2) (1987), 87–154.

[35] E. L. Ince, Ordinary differential equations, Dover Public., New York (1956).

[36] C. G. J. Jacobi, De integratione aequationis differentialis $(A + A'x + A''y)(x\,dy - y\,dx) - (B + B'x + B''y)dy + (C + C'x + C''y)dx = 0$, *Crelle J. für Reine and angew. Math.* **24** (1842), 1–4 ; reprinted in Gesammelte Werke, Band 4, 257-262, Chelsea Public. Comp., New York (1969).

[37] J.-P. Jouanolou, Equations de Pfaff algébriques, *Lect. Notes in Math.* **708**, Springer-Verlag, Berlin (1979)

[38] B. M. Koialovich, Researches on the differential equation $y\,dy - y\,dx = R\,dx$ (in Russian), Sankt Peterburg (1894).

[39] B. M. Koialovich, On the problem of the integration of the differential equation $y\,dy - y\,dx = R(x)\,dx$ (in Russian), in Collection of papers in honour of academician Grave, Gostekhizdat, Moscow (1940), 79–87.

[40] A. N. Kolmogorov, A. P. Jushkevich (editors), Mathematics of XIX century : Chebyshev's ideas in function theory, ordinary differential equations, variational calculus, calculus of finite differences (in Russian), Nauka, Moscow, (1987).

[41] A. N. Korkine, Sur les équations différentielles ordinaires du premier ordre, *C. R. Acad. Sc. Paris* **122** (1896), 1184–1186, errata in C. R. Acad. Sc. Paris **123**, 139; reprinted in [60], vol. 2, 534–536.

[42] A. N. Korkine, Sur les équations différentielles ordinaires du premier ordre, *C. R. Acad. Sc. Paris* **123** (1896), 38–40 ; reprinted in [60] vol. 2, 537–539.

[43] A. N. Korkine, Sur les équations différentielles ordinaires du premier ordre, *Math. Ann.* **48** (1897), 317–364 .

[44] A. N. Korkine, Thoughts about multipliers of differential equations of first degree (in Russian), *Math. Sbornik* **24** (1904), 194–350 and 351–416.

[45] J. Kovačic, An algorithm for solving second order linear homogenous differential equations, *J. Symb. Comp.* **2** (1986), 3–43.

[46] V. V. Kozlov, Integrability and non-integrability in Hamiltonian mechanics (in Russian), *Uspekhi Mat. Nauk.* **38** (1) (1983), 3–67; English translation in *Russian Math Surveys* **38** (1), (1983), 1–76.

[47] S. G. Krein, N. I. Yatskin, Linear differential equations on manifolds (in Russian), Editions of Voronezh University, Voronezh (1980).

[48] S. Lang, Differentiable Manifolds, Springer-Verlag, Berlin, (1985).

[49] K. Ja. Latysheva, On the works of V. P. Jermakov on the theory of differential equations (in Russian), Istoriko-Matematicheskije Issledovanija **9**, Gostekhizdat, Moscow (1956), 691–722.

[50] S. Lie, Gesammelte Abhandlungen, Band 3, 4, B. G. Teubner, Leipzig (1922,1929).

[51] A. Lins Neto, Local structural stability of C^2 integrable forms, *Ann. Inst. Fourier, Grenoble* **27** (2), (1977), 197–225.

[52] N. V. Lokot', Thesis (in Russian, unpublished), Leningrad State Pedagogical Institute (1989).

[53] P. Malliavin, Géométrie différentielle intrinsèque, Hermann, Paris (1972).

[54] D. Morduhai-Boltovskoi, Researches on the integration in finite terms of differential equations of the first order (in Russian), *Communications de la Société Mathématique de Kharkov*, **10** (1906–1909), 34–64 and 231–269 ; english translation of pp. 34–64 by B. Korenblum and M. J. Prelle, *SIGSAM Bulletin* **15** (2), (1981), 20–32.

[55] D. Morduhai-Boltovskoi, On integration of linear differential equations in finite terms (in Russian), Warsaw (1910).

[56] D. Morduhai-Boltovskoi, Sur la résolution des équations différentielles du premier ordre en forme finie, *Rend. Circ. Matem. Palermo* **61** (1937), 49–72.

[57] T. Nagano, Linear differential systems with singularities and an application to transitive Lie algebras, *J. Math. Soc. Japan* **18** (1966), 398–404.

[58] P. J. Olver, Applications of Lie groups to Differential Equations, *Graduate Texts in Math.* **107**, Springer-Verlag (1986).

[59] J. Patera, R. T. Sharp, P. Winternitz, Invariants of real low dimensional Lie algebras, *Journal of Math. Phys.* **17** (6) (1976), 986–994.

[60] P. Painlevé, Œuvres, tomes 1–3, Ed. du CNRS, Paris (1972-1974-1975).

[61] P. Painlevé, Sur les intégrales rationnelles des équations différentielles du premier ordre, *C. R. Acad. Sc. Paris* **110** (1890), 34–36 ; reprinted in Œuvres, tome 2, 220–222.

[62] P. Painlevé, Sur les intégrales algébriques des équations différentielles du premier ordre, *C. R. Acad. Sc. Paris* **110** (1890), 945–948 ; reprinted in Œuvres, tome 2, 233–235.

[63] P. Painlevé, Mémoire sur les équations différentielles du premier ordre, *Ann. Ecole Norm. Sup.* 1ère partie : **8** (1891), 9–58, 103–140 ; 2ème partie : **8** (1891), 201–226, 267–284 and **9** (1891), 9–30 ; 3ème partie : **9** (1892), 101–144, 283–308; reprinted in Œuvres, tome 2, 237–461.

[64] P. Painlevé, Leçons sur la théorie analytique des équations différentielles professées à Stockholm (Septembre, Octobre, Novembre 1895), sur l'invitation de S. M. le Roi de Suède et de Norvége, Ed. Hermann, Paris (1897), reprinted in Œuvres, tome 1, 205–800.

[65] P. Painlevé, Mémoire sur les équations différentielles du premier ordre dont l'intégrale est de la forme $h(x)(y-g_1(x))^{\lambda_1}(y-g_2(x))^{\lambda_2}\cdots(y-g_n(x))^{\lambda_n}=C$, *Ann. Fac. Sc. Univ. Toulouse* (1896), 1–37 ; reprinted in Œuvres, tome 2, 546–582.

[66] E. Picard, Sur un théorème de M. Darboux, *C. R. Acad Sc. Paris* **100** (1885) 618–620 ; reprinted in Ch. E. Picard, Œuvres, tome II, 105–107, Ed. du CNRS, Paris (1979).

[67] V. Poènaru, Singularités C^∞ en Présence de Symétrie, Lect. Notes in Math. **510**, Springer-Verlag, Berlin (1976).

[68] H. Poincaré, Sur l'intégration algébrique des équations différentielles, *C. R. Acad Sc. Paris* **112** (1891) 761–764 ; reprinted in Œuvres, tome III, 32–34, Gauthier-Villars, Paris (1965).

[69] H. Poincaré, Sur l'intégration algébrique des équations différentielles du premier ordre et du premier degré, *Rendic. Circ. Matem. Palermo* **5** (1891) 161–191; reprinted in Œuvres, tome III, 35–58, Gauthier-Villars, Paris (1965).

[70] H. Poincaré, Sur l'intégration algébrique des équations différentielles du premier ordre et du premier degré, *Rendic. Circ. Matem. Palermo* **11** (1897) 193–239; reprinted in Œuvres, tome III, 59–94, Gauthier-Villars, Paris (1965).

[71] M. J. Prelle, M. F. Singer, Elementary first integrals of differential equations, *Trans. Amer. Math. Soc.* **279** (1) (1983), 215–229.

[72] J. F. Price, Lie Groups and Compact Groups, London Math. Soc. Lect. Notes **25**, Cambridge Univ. Press, Cambridge (1977).

[73] A. Ramani, B. Grammaticos, T. Bountis, The Painlevé property and singularity analysis of integrable and non-integrable systems, *Phys. Reports* **180** (1989), 159–245.

[74] F. Schwartz, Symmetries of Differential Equations : From Sophus Lie to Computer Algebra, *SIAM Review* **30** (3) (1988) 450–481.

[75] N. I. Simonov, Euler's applied methods of analysis (in Russian), ed. Gostekhizdat, Moscow (1957).

[76] S. Sternberg, Lectures on Differential Geometry, Prentice Hall (1964), reprinted by Chelsea Public. Comp., New York.

[77] J.-M. Strelcyn, S. Wojciechowski, A method of finding integrals of 3-dimensional dynamical systemes, *Phys. Letters* **133** A (1988) 207–212.

[78] H. J. Sussman, Lie brackets, real analyticity and geometric control, in Differential Geometric Control Theory, R. W. Brockett, R. S. Millman, H. J. Sussman (edit.), Progress in Mathematics **27**, Birkhauser, Basel (1983), 1–116.

[79] E. Tournier (edit.), Computer Algebra and Differential Equations, Acad. Press, New York (1989).

[80] E. Vessiot, Méthodes d'intégrations élémentaires. Etude des équations différentielles ordinaires au point de vue formel, in *Encyclopédie des Sciences Mathématiques Pures et Appliquées*, tome II, vol. 3, fasc. 1, Gauthier-Villars, Paris and B. G. Teubner, Leipzig (1910), 58–170.

[81] E. von Weber, Propriétés générales des sytèmes d'équations aux dérivées partielles. Equations linéaires du premier ordre, in *Encyclopédie des Sciences Mathématiques Pures et appliquées*, tome II, vol. 4, fasc. 1, Gauthier-Villars, Paris and B. G. Teubner, Leipzig (1913), 1–55.

[82] H. Weyl, Classical Groups, Their Invariants and Representations, second edit., Princeton University Press, Princeton (1946).

[83] V. F. Zaitsev, Discret group theoretical analysis of ordinary differential equations (in Russian), *Differentsialnye Uravnienia*, **25** (3) (1989), 379–387; english translation : *Differential Equations*, **25** (1989).

CYCLICITE FINIE DES POLYCYCLES HYPERBOLIQUES

DE CHAMPS DE VECTEURS DU PLAN

MISE SOUS FORME NORMALE.

A. MOURTADA
Université de Bourgogne
Laboratoire de Topologie
CNRS DO 755
B.P. 138, 21004 DIJON Cedex - FRANCE

Sommaire :

INTRODUCTION

Considérons une famille (X_λ) de champs de vecteurs du plan $\mathbf{R}^2, C^\infty$ en $(m,\lambda), m = (x,y)$ et λ dans un voisinage de 0 dans $\mathbf{R}^\wedge, \wedge \in \mathbf{N}^\circ$; et supposons que pour $\lambda = 0, X_o$ admette un polycycle singulier (Γ_k) ayant pour sommets k points de selle hyperboliques

$(P_i)_{i=1,\ldots,k}$; un même sommet peut être compté deux fois (cf. fig.1 a-b) :

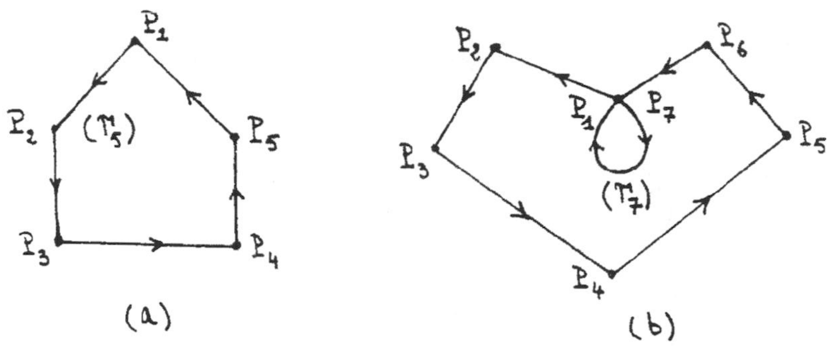

fig. 1

On convient de définir le rapport d'hyperbolicité r d'un point de selle hyperbolique du plan $\mathbf{R}^2$ par : $r = \mid \mu_1/\mu_2 \mid, \mu_1 < 0$ et $\mu_2 > 0$ sont les valeurs propres de la partie linéaire du champ en ce point. On dira que (Γ_k) est de cyclicité $N \in \mathbf{N}$ dans la famille (X_λ) si :

i) $\exists \Omega_o$ voisinage de (Γ_k) et O_0 voisinage de $\lambda = 0$ dans $\mathbf{R}^\Lambda$ tels que $\forall \lambda \in O_0, X_\lambda$ a au plus N cycles limites dans Ω_o ;

ii) $\forall \Omega \subset \Omega_0$ voisinage de (Γ_k) et $\forall O \subset O_0$ voisinage de 0 dans $\mathbf{R}^\Lambda, \exists \lambda \in O$ tel que X_λ ait exactement N cycles limites dans Ω.

Dans cet article, on propose une certaine "forme normale" pour l'application déplacement associée à l'application de retour relative au polycycle perturbé. Donnons d'abord une définition qui précise la terminologie adoptée pour une certaine classe de fonctions.

DEFINITION : *Soient $K \in \mathbf{N}$ et L une fonction de la variable (x, λ), O une partie de $\mathbf{R}^\Lambda$ telle que $0 \in \bar{O}$ et α une fonction de classe C^K en λ sur O, de signe constant et telle que $\mathrm{Lim}_{\substack{\lambda \to 0 \\ \lambda \in O}} \alpha(\lambda) = 0$ et $\mid \alpha(\lambda) \mid \leq \varepsilon/3$ pour tout $\lambda \in O$.Posons $W = \bigcup_{\lambda \in O}]\alpha(\lambda), \varepsilon[\times \{\lambda\}$; et soit ρ une fonction positive, de classe C^K en (x, λ) sur W et telle que $\mathrm{Lim}_{\substack{(x,\lambda) \to (0,0) \\ (x,\lambda) \in W}} \rho(x, \lambda) = 0$; alors on dit que" L* **vérifie les propriétés $(I^K_{\lambda,0})$ par multiplication par ρ sur W"** *si L est de classe C^K en (x, λ) sur W et si $\forall n \leq K$ L vérifie la propriété :*

$$(I^K_{\lambda,0})_n \qquad \qquad Lim_{\substack{(x,\lambda)\to(0,0)\\ (x,\lambda)\in W}} \rho^n . \frac{\partial^n L}{\partial x^n}(x,\lambda) = 0$$

Le résultat principal qu'on démontre dans le §.II.3 est le suivant:

THEOREME 1 .*Soit $K \in \mathbf{N}$, alors il existe $\varepsilon > 0$, $\mathcal{V}$ voisinage de 0 dans $\mathbf{R}^A$, deux fonctions continues et positives sur $\mathcal{V}$: $\rho(\lambda)$ et $\eta(\lambda)$ et des transversales au champ X_λ : $\sigma_i(\lambda), \tau_i(\lambda)$ (cf.fig.2) de classe C^K tels que si on pose:*

$$\mathcal{U}_\lambda =]\rho(\lambda), \varepsilon[, \quad \mathcal{U} = \bigcup_{\lambda\in\mathcal{V}} \mathcal{U}_\lambda \times \{\lambda\} \quad \text{et} \quad V = \bigcup_{\lambda\in\mathcal{V}}]\eta(\lambda), \varepsilon[\times \{\lambda\},$$

l'application déplacement (associée à l'application de retour du polycycle pérturbé) soit définie sur $\mathcal{U}_\lambda$ pour tout $\lambda \in \mathcal{V}$ et soit donnée par: $\forall (x_1, \lambda) \in \mathcal{U}$

$$(1) \quad \Delta(x_1, \lambda) = \Big[\dots \big[x_1{}^{r_1(\lambda)} + b_1(\lambda) \big]^{r_2(\lambda)} + \dots + b_{k-1}(\lambda) \Big]^{r_k(\lambda)} + b_k(\lambda) - \varphi(x_1, \lambda)$$

avec $\varphi(x_1, \lambda) = x_1 [\bar{\alpha}_1(\lambda) + f(x_1, \lambda)]$; x_1 est un paramètre de classe C^K sur $\sigma_1(\lambda)$ et Δ (mesureé sur la transversale $\tau_k(\lambda)$) est de classe C^K en x_1 sur $\mathcal{U}_\lambda$ et toutes ses dériveés sont continues en $(x_1, \lambda) \in \mathcal{U}$. $b_i(\lambda)$ mesure sur la transversale d'entreé du sommet P_{i+1} (avec la convention $k + 1 \equiv 1$) la déformation de la connexion entre les sommets P_i et P_{i+1}. La fonction $\bar{\alpha}_1$ vérifie

$$\forall \lambda \in \mathcal{V} \quad \bar{\alpha}_1(\lambda) > 0.$$

La fonction f est continue sur $]-\varepsilon, \varepsilon[\times \mathcal{V}$,de classe C^K en x_1 sur $]\eta(\lambda), \varepsilon[$ pour tout $\lambda \in \mathcal{V}$ et verifie les propriétés $(I^K_{\lambda,0})$ par multiplication par $(x_1 - \eta(\lambda))$ sur $\mathcal{V}$.

Notons $h_0(x_1, \lambda) = x_1$ et pour $i = 1, \dots, k$ $h_i(x_1, \lambda) = h_{i-1}^{r_i}(x_1, \lambda) + b_i(\lambda)$, h_0 et h_k sont mesurées sur σ_1 et pour $i = 1, \dots, k-1$ h_i est mesurée sur σ_{i+1}, dans (1) l'application $h_i \mapsto h_i^{r_{i+1}}$ pour $i = 0, \dots, k-1$ représente la normalisation de l'application de Dulac le long du coin P_{i+1}, et la translation $h_i^{r_{i+1}} \mapsto h_i^{r_{i+1}} + b_{i+1}$ pour $i = 0, \dots, k-1$ représente la normalisation de la correspondance régulière entre τ_{i+1} et σ_{i+2}.

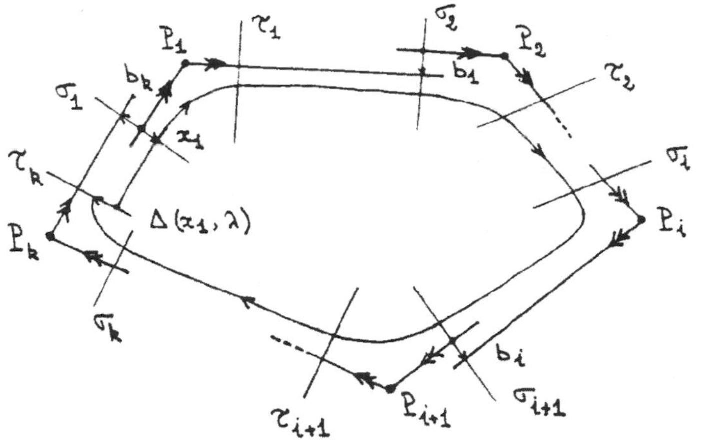

fig. 2

L'idée de la normalisation (1) m'a été suggérée, entre autre, par R. Roussarie que je remercie pour son soutien tout le long de ce travail.

Les cycles limites qui apparaissent au voisinage du polycycle perturbé correspondent aux racines isolées de l'équation en la variable x_1:

(1.1) $$\triangle(x_1, \lambda) = 0 \quad (\text{cf.}(1)).$$

Grâce aux propriétés des fonctions $\bar{\alpha}_1$ et f_0 $\left(\bar{\alpha}_1(0) > 0 \text{ et } Lim_{\substack{x_1 \to 0 \\ \lambda \to 0}} f_0(x_1, \lambda) = 0\right)$ cette équation est équivalente à l'équation plus significative :

(1.2) $$\left[\dots\left[x_1^{r_1(\lambda)} + b_1(\lambda)\right]^{r_2(\lambda)} + \dots + b_{k-1}(\lambda)\right]^{r_k(\lambda)} + b_k(\lambda)\Big] \cdot \left[[\bar{\alpha}_1]^{-1} + f_1\right] = x_1$$

où la fonction f_1 vérifie les mêmes propriétés $\left(I^K_{\lambda.0}\right)$ que la fonction f_0. Cette équation (1.2) montre que le 1er membre est une normalisation de l'application de retour du polycycle par rapport à la transversale σ_1. Or il est facile de voir que pour $\lambda = 0$, cette application s'écrit (cf. par exemple [5] ou §. II.3) :

(1.3) $$\varphi_0(x_1) = x_1^{r_1 r_2 \cdots r_k(0)} \left[[\bar{\alpha}_1(0)]^{-1} + g_0(x_1)\right]$$

où la fonction g_0 vérifie des propriétés que l'on note (I_0^K) et qui sont similaires aux propriétés $\left(I_{\lambda,0}^K\right)$ (cf.§.II.1, Déf.1 et Déf.3). La forme du 1er membre de (1.2) justifie le nom de "forme normale" donné à (1).

Dans [1], on démontre, grâce au théorème ci-dessus, un théorème de finitude de la cyclicité des polycycles hyperboliques génériques en donnant un algorithme pour la résolution de l'équation (1.1).

Avant d'énoncer ce théorème, convenons d'appeler "condition de type C.H." [Condition Hyperbolique] toute condition de la forme:

$$\prod_{j\in I} r_j \neq 1$$

où I est une partie de $\{1,...,k\}$.

THEOREME 2[1] : *sous certaines conditions génériques portant uniquement sur les rapports d'hyperbolicités* $(r_i(0))_{i=1..k}$ *des sommets* $(P_i)_{i=1..k}$, *et contenant toutes les conditions de type C.H., le polycycle* (Γ_k) *est de cyclicité finie dans la famille* (X_λ).

- Si à chaque polycycle (Γ_k) on associe le point $m_k \in \mathbf{R}^k_{+*}$ de coordonnées $m_k = (r_i(0))_{i=1..k}$, [certaines conventions sur le premier point et le sens devant être prises] le mot "générique" du théorème signifie que m_k est dans un ouvert dense de $\mathbf{R}^k_{+*}$.

Si les conditions génériques du théorème 2 sont réalisées, les propriétés $(I^K_{\lambda,0})$ de la fonction f_0 (cf.Définition ci-dessus) sont suffisantes pour négliger le reste f_0 devant la partie principale de Δ (cf.(1)).

Ces conditions génériques ne seront pas toutes connues explicitement à cause de la complexité des relations de récurrence qui les définissent. On montrera cependant qu'elles sont en nombre fini (croissant avec k) et sont de la forme:

$$g_j(r_1, r_2, ..., r_k) \neq 0$$

où g_j est une fonction polynômiale de k variables et à coefficients dans $\mathbf{Z}$. Une majoration significative de la cyclicité maximale est également difficile à établir.

Des résultats plus précis sur le type de ces conditions et sur la cyclicité maximale sont donnés dans [1] et dans [6] où on montre l'existence, pour tout $k\geq 4$, de conditions qui ne sont pas de type C.H.. On traite également, dans ce dernier article, des polycycles à trois et à quatre sommets et on donne une minoration de la cyclicité de certains polycycles (Γ_k) dans les familles génériques.

Le cas $k = 1$ a été étudié par Andronov et al. dans [2], ils ont montré qu'en classe C^1 et sous la condition $r_1(0) \neq 1$; il apparaît au plus un cycle. Le cas $k = 2$ a été traité par L.A.Cherkas dans [3] pour des familles analytiques. Il a montré que sous les conditions $r_1.r_2(0) \neq 1$, $r_1(0) \neq 1$ et $r_2(0) \neq 1$, (Γ_2) est de cyclicité≤ 2. Le résultat du théorème 2 résoud partiellement un problème proposé par J.Sotomayor et repris par R.Roussarie dans [4] concernant la cyclicité des polycycles hyperboliques génériques.

Si on désigne par $K(k)$ le nombre de dérivation intervenant dans l'algorithme de [1], alors dans le théorème 1 ci-dessus, on peut prendre $K = K(k)$ et la famille (X_λ) peut être choisi de classe C^l, l suffisamment grand devant $K(k)$.

I. APPLICATION DE DULAC

Dans ce paragraphe, on donnera à l'application de Dulac relative à un point de selle hyperbolique la structure asymptotique qui nous intéresse dans cette étude. Soit (X_λ) une famille de champs de vecteurs sur $\mathbf{R}^2$ dépendant d'un paramètre $\lambda \in \mathbf{R}^\wedge$. On suppose que cette famille est C^∞ en $(m, \lambda) \in \mathbf{R}^2 \times \mathbf{R}^\wedge$; $m = (x, y)$ et que pour $\lambda \in \mathcal{V}$ où $\mathcal{V}$ est un voisinage de 0 dans $\mathbf{R}^\wedge$, X_λ a un point de selle hyperbolique $P(\lambda)$ de rapport d'hyperbolicité $r(\lambda)$, et que $P(\lambda)$ est la seule singularité du champ X_λ dans un voisinage $\mathcal{U}$ de $P(0)$ dans $\mathbf{R}^2$ pour tout $\lambda \in \mathcal{V}$.

Par conjugaison par un difféomorphisme C^∞ en (m, λ) et multiplication par une fonction C^∞ en (m, λ) de signe constant sur $\mathcal{U} \times \mathcal{V}$, on peut supposer que $P(\lambda) = O = (0,0)$ $\forall \lambda \in \mathcal{V}$ et que les axes de coordonnées sont les variétés invariantes du point de selle O et que dans ces nouvelles coordonnées le champ s'écrit :

$$\textbf{(2)} \quad X_\lambda \begin{cases} \dot{x} &= x \\ \dot{y} &= y \quad [-r(\lambda) + f(x, y, \lambda)] \end{cases}$$

où f est une fonction C^∞ en (x, y, λ) et $f(0,0,\lambda) = 0$ $\forall \lambda \in \mathcal{V}$; (cf. par exemple [7] ou [8] pp. 822-823 où l'on peut voir dans la démonstration (faite dans le cas d'un difféomorphisme individuel) que la conjugaison peut être choisie C^∞ en un paramètre $\lambda \in \mathbf{R}^\wedge$). L'intérêt de l'expression donnée au système (2) ci-dessus est qu'elle permet, par des transformations C^∞ en λ , de définir les trois "invariants" différentiables essentiels d'un point de selle hyperbolique : les variétés stables et instables et le rapport d'hyperbolicité $r(\lambda)$.

I.1 -FORME NORMALE POUR LE SYSTEME (2).

- Comme il sera essentiel d'avoir par rapport au paramètre λ une certaine régularité de toutes les fonctions intervenant dans la mise en équation, on reprend ici la méthode directe de réduction utilisée par R. Roussarie dans [9] pp. 76 dans le cas $r(0) = 1$. On se fixe une base de $\mathbf{R}^2$ correspondante aux coordonnées (x, y) dans lesquelles est écrit le système (2) ; les jets seront pris dans ces coordonnées. Posons :

(2.1) $r_0 = r(0)$ et $\alpha_1(\lambda) = r(\lambda) - r_0$

on a alors le résultat suivant :

PROPOSITION 1 : *On considère le système* (2) *et on se fixe un réel* $a > 0$; *alors, il existe une suite de réels* $(\delta_N)_{N\geq 1}$ *décroissante et des fonctions* C^∞ $\alpha_N(\lambda)$ *définies sur* $W_N = \{\lambda \in \mathcal{V} \; ; \; |\alpha_1(\lambda)| < \delta_N\}$ *avec* $\delta_1 = a$ *telles que :*

1°) *Si* $r_0 = p/q$, *p et q sont deux entiers premiers entre eux,* $M = p + q$: $\forall N \geq 1$ *et* $\forall \lambda \in W_{N+1}$ $J^{N.M+1}X_\lambda(0) \overset{C^\infty}{\sim} x\frac{\partial}{\partial x} - r(\lambda)\, y\frac{\partial}{\partial y} + y\left(\sum_{l=1}^{N} \alpha_{l+1}(x^p y^q)^l\right)\frac{\partial}{\partial y}$

2°) *Si* r_0 *est irrationnel* : $\forall N \geq 1$ *et* $\forall \lambda \in W_N$ $J^N X_\lambda(0) \overset{C^\infty}{\sim} x\frac{\partial}{\partial x} - r(\lambda) y\frac{\partial}{\partial y}$.

La C^∞ –équivalence entre jets d'ordre fixé est la multiplication par des fonctions C^∞ en (x, y, λ) et strictement positives et la conjugaison par difféomorphismes C^∞ en (x, y, λ).

Preuve : il est connu (cf. par exemple [10] pp. 78-80) que si $r_0 = p/q$:

$$J^\infty X_0(0) \overset{C^\infty}{\sim} x\frac{\partial}{\partial x} - r_0 y\frac{\partial}{\partial y} + \sum_{l\geq 1}(x^p y^q)^l . \left(a_l x\frac{\partial}{\partial x} + b_l y\frac{\partial}{\partial y}\right),$$

et si r_0 est irrationnel :

$$J^\infty X_0(0) \overset{C^\infty}{\sim} x\frac{\partial}{\partial x} - r_0 y\frac{\partial}{\partial y},$$

on peut alors appliquer la méthode signalée ci-dessus pour déduire la proposition.

I.2 -UNE STRUCTURE ASYMPTOTIQUE DE L'APPLICATION DE DULAC.

- D'après un théorème de Sternberg (cf. [11]), K étant un entier arbitrairement fixé, il existe un entier $N(K)$ tel que le champ X_λ soit C^K – conjugué à son jet polynômial d'ordre $N(K)$ (la conjugaison peut être choisie C^K en (m, λ)) ; soit d'après la Proposition 1 ci-dessus : $X_\lambda \overset{C^K}{\sim} X_\lambda^{N(K)}$ avec :

$$(3) \quad X_\lambda^{N(K)} = \begin{cases} x\dfrac{\partial}{\partial x} - r(\lambda)y\dfrac{\partial}{\partial y} + y\left(\displaystyle\sum_{l=1}^{\bar{N}(K)} \alpha_{l+1}(x^p y^q)^l\right) \cdot \dfrac{\partial}{\partial y} & \text{si } r_0 = p/q \\ x\dfrac{\partial}{\partial x} - r(\lambda)y\dfrac{\partial}{\partial y} & \text{si } r_0 \text{ est irrationnel} \end{cases}$$

l'entier $\bar{N}(K)$ est donné par $\bar{N}(K) = [(N(K)-1)/M]$, $M = p+q$ (cf. Proposition 1). Par une homothétie, $X_\lambda^{N(K)}$est C^K – équivalent à un champ d'expression similaire définit sur un ouvert contenant le carré $[-1,1] \times [-1,1]$; notons D_λ l'application de Dulac

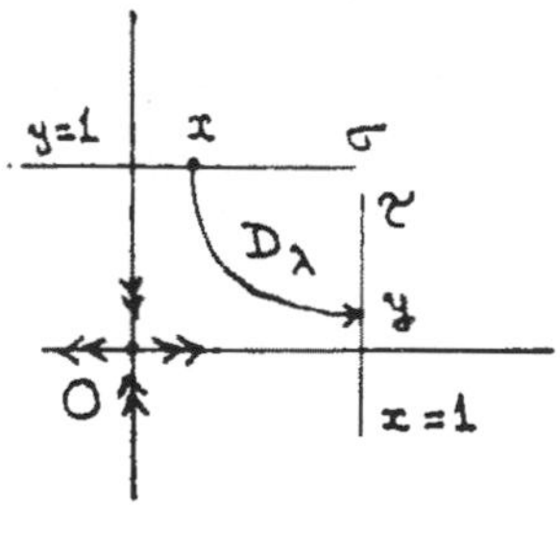

fig. 3

relative au point de selle O, envoyant la demi-transversale positive σ paramétrée par x (cf. fig. 3) sur la demi-transversale positive τ paramétrée par y. On va tout d'abord établir une expression asymptotique de D_λ, essentielle pour la suite de ce travail :

PROPOSITION 2 : *Si dans la Proposition* 1 *on choisit* $a < p/3q$ *dans le cas où* $r_0 = p/q$ *alors il existe* $\mathcal{V}$ *voisinage de* 0 *dans* $\mathbf{R}^\wedge$ *et* $\varepsilon > 0$ *tels que l'application de Dulac* D_λ *pour le système* (3) *s'écrive :*

$$(4) \qquad D_\lambda(x) = x^{r(\lambda)}[1 + d(x,\lambda)] \quad \forall (x,\lambda) \in]0,\varepsilon[\times \mathcal{V},$$

où d *est analytique en* x *sur* $]0,\varepsilon[$ *et est* C^∞ *en* $(x,\lambda) \in]0,\varepsilon[\times\mathcal{V}$; *de plus* d *vérifie la propriété suivante notée* (I_0^∞) :

$$(5)\ (I_0^\infty): \quad \forall n \in \mathbf{N} \quad \mathrm{Lim}_{x\to 0}\, x^n.\frac{\partial^n d}{\partial x^n}(x,\lambda) = 0 \text{ uniformément en } \lambda \in \mathcal{V}.$$

Preuve : la famille $X_\lambda^{N(K)}$ (cf. (3)) étant analytique en (x,y) et en $(\alpha_i)_{i=1..\bar{N}(K)}$, et les α_i étant C^∞ en λ, les propriétés d'analycité et de C^∞ – différentiabilité énoncées sont donc vraies pour D_λ et par conséquent pour d.

Si r_o est irrationnel, d est identiquement nulle et la propriété (I_0^∞) est triviale.

Supposons donc que $r_0 = p/q, p\wedge q = 1$. La démarche qui va suivre s'inspire largement des idées développées par R. Roussarie dans [9] ; dans le 1er quadrant ouvert, faisons le changement de variables suivant :

$$(5.1) \begin{cases} x &= x \\ u &= x^p y^q \end{cases}$$

et posons (cf. (3) et (2.1)) :

$$\bar{\alpha}_1 = -q\alpha_1 \quad \text{et pour } l = 2,...,\bar{N}(K)+1 \quad \bar{\alpha}_l = q\alpha_l,$$

le système (3) est alors transformé dans le système à variables séparées :

$$\begin{cases} \dot{x} &= x \\ \dot{u} &= \sum_{l=1}^{\bar{N}(K)+1} \bar{\alpha}_l.u^l \end{cases}$$

désignons par $u(t,u) = \sum_{j=1}^{\infty} g_j(t,\lambda).u^j$ la solution (analytique pour tout t et pour $\lambda \in W_{\bar{N}(K)+1}$ autour de $u = 0$) de la 2ème ligne de ce système ; posons $\bar{a} = q.a$ (cf. Proposition 1) ; [9] nous fournit alors les estimations :

$\exists C > 1$ tel que $\forall n \in \mathbf{N}$, $\exists C_n > 0$ vérifiant :

$$(5.2)\ \forall j \geq 1 \quad \forall \lambda \in W_{\bar{N}(K)+1} \quad \left|\frac{\partial^n g_j}{\partial t^n}(t,\lambda)\right| \leq C_n.(C.e^{\bar{a}t})^j \ ;$$

on calcule aisément $g_1(t,\lambda) = e^{\bar{a}_1 t}$. Posons : $\tilde{g}_0(t,\lambda) = 1$ et pour $j \geq 1$ $\tilde{g}_j(t,\lambda) = \bar{e}^{\bar{a}_1 t}.g_{j+1}(t,\lambda)$

(5.2) et la formule de Leibniz de dérivation d'un produit nous donnent les estimations suivantes pour les fonctions $\tilde{g}_j$: $\forall n \in \mathbf{N}$, $\exists \tilde{C}_n$ tel que :

(5.3) $\forall j \geq 1, \quad \forall \lambda \in W_{\bar{N}(K)+1} \quad \left|\frac{\partial^n \tilde{g}_j}{\partial t^n}(t,\lambda)\right| \leq \quad \tilde{C}_n.(C.e^{\bar{a}.t})^{j+2}$;

écrivons $u(t,\lambda) = e^{\bar{a}_1 t}.u\left[1 + \sum_{j=1}^{\infty} \tilde{g}_j(t,\lambda).u^j\right]$; le changement de variables (5.1) montre que pour le système (3) la correspondance entre les transversales σ et τ (cf. fig. 3) dans le 1er quadrant ouvert est donnée par :

$$y^q = x^{p-\bar{a}_1}.\left[1 + \tilde{d}(x,\lambda)\right] \text{ pour } x \in [0,\varepsilon] \text{ et } \lambda \in W_{\bar{N}(K)+1}$$

avec

$$\tilde{d}(x,\lambda) = \sum_{j=1}^{\infty} \tilde{g}_j(-Lnx,\lambda).x^{jp}.$$

Démontrons la propriété (I_0^∞) pour la fonction $\tilde{d}$ par récurrence sur n (cf. (5)) : pour $n = 0$, (5.3) nous donne :

$$\forall j \geq 1, \ \forall \lambda \in W_{\bar{N}(K)+1} \quad |\tilde{g}_j(-Lnx,\lambda).x^{jp}| \leq \tilde{C}_0.C^{j+2}.x^{-(j+2)\bar{a}+jp} \ ;$$

choisissons $\bar{a} < p/3$, on a alors :

$$\forall j \geq 1 \qquad -(j+2)\bar{a} + jp > (j-1).2p/3 \geq 0 \ ;$$

la propriété à l'indice $n = 0$ sera donc vraie si on choisit $\varepsilon < C^{-3/2p}$. Supposons la propriété (I_0^∞) vraie jusqu'à l'indice $n - 1 \geq 0$; la formule de Leibniz montre qu'elle sera vraie à l'indice n si et seulement si :

$$\mathrm{Lim}_{x\to 0}\frac{\partial^n}{\partial x^n}[x^n.\tilde{d}(x,\lambda)] = 0 \text{ uniformément en } \quad \lambda \in W_{\bar{N}(K)+1}$$

or $x^n.\tilde{d}(x,\lambda) = \sum_{j=1}^{\infty} \tilde{g}_j(-Lnx,\lambda).x^{jp+n}$; on peut alors utiliser l'estimation de [9] pp. 88 (valable pour un ordre de dérivation n inférieur ou égal aux puissances de x (ici $jp+n \ \ j \geq 1$) :

$\exists M_n > 0, \forall \lambda \in W_{\bar{N}(K)+1} \left|\frac{\partial^n}{\partial x^n}\left[\tilde{g}_j(-Lnx,\lambda).x^{jp+n}\right]\right| \leq \frac{(jp+n)!}{jp!}M_n.C^{j+2}.x^{-(j+2)\bar{a}+jp}$,

on voit donc que pour $\bar{a}$ choisi comme ci-dessus, le même ε vaut.

Maintenant, quitte à diminuer ε et à prendre $\mathcal{V} \subset W_{\tilde{N}(K)+1}$, on peut supposer que :

$$\forall (x, \lambda) \in [0, \varepsilon] \times \mathcal{V} \quad 1 + \tilde{d}(x, \lambda) > 0 ;$$

la fonction d de la proposition est donnée par : $d(x, \lambda) = \left[1 + \tilde{d}(x, \lambda)\right]^{1/q} - 1$ et vérifie aussi la propriété (I_0^∞), ce qui prouve la Proposition 2.

- Conclusion du §.I : on a donc montré que pour la famille de départ X_λ ayant un point de selle hyperbolique $P(\lambda)$ de rapport d'hyperbolicité $r(\lambda)$ (cf. Introduction), si K est un entier fixé, il existe $\varepsilon_1 > 0$, un voisinage $\mathcal{V}$ de 0 dans $\mathbf{R}^\wedge$ et des transversales $\sigma(\lambda)$ et $\tau(\lambda)$ (cf. fig. 4) paramétrées en classe C^K respectivement par x et y et dépendant de façon C^K de λ tels que l'application de Dulac relative au coin $P(\lambda)$ s'écrive :

(6) $y = x^{r(\lambda)}.[1 + d_1(x; \lambda)] \quad \forall (x, \lambda) \in]0, \varepsilon_1[\times \mathcal{V}$,

où d_1 est de classe C^K en (x, λ) et vérifie la propriété suivante notée (I_0^K) :

(7) $(I_0^K) : \forall n \leq K \quad \mathrm{Lim}_{x \to 0} x^n . \frac{\partial^n d_1}{\partial x^n}(x, \lambda) = 0$ uniformément en $\lambda \in \mathcal{V}$.

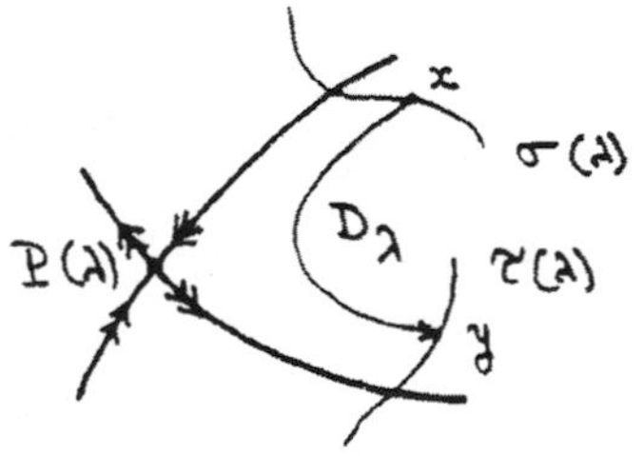

fig. 4

Il est clair que l'on peut écrire aussi :

(8) $x = y^{s(\lambda)}.[1 + d_2(y, \lambda)] \quad \forall (y, \lambda) \in]0, \varepsilon_2[\times \mathcal{V}$;

avec $s(\lambda) = [r(\lambda)]^{-1}$ et d_2 vérifie les mêmes propriétés que d_1.

II -REDUCTION DE LA LOI DE CORRESPONDANCE DU POLYCYCLE

- Cette section sera subdivisée en trois paragraphes. Dans le paragraphe II.1, on donnera des définitions et des lemmes techniques "utiles pour la suite". Dans le paragraphe II.2, on définira un certain vocabulaire lié à la géométrie du polycycle perturbé et on déduira certains résultats. Ces deux premiers paragraphes permettront de rendre plus clairs les calculs intervenant dans le dernier paragraphe II.3 où on procèdera à la réduction de la loi de correspondance du polycycle. Le lecteur peut aller directement à ce dernier paragraphe et se reporter aux deux premiers quand c'est nécessaire.

II.1 -DEFINITIONS ET LEMMES PRELIMINAIRES.

- On donnera d'abord une définition plus précise de la propriété (I_0^K) ci-dessus (cf. (7)) et on définira d'autres propriétés similaires. La plupart des fonctions qu'on rencontrera dans l'étape de mise en équation en II.3 auront une de ces propriétés. On démontrera ensuite quelques conséquences immédiates pour de telles fonctions dans plusieurs Lemmes.

Remarque 0 :

1°) Soit O une partie de $\mathbf{R}^{\wedge}$ et W une partie de $]-\varepsilon,\varepsilon[\times \mathbf{R}^{\wedge}$ du type de celles qu'on rencontrera dans les Définitions 1-4 ; alors on entendra par fonction de classe C^K en λ sur O une fonction qui est continue sur O et de classe C^K sur l'intérieur de O ; de même, on entendra par fonction de classe C^K en (x,λ) sur W une fonction qui est de classe C^K pour tout $(x,\lambda)\in W$ et toutes ses dérivées par rapport à x sont continues sur W et qui est de classe C^K en (x,λ) sur l'intérieur de W.

2°) Il arrivera dans le §.II.3 que l'on exige seulement que les fonctions des Définitions 1-4 soient de classe C^K en x pour tout $(x,\lambda)\in W$et que les dérivées en x soient continues sur W (cf. Déf. 1-4 ci-dessous).

DEFINITION 1 : *Soit H une fonction de la variable (x,λ). On dit que "* H vérifie les propriétés (I_0^K) par multiplication par ρ sur $W=]0,\varepsilon[\times O$ " *(où O est une partie de* $\mathbf{R}^{\wedge}$ *telle que* $0\in\bar{O}$, *et ρ est une fonction positive, de classe C^K en (x,λ) sur W et telle que*

$Lim_{\substack{(x,\lambda)\to(0,0)\\(x,\lambda)\in W}} \rho(x,\lambda) = 0$ *si* H *est de classe* C^K *en* (x,λ) *sur* W *et si* $\forall n \le K$, H *vérifie la propriété suivante qu'on note* $(I_0^K)_n$:

(9) $(I_0^K)_n$: $Lim_{\substack{x\to 0\\x>0}} \rho^n . \frac{\partial^n H}{\partial x^n}(x,\lambda) = 0$ *uniformément en* $\lambda \in O$.

Convention : Si H est une fonction qui vérifie les propriétés (I_0^K) ci-dessus, on convient que :

(10) $H(0,\lambda) = 0 \quad \forall\lambda \in O$.

DEFINITION 2 : *Soient* h *une fonction de la variable* (x,λ) *et* O *une partie de* $\mathbf{R}^\wedge$ *telle que* $0 \in \bar{O}$; *posons* $W =]-\varepsilon,\varepsilon[\times O, W^\circ = W\setminus \{0\} \times O$ *et* $W^1 =]0,\varepsilon[\times O$; *soit* ρ *une fonction positive, de classe* C^K *en* (x,λ) *sur* W^1 *et telle que* $Lim_{\substack{(x,\lambda)\to(0,0)\\(x,\lambda)\in W^1}} \rho(x,\lambda) = 0$, *alors, on dit que* " h vérifie les propriétés (I_1^K) par multiplication par ρ sur W^1" *si* h *est de classe* C^K *en* (x,λ) *sur* W° *et est seulement* C^1 *en* x *pour* $x = 0$ *et si* :

i) $h(0,\lambda) = 0 \quad \forall\lambda \in O$ *et* $Lim_{\substack{x\to 0\\x\neq 0}} \frac{\partial h}{\partial x}(x,\lambda) = 1$ *uniformément en* $\lambda \in O$,

ii) $\forall n \in \{2,...,k\}$, h *vérifie la propriété suivante notée* $(I_1^K)_n$:

(11) $(I_1^K)_n$: $Lim_{\substack{x\to 0\\x>0}} \rho^{n-1}(x,\lambda).\frac{\partial^n h}{\partial x^n}(x,\lambda) = 0$ *uniformémént en* $\lambda \in O$.

- Les propriétés (I_0^K) et (I_1^K) ci-dessus ne sont pas préservées par les divers changements de paramétrages que l'on effectuera dans le paragraphe II.3. On définira ci-dessous des propriétés moins fortes qui seront stables sous ces diverses opérations :

DEFINITION 3 : *Soient* L *une fonction de la variable* (x,λ), O *une partie de* $\mathbf{R}^\wedge$ *telle que* $0 \in \bar{O}$ *et* α *une fonction de classe* C^K *en* λ *sur* O, *de signe constant et telle que* $Lim_{\substack{\lambda\to 0\\\lambda\in O}} \alpha(\lambda) = 0$ *et* $|\alpha(\lambda)| \le \varepsilon/3$ *pour tout* $\lambda \in O$.*Posons* $W = \bigcup_{\lambda\in O}]\alpha(\lambda),\varepsilon[\times\{\lambda\}$; *et soit* ρ *une fonction positive, de classe* C^K *en* (x,λ) *sur* W *et telle que* $Lim_{\substack{(x,\lambda)\to(0,0)\\(x,\lambda)\in W}} \rho(x,\lambda) = 0$; *alors on dit que*" L vérifie les propriétés $(I_{\lambda,0}^K)$ par multiplication par ρ sur W " si L *est de classe* C^K *en* (x,λ) *sur* W *et si* $\forall n \le K$ L *vérifie la propriété* :

(12) $(I_{\lambda,0}^K)_n$: $Lim_{\substack{(x,\lambda)\to(0,0)\\(x,\lambda)\in W}} \rho^n . \frac{\partial^n L}{\partial x^n}(x,\lambda) = 0$

Pour pouvoir effectuer des changements de paramétrages assez réguliers sur une transversale autour de son origine, on aura besoin (au moins) des propriétés suivantes :

DEFINITION 4 : *Soient l une fonction de la variable (x,λ), O et α comme dans la* Déf.3 *ci-dessus. Posons $W =]-\varepsilon,\varepsilon[\times O$, $W^\circ = W \setminus \bigcup_{\lambda\in O}(\alpha(\lambda),\lambda)$ et $W^1 = \bigcup_{\lambda\in O}]\alpha(\lambda),\varepsilon[\times\{\lambda\}$; soit ρ une fonction positive de classe C^K en (x,λ) sur W^1 et telle que $\mathrm{Lim}_{\substack{(x,\lambda)\to(0,0)\\ (x,\lambda)\in W^1}} \rho(x,\lambda) = 0$; alors on dit que "$l$* vérifie les propriétés $(I^K_{\lambda,1})$ par multiplication par ρ sur W^1" *si l est de classe C^K en (x,λ) sur W° et est seulement de classe C^1 en x pour $x = \alpha(\lambda)$ et si :*

i) $l(0,\lambda) = 0 \ \forall\lambda \in O$ *et* $\mathrm{Lim}_{\substack{(x,\lambda)\to(0,0)\\ (x,\lambda)\in W^\circ}} \frac{\partial l}{\partial x}(x,\lambda) = 1,$

ii) $\forall n \in \{2,...,K\}$, *$l$ vérifie la propriété :*

(13) $(I^K_{\lambda,1})_n : \mathrm{Lim}_{\substack{(x,\lambda)\to(0,0)\\ (x,\lambda)\in W^1}} \rho^{n-1} \frac{\partial^n l}{\partial x^n}(x,\lambda) = 0.$

Remarques 1 :

1°) La fonction α des Définitions 3 et 4 peut être identiquement nulle

2°) La deuxième partie de la condition i) des Définitions 2 et 4 permet un certain contrôle en fonction du paramètre de la dérivée première ; elle implique que $\frac{\partial h}{\partial x}(0,\lambda) = 1 \ \forall\lambda \in O$ dans la Déf. 2 et $\mathrm{Lim}_{\substack{\lambda\to 0\\ \lambda\in O}} \frac{\partial l}{\partial x}(\alpha(\lambda),\lambda) = 1$ dans la Déf. 4.

3°) Si dans la Définition 4 ci-dessus, les conditions (13) ne sont vérifiées que pour $(x,\lambda) \in W^2 \subset W^1$, on dira que "$l$ vérifie les propriétés $(I^K_{\lambda,1})$ par multiplication par ρ sur W^2".

- Donnons une dernière définition qui montre l'utilité des fonctions qui vérifient des propriétés (I^K_1) ou $(I^K_{\lambda,1})$:

DEFINITION 5 : *Reprenons les notations de la* Déf. 2 : *on dira que "$\bar{x} = h(x,\lambda)\ \forall(x,\lambda) \in W$"* est un changement de variable (I^K_1) *si la fonction h vérifie les propriétés (I^K_1) (par multiplication par ρ sur W^1(cf.* Déf. 2)). *De même, en reprenant les notations de la* Déf. 4. : *on dira que "$\bar{x} = l(x,\lambda)\ \forall(x,\lambda) \in W$"* est un changement de variable $(I^K_{\lambda,1})$ *si la fonction* l *vérifie les propriétés* $(I^K_{\lambda,1})$ *(par multiplication par* ρ *sur* W^1 *(cf.* Déf. 4)).

- Voici maintenant quelques propriétés immédiates qui découlent des Définitions 1-4. Le résultat fondamental étant contenu dans le Lemme fondamental (Lemme 5) :

LEMME 1 : Soit O une partie de $\mathbf{R}^\wedge$ (espace des paramètres) telle que $0 \in \bar{O}$. Posons $W =]-\varepsilon, \varepsilon[\times O, W^1 =]0, \varepsilon[\times O$ et $W^\circ = W \setminus \{0\} \times O$:

i) Si H est une fonction de la variable $(x, \lambda) \in W^1$ qui vérifie les propriétés (I^K_0) par multiplication par x sur W^1(cf. Déf. 1), alors la fonction h définie sur W^1 par $h(x, \lambda) = x[1 + H(x, \lambda)]$ peut être prolongée sur W en une fonction impaire en x et qui vérifie les propriétés (I^K_1) par multiplication par x sur W^1 (cf. Déf. 2).

ii) Inversement, si h est une fonction de la variable $(x, \lambda) \in W$ qui vérifie les propriétés (I^K_1) par multiplication par x sur W^1 (cf. Déf. 2), alors h s'écrit :$h(x, \lambda) = x[1 + H(x, \lambda)]$ $\forall (x, \lambda) \in W, H$ étant continue en (x, λ) sur W et vérifie les propriétés (I^K_0) par multiplication par x sur W^1 (cf. Déf. 1).

iii) Si L est une fonction de la variable $(x, \lambda) \in W^1$ qui vérifie les propriétés $(I^K_{\lambda,0})$ par multiplication par x sur W^1 (cf. Déf. 3) et si les limites :

(13.1) $$\mathrm{Lim}_{x \to 0} L(x, \lambda) \text{ et } \mathrm{Lim}_{x \to 0}\, x.\frac{\partial L}{\partial x}(x, \lambda)$$

existent $\forall \lambda \in O$, alors la fonction l définie sur W^1 par $l(x, \lambda) = x[1 + L(x, \lambda)]$ peut être prolongée sur W en une fonction impaire en x et qui vérifie les propriétés $(I^K_{\lambda,1})$ par multiplication par x sur W^1 (cf. Déf. 4 et Remarque 1-1°)).

iv) Inversement, si l est une fonction de la variable $(x, \lambda) \in W$ qui vérifie les propriétés $(I^K_{\lambda,1})$ par multiplication par x sur W^1(cf. Déf. 4), alors pour tout $(x, \lambda) \in W$, l s'écrit : $l(x, \lambda) = x[1 + L(x, \lambda)]$ où L est continue en $(x, \lambda) \in W$ et vérifie les propriétés $(I^K_{\lambda,0})$ par multiplication par x sur W^1 (cf. déf. 3 et Remarque 1-1°)).

Preuve :

i) Soient H et h comme dans l'assertion i), définissons sur W la fonction $\tilde{h}$ par :

$$\tilde{h}(x,\lambda) = \begin{cases} h(x,\lambda) & \text{pour } (x,\lambda) \in W^1 \\ 0 & \text{pour } x = 0 \text{ et } \lambda \in O \\ -h(-x,\lambda) & \text{pour } (x,\lambda) \in]-\varepsilon,0[\times O \end{cases}$$

et posons $\frac{\partial \tilde{h}}{\partial x}(0,\lambda) = 1 \ \forall \lambda \in O$. $\tilde{h}$ est impaire en x sur W et est de classe C^K en (x,λ) sur W° ; de plus elle est de classe C^1 en x pour $x = 0$ et vérifie trivialement les conditions i) des propriétés (I_1^K) (cf. Déf. 2). On a :

(13.1') $\forall (x,\lambda) \in W^1$ et $\forall n \in \{2,...,K\}$: $\frac{\partial^n \tilde{h}}{\partial x^n}(x,\lambda) = n.\frac{\partial^{n-1} H}{\partial x^{n-1}} + x.\frac{\partial^n H}{\partial x^n}$,

ce qui montre que $\tilde{h}$ vérifie aussi les conditions (11) des propriétés (I_1^K), la fonction multiplicatrice étant $\rho(x,\lambda) = x$. $\tilde{h}$ vérifie donc les propriétés (I_1^K) par multiplication par x sur W^1.

ii) Posons pour tout $(x,\lambda) \in W^\circ$ $H(x,\lambda) = h(x,\lambda)/x - 1$ et $H(0,\lambda) = 0$; d'après les conditions i) des propriétés (I_1^K) (cf. Déf. 2) et la Remarque 1-2°), H est continue en (x,λ) sur W et est de classe C^K en (x,λ) sur W°, de plus en appliquant le théorème des accroissements finis à h sur $[0,x]$, H vérifie les propriétés $(I_0^K)_0$ et $(I_0^K)_1$ (cf. Déf. 1 (9)) par multiplication par x sur W^1. Comme en i) on a (13.1') ce qui montre que H vérifie les propriétés (I_0^K) par multiplication par x sur W^1.

iii) D'après (13.1), définissons la fonction L par continuité sur $\{0\} \times O$ en posant $L(0,\lambda) = \operatorname{Lim}_{x \to 0^+} L(x,\lambda)$ pour tout $\lambda \in O$ et montrons d'abord que l'existence des deux limites (13.1) entraîne plus précisément que :

(13.2) $\forall \lambda \in O \quad \operatorname{Lim}_{x \to 0} x.\frac{\partial L}{\partial x}(x,\lambda) = 0$

pour cela, on a : $\forall x \in]0,\varepsilon[\ \forall \lambda \in O \ \exists z(x,\lambda) \in]0,x[$ tel que :

$$L(x,\lambda) - L(0,\lambda) = x.\frac{\partial L}{\partial x}(z(x,\lambda),\lambda),$$

l'inégalité $|z(x,\lambda).\frac{\partial L}{\partial x}(z(x,\lambda),\lambda)| \leq |x.\frac{\partial L}{\partial x}(z(x,\lambda),\lambda)|$ et le fait que $\operatorname{Lim}_{x \to 0} z(x,\lambda) = 0$ pour tout $\lambda \in O$ permettent de conclure.

Définissons alors sur W la fonction $\tilde{l}$ par :

$$\tilde{l}(x,\lambda) = \begin{cases} l(x,\lambda) & \text{pour } (x,\lambda) \in W^1 \\ 0 & \text{pour } x = 0 \text{ et } \lambda \in O \\ -l(-x,\lambda) & \text{pour } (x,\lambda) \in]-\varepsilon,0[\times O \end{cases}$$

et posons $\frac{\partial \tilde{l}}{\partial x}(0,\lambda) = 1 + L(0,\lambda)$. On montre facilement, comme pour l'assertion i) ci-dessus, et grâce à (13.1) et (13.2) que $\tilde{l}$ est impaire en x sur W et vérifie les propriétés $(I^K_{\lambda,1})$ par multiplication par x sur W^1.

iv) Posons pour tout $(x,\lambda) \in W^\circ$ $L(x,\lambda) = l(x,\lambda)/x - 1$ et $L(0,\lambda) = \frac{\partial l}{\partial x}(0,\lambda) - 1$ on montre alors, comme pour l'assertion ii) ci-dessus, que L est continue en (x,λ) sur W et vérifie les propriétés $(I^K_{\lambda,0})$ par multiplication par x sur W^1.

Remarque 1.1 : les conditions (13.1) de l'assertion iii) sont essentielles :

a) l'existence de la première des deux limites (13.1) n'est pas une conséquence des propriétés $(I^K_{\lambda,0})$ comme le montre l'exemple : prenons $O =]0,\varepsilon[$ et posons $L(x,\lambda) = \lambda.sin(lnx)$ pour $(x,\lambda) \in W^1 =]0,\varepsilon[\times]0,\varepsilon[$, L vérifie trivialement les propriétés $(I^K_{\lambda,0})$ par multiplication par x sur W^1et $\mathrm{Lim}_{x\to 0} L(x,\lambda)$ n'existe pas pour tout $\lambda \in]0,\varepsilon[$.

b) l'existence de la deuxième des limites (13.1) n'est pas une conséquence des propriétés $(I^K_{\lambda,0})$ et de l'existence de la première limite (13.1) comme le montre l'exemple : $L(x,\lambda) = \lambda\left[(Lnx)^{-1}.sin\left((Lnx)^2\right) + 1\right]$ pour tout $(x,\lambda) \in W^1 =]0,\varepsilon[\times]0,\varepsilon[$, L vérifie les propriétés $(I^K_{\lambda,0})$ par multiplication par x sur W^1 et $\mathrm{Lim}_{x\to 0} L(x,\lambda) = \lambda$ pour tout $\lambda \in]0,\varepsilon[$ et $\mathrm{Lim}_{x\to 0} x.\frac{\partial L}{\partial x}(x,\lambda)$ n'existe pas pour tout $\lambda \in]0,\varepsilon[$

c) de même, l'existence des deux limites (13.1) et les propriétés $(I^K_{\lambda,0})$ n'entraînent pas l'existence de la limite $\mathrm{Lim}_{x\to 0} x^2.\frac{\partial^2 L}{\partial x^2}(x,\lambda)$ pour tout $\lambda \in O$ comme le montre l'exemple : $L(x,\lambda) = \lambda.(Lnx)^{-2}.sin\left((Lnx)^2\right)$,... etc.

- Les lemmes qui suivent, ainsi que le Lemme 1, ne sont établis que dans des cas particuliers qui nous sont utiles pour la suite de l'article. On peut énoncer et démontrer des résultats plus généraux.

LEMME 2 : Soient O, W, W^1 et W° comme dans le lemme 1 :

i) soient H et L des fonctions de la variable $(x,\lambda) \in W^1$qui vérifient respectivement les propriétés (I^K_0) et $(I^K_{\lambda,0})$ par multiplication par x sur W^1. Soient h et l des fonctions de la variable $(x,\lambda) \in W$ qui vérifient respectivement les propriétés (I^K_1) et $(I^K_{\lambda,1})$ par multiplication par x sur W^1 et soit η une fonction de classe C^K en λ sur O (cf. Remarque 0) telle que $\exists a > 0, \forall\lambda \in O : a > \eta(\lambda) > 0$; alors les fonctions composées :

i1) $Ho(\eta.h) : (x,\lambda) \in W^1 \mapsto H(\eta(\lambda).h(x,\lambda),\lambda)$ vérifie les propriétés (I_0^K) par multiplication par x sur W^1.

i2) $Ho(\eta.l)$, $Lo(\eta.h)$ et $Lo(\eta.l)$ (composition comme ci-dessus) vérifient les propriétés $(I_{\lambda,0}^K)$ par multiplication par x sur W^1.

ii) Soient H et L comme en i) et r une fonction de classe C^K en λ sur O telle que $\exists a > 0$, $\exists b > 0, \forall \lambda \in O : r(\lambda) > a$ et $r(\lambda) < b$; alors les fonctions composées :

$$(x,\lambda) \in W^1 \mapsto H(x^{r(\lambda)},\lambda) \text{ et } (x,\lambda) \in W^1 \mapsto L(x^{r(\lambda)},\lambda)$$

vérifient respectivement les propriétés (I_0^K) et $(I_{\lambda,0}^K)$ par multiplication par x sur W^1.

Preuve : aisée, laissée au lecteur; utiliser la formule (13.3) de dérivation d'une fonction composée :

(13.3) $(gof)^{(n)}(x) = \sum_{m=1}^{n} \frac{n!}{m!} . \sum_{i_1+\ldots+i_m=n,i_j\geq 1} \frac{1}{i_1!\ldots i_m!} . g^{(m)}(f(x)).f^{(i_1)}(x)\ldots f^{(i_m)}(x)$

LEMME 3 : Soient O, W, W° et W^1 comme dans le lemme 1 et soit h une fonction de la variable $(x,\lambda) \in W$ qui vérifie les propriétés (I_1^K) (resp. $(I_{\lambda,1}^K)$) par multiplication par x sur W^1 ; alors quitte à réduire ε et O, la fonction $h_\lambda : x \in]-\varepsilon,\varepsilon[\mapsto h(x,\lambda)$ est inversible pour tout $\lambda \in O$ et la fonction l définie pour tout $(y,\lambda) \in W$ par $l(y,\lambda) = h_\lambda^{-1}(y)$ vérifie les propriétés (I_1^K) (resp. $(I_{\lambda,1}^K)$) par multiplication par y sur W^1.

Preuve : h étant de classe C^1 en $x \in]-\varepsilon,\varepsilon[$ pour tout $\lambda \in O$ (cf. Déf. 2 (resp. Déf. 4)) et $Lim_{x\to 0, x\neq 0} \frac{\partial h}{\partial x}(x,\lambda) = 1$ uniformément en $\lambda \in O$ (resp. $Lim_{(x,\lambda)\to(0,0), (x,\lambda)\in W^\circ} \frac{\partial h}{\partial x}(x,\lambda) = 1$), la fonction h_λ est inversible en classe C^1 sur $]-\varepsilon,\varepsilon[$ pour tout $\lambda \in O$ (quitte à réduire ε et O autour du zéro de $\mathbf{R}^\wedge$).Posons alors $l(y,\lambda) = h_\lambda^{-1}(y)$ pour tout $(y,\lambda) \in W$ (on garde la même notation en ε), on peut écrire :

(13.4) $\forall (y,\lambda) \in W \quad h(l(y,\lambda),\lambda) = y,$

h étant de classe C^K en (x,λ) sur W° et de classe C^1 en y sur $]-\varepsilon,\varepsilon[$ pour tout $\lambda \in O$ et $\frac{\partial h}{\partial x}$ étant non nulle pour tout $(x,\lambda) \in W$, l'égalité (13.4) permet de montrer que l est de classe C^K en (y,λ) sur W° et qu'elle vérifie les conditions i) des propriétés (I_1^K) (cf. Déf. 2) (resp. $(I_{\lambda,1}^K)$ (cf. Déf. 4)). La formule (13.3) ci-dessus appliquée à (13.4) pour

$(y, \lambda) \in W^1$ et une récurrence sur $n \in \{2, ..., K\}$ permettent de montrer que l vérifie les conditions (11) (resp. (13)) avec $\rho(y, \lambda) = y$; ce qui prouve le lemme.

LEMME 4 : Soient O, W, W^1 et W° comme dans le lemme 1, soit r une fonction de classe C^K en λ sur O telle que : $(\exists a > 0)(\exists b > 0)(\forall \lambda \in O) : a < r(\lambda) < b$, et posons $s(\lambda) = [r(\lambda)]^{-1}$ pour tout $\lambda \in O$; soit d une fonction de la variable $(x, \lambda) \in W^1$ qui vérifie les propriétés (I_0^K) par multiplication par x sur W^1 (cf. Déf. 1), posons $D(x, \lambda) = x^{r(\lambda)}.[1 + d(x, \lambda)]$ pour tout $(x, \lambda) \in W^1$ et notons D_λ la fonction : D_λ : $x \in]0, \varepsilon[\mapsto D_\lambda(x) = D(x, \lambda)$ (on gardera partout la même notation en ε).

i) Soit $X = h(x, \lambda)$ un changement de variable (I_1^K) défini sur W (cf. Déf. 5) (h vérifie les propriétés (I_1^K) par multiplication par x sur W^1) ; alors dans la variable X, la fonction D s'écrit : $\bar{D}(X, \lambda) = X^{r(\lambda)}.[1 + \bar{d}(X, \lambda)]$ pour tout $(X, \lambda) \in W^1$, la fonction $\bar{d}$ vérifie les propriétés (I_0^K) par multiplication par X sur W^1.

ii) Quitte à réduire ε , la fonction D_λ est inversible pour tout $\lambda \in O$ et la fonction $\tilde{D}$ définie pour tout $(y, \lambda) \in W^1$ par : $\tilde{D}(y, \lambda) = D_\lambda^{-1}(y)$ a pour expression :

$$\tilde{D}(y, \lambda) = y^{s(\lambda)}.[1 + \tilde{d}(y, \lambda)] \; \forall (y, \lambda) \in W^1$$

la fonction $\tilde{d}$ vérifie les propirétés (I_0^K) par multiplication par y sur W^1.

Preuve :

i) D'après le lemme 3, quitte à réduire ε , la fonction h_λ : $x \in]-\varepsilon, \varepsilon[\mapsto h(x, \lambda)$ est inversible pour tout $\lambda \in O$ et la fonction l définie pour tout $(X, \lambda) \in W$ par $l(X, \lambda) = h_\lambda^{-1}(X) = x$ vérifie les propriétés (I_1^K) par multiplication par X sur W^1 et s'écrit d'après le lemme 1 : $l(X, \lambda) = X[1 + L(X, \lambda)]$ pour tout $(X, \lambda) \in W$; la fonction L vérifie les propriétés (I_0^K) par multiplication par X sur W^1. Posons pour tout $(X, \lambda) \in W^1 \bar{D}(X, \lambda) = D(l(X, \lambda), \lambda)$; $\bar{D}$ s'écrit : $\bar{D}(X, \lambda) = X^{r(\lambda)}.[1 + \bar{d}(X, \lambda)]$ et le lemme 2 montre que $\bar{d}$ vérifie les propriétés (I_0^K) par multiplication par X sur W^1.

ii) On a déjà une démonstration géométrique de cette assertion dans le cas où D_λ est l'application de Dulac relative à un point de selle d'un champ de vecteurs du plan $\mathbf{R}^2$ (cf. (6) et (8)). Donnons-en une démonstration analytique dans le cas général : posons $H(x, \lambda) = [1 + d(x, \lambda)]^{s(\lambda)} - 1$ pour $(x, \lambda) \in W^1$, H vérifie trivialement les propriétés (I_0^K) par multiplication par x sur W^1 ; posons aussi pour $(x, \lambda) \in W^1$: $h(x, \lambda) = x[1 + H(x, \lambda)]$ et $h_\lambda(x) = h(x, \lambda)$, d'après le lemme 1 h admet un prolongement $\tilde{h}$ à W impaire en x

et qui vérifie les propriétés (I_1^K) par multiplication par x sur W^1 ; posons pour tout $(x,\lambda) \in W$ $\tilde{h}_\lambda(x) = \tilde{h}(x,\lambda)$, d'après le lemme 3 et quitte à diminuer ε , $\tilde{h}_\lambda$ est inversible pour tout $\lambda \in O$ et la fonction l définie sur W par : $l(y,\lambda) = \tilde{h}_\lambda^{-1}(y)$ vérifie les propriétés (I_1^K) par multiplication par y sur W^1 et s'écrit d'après le lemme 1 : $l(y,\lambda) = y[1+L(y,\lambda)]$ pour tout $(y,\lambda) \in W$, L vérifie les propriétés (I_0^K) par multiplication par y sur W^1. Désignons par f_λ la fonction : $f_\lambda(X) = X^{r(\lambda)}$ pour $X \in \mathbf{R}^{+*}$ et $\lambda \in O$, on vérifie que $\forall(x,\lambda) \in W^1$ $D_\lambda(x) = (f_\lambda o h_\lambda)(x)$ et donc que D_λ est inversible pour tout $\lambda \in O$ et que $D_\lambda^{-1}(y) = \left(\tilde{h}_\lambda^{-1} o f_\lambda^{-1}\right)(y)$; le lemme 2-ii) permet de conclure la preuve.

- Voici maintenant le résultat fondamental qui conduira, grâce à sa forme récursive, à la réduction finale signalée dans l'introduction (cf. (1)). Soit O une partie de $\mathbf{R}^\wedge$ telle que $0 \in \bar{O}$ et soit β une fonction de classe C^K en $\lambda \in O$ (cf. Remarque 0) telle que $\mathrm{Lim}_{\lambda\to 0}\beta(\lambda) = 0$, on suppsose que β est de signe constant sur O au sens suivant :

$$\textbf{(13.5)} \quad \begin{cases} \forall \quad \lambda \in O \ : \ \beta(\lambda) \ > \ 0 & \text{auquel cas on pose} \quad v = + \\ \forall \quad \lambda \in O \ : \ \beta(\lambda) \ \leq \ 0 & \text{auquel cas on pose} \quad v = - \end{cases}$$

soit r une fonction de classe C^K en λ sur O telle que : $(\exists a > 0)(\exists b > 0)(\forall \lambda \in O)$ $a < r(\lambda) < b$; posons pour tout $\lambda \in O$: $s(\lambda) = [r(\lambda)]^{-1}$. Soient η et $\bar{\beta}$ les fonctions :

$$\textbf{(13.6)} \quad \forall \lambda \in O : \eta(\lambda) = \begin{cases} \beta(\lambda) & \text{si} \quad v = + \\ 0 & \text{si} \quad v = - \end{cases} \quad \text{et } \bar{\beta}(\lambda) = [\eta(\lambda)]^{r(\lambda)},$$

soit $\theta > 1$, posons :

$$\textbf{(13.7)} \quad \begin{cases} W =]-\varepsilon,\varepsilon[\times O, W^1 =]0,\varepsilon[\times O, & V = \bigcup\limits_{\lambda\in O}]\eta(\lambda),\varepsilon[\times\{\lambda\}, \\ \mathcal{U} = \bigcup\limits_{\lambda\in O}]\beta(\lambda),\varepsilon[\times\{\lambda\}, & V_\theta = \bigcup\limits_{\lambda\in O}]\theta.\eta(\lambda),\varepsilon[\times\{\lambda\} \end{cases}$$

LEMME 5 (LEMME FONDAMENTAL) : Soit h_1 une fonction de la variable $(x,\lambda) \in W$ (cf. (13.7)) impaire en x et qui vérifie les propriétés $(I_{\lambda,1}^K)$ par multiplication par x sur W^1 (cf. Déf. 4 et (13.7)). Désignons par γ et $\bar{\gamma}$ les fonctions : (cf. (13.5))

$$\textbf{(13.8)} \quad \forall \lambda \in O \ \gamma(\lambda) = h_1(\beta(\lambda),\lambda) \text{ et } \bar{\gamma}(\lambda) = \begin{cases} [\gamma(\lambda)]^{r(\lambda)} & \text{si} \quad v = + \\ 0 & \text{si} \quad v = - \end{cases}$$

Soit "$x = h(y,\lambda)$ $\forall(y,\lambda) \in W$" un changement de variable (I_1^K) (cf. Déf. 5) ; h vérifiant les propriétés (I_1^K) par multiplication par y sur W^1. désignons par h_λ la fonction $h_\lambda(y) = h(y,\lambda)$ $\forall(y,\lambda) \in W$ et posons : (cf. (13.6) et Lemme 3)

(13.9) $\forall \lambda \in O\ \tilde{\beta}(\lambda) = h_\lambda^{-1}(\bar{\beta}(\lambda),$

alors

i) la fonction l définie par :

(13.10) $\forall (x,\lambda) \in W\ l(x,\lambda) = h_1(x - \beta(\lambda),\lambda) + \gamma(\lambda)$

vérifie les propriétés $(I^K_{\lambda\,1})$ par multiplication par $(x - \beta(\lambda))$ sur $\mathcal{U}$ (cf. (13.7)), et par multiplication par x sur V_θ (cf. Remarque 1.3°)). De plus elle s'écrit :

(13.11) $\forall (x,\lambda) \in W\ l(x,\lambda) = x[1 + L(x,\lambda)]$,

où la fonction L vérifie en particulier :

(13.12) $\mathrm{Lim}_{sur\ V}\, L = 0$ et $\mathrm{Lim}_{sur\ V}\, x.\frac{\partial L}{\partial x} = 0$

de plus elle est continue sur W et vérifie les propriétés $(I^K_{\lambda.0})$ par multiplication par $(x - \eta(\lambda))$ sur V (cf. (13.6) et (13.7)) et par multiplication par x sur V_θ (cf. (13.7)) ;

ii) posons pour tout $(y,\lambda) \in W^1$ (cf. (13.7)) :

(13.13) $f(y,\lambda) = \left[l\left([h(y,\lambda)]^{s(\lambda)},\lambda\right)\right]^{r(\lambda)},$

la fonction h_2 définie sur W par :

$$\textbf{(13.14)}\quad \forall \lambda \in O\ h_2(y,\lambda) = \begin{cases} f(y+\tilde{\beta}(\lambda),\lambda) \quad - \quad \bar{\gamma}(\lambda) & \text{si} \quad y \in]0,\varepsilon[\\ 0 & \text{si} \quad y = 0 \\ -h_2(-y,\lambda) & \text{si} \quad y \in]-\varepsilon,0[\end{cases}$$

est impaire en y et vérifie les propriétés $(I^K_{\lambda,1})$ par multiplication par y sur W^1.

Preuve :

i) h_1 étant impaire en x et vérifiant les propriétés $(I^K_{\lambda,1})$ par multiplication par x sur W^1, la fonction l (cf. (13.10) et (13.8)) vérifie trivialement les propriétés $(I^K_{\lambda,1})$ par multiplication par $(x - \beta(\lambda))$ sur $\mathcal{U}$ conformément à la Déf. 4. Pour montrer qu'en plus l vérifie les propriétés $(I^K_{\lambda,1})$ par multiplication par x sur V_θ (conformément à la Remarque 1.3°), considérons les deux cas : si $v = -$ (cf. (13.5)), on a $V_\theta = W^1 \subset \mathcal{U}$ et

$\forall(x,\lambda) \in V_\theta \; 0 \leq x \leq x - \beta(\lambda)$, ce qui prouve l'affirmation ; maintenant si $v = +$, on vérifie que $\forall(x,\lambda) \in V_\theta \; 0 \leq x \leq \hat{\theta}(x - \beta(\lambda))$ avec $\hat{\theta} = \theta.(\theta-1)^{-1}$ et on conclut au résultat.
La relation (13.11) est une application du théorème des accroissements finis à la fonction h_1 sur l'intervalle $(x - \beta(\lambda), -\beta(\lambda))$: $\exists Z(x,\lambda) \in (x - \beta(\lambda), -\beta(\lambda))$ telle que :

$$l(x,\lambda) = x.\frac{\partial h_1}{\partial x}(Z(x,\lambda),\lambda)$$

posons :

(13.15) $L(x,\lambda) = \frac{\partial h_1}{\partial x}(Z(x,\lambda),\lambda) - 1$

la fonction L, qui est aussi définie pour $x \neq 0$ par $L(x,\lambda) = l(x,\lambda)/x - 1$, est continue sur W, est de classe C^1 en x sur W^1 et est de classe C^K en (x,λ) sur V. (13.14) ci-dessus montre que $Lim_{\substack{(x,\lambda)\to(0,0)\\(x,\lambda)\in V}} L(x,\lambda) = 0$ et la relation (13.11) dérivée une fois montre que $Lim_{\substack{(x,\lambda)\to(0,0)\\(x,\lambda)\in V}} x.\frac{\partial L}{\partial x}(x,\lambda) = 0$. Pour montrer les propriétés $(I^K_{\lambda.0})_n$ pour $n \in \{2,...,K\}$ (cf. (12). Déf. 3) on utilise la formule (13.1') appliquée à l et le fait que l vérifie les propriétés $(I^K_{\lambda.1})$ par multiplication par $(x - \beta(\lambda))$ sur $\mathcal{U} \supset V$; de plus si $v = -$, on a $\forall(x,\lambda) \in V : 0 \leq x = x - \eta(\lambda) \leq x - \beta(\lambda)$ et si $v = +$ on a $\forall(x,\lambda) \in V \; 0 \leq x - \eta(\lambda) = x - \beta(\lambda) \leq x$. Maintenant si $\theta > 1$ on a $V_\theta \subset V$ et le fait que l vérifie les propriétés $(I^K_{\lambda.1})$ par multiplication par x sur V_θ et la relation (13.15) ci-dessus prouvent que L vérifie les propriétés $(I^K_{\lambda.0})$ par multiplication par x sur V_θ.

ii) D'après le lemme 1 ii) on peut écrire :

(13.16) $\forall(y,\lambda) \in W \; h(y,\lambda) = y[1 + H(y,\lambda)]$

où H est continue en $(y,\lambda) \in W$ et vérifie les propriétés (I^K_0) par multiplication par y sur W^1 ; de plus on a : $H(0,\lambda) = 0 \; \forall\lambda \in O$ et par dérivation de (13.16) : $\forall\lambda \in O \; \mathrm{Lim}_{y\to 0}\, y.\frac{\partial H}{\partial y}(y,\lambda) = 0$. Grâce à (13.11) et (13.16) écrivons la fonction f (cf. (13.12)) comme suit :

(13.17) $\forall(y,\lambda) \in W^1 \; f(y,\lambda) = y[1 + F(y,\lambda)]$

avec :

(13.18) $\forall(y,\lambda) \in W^1 \; F(y,\lambda) = [1 + H(y,\lambda)].[1 + L([h(y,\lambda)]^{s(\lambda)},\lambda)]^{r(\lambda)} - 1.$

Distinguons maintenant les deux cas $v = +$ et $v = -$ (cf. (13.5)) :

1°) $v = -$: la fonction h_2 (cf. (13.13)) s'écrit pour $(y,\lambda) \in W^1$: $h_2(y,\lambda) = f(y,\lambda)$, (cf. (13.6), (13.8) et (13.9)). Comme L vérifie les propriétés $(I^K_{\lambda\,0})$ par multiplication par x sur $V = W^1$, le lemme 2 i2) et ii) montre que la fonction composée : $(y,\lambda) \in W^1 \mapsto L\left([h(y,\lambda)]^{s(\lambda)},\lambda\right)$ vérifie les propriétés $(I^K_{\lambda,0})$ par multiplication par y sur W^1. (13.11) et (13.14) montrent que les deux limites : $\mathrm{Lim}_{x\to 0} L(x,\lambda)$ et $\mathrm{Lim}_{x\to 0}\, x\frac{\partial L}{\partial x}(x,\lambda)$ existent pour tout $\lambda \in O$. Ceci et ce qui a été dit au sujet de la fonction H (cf. (13.16)) montrent que la fonction F (cf. (13.18)) vérifie les propriétés $(I^K_{\lambda\,0})$ par multiplication par y sur W^1 et que les deux limites : $\mathrm{Lim}_{y\to 0} F(y,\lambda)$ et $\mathrm{Lim}_{y\to 0}\, y.\frac{\partial F}{\partial y}(y,\lambda)$ existent pour tout $\lambda \in O$ et le prolongement (13.13) donne, d'après le lemme 1 iii), une fonction h_2 impaire en y et qui vérifie les propriétés $(I^K_{\lambda\,1})$ par multiplication par y sur W^1.

2°) $v = +$: la fonction f (cf.(13.12) et (13.17)) étant de classe C^1 en y sur W^1, écrivons :

(13.19) $\forall (y,\lambda) \in W^1\ f(y+\tilde{\beta}(\lambda),\lambda) - \bar{\gamma}(\lambda) = y[1 + H_2(y,\lambda)]$

avec :

(13.20) $H_2(y,\lambda) = \frac{\partial f}{\partial y}(Z(y,\lambda),\lambda) - 1$ et $Z(y,\lambda) \in]\tilde{\beta}(\lambda), y + \tilde{\beta}(\lambda)[$;

(il est clair que $\bar{\gamma}(\lambda) = f(\tilde{\beta}(\lambda),\lambda)$ d'après (13.6), (13.8), (13.9) et (13.10)).
Quitte à resteindre O (autour de $0 \in \mathbf{R}^\wedge$), on peut supposer que la fonction h est croissante en y sur W. Posons :

(13.21) $\bar{V} = \bigcup_{\lambda\in O}]\bar{\beta}(\lambda),\varepsilon[\times\{\lambda\}, \tilde{V} = \bigcup_{\lambda\in O}]\tilde{\beta}(\lambda),\varepsilon[\times\{\lambda\}$;

la fonction : $(y,\lambda) \mapsto (h(y,\lambda),\lambda)$ envoie $\tilde{V}$ sur $\bar{V}$ et la fonction $(x,\lambda) \mapsto (x^{s(\lambda)},\lambda)$ envoie $\bar{V}$ sur V. Maintenant, d'après ce qu'on a vu sur L en i) et sur H en (13.16), la fonction F (cf. (13.18)) vérifie trivialement :

$$\mathit{Lim}_{\substack{(y,\lambda)\to(0,0)\\(y,\lambda)\in\tilde{V}}} F(y,\lambda) = 0 \text{ et } \mathit{Lim}_{\substack{(y,\lambda)\to(0,0)\\(y,\lambda)\in\tilde{V}}}\ y.\frac{\partial F}{\partial y}(y,\lambda) = 0$$

ce qui montre, d'après (13.17), (13.19) et (13.20) que : $\mathit{Lim}_{\substack{(y,\lambda)\to(0,0)\\(y,\lambda)\in W^1}} H_2(y,\lambda) = 0$ et $\mathit{Lim}_{\substack{(y,\lambda)\to(0,0)\\(y,\lambda)\in W^1}} y.\frac{\partial H_2}{\partial y}(y,\lambda) = 0$ et qu'en plus, en utilisant (13.19) et le fait que la fonction f est de classe C^1 au point $y = \tilde{\beta}(\lambda)$, les deux limites : $\mathrm{Lim}_{y\to 0} H_2(y,\lambda)$ et $\mathrm{Lim}_{y\to 0}\, y.\frac{\partial H_2}{\partial y}(y,\lambda)$ existent pour tout $\lambda \in O$. Il reste à montrer que :

(13.22) $\forall n \in \{2,...,K\}\ \mathit{Lim}_{\substack{(y,\lambda)\to(0,0)\\(y,\lambda)\in W^1}} y^n.\frac{\partial^n H_2}{\partial y^n}(y,\lambda) = 0$;

pour cela, posons : $\bar{\theta}(\lambda) = \theta^{r(\lambda)}$ et désignons par $\tilde{\theta}$ la fonction de $\lambda \in O$ telle que $\tilde{\theta}(\lambda).\tilde{\beta}(\lambda) = h_\lambda^{-1}(\bar{\theta}(\lambda).\bar{\beta}(\lambda))$, on a : $Lim_{\substack{\lambda \to 0 \\ \lambda \in O}} \tilde{\theta}(\lambda)/\bar{\theta}(\lambda) = 1$. Posons aussi :

(13.23) $\bar{V}_\theta = \bigcup_{\lambda \in O}]\bar{\theta}(\lambda).\bar{\beta}(\lambda), \varepsilon[\times \{\lambda\}$ et $\tilde{V}_\theta = \bigcup_{\lambda \in O}]\tilde{\theta}(\lambda).\tilde{\beta}(\lambda), \varepsilon[\times \{\lambda\}$,

la fonction $(y, \lambda) \mapsto (h(y, \lambda), \lambda)$ envoie $\tilde{V}_\theta$ sur $\bar{V}_\theta$ et la fonction $(x, \lambda) \mapsto (x^{s(\lambda)}, \lambda)$ envoie $\bar{V}_\theta$ sur V_θ (cf. (13.7)). Comme la fonction L (cf. (13.11)) vérifie les propriétés $(I^K_{\lambda,0})$ par multiplication par x sur V_θ, le même raisonnement que dans le cas 1°) ci-dessus (utilisant le lemme 2) montre que la fonction F (cf. (13.18)) vérifie les propriétés $(I^K_{\lambda,0})_n$ pour $n \in \{2, ..., K\}$ (cf. (12). Déf. 3) par multiplication par y sur $\tilde{V}_\theta$ (cf. (13.23)) soit :

(13.24) $\forall n \in \{2, ..., K\}\ Lim_{\substack{(y,\lambda) \to (0,0) \\ (y,\lambda) \in \tilde{V}_\theta}} y^n . \frac{\partial^n F}{\partial y^n}(y, \lambda) = 0.$

Maintenant, en reprenant la formule (13.1') appliquée à l et sans minorer x par $x - \beta(\lambda)$ dans le cas $v = +$, on montre que L vérifie plus précisément :

(13.25) $\forall n \in \{2, ..., K\}\ Lim_{\substack{(x,\lambda) \to (0,0) \\ (x,\lambda) \in V}} x.(x - \beta(\lambda))^{n-1} . \frac{\partial^n L}{\partial x^n}(x, \lambda) = 0.$

Désignons par G la fonction $(x, \lambda) \in W^1 \mapsto G(x, \lambda) = L(x^{s(\lambda)}, \lambda)$ et montrons que G vérifie des propriétés similaires à (13.25) : (cf. (13.23))

(13.26) $\forall n \in \{2, ..., K\}$ $\quad \mathrm{Lim}_{\substack{(x,\lambda) \to (0,0) \\ (x,\lambda) \in \bar{V} \setminus \bar{V}_\theta}} x.(x - \bar{\beta}(\lambda))^{n-1} . \frac{\partial^n G}{\partial x^n}(x, \lambda) = 0$;

pour cela, la formule (13.3) nous donne :

(13.27) $\frac{\partial^n G}{\partial x^n}(x, \lambda) = \sum_{m=1}^{n} * . \frac{\partial^m L}{\partial x^m}(x^{s(\lambda)}, \lambda) . x^{m s(\lambda) - n}$;

où $*$ désigne des fonctions de λ bornées sur O.

Le théorème des accroissements finis montre que $\forall (x, \lambda) \in \bar{V}, \exists Z(x, \lambda) \in]\bar{\beta}(\lambda), x[$ tel que : $x^{s(\lambda)} - [\bar{\beta}(\lambda)]^{s(\lambda)} = s(\lambda).(x - \bar{\beta}(\lambda)).Z^{s(\lambda)-1}$; d'où l'on tire que : (cf. (13.6))

(13.28) $x - \bar{\beta}(\lambda) = r(\lambda).(x^{s(\lambda)} - \beta(\lambda)).Z^{1-s(\lambda)}$;

posons $X = x - \bar{\beta}(\lambda)$,(13.27) et (13.28) donnent :

$$x.X^{n-1} . \frac{\partial^n G}{\partial x^n}(x, \lambda) = \sum_{m=1}^{n} * \ x^s.(x^s - \beta)^{m-1} . \frac{\partial^m L}{\partial x^m}(x^s, \lambda) . \left(\frac{X}{x}\right)^{n-m} . \left(\frac{Z}{x}\right)^{(m-1)(1-s)}$$

on a $\forall(x,\lambda) \in \bar{V}$: $(x-\bar{\beta}(\lambda))/x \leq 1$ et pour $\lambda \in O$ tel que $s(\lambda) \leq 1$ on a $\forall x \geq \bar{\beta}(\lambda)$ $(Z(x,\lambda)/x)^{(m-1)(1-s(\lambda))} \leq 1$ et pour $\lambda \in O$ tel que $s(\lambda) > 1$ on a $\forall x \in]\bar{\beta}(\lambda), \bar{\theta}(\lambda).\bar{\beta}(\lambda)[$ (cf. (13.23)) : $(Z(x,\lambda)/x)^{(m-1)(1-s(\lambda))} \leq [\bar{\theta}(\lambda)]^{-(m-1)(1-s(\lambda))}$; pour $m = 1$, le rapport Z/x ne pose pas de problème et on sait que $\mathrm{Lim}_{\substack{(x,\lambda)\to(0,0)\\(x,\lambda)\in V}} x.\frac{\partial L}{\partial x}(x,\lambda) = 0$; ceci et (13.25) permettent de conclure à (13.26) pour $(x,\lambda) \in \bar{V} \setminus \bar{V}_\theta$. Grâce à la propriété suivante de la fonction h (cf. Déf.2i) : $\mathrm{Lim}_{\,y\to 0}\frac{\partial h}{\partial y}(y,\lambda) = 1$ uniformément en $\lambda \in O$, on montre facilement que la fonction $\tilde{G}(y,\lambda) = G(h(y,\lambda),\lambda)$ (cf. (13.26)) vérifie : $\forall n \in \{2,...,K\}$ $\mathrm{Lim}_{\substack{(y,\lambda)\to(0,0)\\(y,\lambda)\in\tilde{V}\setminus\tilde{V}_\theta}} y.(y-\tilde{\beta}(\lambda))^{n-1}.\frac{\partial^n \tilde{G}}{\partial y^n}(y,\lambda) = 0$, (cf. (13.21) et (13.23)) et donc que la fonction F (cf. (13.18)) vérifie :

(13.29) $\forall n \in \{2,...,K\}$ $\mathrm{Lim}_{\substack{(y,\lambda)\to(0,0)\\(y,\lambda)\in\tilde{V}\setminus\tilde{V}_\theta}} y.(y-\tilde{\beta}(\lambda))^{n-1}.\frac{\partial^n F}{\partial y^n}(y,\lambda) = 0.$

Maintenant, la formule (13.1') appliquée à (13.19) et en utilisant l'expression (13.17) de la fonction f, on obtient :

$$(13.30)\quad \begin{cases} \forall n \in \{2,...,K\} \quad \forall(y,\lambda) \in W^1 : \\ n\dfrac{\partial^{n-1}F}{\partial y^{n-1}}(y+\tilde{\beta}(\lambda),\lambda) + (y+\tilde{\beta}(\lambda)).\dfrac{\partial^n F}{\partial y^n}(y+\tilde{\beta}(\lambda),\lambda) = \\ n\,\dfrac{\partial^{n-1}H_2}{\partial y^{n-1}}(y,\lambda) + y.\dfrac{\partial^n H_2}{\partial y^n}(y,\lambda) \,; \end{cases}$$

Une récurrence sur n dans l'égalité (13.30) qu'on multiplie par y^{n-1} et les résultats (13.24) et (13.29) permettent de conclure à (13.22), ce qui finit la preuve du lemme.

Remarque 2 : le lemme 5 reste vrai si la fonction h ne vérifie que des propriétés $(I^K_{\lambda,1})$ par multiplication par y sur W^1, par contre même si les deux fonctions h_1 et h vérifiaient des propriétés (I^K_1) par multiplication par x (resp. y) sur W^1, la fonction h_2 ne vérifie pas nécessairement des propriétés (I^K_1).

- Voici maintenant un corollaire du Lemme 5, plus fort que le Lemme 5 et qui servira, dans la phase finale de mise en équation dans le § II.3, à donner un sens intrinsèque à certaines fonctions (les b_i) figurants dans la forme normale (1) (cf. Introduction) de l'équation aux cycles limites et à prolonger ces fonctions par continuité sur un voisinage de $0 \in \mathbf{R}^\wedge$:

LEMME 6 : Reprenons les données et les conditions du Lemme 5 en supposant seulement que la fonction h_1 n'est plus impaire en x et en remplacant (13.8) par :

(13.31) $\forall\lambda \in O : \gamma(\lambda) = -h_1(-\beta(\lambda),\lambda)$ et $\bar{\gamma}$ est inchangé.

Alors le résultat i) est inchangé et si dans ii) (cf. Lemme 5) on désigne toujours par f la fonction donnée par (13.12), la fonction $\bar{h}_2$ définie sur W^1 (cf. (13.7)) par :

$$\textbf{(13.32)} \quad \forall (y,\lambda) \in W^1 : \bar{h}_2(y,\lambda) = f(y+\tilde{\beta}(\lambda),\lambda) - \bar{\gamma}(\lambda)$$

admet un prolongement h_2 à W qui vérifie les propriétés $(I^K_{\lambda.1})$ par multiplication par y sur W^1 et qui satisfait à la condition :

$$\textbf{(13.33)} \quad \forall \lambda \in O : h_2(-\tilde{\beta}(\lambda),\lambda) = - \quad \bar{\gamma}(\lambda).$$

Preuve :

i) : grâce à (13.31) la démonstration est la même que celle du Lemme 5.

ii) : reprenons les résultats du Lemme 5 (indépendants du fait que h_1 soit impaire en x) et posons (cf. (13.17) et (13.19)) :

$$\textbf{(13.34)} \quad \forall \lambda \in O : G(\lambda) = \begin{cases} (1+F(\tilde{\beta}(\lambda),\lambda).(1+H_2(\tilde{\beta}(\lambda),\lambda)^{-1} & \text{si} \quad v=+ \\ 1 & \text{si} \quad v=- \end{cases}$$

puis définissons sur W^1 la fonction g : $\forall \lambda \in O$

$$\textbf{(13.35)} \quad g(y,\lambda) = \begin{cases} G(\lambda) - (G(\lambda)-1).(y-\tilde{\beta}(\lambda))^{2K}.(\tilde{\beta}(\lambda))^{-2K} & \text{si } y \in]0,\tilde{\beta}(\lambda)] \\ G(\lambda) & \text{si } y > \tilde{\beta}(\lambda), \end{cases}$$

on vérifie facilement que : $\mathrm{Lim}_{\substack{\lambda \to 0 \\ \lambda \in O}} G(\lambda) = 1$ et que g est de classe C^K en (y,λ) sur W^1 et vérifie : $\mathrm{Lim}_{\substack{\lambda \to 0 \\ \lambda \in O}} g(y,\lambda) = 1$ uniformément en $y \in]0,\varepsilon[$ (de plus la fonction $g-1$ vérifie les propriétés $(I^K_{\lambda,0})$ par multiplication par y sur W^1 (uniformément en $y \in]0,\varepsilon[$)).

Définissons le prolongement h_2 par :

$$\textbf{(13.36)} \quad \forall \lambda \in O : h_2(y,\lambda) = \begin{cases} \bar{h}_2(y,\lambda) & \text{si} \quad y \in]0,\varepsilon[\\ 0 & \text{si} \quad y=0 \\ -g(-y,\lambda). \quad \bar{h}_2(-y,\lambda) & \text{si} \quad y \in]-\varepsilon,0[\end{cases}$$

on montre aisément que h_2 est de classe C^1 en y pour $y=0$ pour tout $\lambda \in O$, qu'elle vérifie les propriétés $(I^K_{\lambda.1})$ par multiplication par y sur W^1 (d'après Lemme 5.ii) Preuve) et que d'après (13.34), (13.35) et le fait que : $\forall \lambda \in O$ $\bar{\gamma}(\lambda) = f(\tilde{\beta}(\lambda),\lambda)$, elle satisfait à (13.33).

II.2 -CONVENTIONS GEOMETRIQUES

- Appliquons la conclusion de la partie I au voisinage de chacun des points de selle $P_i(\lambda)$ $i = 1, ..., k$ de la famille (X_λ) de l'introduction : K étant un entier fixé, il existe un voisinage $\mathcal{V}$ de 0 dans $\mathbf{R}^\wedge$ et pour tout $i = 1, ..., k$, des réels $\varepsilon_{i1} > 0$, $\varepsilon_{i2} > 0$ et des transversales $\sigma_i(\lambda)$ et $\tau_i(\lambda)$ (cf. fig. 5 ci-dessous) paramétrées en classe C^K respectivement par $x_i \in]-\varepsilon_{i1}, \varepsilon_{i1}[$ et $y_i \in]-\varepsilon_{i2}, \varepsilon_{i2}[$ et dépendant de façon C^K de $\lambda \in \mathcal{V}$ tels que l'application de Dulac et son inverse relatives au coin $P_i(\lambda)$ soient données par (6) et (8).

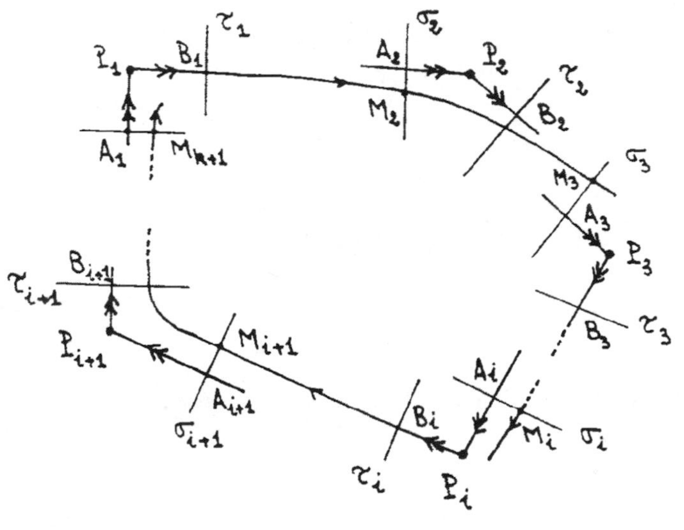

fig. 5

- Pour des raisons qu'on expliquera dans le paragraphe II.3, on est amené à subdiviser le voisinage $\mathcal{V}$ en plusieurs parties (qu'on appellera cônes). On construira ces cônes grâce à la notion de chemin qu'on définit ci-dessous.

DEFINITION 6 : *Soit* $\lambda \in \mathcal{V}$, *on appelle*" chemin partant du sommet P_1 dans le sens positif" *la ligne géométrique obtenue comme réunion de ses k branchements. Pour tout* $i = 1, ..., k$ *le branchement* n°i *a son origine* M_i *sur la transversale* σ_i *et son extrémité* M_{i+1} *sur* σ_{i+1} *(avec la convention* $k+1 \equiv 1$*). Ces k branchements sont définis par récurrence comme suit (cf.* fig. 5) : *pour* $i = 1$, $M_1 = A_1$ *et le branchement* n°1 *est la ligne* $A_1P_1B_1M_2$; *pour* $i \geq 2$ *deux cas se présentent : si* M_i *est d'abscisse* > 0 *sur* σ_i, *le branchement* n°i *est le segment d'orbite* M_iM_{i+1} ; *si* M_i *est d'abscisse* ≤ 0 *sur* σ_i, *le branchement* n°i *est la ligne* $M_iA_iP_iB_iM_{i+1}$.

- Le chemin partant de P_1 dans le sens positif sera aussi appelé "chemin partant de σ_1" sans spécifier le sens. On définit de la même façon "le chemin partant de P_1 dans le sens négatif" (qu'on appellera aussi "chemin partant de τ_1") en commençant par le point B_1 (cf. fig. 5) et en allant dans le sens contraire du champ X_λ. On généralise ensuite cette notion de chemin dans les deux sens (ainsi que les conventions de langage qui vont suivre) à n'importe quel sommet P_i du polycycle.

On dira que le i^e branchement du chemin partant de σ_1 est > 0 (resp. ≤ 0) si son extremité M_{i+1} est d'abscisse > 0 sur σ_i (resp. ≤ 0). Le dernier branchement sera appelé "aboutissement du chemin".

Posons $E = \{+, -\}$; désignons par $v_i \in E$ le signe du i^e branchement du chemin partant de P_j dans le sens $v \in E$ pour un certain $\lambda \in \mathcal{V}$; le $k-uplet$ $\bar{v} = (v_i)_{i=1...k} \in E^k$ sera appelé "signe" du chemin. On définit alors le cône $C(P_j, v, \bar{v})$ par :

(14) $C(P_j, v, \bar{v}) = \{\lambda \in \mathcal{V}$; le chemin partant de P_j dans le sens v a pour signe $\bar{v}\}$

il est clair que :

(15) $\forall j = 1, ..., k \ \forall v \in E \ \bigcup_{\bar{v} \in E^k} C(P_j, v, \bar{v}) = \mathcal{V}$,

et que si $C(P_j, v, \bar{v})$ est non vide, 0 est un point adhérent à $C(P_j, v, \bar{v})$.

II.3 -LA NORMALISATION

- Reprenons le début du paragraphe II.2 : K étant un entier fixé, il existe un voisinage $\mathcal{V}$ de 0 dans $\mathbf{R}^\wedge$ et pour tout $i = 1, ..., k$ des réels $\varepsilon_{i1} > 0$, $\varepsilon_{i2} > 0$ et des transversales $\sigma_i(\lambda)$ et $\tau_i(\lambda)$ (cf. fig. 5) paramétrées en classe C^K respectivement par $x_i \in]-\varepsilon_{i1}, \varepsilon_{i2}[$ et $y_i \in]-\varepsilon_{i2}, \varepsilon_{i2}[$ et dépendant de façon C^K de $\lambda \in \mathcal{V}$ tels que l'application de Dulac $D_{i,\lambda}$ et son inverse relatives au coin $P_i(\lambda)$ soient données par (cf. (6) et (8)) :

$$\textbf{(16.1)} \quad D_{i,\lambda}(x_i) = y_i = x_i^{r_i(\lambda)}.[1 + d_{i1}(x_i, \lambda)] \; \forall (x_i, \lambda) \in]0, \varepsilon_{i1}] \times \mathcal{V} = W_{i1}$$

$$\textbf{(16.2)} \quad D_{i,\lambda}^{-1}(y_i) = x_i = y_i^{s_i(\lambda)}.[1 + d_{i2}(y_i, \lambda)] \; \forall (y_i, \lambda) \in]0, \varepsilon_{i2}] \times \mathcal{V} = W_{i2}$$

où $s_i(\lambda) = [r_i(\lambda)]^{-1}$et d_{i1}(resp. d_{i2}) vérifie les propriétés (I_0^K) par multiplication par x_i (resp. y_i) sur W_{i1}(resp. W_{i2}) (cf. (7) et (9)). Quitte à réduire $\mathcal{V}$, on remplace tous les $\varepsilon_{i1}, \varepsilon_{i2}$ par $\varepsilon > 0$ assez petit.

Notons $\mathcal{R}_{i,\lambda}$ la correspondance régulière entre les transversales τ_i et σ_{i+1} (avec toujours la convention $k+1 \equiv 1$). On peut écrire :

$$\textbf{(17.1)} \quad \mathcal{R}_{i,\lambda}(y_i) = x_{i+1} = \eta_{io}(\lambda) + \psi_i(y_i, \lambda) \; \forall (y_i, \lambda) \in]-\varepsilon, \varepsilon[\times \mathcal{V}$$

$$\textbf{(17.2)} \quad \mathcal{R}_{i,\lambda}^{-1}(x_{i+1}) = y_i = \alpha_{io}(\lambda) + \varphi_i(x_{i+1}, \lambda) \; \forall (x_{i+1}, \lambda) \in]-\varepsilon, \varepsilon[\times \mathcal{V}$$

où α_{io} et η_{io} sont des fonctions de classe C^K en $\lambda \in \mathcal{V}$ telles que :

$$\textbf{(18.1)} \; \alpha_{io}(0) = \eta_{io}(0) = 0 \quad \forall i = 1, ..., k$$

et φ_i (resp. ψ_i) est de classe C^K en (x_{i+1}, λ) (resp. (y_i, λ)) sur $]-\varepsilon, \varepsilon[\times \mathcal{V}$ et est un difféomorphisme en x_{i+1}(resp. y_i) pour tout $\lambda \in \mathcal{V}$ préservant l'origine.
Posons :

$$\textbf{(18.2)} \quad \alpha_{i1}(\lambda) = \frac{\partial \varphi_i}{\partial x_{i+1}}(0, \lambda) \qquad i = 1, ..., k \qquad \lambda \in \mathcal{V} \; ;$$

quitte à resteindre $\mathcal{V}$, on a pour tout $i = 1, ..., k$ et pour tout $\lambda \in \mathcal{V}$:

$$\textbf{(18.3)} \quad \alpha_{i1}(\lambda) > =0.$$

On dira que la déformation (X_λ) du germe X_o le long de (Γ_k) est générique si la dimension de l'espace des paramètres $\wedge$ est $\geq k$ et si les k formes $(D_\lambda \alpha_{io}(0))$ (cf. (17.2)) sont indépendantes. (18.3) montre que cette définition est indépendante du sens de parcours du polycycle.

- Commençons par montrer un premier résultat de réduction valable pour tout $\lambda \in \mathcal{V}$

mais qui dépend de la transversale de départ dans un sens qu'on précisera :

LEMME 7 : Pour tout $i = 1,...,k$, il existe des paramétrages C^K des transversales $\sigma_i(\lambda), \tau_i(\lambda)$ pour tout $\lambda \in \mathcal{V}$ (notés toujours x_i, y_i) dans lesquels l'application de Dulac $D_{i,\lambda}$ et son inverse $D_{i,\lambda}^{-1}$ sont toujours données par les formules (16.1) et (16.2) pour tout $(x_i, \lambda) \in]0, \varepsilon[\times \mathcal{V}$ et pour tout $(y_i, \lambda) \in]0, \varepsilon[\times \mathcal{V}$; et les correspondances $\mathcal{R}_{i,\lambda}^{-1}$ sont données par :

(19.1) $\mathcal{R}_{i,\lambda}^{-1}(x_{i+1}) = y_i = \bar{\alpha}_{io}(\lambda) + x_{i+1}$ pour $i = 1,...,k-1$ et $\forall (x_{i+1}, \lambda) \in]-\varepsilon, \varepsilon[\times \mathcal{V}$

(19.2) $\mathcal{R}_{k,\lambda}^{-1}(x_1) = y_k = \bar{\alpha}_{ko}(\lambda) + \quad \bar{\alpha}_1(\lambda).x_1 \ \forall (x_1, \lambda) \in]-\varepsilon, \varepsilon[\times \mathcal{V}$.

où les fonctions $\bar{\alpha}_{io}$ pour $i = 1,...,k$ sont de classe C^{K-1} en $\lambda \in \mathcal{V}$ et vérifient $\bar{\alpha}_{io}(0) = 0$, et le facteur $\bar{\alpha}_1$ est au moins de classe C^{K-1} en $\lambda \in \mathcal{V}$ et est donné en fonction des α_{i1} (cf. (18.2)) par :

(19.3) $\bar{\alpha}_1(\lambda) = \alpha_{k1}(\lambda). \prod_{j=1}^{k-1} [\alpha_{j1}(\lambda)]^{r_k(\lambda)\, r_{k-1}(\lambda) .\, r_{j+1}(\lambda)}$.

Remarque 3 :

1°) l'inverse du facteur $\bar{\alpha}_1(\lambda)$ ci-dessus (cf. Remarque 4.2°) (19.8) ci-dessous) sera dit "facteur d'hyperbolicité du polycycle par rapport à la transversale σ_1" du fait que la réduction ci-dessus est relative à la transversale σ_1. D'après (18.3) on a :

(19.4) $\forall \lambda \in \mathcal{V} \quad \bar{\alpha}_1(\lambda) > 0$;

2°) dans la définition de la généricité de la déformation (X_λ) donnée ci-dessus, on peut remplacer les (α_{io}) par les $(\bar{\alpha}_{io})$ du lemme comme on le verra dans la preuve.

Preuve : Faisons une première réduction dans laquelle aucune transversale n'est privilégiée:d'après le lemme 4i) et ii), on peut supposer, en posant $\varphi_i(x_{i+1}, \lambda) = \alpha_{i1}(\lambda).g_i(x_{i+1}, \lambda)$ (cf. (17.2) et (18.2)) et en faisant le changement de paramétrage $X_{i+1} = g_i(x_{i+1}, \lambda)$ sur la transversale $\sigma_{i+1}(\lambda)$ pour $i = 1,...,k$ (puis en notant x_i au lieu de X_i) que les correspondances $\mathcal{R}_{i,\lambda}^{-1}$ (cf. (17.2)) s'écrivent :

(19.5) $\forall i = 1,...,k \quad \mathcal{R}_{i,\lambda}^{-1}(x_{i+1}) = y_i = \alpha_{io}(\lambda) + \alpha_{i1}(\lambda).x_{i+1}$

pour tout $(x_{i+1}, \lambda) \in]-\varepsilon, \varepsilon[\times \mathcal{V}$, sans modification dans les formules (16.1) et (16.2). Démontrons le lemme par induction sur i : pour $i = 1$ posons $\bar{x}_2 = \alpha_{11}(\lambda).x_2$ et $\bar{y}_2 = [\alpha_{11}(\lambda)]^{r_2(\lambda)}.y_2$, dans les paramètres $y_1, \bar{x}_2$ la correspondance $\mathcal{R}^{-1}_{1.\lambda}$ (cf. (19.5)) s'écrit $y_1 = \bar{\alpha}_{1o}(\lambda) + \bar{x}_2$ avec $\bar{\alpha}_{1o}(\lambda) = \alpha_{1o}(\lambda)$ et les formules (16.1) et (16.2) ne sont pas modifiées. (19.5) pour $i = 2$ est remplacé par :

$$\bar{y}_2 = [\alpha_{11}(\lambda)]^{r_2(\lambda)}.\alpha_{2o}(\lambda) + \alpha_{21}(\lambda).[\alpha_{11}(\lambda)]^{r_2(\lambda)}.x_3$$

définissons par récurrence les fonctions :

(19.6) $\alpha_1(\lambda) = \alpha_{11}(\lambda)$ et pour $i \in \{2, ..., k\}$ $\alpha_i(\lambda) = \alpha_{i1}(\lambda).[\alpha_{i-1}(\lambda)]^{r_i(\lambda)}$

les paramétrages x_1 et y_1 sur σ_1 et τ_1 étant inchangés, soit $j \in \{2, ..., k-1\}$ et supposons que pour tout $i \in \{2, ..., j\}$ on ait trouvé des paramètres $\bar{x}_i, \bar{y}_i$ sur σ_i, τ_i de la forme $\bar{x}_i = \alpha_{i-1}.x_i$, $\bar{y}_i = [\alpha_{i-1}]^{r_i}.y_i$ tels que les correspondances $\mathcal{R}^{-1}_{i,\lambda}$ s'écrivent : $\bar{y}_i = \bar{\alpha}_{io}(\lambda) + \bar{x}_{i+1}$ pour $i = 2, ..., j-1$ et $\bar{\alpha}_{io}(\lambda) = [\alpha_{i-1}(\lambda)]^{r_i(\lambda)}.\alpha_{io}(\lambda)$; les formules (16.1) et (16.2) étant les mêmes dans les paramètres $\bar{x}_i, \bar{y}_i$ pour $i \in \{2, ..., j\}$. (19.5) pour $i = j$ est remplacé par :

$$\bar{y}_j = [\alpha_{j-1}(\lambda)]^{r_j(\lambda)}.\alpha_{jo}(\lambda) + \alpha_{j1}(\lambda).[\alpha_{j-1}(\lambda)]^{r_j(\lambda)}.x_{j+1}$$

posons alors :

$$\bar{x}_{j+1} = \alpha_j(\lambda).x_{j+1}, \quad \bar{y}_{j+1} = [\alpha_j(\lambda)]^{r_{j+1}(\lambda)}.y_{j+1}$$

et

$$\bar{\alpha}_{jo}(\lambda) = [\alpha_{j-1}(\lambda)]^{r_j(\lambda)}.\alpha_{jo}(\lambda),$$

la correspondance $\mathcal{R}^{-1}_{j,\lambda}$ s'écrit dans les paramètres $\bar{y}_j, \bar{x}_{j+1}$: $\bar{y}_j = \bar{\alpha}_{jo}(\lambda) + \bar{x}_{j+1}$ et les formules (16.1) et (16.2) sont les mêmes dans les paramètres $\bar{x}_{j+1}, \bar{y}_{j+1}$. On achève la preuve du Lemme en posant $\bar{\alpha}_1 = \alpha_k$ (cf. (19.3) et (19.6)).

Remarque 4 :

1°) dans les formules (16.1) et (16.2) écrites dans les paramètres $\bar{x}_i, \bar{y}_i$, les fonctions d_{i1}, d_{i2} sont seulement de classe C^{K-1} en λ .

2°) dans les paramètres du Lemme, les correspondances $\mathcal{R}_{i,\lambda}$ envoyant τ_i sur σ_{i+1} s'écrivent :

(19.7) $\mathcal{R}_{i,\lambda}(y_i) = x_{i+1} = \bar{\eta}_{io}(\lambda) + y_i$ pour $i = 1, ...k-1$ $\forall (y_i, \lambda) \in]-\varepsilon, \varepsilon[\times \mathcal{V}$

(19.8) $\mathcal{R}_{k,\lambda}(y_k) = x_1 = \bar{\eta}_{k_o}(\lambda) + [\bar{\alpha}_1(\lambda)]^{-1} y_k \ \forall (y_k, \lambda) \in]-\varepsilon, \varepsilon[\times \mathcal{V}$

les fonctions $\bar{\eta}_{i_o}$ ayant les mêmes propriétés que les fonctions $\bar{\alpha}_{i_o}$ du Lemme.

Donnons un deuxième résultat de réduction qui permet de définir un deuxième paramètre sur les transversales τ_i par un changement de variable(I_1^K) (cf. Déf. 5). Ce deuxième paramètre servira d'intermédiaire dans les calculs qui suivront :

LEMME 8: Pour tout $i = 1, ..., k$, il existe un changement de variable (I_1^K) (cf. Déf. 5) :

(20.1) $\bar{y}_i = h_i(y_i, \lambda)$

défini sur $]-\varepsilon, \varepsilon[\times \mathcal{V}$ (y_i étant le paramètre sur τ_i donné par le Lemme 7) et dans lequel l'application de Dulac $D_{i,\lambda}$ (cf. (16.1)) s'écrit :

(20.2) $D_{i,\lambda}(x_i) = \bar{y}_i = x_i^{r_i(\lambda)} \ \forall (x_i, \lambda) \in]0, \varepsilon,[\times \mathcal{V}.$

Preuve : D'après (16.2) on a :

$$D_{i,\lambda}^{-1}(y_i) = x_i = y_i^{s_i(\lambda)}[1 + d_{i2}(y_i, \lambda)] \ \forall (y_i, \lambda) \in]0, \varepsilon[\times \mathcal{V}$$

et d_{i2} vérifie les propriétés (I_0^K) par multiplication par y_i sur $]0, \varepsilon[\times \mathcal{V}$. Posons $H_i(y_i, \lambda) = [1 + d_{i2}(y_i, \lambda)]^{r_i(\lambda)} - 1$ pour tout $(y_i, \lambda) \in [0, \varepsilon] \times \mathcal{V}$ (en réduisant ε si nécessaire), H_i vérifie trivialement les propriétés (I_0^K) par multiplication par y_i sur $]0, \varepsilon[\times \mathcal{V}$ et d'après le lemme 1i) la fonction $h_i(y_i, \lambda) = y_i[1 + H_i(y_i, \lambda)]$ peut être prolongée sur $]-\varepsilon, \varepsilon[\times \mathcal{V}$ en une fonction vérifiant les propriétés (I_1^K) par multiplication par y_i sur $]0, \varepsilon[\times \mathcal{V}$. Posons $\bar{y}_i = h_i(y_i, \lambda)$, dans les paramètres $x_i, \bar{y}_i$ $D_{i,\lambda}$ est donnée par (20.2).

- Pour établir le dernier résultat de réduction qui suivra, nous aurons besoin des notations du paragraphe II.2 : comme dans le Lemme 7 gardons σ_1 pour transversale de départ et écrivons (15) pour $j = 1$ et $v = +$ (cf. (14) pour la définition de $C(P_1, v, \bar{v})$) :

(20.3) $\mathcal{V} = \bigcup_{\bar{v} \in E^k} C(P_1, v, \bar{v})$

pour tout $\lambda \in \mathcal{V}$, désignons par $\beta_i(\lambda)$ l'abscisse sur σ_{i+1} (le paramètre étant celui du Lemme 7) de l'extrémité du i^e branchement du chemin partant de P_1 dans le sens v (cf.

Déf. 6, §.II.2) ; comme on le verra dans le lemme qui suit, les fonctions $\beta_i(\lambda)$ sont seulement continues sur $\mathcal{V}$ mais leur restriction à chaque cône $C(P_1, v, \bar{v})$ pour tout $\bar{v} \in E^k$ est de classe C^K. Pour cette raison essentiellement et pour respecter les définitions des propriétés $(I^K_{\lambda,0})$ et $(I^K_{\lambda,1})$ (cf. Déf. 3,4, §.II.1), on établira le lemme sur chaque cône $C(P_1, v, \bar{v})$ seulement. Soit donc $\bar{v} \in E^k$ telle que le cône $C(P_1, v, \bar{v})$ soit non vide et posons :

(21.1) $O_{\bar{v}} = C(P_1, v, \bar{v}), \bar{v} = (v_i)_{i=1 \ldots k}$ et $W =]-\varepsilon, \varepsilon[\times O_{\bar{v}}$

On sait que (d'après (14)) : $\forall \lambda \in O_{\bar{v}}\ v_i.\beta_i(\lambda) \geq 0$ pour tout $i = 1, \ldots, k$.
Dans (20.1) posons $h_{i,\lambda}(y_i) = h_i(y_i, \lambda)$ et désignons par $\bar{\beta}_i$ et $\tilde{\beta}_i$ les fonctions :

(21.2) $\forall \lambda \in O_{\bar{v}}, \forall i \in \{1, \ldots, k-1\}\ \bar{\beta}_i(\lambda) = \begin{cases} [\beta_i(\lambda)]^{r_{i+1}(\lambda)} & \text{si } v_i = + \\ 0 & \text{si } v_i = - \end{cases}$

(21.3) $\forall \lambda \in O_{\bar{v}},\ \forall i \in \{1, \ldots, k-1\}\ \tilde{\beta}_i(\lambda) = h^{-1}_{i+1,\lambda}(\bar{\beta}_i(\lambda))$,

posons aussi :

(21.4) $\forall i \in \{2, \ldots, k\}\ W_i = \bigcup_{\lambda \in O_{\bar{v}}}]\beta_{i-1}(\lambda), \varepsilon[\times \{\lambda\}$ et $W_1 =]0, \varepsilon[\times O_{\bar{v}}$,

LEMME 9 : Soient x_i, y_i les paramètres sur σ_i, τ_i donnés par le lemme 7 et posons :

(21.5) $Y_1 = y_1$ et pour $i = 2, \ldots, k$ $Y_i = y_i - \tilde{\beta}_{i-1}(\lambda)\ \forall \lambda \in O_{\bar{v}}$

(cf. (21.1) et (21.3) ci-dessus), alors il existe :

1°) des changements de paramétrages $(I^K_{\lambda\,1})$ (cf. Déf. 5) :

(21.6) $\hat{Y}_i = \hat{h}_i(Y_i, \lambda)$ pour $i = 1, \ldots, k$ et $(Y_i, \lambda) \in W$

sur les transversales $(\tau_i)_{i=1 \ldots k}$ et qui vérifient les propriétés $(I^K_{\lambda,1})$ par multiplication par Y_i (cf. Déf. 4) sur W_1 (cf. (21.4)) ;

2°) des changements de paramétrages $(I^K_{\lambda,1})$:

(21.7) $\hat{x}_i = l_i(x_i, \lambda)$ pour $i = 2, \ldots, k$ et $(x_i, \lambda) \in W$

sur les transversales $(\sigma_i)_{i=2,\ldots,k}$ et qui vérifient les propriétés $(I^K_{\lambda,1})$ par multiplication par $(x_i - \beta_{i-1}(\lambda))$ sur W_i (cf. (21.1) et (21.4)) ;
tels que si on pose :

(21.8) $\forall\lambda \in O_0\ \forall i \in \{1,...,k-1\}\ \gamma_i(\lambda) = \hat{h}_i(\beta_i(\lambda),\lambda)$

(21.9) $\forall\lambda \in O_0\ \forall i \in \{1,...,k-1\}\ \delta_i(\lambda) = \begin{cases} \gamma_i(\lambda) & \text{si } v_i = + \\ 0 & \text{si } v_i = - \end{cases}$ et $\bar{\gamma}_i(\lambda) = [\delta_i(\lambda)]^{r_{i+1}(\lambda)}$

(21.10) $V_i = \bigcup_{\lambda\in O_0}]\delta_{i-1}(\lambda), \varepsilon[\times\{\lambda\}$

les correspondances $\mathcal{R}_{i,\lambda}$ (cf. (19.7)) s'écrivent :

(22.1) $\forall i \in \{1,...,k-1\}\ \forall(\hat{Y}_i,\lambda) \in W\ : \mathcal{R}_{i,\lambda}(\hat{Y}_i) = \hat{x}_{i+1} = \hat{Y}_i + \gamma_i(\lambda)$

et les applications de Dulac $D_{i,\lambda}$(f. (16.1)) soient données par :

(22.2) $\begin{cases} \forall i \in \{2,...,k\} & \forall(\hat{x}_i,\lambda) \in V_i \ : & D_{i,\lambda}(\hat{x}_i) = \hat{Y}_i = [\hat{x}_i]^{r_i(\lambda)} - \bar{\gamma}_{i-1}(\lambda) \\ \text{et} & \forall(x_1,\lambda) \in W_1 \ : & D_{1,\lambda}(x_1) = \hat{Y}_1 = [x_1]^{r_1(\lambda)} \end{cases}$

Preuve : Par récurrence sur i et en utilisant essentiellement le lemme fondamental (Lemme 5). Démontrons d'abord la 2e ligne de (22.2) : soient x_1 et y_1 les paramètres sur σ_1 et τ_1 donnés par le lemme 7 et soit $\bar{y}_1 = h_1(y_1,\lambda)$ le changement de paramètrage sur τ_1 donné par le Lemme 8 (cf. (20.1)), posons alors d'après (21.5) $\hat{h}_1 = h_1$ et notons $\hat{Y}_1$ au lieu de $\bar{y}_1$, on a immédiatement la 2e ligne de (22.2) d'après (20.2) (cf. Lemme 8) et $\hat{h}_1$ vérifie les propriétés $(I^K_{\lambda,1})$ (puisqu'elle vérifie les propriétés (I^K_1)) par multiplication par Y_1 sur W_1.

Construisons les fonctions $\hat{h}_2$ (cf. (21.6)) et l_2 (cf. (21.7)) et montrons (22.1) pour $i = 1$ et (22.2) pour $i = 2$: la correspondance $\mathcal{R}_{1,\lambda}$ s'écrit dans les paramètres y_1, x_2 du Lemme 7 (cf. (19.7)) : $\mathcal{R}_{1,\lambda}(y_1) = x_2 = \bar{\eta}_{1o}(\lambda) + y_1\ \forall(y_1,\lambda) \in]-\varepsilon,\varepsilon[\times O_0$, ceci montre d'après la définition de la fonction β_1 que :

(23.1) $\forall\lambda \in O_0\ : \beta_1(\lambda) = \bar{\eta}_{1o}(\lambda)$.

Posons : $\forall(x_2,\lambda) \in W : l_2(x_2,\lambda) = \hat{h}_1(x_2-\beta_1(\lambda),\lambda)+\gamma_1(\lambda)$; d'après le lemme 5 i) et (21.8) ci-dessus, la fonction l_2 vérifie les propriétés $(I^K_{\lambda,1})$ par multiplication par $(x_2 - \beta_1(\lambda))$ sur W_1. Faisons sur σ_2 le changement de paramètrage : "$\hat{x}_2 = l_2(x_2,\lambda)\ \forall(x_2,\lambda) \in W$" ; il est clair que dans les paramètres $\hat{Y}_1$, $\hat{x}_2$ la correspondance $\mathcal{R}_{1,\lambda}$ s'écrit :

$$\forall(\hat{Y}_1,\lambda) \in W : \mathcal{R}_{1,\lambda}(\hat{Y}_1) = \hat{x}_2 = \hat{Y}_1 + \gamma_1(\lambda)$$

Posons ensuite :

$$\forall(y_2,\lambda) \in W^1 : f_2(y_2,\lambda) = \left[l_2([h_2(y_2,\lambda)]^{s_2(\lambda)},\lambda)\right]^{r_2(\lambda)}$$

où "$\bar{y}_2 = h_2(y_2,\lambda)\ \forall(y_2,\lambda) \in W$" désigne le changement de paramètrage sur τ_2 donné par le lemme 8 (cf. (20.1)) (on prend plutôt la restriction de ce changement à W) ; d'après (21.5), (21.3) et le Lemme 5 ii) la fonction $\bar{h}_2$ définie sur W^1 par : $\bar{h}_2(Y_2,\lambda) = f_2(Y_2+\tilde{\beta}_1(\lambda),\lambda)-\bar{\gamma}_1(\lambda)$ (cf. (21.9) et l'expression de la fonction l_2 ci-dessus) admet un prolongement $\hat{h}_2$ à W qui vérifie les propriétés $(I^K_{\lambda\ 1})$ par multiplication par Y_2 sur W^1. Faisons sur τ_2 le changement de paramétrage $(I^K_{\lambda\ 1})$: "$\hat{Y}_2 = \hat{h}_2(Y_2,\lambda)\ \forall(Y_2,\lambda) \in W$" (le changement de variable (21.5) n'est qu'un changement d'origine sur τ_2 !) ; pour $(Y_2,\lambda) \in W^1, \hat{Y}_2$ est donné par :

$$\forall(Y_2,\lambda) \in W_1 : \hat{Y}_2 = f_2(Y_2 + \tilde{\beta}_1(\lambda),\lambda) - \bar{\gamma}_1(\lambda)$$

il est clair que dans les paramètres $\hat{x}_2, \hat{Y}_2$ l'application de Dulac $D_{2.\lambda}$ est donnée par : $\forall(\hat{x}_2,\lambda) \in V_2 : D_{2.\lambda}(\hat{x}_2) = \hat{Y}_2 = [\hat{x}_2]^{r_2(\lambda)} - \bar{\gamma}_1(\lambda)$ (cf. (20.1), (20.2), (21.9) et (21.10)), et que le chemin partant de P_1 dans le sens $v = +$ coupe la transversale τ_2 en $Y_2 = 0$ d'après (21.2), (21.3) et (21.5).

Maintenant, supposons que pour tout $i \in \{2,...,j\}$ avec $j \in \{2,....,k-1\}$, on ait construit des changements de paramétrages $(I^K_{\lambda,1})$: "$\hat{Y}_i = \hat{h}_i(Y_i,\lambda)\ \forall(Y_i,\lambda) \in W$" (où Y_i est donné par (21.5)) sur τ_i et "$\hat{x}_i = l_i(x_i,\lambda)\ \forall(x_i,\lambda) \in W$" sur σ_i ($\hat{h}_i$ et l_i vérifiant les propriétés annoncées dans le lemme) tels que pour tout $i \in \{1,...,j-1\}$ la correspondance $\mathcal{R}_{i.\lambda}$ s'écrive dans les paramètres $\hat{Y}_i, \hat{x}_{i+1}$: $\forall(\hat{Y}_i,\lambda) \in W\ \mathcal{R}_{i.\lambda}(\hat{Y}_i) = \hat{x}_{i+1} = \hat{Y}_i + \gamma_i(\lambda)$; et pour tout $i \in \{2,...j\}$ l'application de Dulac $D_{i.\lambda}$ s'écrive dans les paramètres $\hat{x}_i, \hat{Y}_i$: $\forall(\hat{x}_i,\lambda) \in V_i\ D_{i.\lambda}(\hat{x}_i) = \hat{Y}_i = [\hat{x}_i]^{r_i(\lambda)} - \bar{\gamma}_{i-1}(\lambda)$; les fonctions γ_i et $\bar{\gamma}_i$ étant données par (21.8) et (21.9). Supposons de plus que pour tout $i \in \{2,...,j\}$ le chemin partant de P_1 dans le sens $v = +$ coupe la transversale τ_i en $Y_i = 0$ et que pour tout $i \in \{2,...,j-1\}$ on ait la relation suivante entre les fonctions β_i et $\bar{\eta}_{io}$:

$$\textbf{(23.2)}\quad \forall\lambda \in O_v\ \ \bar{\eta}_{io}(\lambda) = \beta_i(\lambda) - \tilde{\beta}_{i-1}(\lambda)$$

(si $j = 2$ prendre la relation (23.1)).

Dans les paramètres y_j, x_{j+1} donnés par le lemme 8, la correspondance $\mathcal{R}_{j.\lambda}$ s'écrit : $\mathcal{R}_{j.\lambda}(y_j) = x_{j+1} = \bar{\eta}_{jo}(\lambda) + y_j\ \forall(y_j,\lambda) \in W$ (cf. (19.7)) et d'après (21.5), elle s'écrit dans les paramètres Y_j, x_{j+1} : $\mathcal{R}_{j.\lambda}(Y_j) = x_{j+1} = \bar{\eta}_{jo}(\lambda) + \tilde{\beta}_{j-1}(\lambda) + Y_j\ \forall(Y_j,\lambda) \in W$. L'hypothèse de récurrence sur le chemin partant de P_1 dans le sens v , la définition 6 et la définition de la fonction β_j montrent que :

$$\textbf{(23.2')}\quad \beta_j(\lambda) = \bar{\eta}_{jo}(\lambda) + \tilde{\beta}_{j-1}(\lambda)\ \forall\lambda \in O_{\bar{v}}$$

et donc que (23.2) ci-dessus est vraie pour tout $i \in \{2,...,k-1\}$.

Posons :

(23.3) $\forall (Y_j, \lambda) \in W : \mathcal{R}_{j,\lambda}(Y_j) = x_{j+1} = \beta_j(\lambda) + Y_j$

(23.4) $\forall (x_{j+1}, \lambda) \in W \; l_{j+1}(x_{j+1}, \lambda) = \quad \hat{h}_j(x_{j+1} - \beta_j(\lambda), \lambda) + \gamma_j(\lambda),$

d'après l'hypothèse de récurrence sur la fonction $\hat{h}_j$, la relation (21.8) (les fonctions $\hat{h}_i$ étant supposées impaires en Y_i sur W) et le lemme 5 i), la fonction l_{j+1} vérifie les propriétés $(I^K_{\lambda,1})$ par multiplication par $(x_{j+1} - \beta_j(\lambda))$ sur W_{j+1}(cf. (21.4)). Faisons sur σ_{j+1} le changement de paramétrage $(I^K_{\lambda,1})$: "$\hat{x}_{j+1} = l_{j+1}(x_{j+1}, \lambda) \; \forall (x_{j+1}, \lambda) \in W$" ; d'après (23.3), (23.4) et (21.6) pour $i = j$, la correspondance $\mathcal{R}_{j,\lambda}$ s'écrit dans les paramètres $\hat{Y}_j, \hat{x}_{j+1}$:

(23.5) $\forall (\hat{Y}_j, \lambda) \in W : \mathcal{R}_{j,\lambda}(\hat{Y}_j) = \hat{x}_{j+1} = \hat{Y}_j + \gamma_j(\lambda).$

Soit "$\bar{y}_{j+1} = h_{j+1}(y_{j+1}, \lambda) \; \forall (y_{j+1}, \lambda) \in W$" le changement de paramétrage (I^K_1) sur τ_{j+1} donné par le Lemme 8 (cf. (20.1)) ; posons :

(23.6) $\forall (y_{j+1}, \lambda) \in W^1 \; f_{j+1}(y_{j+1}, \lambda) = \left[l_{j+1}\left([h_{j+1}(y_{j+1}, \lambda)]^{\sigma_{j+1}(\lambda)}, \lambda\right)\right]^{r_{j+1}(\lambda)}$

d'après (21.5), (21.3), (23.4), (21.8), (21.9) et le lemme 5 ii), la fonction $\bar{h}_{j+1}$ définie sur W^1 par : $\bar{h}_{j+1}(Y_{j+1}, \lambda) = f_{j+1}(Y_{j+1} + \tilde{\beta}_j(\lambda), \lambda) - \bar{\gamma}_j(\lambda)$ admet un prolongement $\hat{h}_{j+1}$ à W (choisi impair en Y_{j+1} pour satisfaire (21.8)) qui vérifie les propriétés $(I^K_{\lambda,1})$ par multiplication par Y_{j+1} sur W^1. Faisons donc sur τ_{j+1} le changement de paramétrage $(I^K_{\lambda,1})$: "$\hat{Y}_{j+1} = \hat{h}_{j+1}(Y_{j+1}, \lambda) \; \forall (Y_{j+1}, \lambda) \in W$", on a :

(23.7) $\forall (Y_{j+1}, \lambda) \in W^1 : \hat{Y}_{j+1} = f_{j+1}(Y_{j+1} + \tilde{\beta}_j(\lambda), \lambda) - \bar{\gamma}_j(\lambda)$

d'après (23.7), (23.6), (20.1), (20.2), (21.9) et (21.10), l'application de Dulac $D_{j+1,\lambda}$ s'écrit dans les paramètres $\hat{x}_{j+1}, \hat{Y}_{j+1}$:

(23.8) $\forall (\hat{x}_{j+1}, \lambda) \in V_{j+1} : D_{j+1,\lambda}(\hat{x}_{j+1}) = \hat{Y}_{j+1} = [\hat{x}_{j+1}]^{r_{j+1}(\lambda)} - \bar{\gamma}_j(\lambda).$

Maintenant, le chemin partant de P_1 dans le sens $v = +$ coupe σ_{j+1}en $x_{j+1} = \beta_j(\lambda)$, donc coupe τ_{j+1} en $y_{j+1} = \tilde{\beta}_j(\lambda)$ d'après (20.1), (20.2), (21.2) et (21.3), et donc coupe τ_{j+1} en $Y_{j+1} = 0$ d'après (21.5). Ceci et (23.2'), (23.5) et (23.8) finit la démonstration par récurrence et prouve le lemme.

Remarque 5 : La réduction (22.2) des applications de Dulac $D_{i,\lambda}$ pour $i \in \{2, ..., k\}$ n'est valable que pour $\hat{x}_i \geq \delta_{i-1}(\lambda)$ pour tout $\lambda \in O_0$ (cf. (21.9) et (21.1)), ceci n'est pas gênant pour la suite puisque dans le cas où $v_{i-1} = +$(i.e $\delta_{i-1}(\lambda) > 0\ \forall\lambda \in O_0$) les points de la transversale σ_i (cf. fig. 5) dont l'abscisse $\hat{x}_i \in [0, \delta_{i-1}(\lambda)[$ ont une trajectoire négative (i.e pour le champ $-X_\lambda$) qui n'aboutit pas sur le côté positif de la transversale de départ σ_1 (cf. déf. 6 pour la définition du chemin partant de σ_1) et donc, il ne peut passer d'orbite fermée par ces points.

- Toujours d'après la définition du chemin partant de P_1 dans le sens $v = +$ (ou partant de σ_1 (cf. Déf. 6)), les seuls points de la transversale τ_i ($i \in \{2,...,k\}$) dont la trajectoire négative aboutit sur le côté positif de σ_1 sont les points d'abscisse $Y_i \geq 0$ (ou $\hat{Y}_i \geq 0$ d'après (22.2) et (21.6)) pour tout $\lambda \in O_0$ (cf. (21.1)). Désignons par T_λ^+ ($\lambda \in O_0$) la loi de correspondance du champ X_λ entre la transversale σ_1 paramétrée par x_1 (cf. Lemme 7 et 9) et la transversale τ_k paramétrée par Y_k (cf. Lemme 9 (21.5)). Le domaine de définition de T_λ^+ sur σ_1 est de la forme :

(24.1) $\forall\lambda \in O_0 : (T_\lambda^+)^{-1}(]0,\varepsilon[) =]\rho(\lambda), \varepsilon[$ (même notation en ε !...)

où ρ est une fonction positive ou nulle de classe C^K en $\lambda \in O_0$ (mais seulement continue sur le voisinage $\mathcal{V}$ (cf. (20.3)) et telle que $Lim_{\lambda \to 0,\ \lambda \in O_0}\ \rho(\lambda) = 0$; posons :

(24.2) $\mathcal{U} = \bigcup_{\lambda \in O_0}]\rho(\lambda), \varepsilon[\times \{\lambda\}$,

on a le corollaire immédiat du Lemme 9 :

LEMME 10 : Posons (cf. Lemme 9 (21.8) et (21.9)) :

(24.3) $\forall\lambda \in O_0 : b_1(\lambda) = \gamma_1(\lambda)$ et pour $i \in \{2,...,k-1\}$ $b_i(\lambda) = \gamma_i(\lambda) - \bar{\gamma}_{i-1}(\lambda)$

et faisons sur τ_k le changement de paramètre $(I^K_{\lambda.1})$: "$\hat{Y}_k = \hat{h}_k(Y_k, \lambda)\ \forall(Y_k, \lambda) \in W$" (cf. (21.6) et (21.1)). Dans les paramètres $x_1, \hat{Y}_k$ la correspondance T_λ^+ définie ci-dessus s'écrit ;

(24.4) $\forall(x_1, \lambda) \in \mathcal{U}$ (cf. (24.2) et (24.1)) :

$$\hat{Y}_k = T_\lambda^+(x_1) = \left[\ldots\left[[[x_1]^{r_1(\lambda)} + b_1(\lambda)]^{r_2(\lambda)} + b_2(\lambda)\right]^{r_3(\lambda)} + \ldots + b_{k-1}(\lambda)\right]^{r_k(\lambda)} - \bar{\gamma}_{k-1}$$

- Désignons maintenant par T_λ^- ($\lambda \in O_0$) la correspondance régulière entre les transversales σ_1 et τ_k pour le champ $-X_\lambda$ (T_λ^- est la restriction à W (cf. (21.1)) de l'application $\mathcal{R}_{k.\lambda}^{-1}$ (cf. (19.2)). Rappelons que (cf. Déf. 6) $\beta_k(\lambda)$ désigne l'abscisse sur σ_1 (paramétrée par x_1) de l'extrémité du $k^{ième}$ branchement (ou aboutissement) du chemin partant de σ_1. Posons comme dans le Lemme 5 : (θ étant un réel > 1)

$$\textbf{(24.5)} \quad \forall\lambda \in O_0 \;\; \eta(\lambda) = \begin{cases} \beta_k(\lambda) & \text{si } v_k = + \\ 0 & \text{si } v_k = - \end{cases} \qquad \text{(cf. (21.1))}$$

$$\textbf{(24.6)} \quad V = \bigcup_{\lambda\in O_0}]\eta(\lambda), \varepsilon[\times\{\lambda\} \text{ et } V_\theta = \bigcup_{\lambda\in O_0}]\theta.\eta(\lambda), \varepsilon[\times \{\lambda\}$$

Posons aussi :

$$\textbf{(24.7)} \quad \forall\lambda \in O_0 \;\; \gamma_k(\lambda) = \hat{h}_k(\bar{\alpha}_1(\lambda).\beta_k(\lambda), \lambda)$$

où $\hat{h}_k$ désigne le changement de paramétrage du Lemme 9 (cf. (21.6)) et $\bar{\alpha}_1$ désigne l'inverse du facteur d'hyperbolicité du polycyle (Γ_k) par rapport à σ_1 (cf. (19.3)). Des Lemmes 5, 7 et 9 on déduit le résultat :

LEMME 11 : Soient x_1 et $\hat{Y}_k$ les paramètres sur σ_1 et τ_k comme dans le Lemme 10. La correspondance T_λ^- définie ci-dessus s'écrit :

$$\textbf{(24.8)} \quad \forall(x_1, \lambda) \in W : \hat{Y}_k = T_\lambda^-(x_1) = x_1.[\bar{\alpha}_1(\lambda) + f(x_1, \lambda)] - \gamma_k(\lambda)$$

(cf. (21.1) pour W, (19.3) pour $\bar{\alpha}_1$ et (24.7) pour γ_k).

La fonction f est seulement continue sur W et vérifie les propriétés $(I_{\lambda.0}^K)$ (cf. Déf. 3) par multiplication par $(x_1 - \eta(\lambda))$ (cf. 24.5)) sur V (cf. (24.6)). De plus elle vérifie les propriétés $(I_{\lambda.0}^K)$ par multiplication par x sur V_θ (cf. (24.6)).

Preuve : La correspondance T_λ^- s'écrit dans les paramètres x_1, Y_k (cf.(19.2) et (21.5))

$$\forall(x_1, \lambda) \in W : Y_k = T_\lambda^-(x_1) = \bar{\alpha}_1(\lambda).x_1 - \bar{\alpha}_1.\bar{\eta}_{ko}(\lambda) - \tilde{\beta}_{k-1}(\lambda)$$

ceci montre d'après la définition de la fonction β_k que :

$$\textbf{(24.9)} \qquad \forall\lambda \in O_0 : \bar{\alpha}_1(\lambda).\bar{\eta}_{k0}(\lambda) = \bar{\alpha}_1(\lambda).\beta_k(\lambda) - \tilde{\beta}_{k-1}(\lambda).$$

Faisons sur σ_1 le changement de paramètre :

$$\textbf{(24.10)} \quad "\hat{x}_1 = \bar{\alpha}_1(\lambda).x_1 \;\; \forall(x_1, \lambda) \in W" \quad \text{(cf. (19.4))}.$$

Ecrivons T_λ^- dans les paramètres $\hat{x}_1, y_k$:

(24.11) $\forall(\hat{x}_1,\lambda) \in W : Y_k = T_\lambda^-(\hat{x}_1) = \hat{x}_1 - \bar{\alpha}_1(\lambda).\beta_k(\lambda)$.

Soit "$\hat{Y}_k = \hat{h}_k(Y_k,\lambda)$ $\forall(Y_k,\lambda) \in W$" le changement de paramètrage $(I^K_{\lambda.1})$ sur τ_k donné par le lemme 9 (cf. (21.6)), et posons :

(24.12) $\forall(\hat{x}_1,\lambda) \in W : l(\hat{x}_1,\lambda) = \hat{h}_k(\hat{x}_1 - \bar{\alpha}_1(\lambda).\beta_k(\lambda),\lambda) + \gamma_k(\lambda)$;

d'après le Lemme 5 i) et (24.7) ci-dessus ($\hat{h}_k$ est supposée impaire en Y_k sur W) la fonction l s'écrit :

(24.13) $\forall(\hat{x}_1,\lambda) \in W : l(\hat{x}_1,\lambda) = \hat{x}_1[1 + L(\hat{x}_1,\lambda)]$

où la fonction L est continue sur W et vérifie les propriétés $(I^K_{\lambda\,0})$ par multiplication par $(\hat{x}_1 - \bar{\alpha}_1(\lambda).\eta(\lambda))$ (cf. (24.5)) sur $\hat{V} = \bigcup_{\lambda\in O_0}]\bar{\alpha}_1(\lambda).\eta(\lambda),\varepsilon[\times\{\lambda\}$ et par multiplication par $\hat{x}_1$ sur $\hat{V}_\theta = \bigcup_{\lambda\in O_0}]\theta.\bar{\alpha}_1(\lambda).\eta(\lambda),\varepsilon[\times\{\lambda\}$ (cf. (24.6)).

D'après (24.10) et (24.13), la fonction l s'écrit dans le paramètre x_1 :

(24.14) $\forall(x_1,\lambda) \in W$ $l(x_1,\lambda) = x_1[\bar{\alpha}_1(\lambda) + f(x_1,\lambda)]$

où la fonction f (d'après (19.4)) est continue sur W et vérifie les propriétés $(I^K_{\lambda.0})$ par multiplication par $(x_1 - \eta(\lambda))$ sur V et par multiplication par x_1 sur V_θ (cf. (24.6)).

(24.11), (24.12) et (24.14) permettent de conclure à (24.8).

- <u>Conclusion de la partie II.3</u> : Fixons σ_1 pour transversale de départ et désignons par $\Delta(.,\lambda)$ l'application déplacement (mesurée sur la transversale τ_k) associée à l'application de retour du polycycle par rapport à la transversale σ_1 (cf. (1) et fig. 2 en Introduction). Considérons la relation (20.3) et choisissons $\bar{v} \in E^k$ $(E = \{+,-\})$, en reprenant les notations et les résultats des Lemmes 9,10 et 11 et en posant :

(24.15) $\forall\lambda \in O_0$ $b_k(\lambda) = \gamma_k(\lambda) - \bar{\gamma}_{k-1}(\lambda)$ (cf. (21.8) et (24.7))

on a : (cf. (24.4) et (24.8))

$$
\textbf{(24.16)}\quad \begin{cases} \forall (x_1,\lambda)\in \mathcal{U} \\ \Delta(x_1,\lambda)=\left[\ldots\left[\left[[x_1]^{r_1(\lambda)}+b_1(\lambda)\right]^{r_2(\lambda)}+b_2(\lambda)\right]^{r_3(\lambda)}+\ldots+b_{k-1}(\lambda)\right]^{r_k(\lambda)} \\ \qquad +\, b_k(\lambda)-x_1[\bar{\alpha}_1(\lambda)+f(x_1,\lambda)] \end{cases}
$$

* La signification du terme $\bar{\alpha}_1$ a été vue en Introduction et dans la Remarque 3. 1°) et son expression est donnée par le Lemme 7 (cf. (19.3)).

* Les fonctions b_i (cf. (24.3) et (24.15)) sont liées aux fonctions $\bar{\eta}_{io}$ (cf.(19.7) et (19.8)) qui ont une signification intrinsèque que les fonctions b_i n'ont pas toutes suivant le cône O_v choisi (cf. (21.1)). Ceci est dû à la façon arbitraire dont on a prolongé les fonctions $\hat{h}_i$ du Lemme 9 (cf. (21.6)) (par symétrie par rapport à $Y_i = 0$ pour la clarté du texte). Le Lemme 6 permet de remédier à cela en remplaçant dans le Lemme 9 (21.8) par : (cf. (13.31))

$$
\textbf{(24.17)}\quad \forall\lambda\in O_v\ \forall i\in\{2,\ldots,k-1\}\ \gamma_i(\lambda)=-\hat{h}_i(-\beta_i(\lambda),\lambda),
$$

et en remplaçant aussi (24.7) par :

$$
\textbf{(24.18)}\quad \forall\lambda\in O_v\ \gamma_k(\lambda)=-\hat{h}_k(-\bar{\alpha}_1(\lambda).\beta_k(\lambda),\lambda)\ ;
$$

on applique alors le Lemme 6 pour construire des prolongements $\hat{h}_i$ dans le Lemme 9 qui satisfont à : (cf. (13.33))

$$
\textbf{(24.19)}\quad \forall\lambda\in O_v\ \forall i\in\{2,\ldots,k-1\}\ \hat{h}_i(-\tilde{\beta}_{i-1}(\lambda),\lambda)=-\bar{\gamma}_{i-1}(\lambda)
$$

(cf. (21.3) et (21.9)).

D'après (23.1), (23.2), (24.9), (24.3), (24.15) et (24.17), (24.18) et (24.19) ci-dessus, on a :

$$
\forall\lambda\in O_v :\begin{cases} & b_1(\lambda) = \hat{h}_1(\bar{\eta}_{10}(\lambda),\lambda)\ (\hat{h}_1 \text{ est impaire en } Y_1) \\ \forall i\in\{2,\ldots,k-1\} & b_i(\lambda) = -(\hat{h}_i(-\beta_i(\lambda),\lambda)-\hat{h}_i(-\tilde{\beta}_{i-1}(\lambda),\lambda)) \\ \text{et} & b_k(\lambda) = -(\hat{h}_k(-\bar{\alpha}_1(\lambda).\beta_k(\lambda),\lambda)-\hat{h}_k(-\tilde{\beta}_{k-1}(\lambda),\lambda)) \end{cases}
$$

et donc d'après la 2e partie de la condition i) des propriétés $(I^K_{\lambda,1})$ (cf. Déf.4), il existe pour tout $i\in\{1,\ldots,k\}$ des fonctions a_i continues en λ sur O_v telles que :

$$
\textbf{(24.20)}\quad \forall\lambda\in O_v :\begin{cases} b_i(\lambda) = a_i(\lambda).\bar{\eta}_{io}(\lambda) \text{ pour } i=1,\ldots,k-1 \\ b_k(\lambda) = a_k(\lambda).\bar{\alpha}_1(\lambda).\bar{\eta}_{k0}(\lambda) \end{cases}
$$

avec :

(24.21) $\forall i = 1, ..., k \qquad Lim_{\substack{\lambda \to 0 \\ \lambda \in O_0}} a_i(\lambda) = 1.$

Ainsi pour tout $i \in \{1, ..., k\}$ et pour tout $\lambda \in O_0$, $b_i(\lambda)$ mesure sur la transversale σ_{i+1} (à un facteur près) la déformation de la connexion entre les sommets P_i et P_{i+1} $(k+1 \equiv 1)$ (cf. Remarque 4.2°).

* Le reste f a été décrit dans le Lemme 11 (cf. (24.8)).

Maintenant, en utilisant la restriction signalée dans la Remarque 0.2°) (on exige seulement la continuité par rapport au paramètre λ) et les résultats élémentaires de la preuve du Lemme 6 (cf. (13.34), (13.35) et (13.36)), il est clair qu'on peut prolonger toutes les fonctions ci-dessus (et à partir des données du Lemme 9 (cf. (21.2)) par continuité par rapport à λ sur le voisinage $\mathcal{V}$ (cf. (20.3)), les changements de paramétrage $\hat{h}_i$ et l_i du Lemme 9 vérifieraient alors les propriétés $(I^K_{\lambda\ 1})$ conformément à la Remarque 0.2°) et on peut énoncer :

THEOREME : *Soit $\mathcal{V}$ le voisinage de 0 dans $\mathbf{R}^\wedge$ donné par* (20.3). *Prolongeons la relation* (24.2) *à $\mathcal{V}$ en posant :*

(25.1) $\mathcal{U} = \bigcup_{\lambda \in \mathcal{V}}]\rho(\lambda), \varepsilon[\times \{\lambda\}$

la fonction ρ étant donnée par (24.1). *Désignons par $\beta_k(\lambda)$ pour tout $\lambda \in \mathcal{V}$, l'abscisse sur la transversale σ_1 (cf.* fig. 2) *(paramétrée par x) de l'extrémité de l'aboutissement du chemin partant du sommet P_1 dans le sens $v = +$* (cf. §.II.2) *et prolongeons les relations* (24.5) *et (24.6) à $\mathcal{V}$ en posant :*

(25.2) $\forall \lambda \in \mathcal{V} : \eta(\lambda) = \begin{cases} \beta_k(\lambda) & \text{si } \beta_k(\lambda) > 0 \\ 0 & \text{si } \beta_k(\lambda) \leq 0 \end{cases}$

(25.3) $V = \bigcup_{\lambda \in \mathcal{V}}]\eta(\lambda), \varepsilon[\times \{\lambda\}$ et $V_\theta = \bigcup_{\lambda \in \mathcal{V}}]\theta.\eta(\lambda), \varepsilon[\times \{\lambda\}$,

alors, l'application déplacement $\triangle(., \lambda)$ (associée à l'application de retour du polycycle par rapport à σ_1) admet $\mathcal{U}$ pour domaine de définition et est donnée par :

(25.4)
$$\begin{cases} \forall (x_1, \lambda) \in \mathcal{U} \\ \triangle(x_1, \lambda) = \left[... \left[\left[[x_1]^{r_1(\lambda)} + b_1(\lambda) \right]^{r_2(\lambda)} + b_2(\lambda) \right]^{r_3(\lambda)} + ... + b_{k-1}(\lambda) \right]^{r_k(\lambda)} \\ \qquad + b_k(\lambda) - x_1 [\bar{\alpha}_1(\lambda) + f(x_1, \lambda)] \end{cases}$$

avec :

* $\forall i \in \{1,...,k\}$ $r_i(\lambda)$ *désigne le rapport d'hyperbolicité du sommet* P_i *(cf. Introduction),*

* $\bar{\alpha}_1$ *est le facteur d'hyperbolicité du polycycle par rapport à la transversale* σ_1 *(cf.*(19.3) *et Remarque* 3.1°))

* $\forall i \in \{1,...,k\}$ b_i *est une fonction continue sur* $\mathcal{V}$ *et* $b_i(\lambda)$ *mesure pour tout* $\lambda \in \mathcal{V}$ *la déformation de la connexion entre les sommets* P_i *et* P_{i+1} *(avec la convention* $k+1 \equiv 1$*). Les fonctions* b_i *sont lisibles sur les transversales* σ_{i+1} *et sont données par* (24.20) *que l'on prolonge à* $\mathcal{V}$.

* *La fonction* f *est définie et continue sur* $]-\varepsilon,\varepsilon[\times\mathcal{V}$ *et vérifie les propriétés* $(I^K_{\lambda,0})$ (*cf.* Déf. 3, II.1) *et au sens de la Remarque* 0.2°) *par multiplication par* $(x-\eta(\lambda))$ (*cf.* (25.2)) *sur* V (*cf.* (25.3)) *et par multiplication par* x *sur* V_θ.

Remarque 6 :

1°) le fait que la fonction "reste" f vérifie les propriétés $(I^K_{\lambda,0})$ par multiplication par x sur V_θ nous servira essentiellement dans l'article [6] où on etudie certains cas particuliers de polycycles.

2°) l'application déplacement $\triangle$ (cf. (25.4)) est de classe C^K en x sur son domaine de définition $\mathcal{U}$ et ses dérivées par rapport à x sont continues en $(x,\lambda) \in \mathcal{U}$; K étant l'entier fixé dans la conclusion de la section I.

REFERENCES

[1] A. Mourtada:" *Cyclicité finie des polycycles hyperboliques de champs de vecteurs du Plan. - Algorithme de Finitude*" (Preprint-Dijon).

[2] A.A. Andronov, E.A. Leontovich, I.I. Gordon and A.G. Maier : " *Theory of bifurcation of Dynamical Systems on the Plane*". Israel Program of Scientific Translations, Jerusalem, 1971.

[3] L.A. Cherkas : "*Structure of a successor function in the neighborhod of separatrix of a perturbed analytic autonomous system in the Plane*". Translated from differentsial'nye Uraneniya, vol. 17, n° 3, March. 1981, pp. 469-478.

[4] R. Roussarie : "*A note on Finite Cyclicity Property and Hilbert's 16th Problem*". Dynamical Systems (Proc. Chilean Symp., Valparaiso 1986) (Lecture Notes in Mathematics 1331) ed R. Barmon, R. Labarca and J. Palis Jr (Berlin : Springer), pp 161-168.

[5] H. Dulac : "*Sur les cycles limites*". Bull. soc. Math. France, 51, (1923), pp 45-188.

[6] A. Mourtada : "*Polycycles Hyperboliques Génériques à trois et quatre sommets*",(Preprint-Dijon).

[7] S. Sternberg : "*On the behaviour of invariant curves near a hyperbolic point of a surface transformation*". American Journal of Mathematics, vol. 75 (1955), pp. 526-534.

[8] S. Sternberg : "*Local contractions and a theorem of Poincaré*". American Journal of Mathematics, vol. 79 (1957), pp. 809-824.

[9] R. Roussarie : "*On the number of limit cycles which appear by perturbation of separatrix loop of planar vector fields*". BOL. SOC. BRAS. MAT., vol. 17 n° 2 (1986), pp 67-101.

[10] R. Roussarie : "*Modèles locaux des champs et de formes*". Société Mathématique de France, Astérisque 30, 1975.

[11] S. Sternberg :"*On the Structure ol local homeomorphism of euclidean n − space* III". American Journal of Math., vol 81 (1959), pp. 578-604.

Bifurcation of Limit Cycles

L.M. Perko
Department of Mathematics, Northern Arizona University
Flagstaff, AZ 86011

We consider the local bifurcation and global behavior of one-parameter families of limit cycles of a planar analytic system

$$\dot{x} = f(x,\lambda) \qquad (1_\lambda)$$

depending on a parameter $\lambda \in \mathbf{R}$. One-parameter families of limit cycles can bifurcate from multiple limit cycles, cycles belonging to a period annulus, separatrix cycles or nonhyperbolic critical points of (1_λ). We restrict ourselves to the first three possibilities and present some basic results on the local bifurcation of one-parameter families of limit cycles from multiple limit cycles, separatrix cycles and from centers in Section 1. The global behavior of any one-parameter family of limit cycles of (1_λ) is discussed in Section 2.

For the most part, this paper presents a survey of previously established results; however, it also includes some new results on the global behavior of one-parameter families of limit cycles of (1_λ). See the Generalized Planar Termination Principle and its extension in Section 2. Several interesting examples which illustrate the general theory are also presented in this paper.

1. Local Bifurcations

Assume that the planar analytic system (1_{λ_0}) has a cycle

$$\Gamma_0 : x = \gamma_0(t)$$

of minimum period T_0. In this section we study the local bifurcation of one-parameter families of limit cycles of (1_λ) from Γ_0. We begin by discussing the displacement function and how it can be used to measure the rate of growth of a simple limit cycle with respect to λ. We then present the theory of local bifurcations from multiple limit cycles, separatrix cycles and cycles belonging to a period annulus of (1_λ).

1.1 The Displacement Function

The main analytical tool that is used in the study of one-parameter families of limit cycles of (1_λ) is the Poincaré map $h(n,\lambda)$ or the **displacement function for** (1_λ)

$$d(n,\lambda) = h(n,\lambda) - n$$

where n is the signed distance along a normal line ℓ_{t_0} to Γ_0 at a point $\gamma(t_0) \in \Gamma_0$; cf. Figure 1.

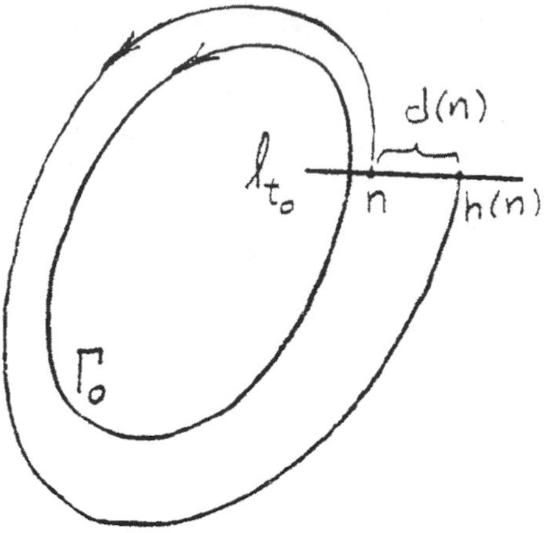

Figure 1: The Poincaré map, h(n), and the displacement function, d(n), for (1_λ) in a neighborhood of the cycle Γ_0.

If we wish to explicitly note the dependence of the displacement function on t_o, i.e., on the point $\gamma_o(t_o) \in \ell_{t_o} \cap \Gamma_o$, we write $d(n,\lambda,t_o)$. The existence and analytic properties of $d(n,\lambda,t_o)$ are established for example in [1]. See Lemma 1.1 and Remark 1.1 in [2]. For convenience in notation we define the function $f_o(t) = f(\gamma_o(t),\lambda_o)$ and let $\nabla \cdot f_o(t)$ denote the function $\nabla \cdot f(\gamma_o(t),\lambda_o)$. We also define ω_o to be ± 1 when the cycle Γ_o is positively or negatively oriented respectively and define σ_o to be ± 1 when Γ_o is a limit cycle which is unstable or stable on its exterior respectively. The following result is classical; cf. [1], p. 383.

Lemma 1: For $\delta > 0$, $|n| < \delta$ and $|\lambda - \lambda_o| < \delta$, let $d(n,\lambda)$ be the displacement function for (1_λ). Then the derivative $d_n(0,\lambda_o)$ is independent of the point $\gamma_o(t_o)$ on Γ_o and

$$d_n(0,\lambda_o) = \exp \int_o^{T_o} \nabla \cdot f_o(t)dt - 1.$$

Definition 1: If $d(0,\lambda_o) = d_n(0,\lambda_o) = d_n^{(2)}(0,\lambda_o) = \ldots = d_n^{(m-1)}(0,\lambda_o) = 0$ and $d_n^{(m)}(0,\lambda_o) \neq 0$ then, for $m > 1$, Γ_o is called a **multiple limit cycle of multiplicity** m. If $m = 1$, i.e., if $d(0,\lambda_o) = 0$ and $d_n(0,\lambda_o) \neq 0$, then Γ_o is called a **simple limit cycle** or **hyperbolic limit cycle**.

Remark 1: It follows from Lemma 1 that Γ_o is a simple limit cycle of (1_{λ_o}) iff

$$\int_o^{T_o} \nabla \cdot f_o(t)dt \neq 0$$

and if Γ_o is a simple limit cycle then σ_o is equal to the sign of the above integral. It follows from Theorem 42 in [1] that the multiplicity m of Γ_o is independent of the point $\gamma_o(t_o)$. And for analytic systems

$$d(0,\lambda_o) = d_n(0,\lambda_o) = d_n^{(2)}(0,\lambda_o) = \ldots = 0$$

iff Γ_o is a cycle which belongs to a continuous band of cycles of (1_{λ_o}); cf. [1].

Just as the derivative $d_n(0,\lambda_o)$ can be computed in terms of the vector field f and the function $\gamma_o(t)$, so can the derivative $d_\lambda(0,\lambda_o,t_o)$. This is an important result: $d_\lambda(0,\lambda_o,0)$ is given by equation (36) on p. 384 in [1] and $d_\lambda(0,\lambda_o,t_o)$ is given by the next lemma proved in [3] by the author.

Lemma 2: For $\delta > 0$, $|n| < \delta$, $|\lambda - \lambda_o| < \delta$ and $t_o \in \mathbf{R}$, let $d(n,\lambda,t_o)$ denote the displacement function for the system (1_λ) along the normal line ℓ_{t_o} to Γ_o. Then

$$d_\lambda(0,\lambda_o,t_o) = -\frac{\omega_o \exp \int_0^{t_o+T_o} \nabla\cdot f_o(t)dt}{|f_o(t_o)|} \int_{t_o}^{t_o+T_o} \frac{f \wedge f_\lambda(\gamma_o(t),\lambda_o)}{\exp \int_0^t \nabla\cdot f_o(s)ds}\, dt$$

where the wedge product of two vectors $x = (x_1,x_2)$ and $y = (y_1,y_2)$ is defined by $x \wedge y = x_1 y_2 - y_1 x_2$.

We next show how the formulas in Lemmas 1 and 2 can be used to compute the rate of growth of a simple limit cycle with respect to the parameter λ. If Γ_o is a simple limit cycle, then at any point $\gamma_o(t_o)$ on Γ_o, $d_n(0,\lambda_o,t_o) \neq 0$ and it follows from the implicit function theorem that for each $t_o \in \mathbf{R}$ the curve $d(n,\lambda,t_o) = 0$ is described by a function $n(\lambda,t_o)$. The rate of growth of the limit cycle Γ_o with respect to λ at the point $\gamma_o(t_o) \in \Gamma_o$ is determined by

$$\frac{\partial n}{\partial \lambda}(\lambda_o,t_o) = -\frac{d_\lambda(0,\lambda_o,t_o)}{d_n(0,\lambda_o)};$$

cf. equation (3.17) in [4]. It follows from Lemmas 1 and 2 that

$$\operatorname{sgn}\left[\frac{\partial n}{\partial \lambda}(\lambda_o,t_o)\right] = \omega_o\, \sigma_o\, \mu(t_o)$$

where $\mu(t_o)$ is the sign of the **Melnikov function**

$$M(t_o) = \int_{t_o}^{t_o+T_o} \frac{f \wedge f_\lambda(\gamma_o(t),\lambda_o)}{\exp \int_0^t \nabla\cdot f_o(s)ds}\, dt.$$

And this leads to the following result:

Theorem 1: If Γ_o is a simple limit cycle of (1_{λ_o}) then Γ_o belongs to a unique, one-parameter family of limit cycles Γ_λ of (1_λ) and at any point $\gamma_o(t_o)$ on Γ_o, increasing the parameter λ will cause the limit cycle Γ_λ to expand or contract along the normal line ℓ_{t_o} iff $\omega_o\, \sigma_o\, \mu(t_o) = \pm 1$ respectively.

Since, by definition, any one-parameter family of rotated vector fields $f(x,\lambda)$ satisfies $f \wedge f_\lambda(x,\lambda) > 0$, cf. [4] or [5], Theorem 7 in [4] and Theorem D in [5] on the monotone growth of limit cycles follow as corollaries of Theorem 1 above. In general, limit cycles do not expand or contract monotonically with the parameter as is illustrated by the following example in [2].

Example 1: The one-parameter family of limit cycles for the system

$$\dot{x} = -y + x\left[(x - \lambda)^2 + y^2 - 1\right]$$
$$\dot{y} = x + y\left[(x - \lambda)^2 + y^2 - 1\right]$$

is shown in Figure 2 for various values of the parameter $\lambda \in [0,1)$. The corresponding limit cycles for $\lambda \in (-1,0]$ can be obtained by reflecting those in Figure 2 about the origin. There are Hopf bifurcations at the origin for $\lambda = \pm 1$.

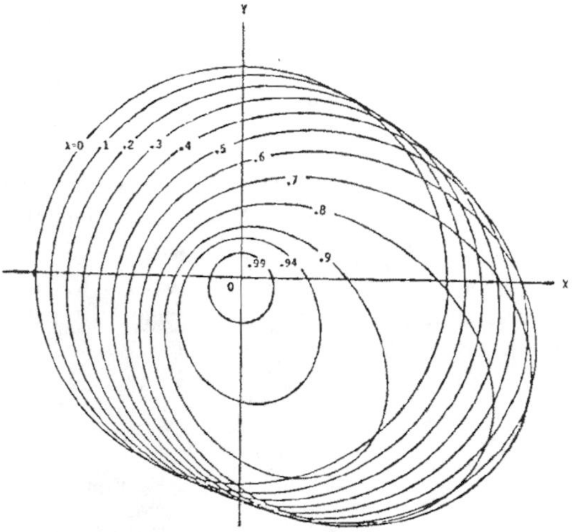

Figure 2: A one-parameter family of limit cycles which exhibits non-monotonic growth.

1.2 Bifurcation at Nonsingular Multiple Limit Cycles

In this section we distinguish between singular and nonsingular, multiple limit cycles of (1_λ) and show that the bifurcation theory for nonsingular, multiple limit cycles is exactly the same as the bifurcation theory for multiple limit cycles belonging to a one-parameter family of rotated vector fields as described in [4] or [5]. In particular, the only bifurcation that can occur at a nonsingular, multiple limit cycle of (1_λ) is the saddle node bifurcation and locally the resulting hyperbolic stable and unstable limit cycles expand and contract monotonically as the parameter λ varies in a certain sense described in Theorem 3 below. The theory in this section is based on the following theorem whose proof is a simple consequence of Lemma 2; cf. [3].

Theorem 2: Under the hypotheses of Lemma 2, if Γ_0 is a multiple limit cycle of (1_{λ_0}) then

$$d_\lambda(0,\lambda_0,t_0) = \left[\frac{|f_0(0)|}{|f_0(t_0)|} \exp\int_0^{t_0} \nabla\cdot f_0(t)dt \right] d_\lambda(0,\lambda_0,0).$$

Corollary: If Γ_o is a multiple limit cycle of (1_{λ_o}), then $d_\lambda(0,\lambda_o,0) = 0$ iff for all $t_o \in \mathbf{R}$ $d_\lambda(0,\lambda_o,t_o) = 0$.

Definition 2: The limit cycle Γ_o is a **singular, multiple limit cycle of** (1_{λ_o}) if $d_n(0,\lambda_o) = d_\lambda(0,\lambda_o) = 0$. If $d_n(0,\lambda_o) = 0$ and $d_\lambda(0,\lambda_o) \neq 0$ then Γ_o is a **nonsingular, multiple limit cycle of** (1_{λ_o}). If Γ_o is a nonsingular, multiple limit cycle of (1_{λ_o}) then for all $t_o \in \mathbf{R}$, $\mu(t_o) \equiv \mu_o = \pm 1$.

Remark 2: It follows from Lemmas 1 and 2 that Γ_o is a nonsingular, multiple limit cycle of (1_{λ_o}) iff

$$\int_0^{T_o} \nabla \cdot f_o(t)dt = 0$$

and

$$\int_0^{T_o} e^{-\int_0^t \nabla \cdot f_o(s)ds} f \wedge f_\lambda(\gamma_o(t),\lambda_o)dt \neq 0.$$

Note that since $f \wedge f_\lambda(x,\lambda) > 0$ in a one-parameter family of rotated vector fields, any multiple limit cycle which belongs to a one-parameter family of rotated vector fields is a nonsingular, multiple limit cycle. The next theorem follows immediately from the implicit function theorem (since $d_\lambda(0,\lambda_o) \neq 0$ for a nonsingular, multiple limit cycle) and it generalizes Theorems 71 and 72 in [1], Theorem 8 in [4] and Theorem F in [5].

Theorem 3: If Γ_o is a nonsingular, multiple limit cycle of (1_{λ_o}), then Γ_o belongs to a unique, one-parameter family of limit cycles of (1_λ) and

(1) if the multiplicity of Γ_o is odd, then the family expands or contracts monotonically iff $\omega_o \, \sigma_o \, \mu_o(\lambda - \lambda_o)$ increases or decreases respectively and

(2) if the multiplicity of Γ_o is even, then Γ_o bifurcates into a simple stable limit cycle and a simple unstable limit cycle which expand and contract monotonically as $\omega_o \, \sigma_o \, \mu_o(\lambda - \lambda_o)$ increases and Γ_o disappears as $\omega_o \, \sigma_o \, \mu_o(\lambda - \lambda_o)$ decreases.

Example 2: The analytic system

$$\dot{x} = -y - x[\lambda - (r^2 - 1)^2]$$
$$\dot{y} = x - y[\lambda - (r^2 - 1)^2]$$

has a nonsingular, multiple limit cycle of multiplicity two, at the bifurcation value $\lambda = 0$, given by $\gamma_o(t) = (\cos t, \sin t)$. The bifurcation diagram for this system is shown in Figure 3; cf. [6]. There is a Hopf bifurcation at the origin at the bifurcation value $\lambda = 1$. Note that this system defines a one-parameter family of rotated vector fields with parameter $\lambda \in \mathbf{R}$.

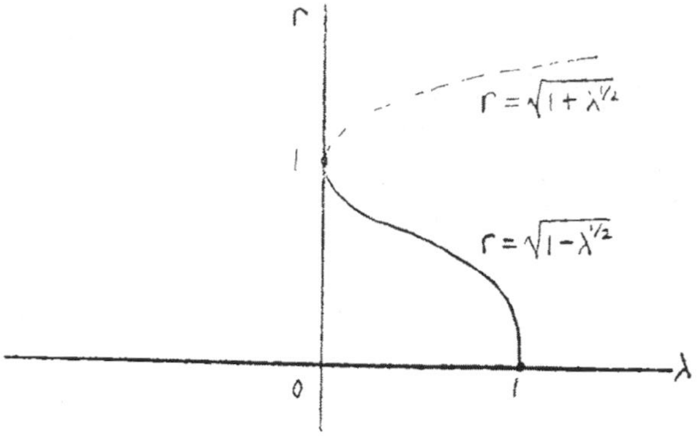

Figure 3: The bifurcation diagram for the system in Example 2.

1.3 Bifurcation at Singular Multiple Limit Cycles

If Γ_0 is a nonsingular, multiple limit cycle of (1_{λ_0}), then no matter how large its multiplicity there is only one branch of $d(n,\lambda) = 0$ passing through the point $(0,\lambda_0)$. On the other hand, if Γ_0 is a singular, multiple limit cycle of (1_{λ_0}) of multiplicity m, then the Weierstrass preparation theorem implies that there may be no branches of $d(n,\lambda) = 0$ passing through $(0,\lambda_0)$ or there may be as many as m branches of $d(n,\lambda) = 0$ passing through $(0,\lambda_0)$. In the former case Γ_0 is called an **isolated limit cycle** of the one-parameter family of vector fields (1_λ). For example the system

$$\dot{x} = -y + x\Psi(r,\lambda)$$
$$\dot{y} = x + y\Psi(r,\lambda) \tag{2}$$

with $\Psi(r,\lambda) = (r^2 - 1)^2 + \lambda^2$ has an isolated limit cycle for $\lambda = o$ which is a circle of radius one. The above system with

$$\Psi(r,\lambda) = (r - 1)[\lambda - (r^2 - 1)^2][1 + \lambda - r^2][1 - \lambda - r^2]$$

has four one-parameter families of limit cycles bifurcating from a singular, multiple limit cycle of multiplicity four at the bifurcation value $\lambda = 0$. The bifurcation diagram is shown in Figure 4.

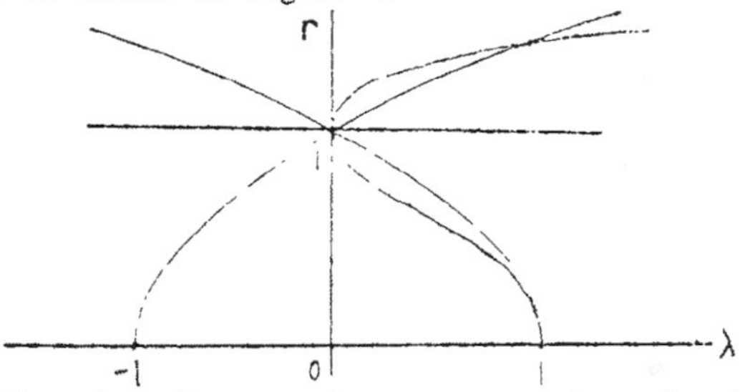

Figure 4: The bifurcation diagram of a system with a singular, multiple limit cycle of multiplicity four at $\lambda = 0$.

As in [2], Puiseux series can be used to analyze the branches $n(\lambda,t_0)$ of $d(n,\lambda,t_0) = 0$ passing through the point $(0,\lambda_0)$; cf., eg. [7]. If Γ_0 is a singular, multiple limit cycle of (1_{λ_0}) which belongs to a one-parameter family of limit cycles Γ_λ of (1_λ), then Γ_λ is defined by a branch $n(\lambda,t_o)$ of $d(n,\lambda,t_o) = 0$ where $d(n,\lambda,t_o)$ is the displacement function for (1_λ) along ℓ_{t_o} and $n(\lambda,t_o)$ can be expanded in a Puiseux series

$$n(\lambda,t_o) = (\sigma\lambda)^{k/m} \sum_{i=0}^{\infty} a_i(t_o)(\sigma\lambda)^{i/m} \tag{3}$$

which converges for $0 \leq \sigma\lambda < \sigma\delta$ where σ is ± 1, $a_o(t_o) \neq 0$ except possibly at finitely many $t_o \in [0,T_o)$, k and m are unique, relatively prime positive integers and δ is some positive constant. In this case, Γ_o is said to be a multiple limit cycle of (1_{λ_0}) which belongs to a one-parameter family of limit cycles Γ_λ of (1_λ) of **reduced multiplicity** m. Using the above Puiseux series, the following theorem was proved in [3].

Theorem 4: Suppose that Γ_o is a singular, multiple limit cycle of (1_{λ_0}) which belongs to a one-parameter family of limit cycles Γ_λ of (1_λ), corresponding to a branch $n(\lambda,t_o)$ of $d(n,\lambda,t_o) = 0$, of reduced multiplicity m. Then there is a $\delta > 0$ such that $n(\lambda,t_o)$ can be expanded in a Puiseux series (3) which converges for $0 \leq \sigma\lambda < \sigma\delta$ and $\sigma = \pm 1$; furthermore,

(1) if m is even then Γ_o bifurcates into a simple stable limit cycle and a simple unstable limit cycle belonging to the family Γ_λ as $\sigma\lambda$ increases and Γ_o disappears as $\sigma\lambda$ decreases,

(2) if m is odd and k is odd then the limit cycles in the family Γ_λ expand or contract along ℓ_{t_o} as $\sigma\lambda$ increases according to whether $a_o(t_o)$ is positive or negative respectively, and

(3) if m is odd and k is even then the limit cycles in the family Γ_λ expand or contract along ℓ_{t_o} as λ increases in $(0,\delta)$ or as λ decreases in $(-\delta,0)$ according to whether $a_o(t_o)$ is positive or negative respectively.

It was also shown in [3] that in the first case in Theorem 4, the stable and unstable hyperbolic limit cycles, resulting from the saddle-node type of bifurcation at $\lambda = \lambda_o$, expand and contract monotonically (along all normal lines ℓ_{t_o} to Γ_o) as the parameter λ varies monotonically in a neighborhood of λ_o.

1.4 Bifurcation from a Center

If the cycle Γ_o belongs to a continuous band of cycles or period annulus $\mathfrak{a}$ of (1_{λ_0}), it was shown in [8] that if the components of $f(x,\lambda_o)$ are relatively prime analytic functions of $x = (x_1,x_2)$, then the inner and outer boundaries of $\mathfrak{a}$ consist of either a single critical point or graphic of (1_{λ_0}). If the components of $f(x,\lambda_o)$ are polynomials in $x = (x_1,x_2)$, then the inner and outer boundaries of $\mathfrak{a}$ consist of either a single critical point or generalized graphic (defined in Section 2.1) on the Poincaré sphere.

It might be thought that in this case it is possible to have an infinite number of one-parameter families bifurcating from Γ_o as λ varies from λ_o, but it was shown in the appendix in [2] that this, in fact, is not possible. Since we wish to study the bifurcation of one-parameter families of limit cycles from Γ_o with small variations of the parameter λ from λ_o, we let $\lambda = \lambda_o + \varepsilon\mu$ where ε is a small parameter and $\mu \in \mathbf{R}$. If we wish to include more than the one-parameter μ, we let $\mu \in \mathbf{R}^n$. The system (1_λ) can then be written in the form

$$\dot{x} = f(x) + \varepsilon g(x,\varepsilon,\mu). \qquad (4_\mu)$$

We assume that the unperturbed system

$$\dot{x} = f(x)$$

has a period annulus $\mathcal{A} = \bigcup_\alpha \Gamma_\alpha$ where the one-parameter family of periodic orbits Γ_α with period T_α is given by

$$\Gamma_\alpha : x = \gamma_\alpha(t)$$

and where the parameter α is equal to the arc length along a Poincaré section Σ normal to the one-parameter family of periodic orbits Γ_α. We let $d(\alpha,\varepsilon,\mu)$ denote the displacement function for the system (4_μ) along Σ. The following lemmas are proved in [9]; also cf. Theorem 2.3 in [10].

Lemma 3: Let $d(\alpha,\varepsilon,\mu)$ be the displacement function for the system (4_μ) along Σ. Then

$$d_\varepsilon(\alpha,0,\mu) = -\frac{\omega_0}{|f(\gamma_\alpha(0))|} M(\alpha,\mu)$$

where the Melnikov function

$$M(\alpha,\mu) = \int_0^{T_\alpha} \frac{f \wedge g(\gamma_\alpha(t),0,\mu)}{\exp \int_0^t \nabla \cdot f(\gamma_\alpha(s))ds}\, dt.$$

Lemma 4: Under the assumptions of Lemma 3, if $M(\alpha_0,\mu_0) = \ldots = M_\alpha^{(k-1)}(\alpha_0,\mu_0) = 0$ then

$$\partial_\varepsilon d_\alpha^{(k)}(\alpha_0,0,\mu_0) = -\frac{\omega_0}{|f(\gamma_\alpha(0))|} M_\alpha^{(k)}(\alpha_0,\mu_0).$$

The next theorem, proved in [9], illustrates the use of the Melnikov function $M(\alpha,\mu)$, defined in Lemma 3, in establishing the bifurcation of limit cycles from a center.

Theorem 5: Suppose that the analytic system (4_μ) has a period annulus $\mathcal{A} = \bigcup_\alpha \Gamma_\alpha$ and that

$$M(\alpha_0,\mu_0) = 0, \quad M_\alpha(\alpha_0,\mu_0) \neq 0.$$

Then for all sufficiently small $\varepsilon \neq 0$, the system (1_{μ_0}) has a unique hyperbolic limit cycle in an $O(\varepsilon)$ neighborhood of the cycle Γ_{α_0}. If $M(\alpha_0,\mu_0) \neq 0$, then for sufficiently small $\varepsilon \neq 0$ (1_{μ_0}) has no cycle in an $O(\varepsilon)$ neighborhood of Γ_{α_0}. Furthermore, if

$$M(\alpha_0,\mu_0) = M_\alpha(\alpha_0,\mu_0) = \ldots = M_\alpha^{(m-1)}(\alpha_0,\mu_0) = 0,\ M_\alpha^{(m)}(\alpha_0,\mu_0) \neq 0,\ M_{\mu_1}(\alpha_0,\mu_0) \neq 0,$$

then for all sufficiently small $\varepsilon \neq 0$, there exists a $\mu_\varepsilon = \mu_0 + O(\varepsilon)$ such that the system (1_{μ_ε}) has a unique limit cycle of multiplicity m in an $O(\varepsilon)$ neighborhood of Γ_{α_0}.

Example 3: Using Theorem 5, it was shown in [9] that for $\mu_2\mu_3 < 0$, $0 < \mu_1\mu_3 < 9\mu_2^2/40$, and sufficiently small $\varepsilon \neq 0$, the system

$$\dot{x} = -y + \varepsilon(\mu_1 x + \mu_2 x^3 + \mu_3 x^5)$$
$$\dot{y} = x$$

has exactly two hyperbolic limit cycles of radii $r = \sqrt{\dfrac{-3\mu_2 \pm \sqrt{9\mu_2^2 - 40\mu_1\mu_3}}{5\mu_3}}$ and that for $\mu_2\mu_3 < 0$, there exists a $\mu_1 = 9\mu_2^2/40\mu_3 + O(\varepsilon)$ such that for sufficiently small $\varepsilon \neq 0$ this system has a unique semi-stable limit cycle of multiplicity two of radius $r = \sqrt{3|\mu_2|/5|\mu_3|}$.

The next theorem once again illustrates the power of the Poincaré map or the displacement function in studying the bifurcation of limit cycles. The first part of this theorem is essentially Lemma 2.1 in [10] and the second part is Theorem 5 in [9].

Theorem 6: Suppose that the analytic system (4_μ) has a period annulus $\mathcal{a} = \cup_\alpha \Gamma_\alpha$ and that $d(\alpha,\varepsilon,\mu)$ is the displacement function for (4_μ) along Σ. If for some $\delta > 0$ and all $|\alpha| < \delta$

$$d(\alpha,0,\mu_o) \equiv d_\varepsilon(\alpha,0,\mu_o) \equiv \ldots \equiv d_\varepsilon^{(k-1)}(\alpha,0,\mu_o) \equiv 0,$$
$$d_\varepsilon^{(k)}(\alpha_o,0,\mu_o) = 0,\ \partial_\varepsilon^{(k)} d_\alpha(\alpha_o,0,\mu_o) \neq 0,$$

then for all sufficiently small $\varepsilon \neq 0$, the system (1_{μ_o}) has a unique hyperbolic limit cycle in an $O(\varepsilon)$ neighborhood of the cycle Γ_{α_o}. Furthermore, if for some $\delta > 0$ and all $|\alpha| < \delta$

$$d(\alpha,0,\mu_o) \equiv d_\varepsilon(\alpha,0,\mu_o) \equiv \ldots \equiv d_\varepsilon^{(k-1)}(\alpha,0,\mu_o) \equiv 0,$$
$$d_\varepsilon^{(k)}(\alpha_o,0,\mu_o) = \partial_\alpha d_\varepsilon^{(k)}(\alpha_o,0,\mu_o) = \ldots = \partial_\alpha^{(m-1)} d_\varepsilon^{(k)}(\alpha_o,0,\mu_o) = 0,$$
$$\partial_\alpha^{(m)} d_\varepsilon^{(k)}(\alpha_o,0,\mu_o) \neq 0 \text{ and } \partial_{\mu_1} d_\varepsilon^{(k)}(\alpha_o,0,\mu_o) \neq 0,$$

then for all sufficiently small $\varepsilon \neq 0$ there is a $\mu_\varepsilon = \mu_o + O(\varepsilon)$ such that the system (1_{μ_ε}) has a unique limit cycle of multiplicity m in an $O(\varepsilon)$ neighborhood of the cycle Γ_{α_o}.

Example 4: Using Theorem 6, it was shown in [9] that given A, B, C > 0, the quadratic system

$$\dot{x} = -y + \varepsilon^6 ax + 8\varepsilon^5 bxy + \varepsilon\left[y^2 + 8\left(\frac{d}{25}\right)^{1/3} xy - 2x^2\right]$$
$$\dot{y} = x + \varepsilon^6 ax - 12\varepsilon^3 \frac{c}{(5d)^{1/3}} xy + \varepsilon\left[4\left(\frac{d}{25}\right)^{1/3}(x^2 - y^2) - xy\right]$$

with $\varepsilon \neq 0$ sufficiently small, $a = ABC$, $b = AB + AC + BC$, $c = A + B + C$ and $d = 1$ has exactly three hyperbolic limit cycles asymptotic to circles of radii $r = \sqrt{A}, \sqrt{B}$ and $\sqrt{C}$

as $\varepsilon \to 0$; and for $ab > 0$ and $c = b^2/3a$, there is a $d = c^2/3b + O(\varepsilon)$ such that for sufficiently small $\varepsilon \neq 0$ this system has a unique limit cycle of multiplicity three asymptotic to the circle of radius $r = \sqrt{3|a|/|b|}$ as $\varepsilon \to 0$. These results follow immediately from the fact, shown in [9], that with $\mu = (a,b,c,d)$ the displacement function for this system along the x-axis $d(x,\varepsilon,\mu) = 2\pi\varepsilon^6 x(a - bx^2 + cx^4 - dx^6) + O(\varepsilon^7)$. The above results are illustrated by the numerical results shown in Figures 5 and 6 below.

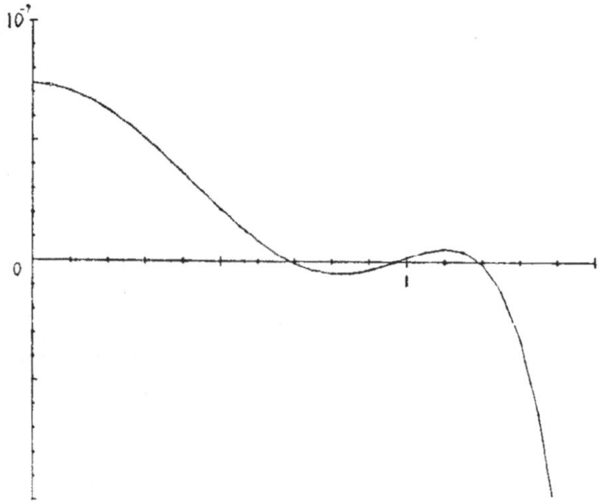

Figure 5: The function d(x)/x for the system in Example 4 with $a = 3/4$, $b = 11/4$, $c = 3$, $d = 1$ and $\varepsilon = .05$.

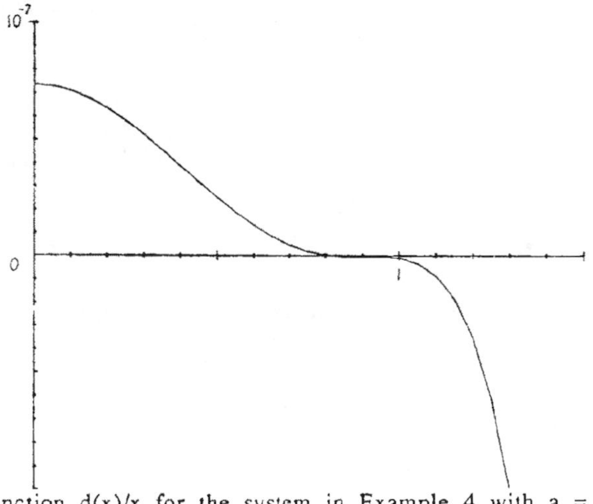

Figure 6: The function d(x)/x for the system in Example 4 with $a = 3/4$, $b = 3\sqrt{3}/2$, $c = 3$, $d = 2\sqrt{3}$ and $\varepsilon = .05$.

1.5 Bifurcation from a Separatrix Cycle

We end this section on local bifurcations with some recent results concerning the bifurcation of limit cycles from separatrix cycles or graphics (defined in the next section) and the persistence of separatrix cycles under perturbations. We assume that the unperturbed system

$$\dot{x} = f(x)$$

has a continuous band of cycles, Γ_α, on the interior (or exterior) of a separatrix cycle

$$\Gamma_0 : x = \gamma_0(t)$$

and that $\Gamma_\alpha \to \Gamma_0$ as $\alpha \to 0^+$.

The Melnikov function, $M(\alpha,\mu)$, for the perturbed system (4_μ) along the cycle Γ_α of the unperturbed system is defined in Lemma 3 of the previous section. We define the Melnikov function for (4_μ) along the separatrix cycle Γ_0 by

$$M(\mu) = \int_{-\infty}^{\infty} \frac{f \wedge g(\gamma_0(t),0,\mu)}{\exp \int_0^t \nabla \cdot f(\gamma_0(s))ds} \, dt.$$

The next theorem summarizes some recent results of the author on the persistence of the separatrix cycle Γ_0 and some results of Dingjun, Moan and Deming on the bifurcation of a limit cycle from a separatrix cycle under perturbation; cf. Theorems 6 and 7 in [9].

Theorem 7: Suppose that the analytic system (4_μ) with $\varepsilon = 0$ has a period annulus $\mathcal{a}$ on the interior or exterior of a separatrix cycle or graphic Γ_0, that

$$M(\mu_0) = 0, \; M'(\mu_0) \neq 0, \text{ and that } M_\alpha(\alpha,\mu_0) \neq 0$$

for all sufficiently small $\alpha > 0$. Then with

$$\sigma = \text{sgn}[M'(\mu_0)\, M_\alpha(\alpha,\mu_0)],$$

it follows that for all sufficiently small $\varepsilon = 0$ there is a $\mu_\varepsilon = \mu_0 + 0(\varepsilon)$ such that for all sufficiently small $|\mu - \mu_\varepsilon|$

(1) if $\sigma(\mu - \mu_\varepsilon) < 0$, there is a unique limit cycle of (4_μ) in an $0(\varepsilon)$ neighborhood of Γ_0

(2) if $\mu = \mu_\varepsilon$, there is a unique separatrix cycle of (4_{μ_ε}) in an $0(\varepsilon)$ neighorhood of Γ_0 and

(3) if $\sigma(\mu - \mu_\varepsilon) > 0$, there is no cycle or separatrix cycle of (4_μ) in an $0(\varepsilon)$ neighorhood of Γ_0.

If Γ_0 is a graphic which consists of homoclinic and/or heteroclinic loops Γ_1, Γ_2, ... Γ_k, then we must replace $M(\mu)$ in Theorem 7 by the compound Melnikov function

$$M(\mu) = M_1(\mu) + M_2(\mu) + \ldots + M_k(\mu)$$

where $M_j(\mu)$ is the Melnikov function computed along Γ_j for $j = 1, \ldots k$. We illustrate the use of Theorems 5 and 7 in the next two examples; cf. Sections 4 and 5 in [9].

Example 5: Consider the perturbed Lienard system

$$\dot{x} = y + \varepsilon(ax + bx^2 + cx^3)$$
$$\dot{y} = -x + x^3.$$

The unperturbed system with $\varepsilon = 0$ has a separatrix cycle consisting of two heteroclinic loops $\Gamma_0^{\pm}$; cf. Figure 7.

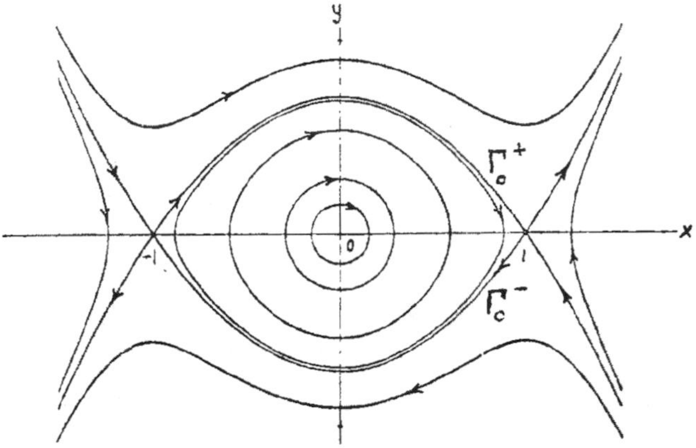

Figure 7: The phase portrait for the unperturbed system in Example 5 with $\varepsilon = 0$.

Using Theorems 5 and 7 and the theory of rotated vector fields, it was shown in [9] that for $b \in \mathbf{R}$, $c < 0$ (or $c > 0$) and all sufficiently small $\varepsilon \neq 0$, the above system has at most one limit cycle and there is a one-parameter family of limit cycles of the above system which is born in a Hopf bifurcation at $a = 0$ and which expands monotonically as a increases (or as a decreases) until it intersects the saddles at $\pm(1, -\varepsilon(a \pm b + c))$ and forms a separatrix cycle in an $O(\varepsilon)$ neighborhood of $\Gamma_0^+ \cup \{0\} \cup \Gamma_0^-$ at some value of $a = a_\varepsilon = -\frac{3}{5}c + O(\varepsilon)$.

Example 6: Consider the perturbed Lienard system

$$\dot{x} = y + \varepsilon(ax + bx^2 + cx^3)$$
$$\dot{y} = x - x^3.$$

The unperturbed system with $\varepsilon = 0$ has a compound separatrix cycle or graphic consisting of two homoclinic loops $\Gamma_0^{\pm}$; cf. Fig. 8.

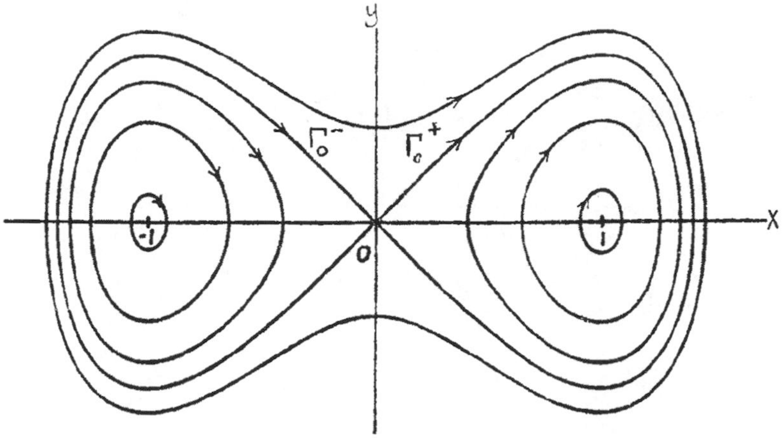

Figure 8: The phase portrait for the unperturbed system in Example 6 with $\varepsilon = 0$.

Using Theorems 5 and 7 and the theory of rotated vector fields, it was shown in [9] that for $b \in \mathbf{R}$, $c > 0$ and all sufficiently small $\varepsilon \neq 0$, there is a one-parameter family of limit cycles of the above system around the critical point $(1, -\varepsilon(a + b + c))$ which is born in a Hopf bifurcation at $a = -3c$ and which expands monotonically as a increases until it intersects the saddle at the origin and forms a separatrix cycle Γ_ε^+ in an $O(\varepsilon)$ neighborhood of Γ_0^+ at some value of $a = a_\varepsilon = -\frac{12}{5}c + O(\varepsilon)$. By symmetry there is also a one-parameter family of limit cycles around the critical point $(-1, \varepsilon(a - b + c))$ which expands as a increases and intersects the saddle at the origin and forms a separatrix cycle Γ_ε^- in an $O(\varepsilon)$ neighorhood of Γ_0^- at the bifurcation value $a = a_\varepsilon$. Now as a increases from a_ε, a one-parameter family of limit cycles bifurcates from the compound separatrix cycle $\Gamma_\varepsilon^+ \cup \{0\} \cup \Gamma_\varepsilon^-$ and it expands monotonically as a increases to the bifurcation value a_ε^* where a saddle-node bifurcation occurs. (We have numerically determined that $a_\varepsilon^* = -kc + O(\varepsilon)$ where $k = 2.256 \cdots$.) Furthermore, for all $a < a_\varepsilon^*$, there is another branch of this one-parameter family of limit cycles (as discussed in [2]) which contracts monotonically from infinity as a increases to a_ε^*. At $a = a_\varepsilon^*$ there is a semistable limit cycle of multiplicity two of the system in Example 6 and for $a > a_\varepsilon^*$, that system has no limit cycles or separatrix cycles. The following numerically computed limit cycles and separatrix cycles illustrate these results.

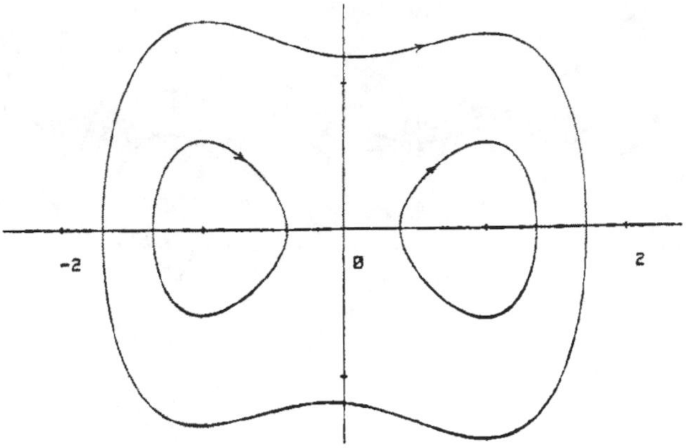

a = -.9

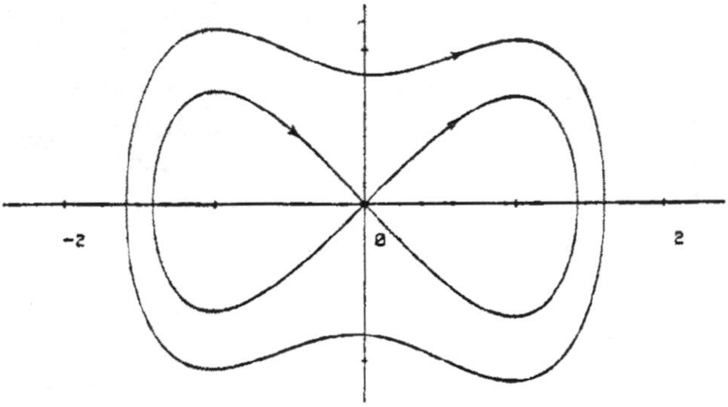

$a = a_\varepsilon \cong -.8$

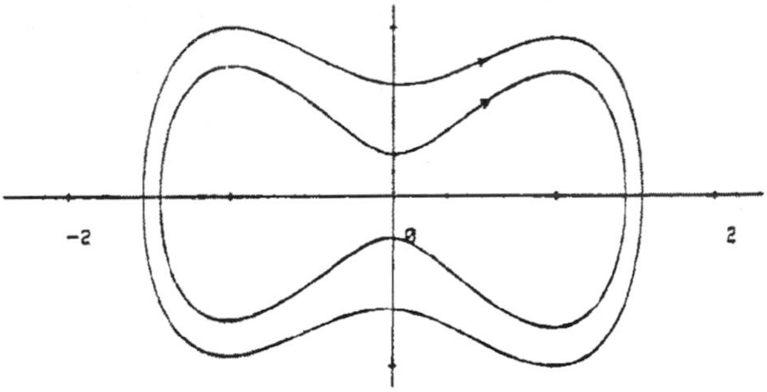

a = -.77

Figure 9: Some limit cycles and separatrix cycles for the system in Example 6 with $\varepsilon = .1$, $c = 1/3$, $a = -.9$, $a = a_\varepsilon \cong -.8$, and $a = -.77 < a_\varepsilon^* \cong -.752$.

2. Global Bifurcations

In this section, we describe the global behavior of any one-parameter family of limit cycles of (1_λ). Using Puiseux series, it was shown in [2] that when f(x,λ) is an analytic function, a one-parameter family of limit cycles of (1_λ) can be continued through any bifurcation in a unique way. This is not generally possible even for C^∞ systems. It was also shown in [2] that any one-parameter family of limit cycles of (1_λ) is contained in a unique, maximal one-parameter family of limit cycles of (1_λ) which is either open or cyclic. And the Planar Termination Principle in [2] describes how an open family terminates. After reviewing some of the basic terminology and results in [2], we establish a generalization of the Planar Termination Principle and show how it can be extended to a global result on the Poincaré sphere.

2.1 The Planar Termination Principle

We first of all recall some of the basic definitions in [2] and generalize the definition of a graphic. A **separatrix cycle** of (1_λ) is a simple closed curve which consists of a finite number of critical points and compatibly oriented trajectories of (1_λ). A **graphic** or **compound separatrix cycle** of (1_λ) is a connected union of a finite number of compatibly oriented separatrix cycles of (1_λ). A precise definition of a graphic can be found in [8] or [11]. In order to treat systems (1_λ) where the components of the vector field f(x,λ) are not necessarily relatively prime, we need the following:

Definition 3: A **generalized graphic** of (1_λ) is a connected union of a finite number of critical points, arcs of critical points and compatibly oriented trajectories of (1_λ).

The idea of a generalized graphic was introduced in [12]. While the Poincaré map is always defined on one side or the other of a separatrix cycle or graphic of (1_λ), this need not be the case for a generalized graphic of (1_λ).

As in [2], a one-parameter family of limit cycles S is said to be **open** if any path $\gamma : (\beta_0,\beta_1) \to \mathbf{R}\times\mathbf{R}^2$ which generates S is defined on an open interval. And, roughly speaking, S is said to be a **cyclic family** if topologically it is a two-dimensional torus; i.e., it has a closed loop bifurcation diagram. These notions are made precise in [2]. The system (2) with $\Psi(r,\lambda) = (r^2 - 2)^2 + \lambda^2 - 1$ furnishes an example of a cyclic family. Another example is furnished by the last example in Section 4 of [10]; cf. Figure 1 in [10]. If we assume that the system (1_λ) has a one-parameter family of limit cycles Γ_β where the parameter $\lambda = \lambda(\beta)$ then, as in [2] **the limit set of** Γ_β as $\beta \to \beta_0$ is defined as

$$\lim_{\beta\to\beta_0} \Gamma_\beta = \{x\in\mathbf{R}^2 \mid \exists \text{ a sequence } \beta_n \to \beta_0 \text{ and a sequence } x_n\in\Gamma_{\beta_n} \text{ with } x_n \to x \text{ as } n \to \infty\}.$$

With these definitions in mind, we can now state the main result in [2].

Theorem 8 (The Planar Termination Principle): If (1_λ) is a relatively prime, planar, analytic system, then any one-parameter family of limit cycles of (1_λ) is contained in a unique, maximal, one-parameter family of limit cycles S of (1_λ) and S is either open or cyclic. If S is open at an endpoint β_0 of the domain of γ (where γ is a path that generates S) then as $\beta \to \beta_0$ either the parameter $\lambda(\beta)$ becomes unbounded, the orbits Γ_β become unbounded, or $\lambda(\beta) \to \lambda_0$ and the family terminates at a critical point or graphic of (1_{λ_0}), i.e. $\lim_{\beta\to\beta_0} \Gamma_\beta$ is either a critical point or graphic of (1_{λ_0}).

Corollary: Under the hypotheses of the above theorem, the periods T_β of the one-parameter family of periodic orbits Γ_β become unbounded as $\beta \to \beta_0$ only if Γ_β approaches a degenerate critical point or graphic of (1_{λ_0}) as $\beta \to \beta_0$.

Example 7: The system

$$\dot{x} = -\lambda y + x(r^2 - \lambda^2)$$
$$\dot{y} = \lambda x + y(r^2 - \lambda^2)$$

has a one-parameter family of limit cycles, $\gamma_\lambda(t) = \lambda(\cos\lambda t, \sin\lambda t)$, which approaches a degenerate critical point at the origin as the parameter $\lambda \to 0$. In this case, the periods $T_\lambda = 2\pi/\lambda \to \infty$ as $\lambda \to 0$.

Even though the periods T_λ always increase without bound when a one-parameter family of periodic orbits Γ_λ approaches a graphic in $\mathbf{R}^2$, it is not always the case that the periods T_λ necessarily become unbounded as Γ_λ approaches a degenerate critical point. Also, some preliminary results of A. Gasull indicate that it is possible for the periods T_λ to remain bounded as the periodic orbits Γ_λ approach a graphic on the Poincaré sphere with no finite critical points.

Remark 3: Wintner's Principle of Natural Termination for analytic systems in $\mathbf{R}^n$ necessarily includes terminations at period-doubling bifurcations (which cannot occur in $\mathbf{R}^2$) and terminations as $T_\beta \to \infty$ which can occur in more complex ways than described in the above corollary, e.g., at a "homoclinic explosion." Wintner established his Principal of Natural Termination, using Puiseux series, in 1930 after years of studying the termination of various one-parameter families of periodic orbits in the restricted three-body problem computed at the Copenhagen Observatory; cf. [13].

2.2 The Generalized Planar Termination Principle

If we remove the hypotheses that the components of the vector field $f(x,\lambda)$ be relatively prime, then the Planar Termination Principle remains valid as long as we replace "graphic" by "generalized graphic" in the statement of Theorem 8. This follows since Lemmas 2.1 - 2.4 in [2] as well as their proofs hold in this case; however, Lemma 2.5 in [2] must be replaced by the following:

Lemma 5: If (1_λ) is a planar, analytic system then the limit set

$$C_0 = \lim_{\beta\to\beta_0} \Gamma_\beta$$

contains at least one critical point of (1_{λ_0}) and C_0 contains at most a finite number of isolated critical points of (1_{λ_0}); any nonisolated critical point of (1_{λ_0}) in C_0 belongs to a continuous arc of critical points of (1_{λ_0}) and there are at most a finite number of such arcs of critical points in C_0.

The first part of this lemma is proved in exactly the same way as Lemma 2.5 in [2]. If x_0 is a nonisolated critical point of (1_{λ_0}), then for any $\delta > 0$ there is a critical point x_s of (1_{λ_0}) with $|x_0 - x_s| < \delta$. Now if $x(t,\tilde{x},\beta)$ is the trajectory of $(1_{\lambda(\beta)})$ through the point $\tilde{x}$ at $t = 0$, then as in the proof of Lemma 2.4 in [2], it can be shown that for some $\delta > 0$ and each $s \in (-\delta,\delta)$ there are sequences $t_n(s)$ and $x_n(s) \in \Gamma_{\beta_n}$ such that $x(t_n(s),x_n(s),\beta_n) \to x_s$ uniformly for all $s \in (-\delta,\delta)$ as $n \to \infty$. And this implies that $x_0 \in C_0$ belongs to a continuous arc of critical points of (1_{λ_0}). The last part of Lemma 5 follows as in the proof of Theorem 1 in [8]. We then have the following generalization of Theorem 8.

Theorem 9 (The Generalized Planar Termination Principle): If (1_λ) is a planar, analytic system, then any one-parameter family of limit cycles of (1_λ) is contained in a unique, maximal, one-parameter family of limit cycles S of (1_λ) and S is either open or cyclic. If S is open at β_0 then as $\beta \to \beta_0$ either the parameter $\lambda(\beta)$ becomes unbounded, the orbits Γ_β become unbounded, or $\lambda(\beta) \to \lambda_0$ and the family terminates at a critical point or generalized graphic of (1_{λ_0}).

We next extend this result to a global result on the Poincaré sphere. To do this we assume that the components $P(x_1,x_2,\lambda)$ and $Q(x_1,x_2,\lambda)$ of $f(x,\lambda)$ are polynomials of degree m_λ and we must also assume that there is a $\delta > 0$ such that the degree $m_\lambda \equiv m$, a constant, for all $\lambda \in (\lambda_0 - \delta,\lambda_0)$ or $\lambda \in (\lambda_0,\lambda_0 + \delta)$.

Theorem 10 (The Extended Planar Termination Principle): Every one-parameter family of limit cycles of a polynomial system (1_λ) of degree m_λ is contained in a unique, maximal one-parameter family of limit cycles S of (1_λ) and S is either open or cyclic. If S is open at β_0 then as $\beta \to \beta_0$ either the parameter $\lambda(\beta)$ becomes unbounded or $\lambda(\beta) \to \lambda_0$ and if for some $\delta > 0$ $m_\lambda \equiv m$ for all $\lambda \in (\lambda_0 - \delta,\lambda_0)$ or $\lambda \in (\lambda_0,\lambda_0 + \delta)$ then the family terminates at a critical point or generalized graphic of the vector field on the Poincaré sphere defined by

$$\begin{vmatrix} dx & dy & dz \\ x & y & z \\ P_0(x,y,z) & Q_0(x,y,z) & 0 \end{vmatrix} = 0 \tag{5}$$

where $P_0(x,y,z) = z^m P(\frac{x}{z},\frac{y}{z},\lambda_0)$ and $Q_0(x,y,z) = z^m Q(\frac{x}{z},\frac{y}{z},\lambda_0)$.

We illustrate this theorem with an example from the work of Rothe and Shafer on predator-prey systems [14].

Example 8: As in [14], the system

$$\dot{x} = x[(1 - ax)(4 + 3x) - 4y]$$
$$\dot{y} = y(x - 4)$$

has a critical point at (4, 4 - 16a) and a limit cycle around this critical point in the first quadrant for the parameter $a \in (0,a^*)$ where $a^* = 12/112$. The global phase portrait in the first octant of the Poincaré sphere is shown in Figure 10. As the parameter $a \to 0^+$, the limit cycle expands to a generalized graphic on the Poincaré sphere consisting of trajectories along the positive x and y axes, a critical point at the origin and an arc of critical points along the equator of the Poincaré sphere as shown in Figure 10. In terms of local coordinates at the point (1,0,0) on the Poincaré sphere, cf. [6], the system (5) for this example becomes

$$\dot{y} = -y[(z - a)(4z + 3) - 4yz + z(4z - 1)] \to -yz(8z - 4y + 2)$$
$$\dot{z} = -z[(z - a)(4z + 3) - 4yz] \to -z^2(4z - 4y + 3)$$

as $a \to 0^+$. We see that for a = 0 the equator of the Poincaré sphere consists of critical points. Dividing the above system with a = 0 through by z gives a reduced system with a saddle at y = 1/2, z = 0.

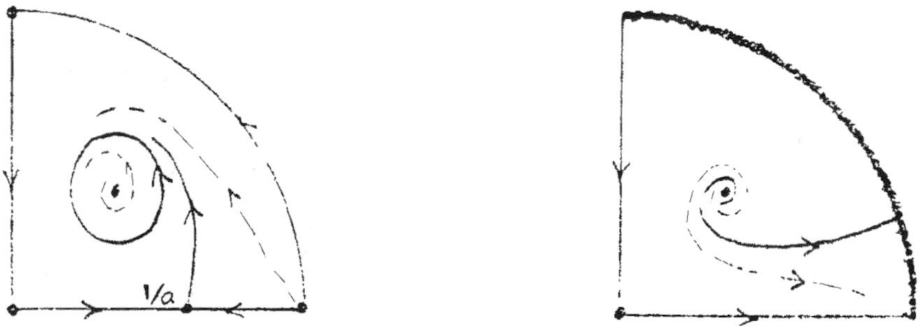

$0 < a < a^*$ $\qquad$ $a = 0$

Figure 10: The first octant global phase portrait for the system in Example 8 with $0 \le a < a^*$.

References

1. A.A. Andronov et al., *Theory of bifurcations of dynamical systems on a plane*, Kefer Press, Jerusalem (1971).
2. L.M. Perko, Global families of limit cycles of planar analytic systems, *Trans. A.M.S.*, to appear (1990).
3. L.M. Perko, Bifurcation of limit cycles: geometric theory, submitted to *Proc. A.M.S.* (1990).
4. G.F.D. Duff, Limit cycles and rotated vector fields, *Annals of Math.*, 67 (1953) 15-31.
5. L.M. Perko, Rotated vector fields and the global behavior of limit cycles for a class of quadratic systems in the plane, *J. Diff. Eq.*, 18 (1975) 63-86.

6. L.M. Perko, *Differential equations and dynamical systems*, Textbooks in Applied Math. Series, Springer Verlag, New York (1990).
7. S.N. Chow and J.K. Hale, *Methods of bifurcation theory*, Comprehensive Studies in Mathematics, Springer Verlag, New York (1982).
8. L.M. Perko, On the accumulation of limit cycles, *Proc. A.M.S.*, 99 (1987) 515-526.
9. T.R. Blows and L.M. Perko, Bifurcation of limit cycles from centers, submitted to *J. Diff. Eq.* (1990).
10. C. Chicone and M. Jacobs, Bifurcation of limit cycles from quadratic isochrones, to appear in *J. Diff. Eq.* (1991).
11. J.P. Francoise and C.C. Pugh, Keeping track of limit cycles, *J. Diff. Eq.*, 65 (1986) 139-157.
12. Y. Kuang, Finiteness of limit cycles of planar autonomous systems, *Applicable Analysis* 32 (1989) 253-264.
13. A. Wintner, Beweis des E. Stromgrenschen dynamischen Abschlusprinzipsder periodichen Bahngruppen im restringierten Dreikorperproblem, *Math. Zeit*, 34 (1931) 321-349.
14. F. Rothe and D.S. Shafer, Bifurcation in a quartic polynomial system arising in biology, this proceedings (1990).

Universal unfolding of a singularity of a symmetric vector field with 7-jet C^∞-equivalent to $y\, \partial/\partial x + (\pm x^3 \pm x^6 y)\, \partial/\partial y$ [1]

by Christiane Rousseau

Département de Mathématiques et de Statistique et Centre de Recherches Mathématiques, Université de Montréal, C.P. 6128, Succursale A, Montréal, Québec, H3C 3J7, Canada.

Abstract.
In this paper we study the codimension 4 singularity at the origin for symmetric vector fields with nilpotent linear part and 7-jet C^∞-equivalent to $y\, \partial/\partial x + (ax^3 + bx^6 y)\, \partial/\partial y$, $a,b \neq 0$. For this we introduce the universal unfolding of the singularity and derive its bifurcation diagram. The methods are classical and make an extensive use of properties of elliptic integrals.

1. Introduction.

This paper is part of a systematic study of low codimension singularities of vector fields. As noted by Arnold in [2], singularities of codimension $\leq k$ appear generically in k-parameter families of vector fields, and are interesting for the applications to phenomena depending on k parameters. We consider here the particular case of vector fields symmetric about the origin, and codimension 4 singularities with nilpotent linear part. The corresponding codimension 2 case was studied by Carr [3] and Horozov [7], and the codimension 3 case appears in [10].

It was first pointed by Arnold [1] and Takens [15] that the study of bifurcations of symmetric vector fields under a rotation of order q gives information about Hopf bifurcation of a fixed point of a diffeomorphism F of the plane with eigenvalues $e^{\pm 2\pi i p/q}$. Through the Poincaré return map F, this gives information on the loss of stability of a periodic orbit of a vector field in $\mathbb{R}^n$, when two conjugate eigenvalues of F cross the unit circle. For $q = 2$, in the generic case, the jacobian matrix of F at the fixed point is a Jordan block with eigenvalue -1. The map F^2 then can be identified to arbitrarily high order to the flow map of a symmetric vector field with a singular point at the origin having nilpotent linear part ([1], [4], [13], [15]). More precisely the bifurcation diagram of a singularity of a symmetric vector field with a double-zero eigenvalue gives the "large scale" bifurcation diagram for the fixed point of the diffeomorphism F.

We study here singularities of symmetric vector fields, with 3-jet at the origin given by

[1] This work was supported by NSERC and FCAR.

$$\dot{x} = y \qquad (1.1)$$
$$\dot{y} = \eta_1 x^3 \ , \quad \eta_1 = \pm 1.$$

The topological type of these singularities is determined by their 3-jet [15] (saddle type for $\eta_1 = 1$ and focus type for $\eta_1 = -1$). Such singularities of finite codimension have normal form [16]

$$\dot{x} = y \qquad (1.2)$$
$$\dot{y} = \eta_1 x^3 + \eta_2 x^{2(n-1)} y + o(|x,y|^{2n-1}), \quad \eta_2 = \pm 1,$$

and proposed universal unfoldings

$$\dot{x} = y \qquad (1.3)$$
$$\dot{y} = \varepsilon_1 x + \eta_1 x^3 + \varepsilon_2 y + \varepsilon_3 x^2 y + \ldots + \varepsilon_n x^{2(n-2)} y + \eta_2 x^{2(n-1)} y + o(|x,y|^{2n-1}).$$

A family in general position with the singularity (1.2) can be brought to the form (1.3) with the remainder depending on ε. These systems are studied as perturbations of Hamiltonian vector fields, and limit cycles correspond to zeros of elliptic integrals. When $\eta_1 = 1$, the singularity is of *saddle type* . It has been shown in this case by Petrov [11] that the maximum number of zeros of the elliptic integrals is n-1. The bifurcation diagram being the same as for the cusp of order n (it is described in detail in [9] for the cusp of order 4), the result will be stated without proof. When $\eta_1 = -1$, the singularity is of *focus type* . The maximum number of zeros of the elliptic integrals is $3[(n-1)/2] + 1 + (-1)^n$ [14] (which gives 5 when n = 4). We discuss completely the bifurcation diagram in the case $\eta_1 = -1$.

In the space of germs of C^∞ vector fields at the origin with a singular point at the origin we have the strata:

- Σ_0, the set of germs for which the linear part has eigenvalues with non-zero real part;
- Σ_{1H+} (resp. Σ_{1H-}), the set of germs for which the linear part has a pair of pure imaginary eigenvalues and a positive (resp. negative) radial cubic component in the normal form;
- $\Sigma_{1P\pm\pm}$, the set of germs with a generic pitchfork bifurcation, i.e. a zero eigenvalue, a positive (negative) eigenvalue, and a positive (negative) cubic term in the normal form on the center manifold;
- $\Sigma_{2H\pm}$, the set of germs for which the linear part has a pair of pure imaginary eigenvalues, a zero radial cubic component and a positive/negative radial component of degree 5 in the normal form;
- Σ_{2S+} , Σ_{2S-} (a > 0, "saddle case"), Σ_{2F+} and Σ_{2F-} (a < 0, "focus case"), the set of germs with 3-jet C^∞-equivalent to

$$\dot{x} = y$$
$$\dot{y} = ax^3 + bx^2y \ , \quad a,b \neq 0.$$

For b > 0 (resp. < 0), the saddle/focus is repelling (resp. attracting);

- $\Sigma_{3H\pm}$, the set of germs for which the linear part has a pair of pure imaginary eigenvalues, zero radial components of degree 3 and 5, and a positive/negative radial component of degree 7 in the normal form;
- Σ_{3S+} , Σ_{3S-} (a > 0, "saddle case"), Σ_{3F+} and Σ_{3F-} (a < 0, "focus case"), the set of germs with 5-jet C^∞-equivalent to

$$\dot{x} = y$$
$$\dot{y} = ax^3 + bx^4y \ , \quad a,b \neq 0,$$

depending whether b is positive or negative;

- Σ_{4S+}, Σ_{4S-} ($a > 0$, "saddle case"), Σ_{4F+} and Σ_{4F-} ($a < 0$, "focus case"), the set of germs with 7-jet C^∞-equivalent to

$$\dot{x} = y$$
$$\dot{y} = ax^3 + bx^6y \;,\quad a,b \neq 0,$$

depending whether b is positive or negative;

In this paper we give evidence for the following theorem:

Theorem 1.1. A local 4-parameter family X_λ of symmetric vector fields at $(0,0) \in \mathbb{R}^2 \times \mathbb{R}^4$ cutting one of the strata Σ_{4S+}, Σ_{4S-}, Σ_{4F+} and Σ_{4F-} transversally is fiber-C^0 equivalent (definition given in [5] and [6]) to

$$\dot{x} = y \tag{1.4}$$
$$\dot{y} = \varepsilon_1 x + \varepsilon_2 y + \varepsilon_3 x^2 y + \varepsilon_4 x^4 y + \eta_1 x^3 + \eta_2 x^6 y, \quad \eta_i = \pm 1.$$

Similar theorems for i-parameter families cutting one of the strata $\Sigma_{iS\pm}$, $\Sigma_{iF\pm}$, $i = 2,3$ transversally follow from results of [3], [7], [10].

The paper is organized in the following way. In Section 2 we give elementary properties of system (1.4). In Section 3 we study the elliptic integrals which will be needed in the rest of the paper. In Section 4 we give the bifurcation diagram of (1.4). Finally in Section 5 we derive the proof of Theorem 1.1.

Notation for all bifurcation curves (surfaces or hypersurfaces) in the text and in all figures:

- (H) (resp. (H_2), etc...) denotes Hopf bifurcation, (resp. Hopf bifurcation of order 2, etc...)
- (HL) (resp. (HL_2), etc...) denotes double homoclinic loop bifurcation, (resp. double homoclinic loop bifurcation of order 2, etc...)
- ($2C_e$), ($3C_e$), ($4C_e$) (resp. ($2C_i$), ($3C_i$)) denotes bifurcation of external (large) limit cycle (resp. pairs of internal (small) limit cycles) of multiplicity 2, 3, 4 (resp. 2, 3). We will abbreviate to (2C) or (3C) when the context is clear.
- (P) denotes pitchfork bifurcation.
- (DZ) (resp. (DZ_3)) denotes double-zero eigenvalue bifurcation (resp. double-zero eigenvalue bifurcation of codimension 3).
- Intersections of the previous bifurcations will be denoted by (H,HL), (H,2C), etc...
- In all bifurcation diagram figures, the large limit cycles in each region are drawn surrounding a number: the number of pairs of small limit cycles.

2. Elementary properties of (1.4).

We study the case $\eta_2 = -1$ (attracting focus). The case $\eta_2 = 1$ (repelling focus) follows using $(x,y,t,\varepsilon_1,\varepsilon_2,\varepsilon_3,\varepsilon_4) \mapsto (x,-y,-t,\varepsilon_1,-\varepsilon_2,-\varepsilon_3,-\varepsilon_4)$.

Singular points of (1.4) (for $\eta_1 = \eta_2 = -1$).

For $\varepsilon_1 < 0$, the system has a unique singular point at (0,0). The singular point undergoes a Hopf bifurcation for $\varepsilon_2 = 0$. The Hopf bifurcation is of order 2 (resp. 3) for $\varepsilon_3 = 0$ (resp. $\varepsilon_3 = \varepsilon_4 = 0$). The region with three limit cycles is located inside $\varepsilon_2 > 0$, $\varepsilon_3 < 0$, $\varepsilon_4 > 0$.

The system has a pitchfork bifurcation at $\varepsilon_1 = 0$, $\varepsilon_2 \neq 0$. For $\varepsilon_1 = \varepsilon_2 = 0$, we have a double-zero eigenvalue bifurcation. The generic codimension 2 cases (DZ) occur for $\varepsilon_3 \neq 0$, separated by the codimension 3 cases (DZ_3) occuring for $\varepsilon_3 = 0$, $\varepsilon_4 \neq 0$. The bifurcation diagram of (DZ) (resp. (DZ_3)) is found in [3], [7] (resp [10] and Figure 1) for $\varepsilon_3 < 0$ (resp. $\varepsilon_4 < 0$) and the other cases are derived through $(x,t,\varepsilon_2,\varepsilon_3,\varepsilon_4) \longmapsto (-x,-t,-\varepsilon_2,-\varepsilon_3,-\varepsilon_4)$). When $\varepsilon_4 > 0$ the origin in (DZ_3) is surrounded by a limit cycle. Hence there is an additional large limit cycle in each region of the bifurcation diagram.

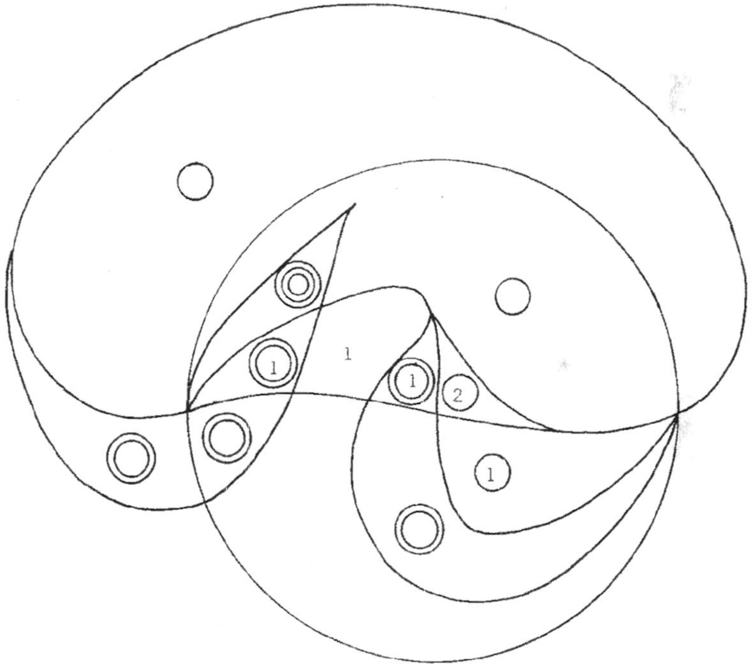

Figure 1. Bifurcation diagram of (DZ_3).

For $\varepsilon_1 > 0$, system (1.4) has three singular points. The origin is a saddle, and the two singular points $q_\pm = (\pm\sqrt{\varepsilon_1}, 0)$ are nodes or foci. There is a Hopf bifurcation at the points $q_\pm$ for $\varepsilon_2 + \varepsilon_3\varepsilon_1 + \varepsilon_4\varepsilon_1{}^2 - \varepsilon_1{}^3 = 0$. We can study it by the classical method of Lyapunov coefficients. An alternative approach is to study it by means of elliptic integrals. This method has the advantage that it can be used also to study the double homoclinic loop bifurcation, and the bifurcations of multiple limit cycles. It proceeds in the following way: for $\varepsilon_1 > 0$, we make the change of coordinates

$$x = \delta u \qquad \varepsilon_1 = \delta^2$$
$$y = \delta^2 v \qquad \varepsilon_2 = \delta^6 \mu_1 \tag{2.1}$$
$$\tau = \delta t \qquad \varepsilon_3 = \delta^4 \mu_2$$
$$\varepsilon_4 = \delta^2 \mu_3$$

and the system becomes

$$u' = v \tag{2.2}$$
$$v' = u - u^3 + \delta^5 (\mu_1 v + \mu_2 u^2 v + \mu_3 u^4 v - u^6 v)$$

which is studied as a perturbation of a Hamiltonian system (for $\delta = 0$) with Hamiltonian function

$$H(u,v) = v^2/2 - u^2/2 + u^4/4. \tag{2.3}$$

Level curves of H are drawn in Figure 2.

Limit cycles are approximately obtained as zeros of the Sotomayor-Melnikov function

$$M(h) = \int_{H=h} (\mu_1 v + \mu_2 u^2 v + \mu_3 u^4 v - u^6 v)\, du. \tag{2.4}$$

Pairs of small limit cycles correspond to values $h \in [-1/4, 0)$, and large limit cycles to values $h > 0$.

M(h) is everywhere differentiable except at $h = 0$ where we have a double homoclinic loop. In the neighborhood of $h = 0$ an additional discussion is needed since M is the principal part of the Poincaré return in the neighborhood of a singular polycycle. The only complete results about singular polycycles were obtained by Roussarie [12] for singular monocycles. Because of the symmetry of the system, our analysis can be reduced to Roussarie's case of a unique saddle using a Poincaré "half-return" map, with principal part equal to M/2 (details in [8] and [14]).

We call

$$I_i = I_i(h) = \int_{H=h} u^{2i} v\, du. \tag{2.5}$$

Using

$$I_2 = 4h\, I_0/7 + 8\, I_1/7, \qquad I_3 = 16h\, I_0/21 + (32 + 28h)\, I_1/21, \tag{2.6}$$

we get

$$M(h) = [\mu_1 + 4h\mu_3/7 - 16h/21]\, I_0 + [\mu_2 + 8\mu_3/7 - (32+28h)/21]\, I_1. \tag{2.7}$$

Hence zeros of correspondSince $I_0 > 0$ except at $h = -1/4$, we study zeros of

$$\overline{M}(h) = [\mu_1 + 4h\mu_3/7 - 16h/21] + [\mu_2 + 8\mu_3/7 - (32+28h)/21]\, P, \tag{2.8}$$

where $P = P(h) = I_1/I_0$.

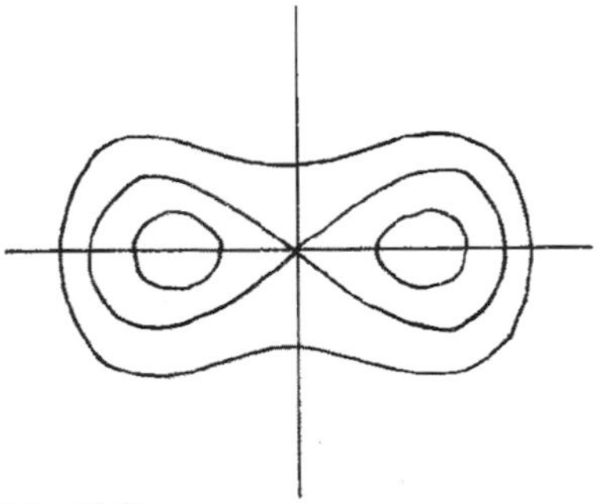

Figure 2. Level curves of H in (2.3)

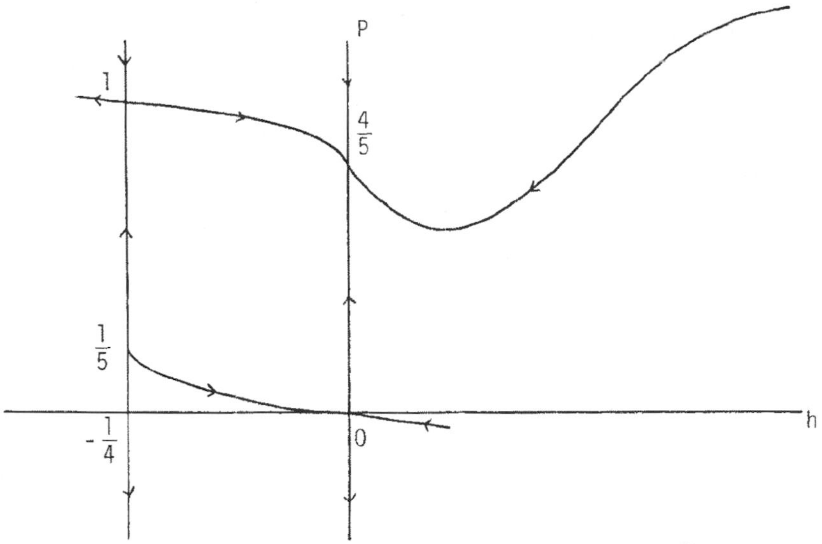

Figure 3. Graph of P(h) as a uion of trajectories of (3.2).

3. Properties of the function P(h)

Proposition 3.1. i) P satisfies a Ricatti equation

$$4h(1 + 4h)P' = 5P^2 + 8hP - 4h - 4P \tag{3.1}$$

ii) The graph of P(h) is the union of trajectories of the vector field (Figure 3)

$$\dot{h} = - 4h(1 + 4h) \tag{3.2}$$
$$\dot{P} = - 5P^2 - 8hP + 4h + 4P,$$

namely the saddle point (-1/4,1), its unstable manifold going to the node (0,4/5), the node (0,4/5) and a trajectory going to the node (0, 4/5) for $h > 0$.

iii) $P > 1/2$ (3.3)

iv) $P(-1/4) = 1$, $P'(-1/4) = -1/2$, $P''(-1/4) = -7/8$, $P'''(-1/4) = -609/128$. (3.4)

v) Around $h = 0$, P(h) has an asymptotic expansion

$$P(h) = c_0 + c_1 h \ln|h| + c_2 h + c_3 h^2 \ln|h| + \dots, \tag{3.5}$$

with $c_0 = 4/5$, $c_1 > 0$.

iv) Near $h = 0$

$$P(h) - 4/5 \sim c_1 h \ln|h|,\ P'(h) \sim c_1 \ln|h|,\ P''(h) \sim c_1/h,\ P'''(h) \sim - c_1/h^2,\ P^{iv}(h) \sim 2c_1/h^3. \tag{3.6}$$

v) At infinity

$$P(h) \sim k h^{1/2},\ P'(h) \sim k/2\ h^{-1/2},\ P''(h) \sim - k/4\ h^{-3/2}, \tag{3.7}$$
$$P'''(h) \sim 3k/8\ h^{-5/2},\ P^{iv}(h) \sim -15k/16\ h^{-7/2},$$

where k is a positive constant.

vi) $2h\ (1+4h)\ P'' = P'(5P - 12h - 4) + 4P - 2.$ (3.8)

vii) $2h\ (1+4h)\ P''' = P'(5P' - 8) + P''(5P - 28h - 6).$ (3.9)

viii) $2h\ (1+4h)\ P^{iv} = P''(15P' - 36) + P'''(5P - 44h - 8).$ (3.10)

x) $P', P'', P''' < 0$ for $h \in [-1/4,0]$.

Proof. It can be found in [3], [10] and [14] or is obvious. ∎

The following method is used to prove many properties of P and its derivatives. Using (3.1) and (3.8)-(3.10) these properties are transformed into polynomial equalities or inequalities in the variables h and P which are proved by considering the relative position of the graph of P(h) and some algebraic curves in (h,P)-plane.

Proposition 3.2. i) P', P'', P''' have exactly one positive zero $h_1 < h_2 < h_3$ respectively

$$P'(h_1) = P''(h_2) = P'''(h_3) = 0. \tag{3.11}$$

ii) $$N_1 = 3P''^2 - 2P'P''' < 0 \text{ on } [-1/4,0]. \tag{3.12}$$

iii) $N_1 = 3P''^2 - 2P'P'''$ has exactly two positive zeros h_4 and h_5. At these points

$$N_1' = 2(2P''P''' - P'P^{iv}) \neq 0. \tag{3.13}$$

iv) There are exactly two positive zeros h_6 and h_7 of the function

$$N_2(h) = 4(1 - P) + P'(1 + 4h). \tag{3.14}$$

v) There is exactly one positive zero h_8 of the function

$$N_3(h) = 4 - 5P + 5hP'. \tag{3.15}$$

vi) The points h_1-h_8 satisfy

$$0 < h_6,\ h_4 < h_1 < h_2 < h_3 < h_5 < h_8 < h_7. \tag{3.16}$$

Proof. i) The existence of h_2, h_3, h_4 follows from the asymptotic behaviour of P', P'', P''' at $h = 0$ and at $h = +\infty$. The unicity goes in the following way. If $P'(h^*) = 0$, then from (3.3) and (3.8) we deduce that necessarily $P''(h^*) > 0$.
Similarly if $P''(\hat{h}) = 0$ and $P''(h) > 0$ for $0 < h < \hat{h}$ we deduce that necessarily $P'''(\hat{h}) \leq 0$. If $P'''(\hat{h}) = 0$ then we get $P^{(n)}(\hat{h}) = 0$ for all $n \geq 2$. This is impossible since P is not analytic near $h = 0$. Then $P'''(\hat{h}) < 0$. By (3.9), $0 < P'(\hat{h}) < 8/5$ and P' is decreasing for $h > \hat{h}$. Hence, if there is another zero $\hat{h}_1$ of P'' it must also satisfy $P'''(\hat{h}_1) < 0$.
If we now suppose $P'''(\tilde{h}) = 0$, from (3.6) and (3.10) we get $P''(\tilde{h}) < 0$ and $P^{iv}(\tilde{h}) > 0$.
ii) We first show that if $h \in [-1/4,0]$, then $N_1(h) \neq 0$. For this purpose, we suppose $N_1(h^*) = 0$, and we show that necessarily $N_1'(h^*) = 2(2P''P''' - P'P^{iv}) < 0$. This is a contradiction to $N_1(-1/4) < 0$. Using (3.9) and (3.10) we can write

$$\begin{aligned} 2h(1+4h)(2P''P''' - P'P^{iv}) &= 2P''^2(5P-28h-6) + 2P'P''(5P'-8) - P'P'''(5P-44h-8) \\ &\quad - P'P''(15P'-36) \\ &= 2P''^2(5P-28h-6) + P'P''(20-5P') - P'P'''(5P-44h-8). \end{aligned} \tag{3.17}$$

$$\begin{aligned} 2h(1+4h)(2P''P''' - P'P^{iv})|_{h=h^*} &= P''^2[2(5P-28h-6) - 3(5P-44h-8)/2] + P'P''(20-5P') \\ &= 5P''[P''(P+4h) + 2P'(4-P')]/2 > 0. \end{aligned} \tag{3.18}$$

The result follows from $P', P'', P''' < 0$ for $h \in [-1/4,0]$ and $P + 4h \geq 0$ (since it is zero at $h = -1/4$, positive at $h = 0$ and the derivative is first positive, then negative).
iii) The existence of two zeros of N_1 for $h > 0$ follows from

$N_1(h) < 0$ near $h = 0$, using the asymptotic expansion of P(h): $N_1(h) \sim c_1{}^2/h^2\,[3+2\ln|h|]$,

$N_1(h) > 0$ at the points $P'(h) = 0$ and $P''(h) = 0$,

$N_1(h) < 0$ near infinity, using $N_1(h) \sim -3k^2h^{-3}/16$.

Using (3.18) we show that if $N_1(h^*) = 0$ for $h^* > 0$ and $P'(h^*) < 0$ (resp. > 0), then

$N'_1(h^*) > 0$ (resp. < 0).

Let $E = P''(P+4h) + 2P'(4-P')$ be the quantity appearing in (3.18). Our conclusion follows if we show $E > 0$.

$$2h(1+4h)E = P'(5P-12h-4)(P+4h) + (4P-2)(P+4h) - P'(5P^2+8hP-4h-4P) + 16h(1+4h)P'$$

$$= 4h(1+4h)P' + 2(2P-1)(P+4h) \qquad (3.19)$$

$$= 9P^2 + 24hP - 6P - 12h = 12(2P-1)h + 3P(3P-2). \qquad (3.20)$$

By (3.19) $E > 0$ in the region $P' > 0$. In the region $P' < 0$ we use the fact that

$$Q_1(h) = 16/3\, h^3(1+4h)^3 N_1(h) \qquad (3.21)$$

$$= 75P^4 - 80hP^3 - 64h^2P^2 - 160P^3 + 104hP^2 + 64h^2P + 120P^2 - 16h^2 - 32P - 16h$$

$$= -16(2P-1)^2h^2 + 8(2P-1)(2+4P-5P^2)h + P(5P-4)(15P^2-20P+8)$$

$$= A(P)h^2 + B(P)h + C(P),$$

using (3.1), (3.8) and (3.9). We consider the algebraic curve $Q_1 = A(P)h^2 + B(P)h + C(h) = 0$, by slicing it with lines $P = Cst$. The discriminant

$$\Delta = B(P)^2 - 4A(P)C(P) = 256(2P-1)^2(25P^4 - 50P^3 + 29P^2 - 4P + 1) > 0, \qquad (3.22)$$

everywhere for $P > 1/2$ since $R(P) = 25P^4 - 25P^3 + 29P^2 - 4P + 1$ has no real roots (because $R'(P) = 2(2P-1)(25P^2 - 25P + 2)$ has all its roots inside $[0,1]$ and $R(P) - (2P-1)R'(P)/8 = (2 + 17P - 17P^2)/4 > 0$ for $P \in [0,1]$, i.e. $R|_{R'=0} > 0$).

For $1/2 < P < 4/5$ the algebraic curve $Q_1 = 0$ has two branches $h = h(P)$ in the region $h > 0$. Solving $E = 0$ for h in (3.20) and replacing in (3.21) we get

$$Q_1|_{E=0} = 24P(P-1)(2P-1)^2 < 0. \qquad (3.23)$$

Hence there is no intersection point of the algebraic curves $E = 0$ and $Q_1 = 0$. At $P = 4/5$ the point on $E = 0$ has a negative h-coördinate and the points on $Q_1 = 0$ have positive (or zero) h-coordinates. It follows that the curve $E = 0$ is located to the left of the two branches of $Q_1 = 0$. Therefore $E|_{Q_1=0} > 0$.

iv) A zero h_0 of the function $N_2(h) = 4(1-P) + P'(1+4h)$ corresponds to a point of $P(h)$ such that the tangent line to $P(h)$ at the point h_0 passes through the point $(-1/4,1)$. Using the sign of P' and P'' we can see graphically that there are two such points.

v) Similarly a zero h_0 of $N_3(h) = 4 - 5P + 5hP'$ corresponds to a point h_0 such that the tangent line to $P(h)$ at $h = h_0$ passes through the point $(0, 4/5)$.

vi) $h_8 < h_7$ follows from consideration of the tangent lines to $P(h)$. It is enough to show $h_5 < h_8$. For this we consider $Q_1(h)$ given in (3.21) and

$$4(1+4h)N_3(h) = 25P^2 - 40hP - 40P + 44h + 16. \qquad (3.24)$$

If we put $N_3(h) = 0$ in $Q_1(h)$, i.e. $h = (25P^2 - 40P + 16)/(40P - 44)$, we get

$$Q_1(h)|_{h=h_8} = -4(5P-4)(5P-1)(P-1)(20P^2 - 45P + 28)/(10P-11)^2|_{h=h_8} < 0 \qquad (3.25)$$

since, from the graph of the hyperbola $N_3(h) = 0$, necessarily $P(h_8) > 11/10$. ■

<u>Remark</u>. A proof that $h_4 < h_6$ (which is satisfied numerically) would follow from a proof that $P(h) > 3/4$ for all positive h, which is also satisfied numerically, but which we do not know how to prove. We have made all our proofs independent of that fact.

<u>Memo 3.3 (on the points $h_i > 0$).</u> (3.26)

- $P'(h_1) = 0$, ($h_1 \simeq 0.09$), $P(h_1) < 4/5$.
- $P''(h_2) = 0$, ($h_2 \simeq 0.424$).

- $P'''(h_3) = 0$, ($h_3 \simeq 0.765$).
- $N_1(h_4) = N_1(h_5) = 0$, $h_4 < h_5$, with $N_1 = 3P''^2 - 2P'P'''$, ($h_4 \simeq 0.017$, $h_5 \simeq 0.905$), $P(h_4) < 4/5$.
- $N_2(h_6) = N_2(h_7) = 0$, $h_6 < h_7$, with $N_2 = 4(1 - P) + P'(1 + 4h)$, ($h_6 \simeq 0.019$, $h_7 \simeq 5.66$), $P(h_6) < 4/5$, $P(h_7) > 1$.
- $N_3(h_8) = 0$, with $N_3 = 4 - 5P + 5hP'$, ($h_8 \simeq 1.87$), $P(h_8) > 4/5$.
- $N_4(h_9) = N_4(h_{10}) = 0$, $h_9 < h_{10}$, with $N_4 = hP''(4 - 5P) + 2P'(4 - 5P + 5hP')$, ($h_9 \simeq 0.028$, $h_{10} \simeq 4.3$); N_4 is related to v_1 introduced in Section 4.
- $N_5(h_{11}) = 0$, with $N_5 = P''(1 + 4h)(1 - P) + P'[8(1 - P) + 2P'(1 + 4h)]$, ($h_{11} \simeq 13$); N_5 is related to v_3 introduced in Section 4.
- $N_6(h_{12}) = 0$, with $N_6 = -(1 + 4h)^2P' - 32(P - 1)^2$, ($h_{12} \simeq 0.008$); N_6 is related to v_4 introduced in Section 4.

4. Bifurcation diagram of 2.2.

Proposition 4.1.[14] The function $\overline{M}(h)$ has at most five zeros $h \geq -1/4$ for all values of μ_1, μ_2, μ_3. The maximum number of positive zeros is 4, and the maximum number of zeros inside $[-1/4,0]$ is 3.

Proposition 4.2. i) The derivatives of $\overline{M}(h)$ are given by

$$\overline{M}(h) = [\mu_1 + 4h\mu_3/7 - 16h/21] + [\mu_2 + 8\mu_3/7 - (32+28h)/21]\, P = A(h) + B(h)\, P, \tag{4.1}$$

$$\overline{M}'(h) = 4\mu_3/7 - 16/21 - 4P/3 + [\mu_2 + 8\mu_3/7 - (32+28h)/21]\, P' \tag{4.2}$$

$$= 4\mu_3/7 - 16/21 - 4P/3 + B(h)\, P',$$

$$\overline{M}''(h) = -8P'/3 + [\mu_2 + 8\mu_3/7 - (32+28h)/21]\, P'' = -8P'/3 + B(h)\, P'', \tag{4.3}$$

$$\overline{M}'''(h) = -4P'' + [\mu_2 + 8\mu_3/7 - (32+28h)/21]\, P''' = -4P'' + B(h)\, P''', \tag{4.4}$$

$$\overline{M}^{iv}(h) = -16P'''/3 + [\mu_2 + 8\mu_3/7 - (32+28h)/21]\, P^{iv} = -16P'''/3 + B(h)\, P^{iv}. \tag{4.5}$$

ii) A limit cycle of order 2 (resp. 3) occurs for

$$\overline{M}(h) = \overline{M}'(h) = 0,\ \overline{M}''(h) \neq 0, \tag{4.6}$$

and a limit cycle of order 3 for

$$\overline{M}(h) = \overline{M}'(h) = \overline{M}''(h) = 0,\ \overline{M}'''(h) \neq 0. \tag{4.7}$$

This gives a smooth surface (curve) in μ-space.
There is no small limit cycle of order 4. There are two points in μ-space corresponding to large limit cycles of order 4.

Proof. ii) For cycles of multiplicity 4 we need to consider the equations $\overline{M}(h) = \overline{M}'(h) = \overline{M}''(h) = \overline{M}'''(h) = 0$. Since $\overline{M}''(h) = \overline{M}'''(h) = 0$ gives

$$N_1(h) = 3P''(h)^2 - 2P'(h)P'''(h) = 0, \tag{4.8}$$

there are no small limit cycle of multiplicity 4 and two large limit cycles of order 4 by Proposition 3.2 ii) and iii). ■

Proposition 4.3. i) Hopf bifurcation at q_+ occurs at $\mu_1 + \mu_2 + \mu_3 - 1 = 0$. It is of order 2 under the additional condition $\mu_2 = -3$ and of order 3 at $(\mu_1,\mu_2,\mu_3) = (-1,-3,5)$. At this point the weak focus of order 3 is attracting.

ii) Double homoclinic loop through the origin occurs for $\overline{M}(0) = 0$, with

$$\overline{M}(0) = \mu_1 + 4\mu_2/5 + 32\mu_3/35 - 128/105. \qquad (4.9)$$

It is of order 2 if $\mu_1 = 0$ and of order 3 at $(\mu_1,\mu_2,\mu_3) = (0, -32/15, 16/5)$, with bifurcation diagram given in Figure 4.

iii) Bifurcation (HL_2,H) occurs at (0,-8/3, 11/3).

iv) Bifurcation (HL,H_2) occurs at (-4/9,-3,40/9).

Proof. i) Hopf bifurcation occurs for $\overline{M}(-1/4) = 0 = \mu_1 + \mu_2 + \mu_3 - 1$. Hopf bifurcation of order 2 (resp.3) occurs for $\overline{M}'(-1/4) = -(\mu_2 + 3)/2 = 0$, $\overline{M}''(-1/4) = -7\mu_2/8 - \mu_3 + 19/8 \neq 0$ (resp. $\overline{M}'(-1/4) = \overline{M}''(-1/4) = 0$, $\overline{M}'''(-1/4) = -15/4 < 0$).

ii) A theoretical discussion of how to reduce the study of the bifurcation diagram of the double homoclinic loop to the study of the zeros of $\overline{M}$ can be found in [8] and [14] (using the techniques of [12]).

From the expansion (3.5) for P(h) we deduce an asymptotic expansion for $\overline{M}(h)$ in the neighborhood of $h = 0$

$$\overline{M}(h) = \alpha_0 + \alpha_1 h \ln|h| + \alpha_2 h + \alpha_3 h^2 \ln|h| + \ldots, \qquad (4.10)$$

with

$$\begin{aligned} \alpha_0 &= \mu_1 + 4\mu_2/5 + 32\mu_3/35 - 128/105, \\ \alpha_1 &= (\mu_2 + 8\mu_3/7 - 32/21)\, c_1, \\ \alpha_2 &= 4\mu_3/7 - 64/35 + (\mu_2 + 8\mu_3/7 - 32/21)\, c_2, \\ \alpha_3 &= -4c_1/3 + (\mu_2 + 8\mu_3/7 - 32/21)\, c_3. \end{aligned} \qquad (4.11)$$

The condition $\alpha_1 = 0$ gives $\mu_2 + 8\mu_3/7 - 32/21 = 0$ (which together with $\alpha_0 = 0$ yields $\mu_1 = 0$). It is also clear that if $\alpha_0 = \alpha_1 = \alpha_2 = 0$, then $\alpha_3 < 0$. ■

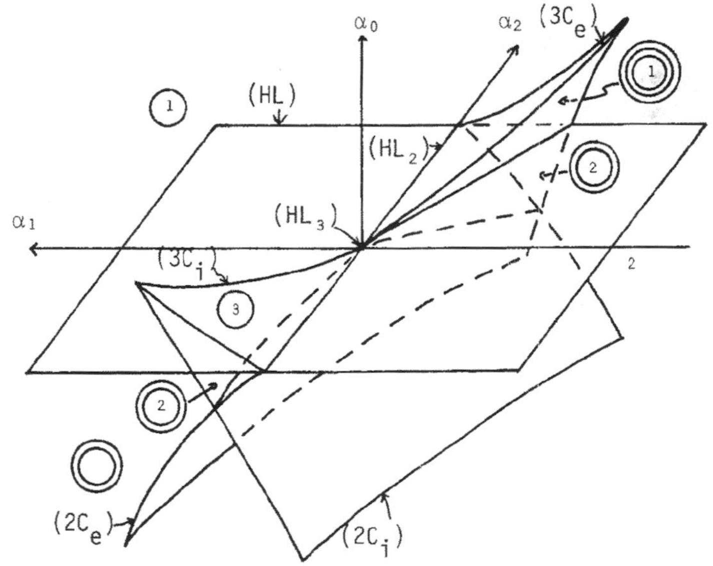

Figure 4. Bifurcation diagram of the double homoclinic loop of order 3.

The key to the understanding of the bifurcation diagram comes from the surface of double limit cycle. It is a ruled surface which is therefore well described by its intersection curves (H,2C) and (HL,2C) with the planes (H) and (HL). The surface is singular on a curve (3C) along which we have a triple limit cycle. All these curves are best studied if we change to ν-coordinates, where ν_1 (resp. ν_2, ν_3, ν_4) are zero on (HL) (resp. (HL_2), (H), (H_2)).

Proposition 4.4. The curve of triple limit cycles is given by the equations

$$\nu_1 = 4[hP''(4 - 5P) + 2P'(4 - 5P + 5hP')]/(15P'') \quad (4.12)$$

$$\nu_2 = [- 4hPP'' + 8P'(hP' - P)]/(3P'') \quad (4.13)$$

$$\nu_3 = \{P''(1 + 4h)(1 - P) + P'[8(1 - P) + 2P'(1 + 4h)]\}/(3P''), \quad (4.14)$$

in the coordinates

$$\begin{aligned} \nu_1 &= \mu_1 + 4\mu_2/5 + 32\mu_3/35 - 128/105 \\ \nu_2 &= \mu_1 \\ \nu_3 &= \mu_1 + \mu_2 + \mu_3 - 1. \end{aligned} \quad (4.15)$$

- For $h \in [-1/4,0]$, the curve joins (H_3) ($h = -1/4$) to (HL_3) ($h = 0$).
- For increasing $h > 0$, it continues in the region $\nu_1 < 0$,
- passes through a limit cycle of order 4 at $h = h_4$,
- then crosses $\nu_1 = 0$ transversally at $h = h_9$ with $\nu_2 > 0$,
- and goes to infinity when $P''(h_2) = 0$, in the region $\nu_1,\nu_3 > 0$.
- It reappears from infinity in the region ν_1, $\nu_3 < 0$,
- passes through a limit cycle of order 4 at $h = h_5$.
- crosses the plane $\nu_1 = 0$ transversally at $h = h_{10}$, with $\nu_2 > 0$,
- crosses the plane $\nu_3 = 0$ transversally at $h = h_{11}$, with $\nu_4 = \mu_2 + 3 < 0$,
- and goes to infinity when h goes to infinity in the region $\nu_1,\nu_3 > 0$.

(The sign of ν_2 (resp. ν_4) determines on which side of the lines (HL_2) (resp. (H_2)) the curve (3C) crosses the (H)-plane (resp. (HL)-plane)).

Proof. We can study the sign of ν_1 and ν_3 near $h = 0$ (using the asymptotic expansion of P), near $P''(h) = 0$ and at infinity. This gives the existence of the crossings of (3C) with $\nu_1 = 0$ and $\nu_3 = 0$.

Transversality of the crossings follows from $\nu_1' \neq 0$ (resp. $\nu_3' \neq 0$) when $\nu_1 = 0$ (resp. $\nu_3 = 0$). Indeed ν_1' and ν_3' are given by

$$\nu_1' = 4/(15P''^2)\ (4 - 5P + 5hP')(3P''^2 - 2P'P'''), \quad (4.16)$$

$$\nu_3' = 1/(3P''^2)\ [4(1 - P) + P'(1 + 4h)](3P''^2 - 2P'P'''). \quad (4.17)$$

Crossings with $\nu_1 = 0$. We first describe $\nu_1(h)$ for increasing h. ν_1' has three zeros $h_4 < h_5 < h_8$, and ν_1 passes at infinity at $h = h_2$ (the h_i are defined in Memo 3.3). Since $\nu_1 = 0$ at $h = 0$, and ν_1 is first decreasing, the curve cannot cross $\nu_1 = 0$ before going through a zero of ν_1', namely $h = h_4$. Then ν_1 is increasing to $+\infty$ until $h = h_2$. The curve reappears near $\nu_1 = -\infty$, and increases everywhere except in the interval (h_5, h_8). We have a unique crossing at $h = h_{10}$, with $\nu_1'(h_{10}) > 0$, if we show $\nu_1(h_5) < 0$. It will then follow that $h_{10} > h_8$.

For this we use $Q_1(h)$ in (3.21) and

$$\begin{aligned} Q_2(h) &= 15/2\ P''(1 + 4h)^2\nu_1 = 25P^3 - 40hP^2 - 45P^2 + 76hP + 30P - 28h - 8 \\ &= - [4(2P - 1)(5P - 7)]h + [(5P - 4)(5P^2 - 5P + 2)]. \end{aligned} \quad (4.18)$$

The conclusion follows by comparing the algebraic curves $Q_1 = 0$ and $Q_2 = 0$ in the (h,P)-plane. For fixed P, $Q_1 = 0$ has two positive h-roots for $P < 4/5$, and one for $P > 4/5$. Then h_4 is the intersection point of the left branch with P(h), and h_5 is the intersection of the right branch in $P < 4/5$ which becomes the unique branch in $P > 4/5$. $Q_2 = 0$ has no branch for $h > 0$ and $4/5 < P < 7/5$, one branch for $P > 7/5$ (on which $h \longrightarrow +\infty$ when $P \longrightarrow 7/5$), and one branch for $1/2 < P < 4/5$ (on which $h \longrightarrow 0$ when $P \longrightarrow 4/5$). We first show that $Q_1 = 0$ and $Q_2 = 0$ have no intersection point in $h > 0$, $P > 1/2$. This follows from solving $Q_2 = 0$ for h,

$$h = [(5P - 4)(5P^2 - 5P + 2)]/[4(2P - 1)(5P - 7)] \tag{4.19}$$

and replacing in Q_1,

$$Q_1|_{Q_2=0} = [40(P - 1)^2(2P - 1)(1 - 5P)(5P - 4)]/(5P - 7)^2 \neq 0, \tag{4.20}$$

since there is no positive h with $Q_2 = 0$ and $P = 4/5, 1$.

Then the curve $Q_2 = 0$ lies to the right of $Q_1 = 0$ in the region $h > 0$, $P > 7/5$. Hence we get that for $P > 4/5$, if $Q_1 = 0$, then $Q_2 > 0$. Similarly, in the region $P \leq 4/5$, $Q_2 = 0$ lies to the left of the right branch of $Q_1 = 0$, and on the right branch of $Q_1 = 0$ we have $Q_2 > 0$. In all cases we get that when $N_1(h_5) = 0$, then $v_1 < 0$, i.e. $v_1(h_5) < 0$.

It follows from the discussion that $P(h_{10}) > 7/5$.

Proof that $v_2 > 0$ at the crossings with $v_1 = 0$. It follows from

$$v_2|_{v_1=0} = 0 = 32/3\ hP'^2/[(4 - 5P)P''], \tag{4.21}$$

and the fact that at both crossings we have $(4 - 5P)P'' > 0$.

Crossings with $v_3 = 0$. Similarly we describe $v_3(h)$ for positive increasing h. Near $h = 0$ we have v_3, $v_3' > 0$. v_3' has four zeros h_4, $h_6 < h_5 < h_7$. For h just below h_2 $v_3 > 0$. We want to show that $v_3 > 0$, for $0 < h < h_2$. For this it is enough to show that $v_3(h_4)$, $v_3(h_6) > 0$. For h just above h_2, v_3 is negative. It increases until $h = h_5$, decreases until $h = h_7$, and increases then to a positive value for $h \longrightarrow +\infty$. We will show that we have a unique crossing at $h = h_{11}$ for which $v_3' > 0$ if we show that $v_3(h_5) < 0$. It will then follow that $h_{11} > h_7$. The proofs that $v_3(h_4) > 0$ and $v_3(h_5) < 0$ go together: we prove that $P''v_3|_{Q_1=0} > 0$. For this we use

$$\begin{aligned} Q_3(h) &= 24P''h^2(1 + 4h)v_3 \qquad (4.22)\\ &= 20hP^3 - 32h^2P^2 + 25P^3 + 8hP^2 + 80h^2P - 40P^2 - 28hP - 32h^2 + 16P + 8h \\ &= -16(2P - 1)(P - 2)h^2 + 4(5P^3 + 2P^2 - 7P + 2)h + P(5P - 4)^2 \\ &= a(P)h^2 + b(P)h + c(P), \end{aligned}$$

which we consider for values of $P > 1/2$ (because of (3.3)).

Our conclusion will follow if we prove that $Q_3 > 0$ at the points where $Q_1 = 0$.

First for $P < 2$, a(P), c(P) > 0. b(P) has three real roots since $b(0) > 0$, $b(1/2) < 0$, $b(4/5) > 0$. Since $b(3/4) < 0$ the highest root P^* satisfies $3/4 < P^* < 4/5$. Then $b(P) > 0$ if $P > P^*$. Also the discriminant $\Delta = b^2 - 4ac$ is given by

$$\Delta = 16(P - 1)^2(25P^4 + 270P^3 - 371P^2 + 108P + 4). \tag{4.23}$$

The quantity

$$q(P) = 25P^4 + 270P^3 - 371P^2 + 108P + 4 \tag{4.24}$$

has four real roots since $q(-1) < 0$, $q(1/2) > 0$, $q(3/4) < 0$, $q(4/5) > 0$. Let $\hat{P} < \tilde{P}$ be the highest roots.

$\Delta < 0$ for $\hat{P} < P < \tilde{P}$. We have that $P^* < \hat{P}$, since

$$q(P)|_{P=P^*} = (5P + 52)b(P^*) - 2(220P^{*2} - 231P^* + 50) = -2(220P^{*2} - 231P^* + 50) < 0, \tag{4.25}$$

for $P > 3/4$.
It follows that in the region $h > 0$, $\hat{P} < P < 2$, there are no points of the algebraic curve $Q_3 = 0$. Hence, $Q_3 > 0$ in that region.
We now consider the case $P \notin (\hat{P},2)$. For $P > 2$ a unique branch of $Q_3 = 0$ appears in $h > 0$ (on it h goes to $+\infty$ when P approaches 2). For $P < \hat{P}$ two branches of $Q_3 = 0$ start at a double point for $P = \hat{P}$.
It was shown before (in the discussion on v_1) that $Q_1 = 0$ has a unique branch for $h > 0$ and $P > 2$ and two branches for $P < 4/5$. We show that there is no intersection point between $Q_1 = 0$ and $Q_3 = 0$ in the region $h > 0$. Then, for $P > 2$, the branch of $Q_3 = 0$ is always to the right of the branch $Q_1 = 0$. Also for $P < \hat{P}$, the two branches of $Q_3 = 0$ are not separated by a branch of $Q_1 = 0$. We get the conclusion in all cases.
Suppose that there is an intersection point. Then

$$(2P - 1)Q_3 - (P - 2)Q_1 = 0 = -75P^5 + 120hP^4 + 360P^4 - 268hP^3 - 545P^3 + 144hP^2 + 344P^2 + 60hP - 80P - 40h. \tag{4.26}$$

Solving this for h

$$h = [P(5P - 4)(15P^3 - 60P^2 + 61P - 20)]/[4(2P - 1)(15P^3 - 26P^2 + 5P + 10)], \tag{4.27}$$

(the denominator is positive for $P > 0$) and replacing in $Q_1 = 0$, we get

$$Q_1|_{(4.27)} = [840P^2(5P - 4)(P - 1)^4(5P - 1)(2P - 1)]/(15P^3 - 26P^2 + 5P + 10)^2 \neq 0. \tag{4.28}$$

To prove $v_3(h_6) > 0$, we remark that (N_2 is given in (3.14))

$$4hN_2(h) = 5P^2 - 8hP - 4P + 12h. \tag{4.29}$$

Solving $4hN_2 = 0$ for h and replacing in Q_3 (given in (4.22)) we get

$$Q_3|_{N_2=0} = 42\,P(5P - 4)(P - 1)^3/(2P - 3)^2 > 0. \tag{4.30}$$

Proof that $v_4 = \mu_2 + 3 < 0$ when $v_3 = 0$. We can verify that

$$v_4 = (-105v_1 + 9v_2 + 96v_3)/12 = [P''(9 + 4h - 8P) + 8P'(1 + 2P')]/(3P'') \tag{4.31}$$

Then

$$v_4|_{v_3=0} = 2/3\;P'[P'(1 + 4h)^2 + 32(P - 1)^2]/[P''(P - 1)(1 + 4h)], \tag{4.32}$$

and $v_4 < 0$ when $v_3 = 0$ follows from $P' > 0$, $P'' < 0$ and $P > 1$.
Proof that $h_{10} < h_{11}$. Since $h_8 < h_{10}$ and $h_7 < h_{11}$, the conclusion follows if we show $h_{10} < h_7$. Since $N_2(h_7) = 0$, h_7 is a zero of (4.29). Similarly h_{10} is a solution of $Q_2 = 2(1+ 4h)^2N_4 = 0$, with Q_2 given in (4.18). Solving $4hN_2 = 0$ for h in (4.29) and replacing in Q_2 we get

$$Q_2|_{h=h_7} = -6(5P - 4)(P - 1)^2/(2P - 3) < 0, \tag{4.33}$$

since there are no points of $N_2 = 0$ for $4/5 < P < 3/2$ and $h > 0$. We get also $P(h_7) > 3/2$. ■

<u>Proposition 4.5</u>. The curve (2C) inside the (H)-plane has the shape described in Figure 5, in the coordinates

$$v_1 = \mu_1 + 4\mu_2/5 + 32\mu_3/35 - 128/105, \qquad v_4 = \mu_2 + 3. \tag{4.34}$$

The two points where the curve goes to infinity correspond to the two points where the tangent line to P(h) passes through the point (-1/4,1). There is one point of triple limit cycle on the branch corresponding to the highest values of h. The intersection of the curve with (H_2) is transversal.

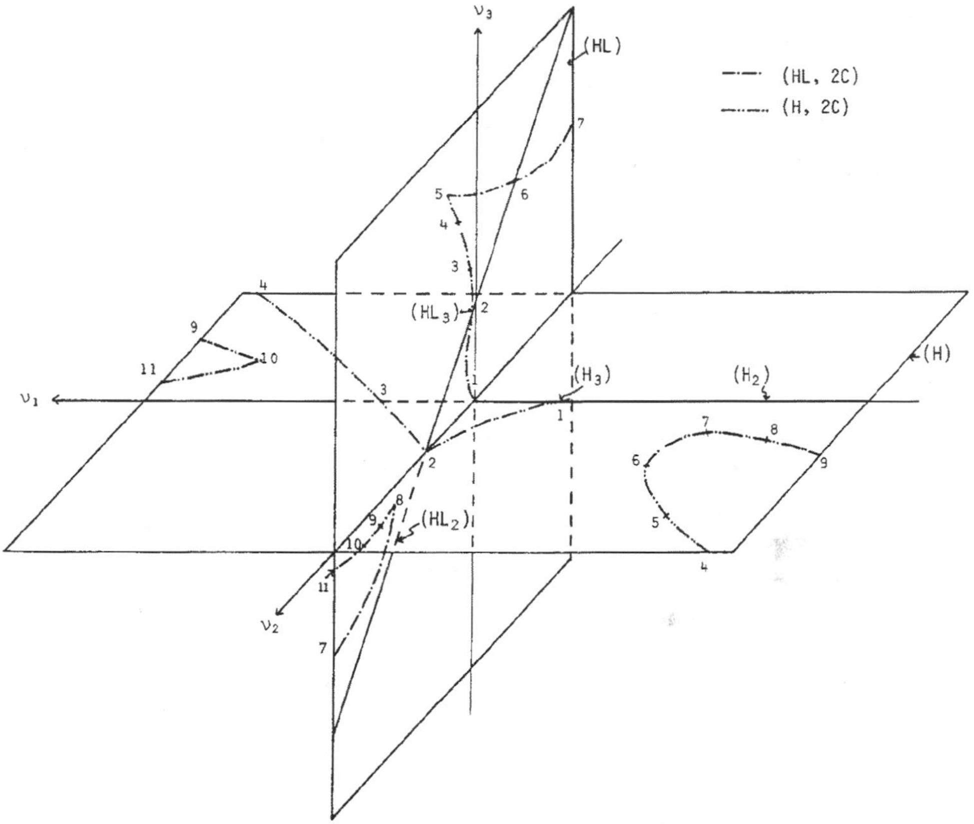

Figure 5. Curves (2C) inside (H)- and (HL)-planes.

Proof. The equation of (2C) inside (H) is given by the equations $\overline{M}(h) = \overline{M}'(h) = 0$ and $\nu_3 = \mu_1 + \mu_2 + \mu_3 - 1 = 0$. This gives the equations

$$\nu_1 = 4/15\ [h(1 + 4h)P' + (5P - 4)(P - 1)]/[4(P - 1) - P'(1 + 4h)] \qquad (4.35)$$

$$= 1/15\ [25P^2 + 8hP - 4h - 40P + 16]/[4(P - 1) - P'(1 + 4h)],$$

$$\nu_4 = 1/3\ [-(1 + 4h)^2P' - 32(P - 1)^2]/[4(P - 1) - P'(1 + 4h)]. \qquad (4.36)$$

The derivatives of ν_1 and ν_4 with respect to h are given by (since $N_2 = 4(P - 1) - P'(1 + 4h)$)

$$\nu_1' = 4/(15N_2^2)(4 - 4h - 5P)\ [P''(1 + 4h)(1 - P) + 2P'(4(1 - P) + P'(1 + 4h))], \qquad (4.37)$$

$$\nu_4' = 4/(3N_2^2)(8P + 4h - 7)\ [P''(1 + 4h)(1 - P) + 2P'(4(1 - P) + P'(1 + 4h))]. \qquad (4.38)$$

The quantity into brackets is the numerator of ν_3 in (4.14). It has the unique zero h_{11} ($h_{11} > h_7$), which corresponds to a point of (3C) studied in Proposition 4.4. The curve has a cusp point at $h = h_{11}$. ν_1' and ν_4' both have an additional zero at $4 - 4h - 5P = 0$ and at $7 - 4h - 8P = 0$ respectively.

For $h > 0$ the numerator of ν_1 is always positive: this comes from the relative position of the hyperbola $25P^2 + 8hP - 4h - 40P + 16 = 0$ and the graph of P(h). Hence ν_1 changes sign only twice (by passing through infinity) when the denominator changes sign at the two points h_6 and h_7. ν_4 also changes sign twice when the denominator is zero. The numerator of ν_4 can have no zero for $P' > 0$. It has one zero at $h = h_{12}$ for $P' < 0$, since it is $+\infty$ at $h = 0$ and negative at $P' = 0$. We want to show that $h_{12} < h_6$. This follows if we let

$$N_4(h) = 4h[-(1+4h)^2P' - 32(P-1)^2] = 16(1-2P)h^2 - 4(31 - 66P + 37P^2)h + P(4-5P) \tag{4.39}$$

and we evaluate

$$N_4|_{N_2=0} = -84P(5P-4)(P-1)^3/(2P-3)^2 < 0, \tag{4.40}$$

where N_2 is given in (3.14) and (4.29).
Also

$$N_4|_{h=(7-8P)/4} = 168(P-1)^3 < 0. \tag{4.41}$$

The line $7 - 4h - 8P = 0$ lies above the left branch of the hyperbola $N_2 = 0$, yielding that the point at which $v_4' = 0$ lies between h_6 and h_7 and satisfies $N_4 > 0$, i.e. $v_4 < 0$. This finishes the proof that h_{12} is the only point at which $v_4 = 0$ and that at this point $v_4' < 0$.
In the v-coordinates

$$\overline{M}(h) = -4/9 + 16h/9 + 35v_1/3 - 20hv_1/3 + 4v_4/3 - 4hv_4/3$$
$$+ P[5/9 - 4h/3 - 40v_1/3 - 5v_4/3], \tag{4.42}$$

$$\overline{M}'(h) = 16/9 - 20v_1/3 - 4v_4/3 - 4P/3 + P'[5/9 - 4h/3 - 40v_1/3 - 5v_4/3]. \tag{4.43}$$

Under the hypotheses $\overline{M}(h) = \overline{M}'(h) = 0$ we have

$$\overline{M}''(h) = -4/3\ \{2P'[4(P-1) - P'(1+4h)] + P''(P-1)(1+4h)\}/[4(P-1) - P'(1+4h)]. \tag{4.44}$$

$\overline{M}''(h)$ changes sign three times, twice when the denominator is zero (i.e. when the curve passes through infinity, and once for $h = h_{11}$. It follows in this case from Proposition 4.4 that $\overline{M}'''(h_{11}) < 0$. ■

Proposition 4.6. The curve (2C) inside the (HL)-plane has the shape described in Figure 5, in the coordinates

$$v_2 = \mu_1, \tag{4.45}$$
$$v_3 = \mu_1 + \mu_2 + \mu_3 - 1.$$

The point where the curve goes to infinity corresponds to the point where the tangent line to $P(h)$ passes through the point $(0,4/5)$. There are two points of triple limit cycle, one for $P' < 0$, one for $P'' < 0$. The intersection of the curve with the line (HL_2) is transversal.

Proof. The equation of (2C) inside (HL) is given by the equations $\overline{M}(h) = \overline{M}'(h) = 0$ and $v_1 = \mu_1 + 4\mu_2/5 + 32\mu_3/35 - 128/105 = 0$. This gives the equations

$$v_2 = -16/3\ h^2P'/[4 - 5P + 5hP'], \tag{4.46}$$

$$v_3 = 1/3\ [h(1+4h)P' + (5P-4)(P-1)]/[4 - 5P + 5hP'] \tag{4.47}$$
$$= 1/12\ [25P^2 + 8hP - 4h - 40P + 16]/[4 - 5P + 5hP'].$$

The derivatives of v_2 and v_3 with respect to h are given by (since $N_3 = 4 - 5P + 5hP'$)

$$v_2' = -16/(3N_3^2)\ h\ [hP''(4-5P) + 2P'(4 - 5P + 5hP')], \tag{4.48}$$

$$v_3' = 1/(3N_3^2)\ (5P + 4h - 4)\ [hP''(4-5P) + 2P'(4 - 5P + 5hP')]. \tag{4.49}$$

The quantity into brackets is the numerator of v_1 in (4.12). It has two zeros h_9 and h_{10}, which correspond to points of (3C) studied in Proposition 4.4, and are cusp points for (2C) in (HL). v_3' has an additional zero at $5P + 4h - 4 = 0$. For $h > 0$ the numerator of v_3 is always positive (it is the same as in v_1 in (4.35)). The denominator changes sign only once at $h = h_8$, corresponding to sign changes for v_2 and v_3. v_2 also changes sign when $P' = 0$, i.e. $h = h_1$. At this point $v_2' < 0$.

In the ν-coordinates

$$\overline{M}(h) = 4h/3 - 4h\nu_3 + \nu_2 - h\nu_2 - P[4h/3 + 5\nu_2/4], \tag{4.50}$$

$$\overline{M}'(h) = 4/3 - 4\nu_3 - \nu_2 - 4P/3 - P'[4h/3 + 5\nu_2/4]. \tag{4.51}$$

Under the hypotheses $\overline{M}(h) = \overline{M}'(h) = 0$ we have

$$\overline{M}''(h) = 4/3\ \{- 2P'(4 - 5P + 5hP') + P''h(5P - 4)\}/[4 - 5P + 5hP']. \tag{4.52}$$

$\overline{M}''(h)$ changes sign three times, once when the denominator is zero (i.e. the curve passes through infinity), and twice for $h = h_9$ and $h = h_{10}$. It follows from Proposition 4.4 that $\overline{M}'''(h_9)$, $\overline{M}'''(h_{10}) < 0$. ■

Proposition 4.7. Figure 5 contains the two curves (H,2C) and (HL,2C) as parametrized by values of h. This gives that the ruled surface of double limit cycles (2C) generated by the lines $\overline{M}(h) = \overline{M}'(h) = 0$ is as in Figure 6. In particular we get that the surface is connected. The surface is transversal to the planes (H) and (HL), except along (H_2), (HL_2) and (H,HL).
Proof. The points 1-11 in Figure 5 correspond to the values of h

$$-1/4 < 0 < h_{12} < h_6 < h_9 < h_1 < h_8 < h_{10} < h_7 < h_{11} < +\infty, \tag{4.53}$$

where the h_i have been defined in Memo 3.3.
The only inequality which remains to be proved is $h_6 < h_9$. For this we use (4.29) and (4.18) and we consider

$$Q_2|_{N_2=0} = - 6\ (5P - 4)(P - 1)^2/(2P - 3) < 0, \tag{4.54}$$

for $P < 4/5$, from which the inequality follows.

The transversality follows from the the fact that we have a ruled surface. ■

Theorem 4.8. The bifurcation diagram of system (2.2) in μ-space is given in Figure 6.
Proof. It follows from the preceding propositions, and the bifurcation diagram around the double homoclinic loop bifurcation of codimension 3 (Figure 4). ■

5. Sketch of Proof of Theorem 1.1.

We first derive the bifurcation diagram of system 1.4.

Theorem 5.1. The bifurcation diagram of (1.4) for $\eta_1 = -1$ is a cone. We describe its intersection with a small 3-sphere S_3 in ε-space. On a closed half 3-sphere S_3^+ ($\varepsilon_1 \geq 0$) the system has three singular points (with multiplicity). The closed half 3-sphere S_3^+ can be identified to a closed 3-ball $B_3 \subset \mathbb{R}^3$, $\mathbb{R}^3$ being identified with S_3 minus a point outside the bifurcation diagram in the region $\varepsilon_1 < 0$. The bifurcation diagram inside B_3 is given in Figure 6. The bifurcation diagram on the boundary of B_3 which we denote S_2 is given in Figure 7. The Roman numerals in Figure 6 indicate to which (relative) codimension 2 bifurcation points on S_2 are connected the codimension 2 bifurcation curves inside B_3. Outside B_3 the system has a generic Hopf bifurcation of order 3, with the related surface (curve) of double (triple) limit cycle. The surface (resp. curve) of double (resp. triple) limit cycle enters B_3 as shown in Figure 7.

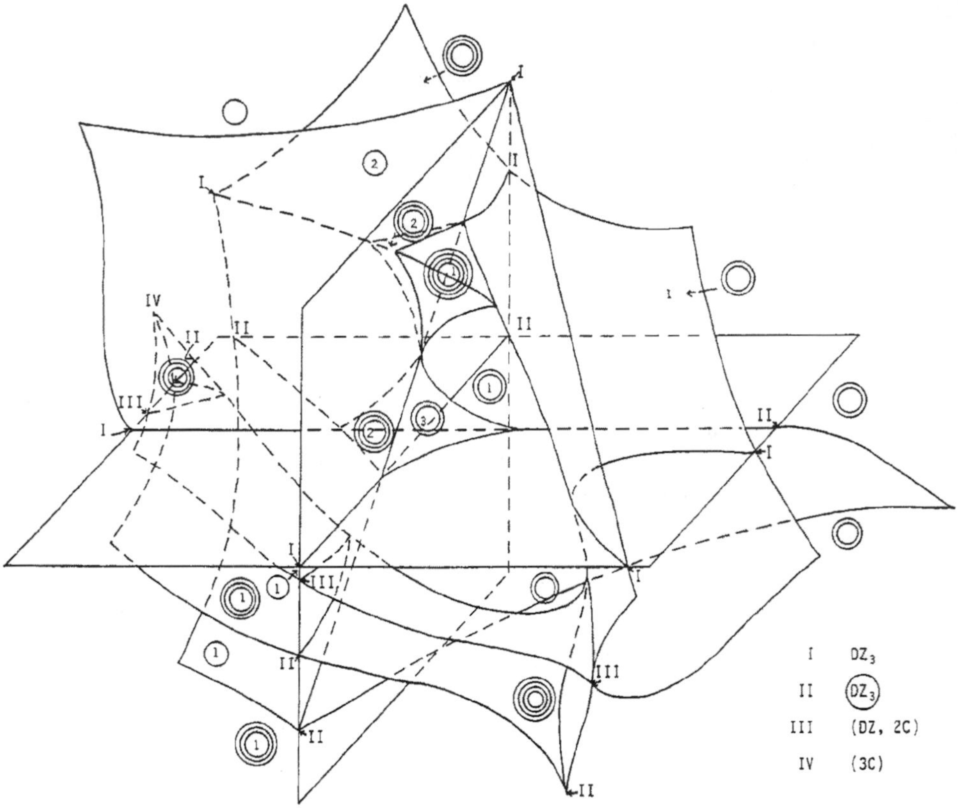

Figure 6. Bifurcation diagram of (2.2).

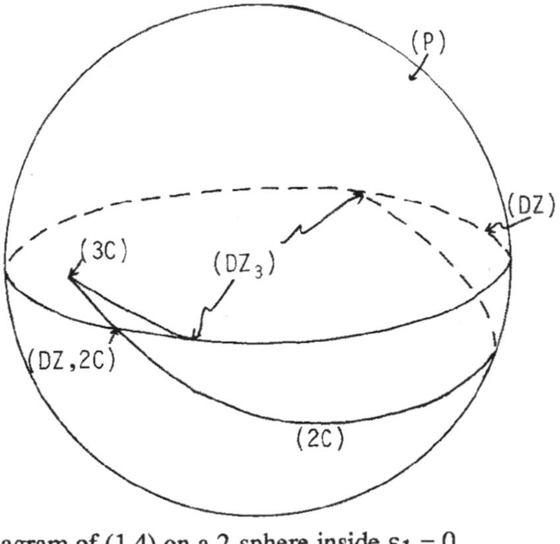

Figure 7. Bifurcation diagram of (1.4) on a 2-sphere inside $\varepsilon_1 = 0$

Proof. The bifurcation diagram on S_3 is described as a union of four cones around each of the coordinates axes:
- a cone C_1 around the ε_1-axis constructed on an arbitrary compact in $(\varepsilon_2,\varepsilon_3,\varepsilon_4)$-space;
- a cone C_2 around the ε_2-axis constructed on the product of a small neighborhood of 0 in ε_1-space with an arbitrary compact in $(\varepsilon_3,\varepsilon_4)$-space;
- a cone C_3 around the ε_3-axis constructed on the product of a small neighborhood of 0 in $(\varepsilon_1,\varepsilon_2)$-space with an arbitrary compact in ε_4-space;
- a cone C_4 around the ε_4-axis constructed on a small neighborhood of 0 in $(\varepsilon_1,\varepsilon_2,\varepsilon_3)$-space.

The construction of C_1 follows from (2.1) and Section 4.

The construction of C_2, C_3 and C_4 follows from the study of the bifurcation diagram of (1.4) on $\varepsilon_1 = 0$. The system has a pitchfork bifurcation (P), except on $\varepsilon_1 = \varepsilon_2 = 0$, where we have a bifurcation of double-zero eigenvalue (DZ). This bifurcation is degenerate at two points: $\varepsilon_1 = \varepsilon_2 = \varepsilon_3 = 0$, $\varepsilon_4 \neq 0$. In case $\varepsilon_4 < 0$ (resp. > 0) we recover the bifurcation diagram for an attracting (DZ_3) ([10] and Figure 1) (resp. the bifurcation diagram for a repelling (DZ_3) where everything is surrounded by a large attracting limit cycle).

The bifurcation diagram of (1.4) on $\varepsilon_1 = 0$ is studied by transforming the system into a near-Hamiltonian system

$$\begin{aligned} u' &= v \\ v' &= -u^3 + \delta^5(\mu_1 v + \mu_2 u^2 v + \mu_3 u^4 v - u^6 v) \end{aligned} \tag{5.1}$$

through

$$\begin{aligned} x &= \delta u & \varepsilon_1 &= \delta^2 \mu_0 \\ y &= \delta^2 v & \varepsilon_2 &= \delta^6 \mu_1 \\ \tau &= \delta t & \varepsilon_3 &= \delta^4 \mu_2 \\ & & \varepsilon_4 &= \delta^2 \mu_3. \end{aligned} \tag{5.2}$$

For $\mu_0 = 0$, we define similarly the elliptic integrals. The new function P is given by $P(h) = k\,h^{1/2}$ [10], and zeros of Sotomayor-Melnikov function (corresponding to large limit cycles) are zeros of

$$\overline{M}(h) = \mu_1 + 4\mu_3/3\,h + (\mu_2 - 4h/3)P. \tag{5.3}$$

The bifurcation diagram for these zeros can be obtained easily (it is the bifurcation diagram of the zeros of a cubic polynomial in $h^{1/2}$), and the "structurally stable behaviour" (consisting of hyperbolic large limit cycles or generic bifurcations of double or triple limit cycles) remains in a small neighborhood of each point. By the universality of the unfolding for a point of pitchfork bifurcation there can be no small limit cycle in a sufficiently small perturbation of such a point.

To construct the cone C_2, we let $\mu_0 = 0$, scale $\mu_1 = \pm 1$, and take (μ_2,μ_3) in an arbitrary compact. The only bifurcations that occur are generic bifurcations of double and triple limit cycles. So we can derive the bifurcation diagram for μ_0 in a small neighborhood of 0.
Similarly to construct C_3 we let $\mu_0 = \mu_1 = 0$, scale $\mu_2 = \pm 1$ and take μ_3 in an arbitrary compact. The bifurcation diagram contains (DZ) bifurcation and its intersection with a generic double limit cycle. This gives the bifurcation diagram for (μ_0,μ_1) in a small neighborhood of zero.

Finally to construct C_4 we let $\mu_0 = \mu_1 = \mu_2 = 0$, and scale $\mu_3 = \pm 1$. We have a generic (DZ_3) bifurcation. When $\mu_3 = +1$, we also have the existence of a hyperbolic attracting limit cycle. This gives us the bifurcation diagram for (μ_0,μ_1,μ_2) in a small neighborhood of 0.

The final point is that results are valid in a **fixed** neighborhood V of the origin in (x,y)-space. Up to now all results are only valid in a domain V_δ in (x,y)-space depending on δ. Exactly as in [6] we can show that trajectories starting on ∂V will enter in ∂V_δ. ∎

Theorem 5.2. The bifurcation diagram of (1.4) ($\eta_1 = +1$) is a cone. We describe its intersection with a small 3-sphere S_3 in ε-space. Outside a closed half 3-sphere S_3^- the system has no singular points. Identifying S_3^- with a closed 3-ball B_3, the bifurcation diagram inside B_3 is given in Figure 8 in μ-coordinates (obtained as in (5.2) with $\mu_0 = -1$). The bifurcation diagram on the boundary $\partial B_3 = S_2$ is given by a circle on which we have (DZ) with two points (DZ_3) on it.

Proof. Same as in [9] and [10]. ∎

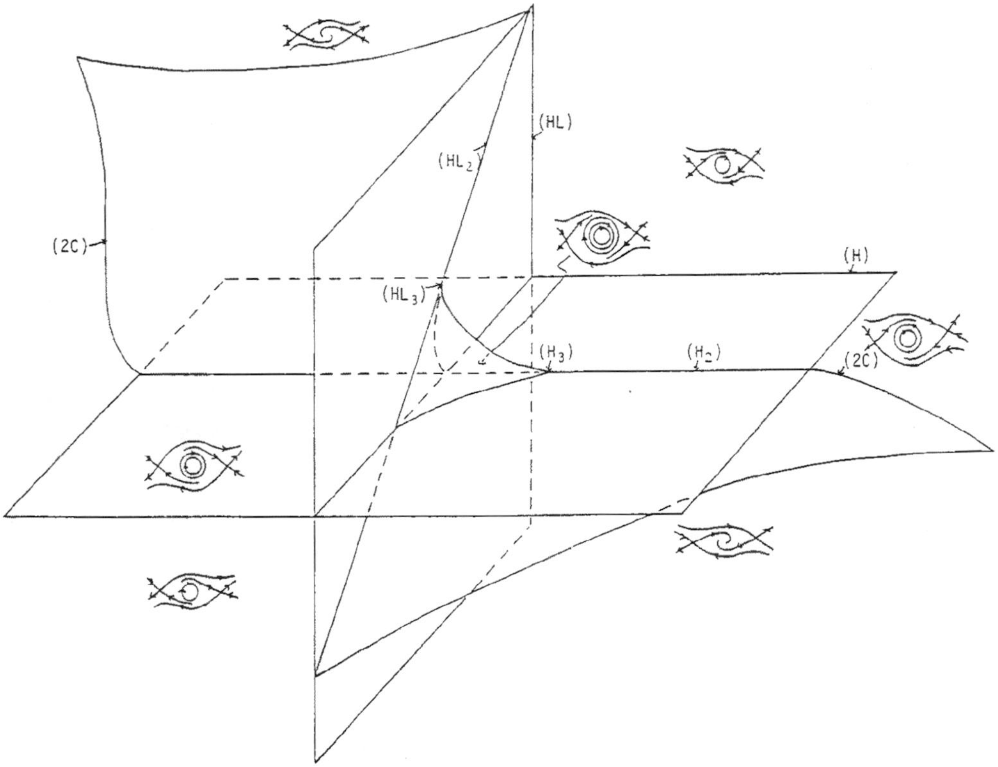

Figure 8. Bifurcation diagram in μ-coordinates for (1.4) with $\eta_1 = +1$.

Theorem 5.3. The bifurcation diagram of (1.4) can be seen as a transition from the attracting (DZ_3) to the repelling (DZ_3), through an attracting (DZ_4). Cutting properly the bifurcation diagram of (DZ_4) the transition is given by the sequence of 2-dimensional bifurcation diagrams appearing in Figures 9 and 10 for the focus ($\eta_1 = -1$) and saddle ($\eta_1 = +1$) cases respectively.

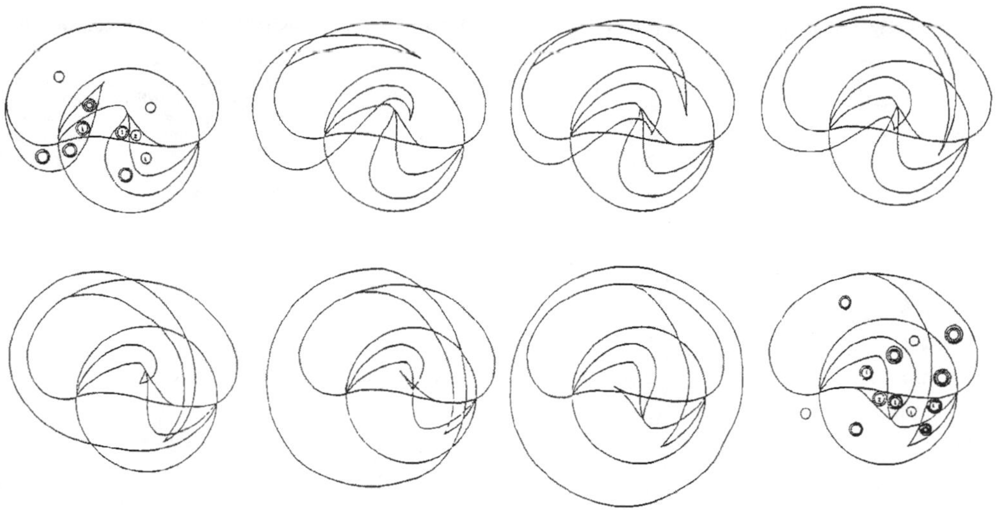

Figure 9. Transition from the attracting (DZ_3) of focus type to the repelling (DZ_3) of focus type through an attracting (DZ_4) of focus type.

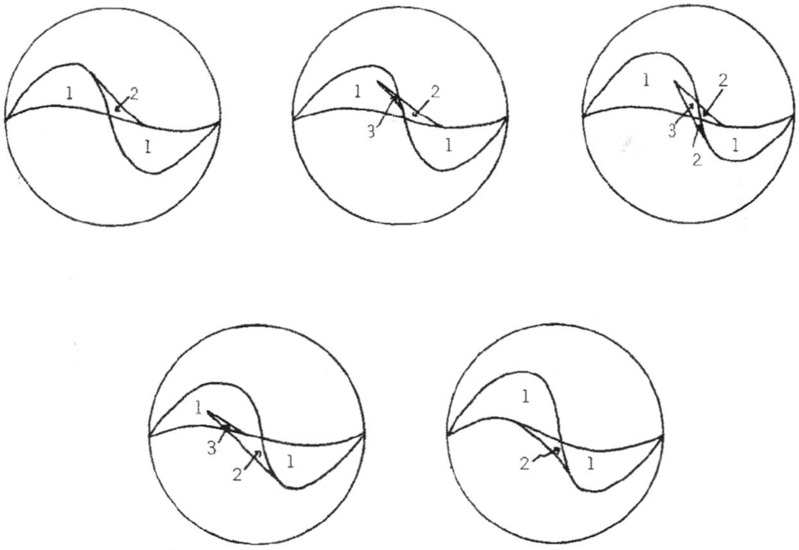

Figure 10. Transition from the attracting (DZ_3) of saddle type to the repelling (DZ_3) of saddle type through an attracting (DZ_4) of saddle type.

<u>Ideas of the proof of Theorem 1.1</u>. The complete proof of the theorem is long and technical, but all the techniques are exactly the same as in [5] and [6]. So we sketch only the ideas.

i) A 4-parameter family X_λ of vector fields cutting Σ_{4F-} transversally can be brought (allowing division by a positive function) to the normal form

$$\dot{x} = y \tag{5.1}$$
$$\dot{y} = \varepsilon_1(\lambda)x - x^3 + y[\varepsilon_2(\lambda) + \varepsilon_3(\lambda)x^2 + \varepsilon_4(\lambda)x^4 - x^6h(x,\lambda)] + y^2Q(x,y,\lambda),$$

where $h(x,\lambda) = 1 + o(|x,\lambda|)$, $Q(x,\lambda) = o(|x,y,\lambda|^N)$, $N \geq 5$ and

$$D(\varepsilon_1,\varepsilon_2,\varepsilon_3,\varepsilon_4)/D(\lambda_1,\lambda_2,\lambda_3,\lambda_4)\ (0) \neq 0. \tag{5.2}$$

(This is the transversality condition.)

ii) The study of singular points and Hopf bifurcations can be done as for system (1.4).

iii) In the same way we show that the bifurcation diagrams of (1.4) and (5.1) are topologically the same. This follows from the fact that all bifurcations occuring in (1.4) are generic, and all intersections of bifurcation curves are transversal. Hence these remain in (5.1) which can be viewed after change of coordinates and parameters as a small perturbation of (1.4).

iv) The topological type of X_λ in a fixed neighborhood of $0 \in \mathbb{R}^2$ is constant in each connected component of the complement of the bifurcation set, and in each stratum of the bifurcation set (surfaces, curves or points). It is the same as that of (1.4) for suitable ε_i's. The topological type of X_λ (in the hyperbolic case) depends only on the numbers of small and large limit cycles, and the stability of the singular points $q_\pm$. The families X_λ and (1.4) then are fiber-C^0 equivalent (since we do not know if the equivalence depends continuously on the parameters). ■

<u>Acknowledgements</u>. The author wants to thank Freddy Dumortier, Dana Schlomiuk and Jorge Sotomayor for helpful suggestions and/or stimulating discussions.

<u>References</u>.

[1] V. I. Arnold, *Loss of stability of self-oscillations close to resonances and versal deformations of equivariant vector fields,* Funct. Anal. Appl., <u>11</u>, (1977), 1-10.

[2] V. I. Arnold, *Geometrical methods in the theory of ordinary differential equations,* Springer Verlag, New York, Heidelberg, Berlin, 1983 (Russian original 1978).

[3] J. Carr, *Applications of centre manifold theory,* Springer Verlag, New York, Heidelberg, Berlin, 1981.

[4] S. N. Chow, C. Li and D. Wang, *Center manifolds, normal forms and bifurcations of vector fields,* book, in preparation.

[5] F. Dumortier, R. Roussarie and J. Sotomayor, *Generic 3-parameter families of vector fields on the plane, unfolding a singularity with nilpotent linear part. The cusp case of codimension 3,* Ergodic theory and dynamical systems, **7**, (1987), 375-413.

[6] F. Dumortier, R. Roussarie and J. Sotomayor, *Generic 3-parameter families of planar vector fields, unfoldings of saddle, focus and elliptic singularities with nilpotent linear parts,* preprint, 1989.

[7] E. I. Horozov, *Versal deformations of equivariant vector fields under symmetries of order 2 and 3,* Trudy Seminar Petrovskii, **5**, (1979), 163-192.
[8] A Jebrane and R. Mourtada, *Cyclicité finie des lacets doubles,* preprint (Dijon), 1990.
[9] C. Li and C. Rousseau, *A system with three limit cycles appearing in a Hopf bifurcation and dying in a homoclinic bifurcation: the cusp of order 4,* J. Differential Equations, **79**, (1989), 132-167.
[10] C. Li and C. Rousseau, *Codimension 2 symmetric homoclinic bifurcations and application to 1:2 resonance,* to appear in Canadian J. Math.
[11] G. S. Petrov, *Elliptic integrals and their nonoscillation,* Funct. Anal. Appl., **20**, (1986), 37-40.
[12] R. Roussarie, *On the number of limit cycles which appear by perturbation of separatrix loop of planar vector fields,* Bol. Soc. Bras. Mat., **17**, 67-101.
[13] C. Rousseau, *Codimension 1 and 2 bifurcations of fixed points of diffeomorphisms and of periodic orbits of vector fields,* to appear in Annales Mathématiques du Québec.
[14] C. Rousseau and H. Zoladek, *Zeroes of complete elliptic integrals in real domain,* preprint, 1989.
[15] F. Takens, *Forced oscillations and bifurcations,* in Applications of global analysis I, Comm. Math. Ins. Rijksuniveersiteit Utrecht, (1974), 1-59.
[16] F. Takens, *Singularities of vector fields,* Publ. Math. I.H.E.S., **43**, (1974), 47-100.

Bifurcation in a Quartic Polynomial System Arising in Biology

by

Franz Rothe

and

Douglas S. Shafer

Mathematics Department
University of North Carolina at Charlotte
Charlotte, North Carolina 28223 USA

1. Introduction.

The generalized Gause model for the interaction of species of prey and predators, with respective population densities $x(t)$ and $y(t)$ is ([4]):

$$\begin{aligned} \dot{x} &= xg(x) - p(x)y \\ \dot{y} &= q(x)y - \delta y \end{aligned} \tag{1}$$

where for biological realism one requires that all functions be continuous, $g(x)$ positive at zero but decreasing, $p(x)$ vanishing at 0 but positive otherwise, $q(x)$ with the same properties as $p(x)$, and $\delta > 0$. If $p(x)$ is allowed to increase, then decrease, system (1) then models "group defense" of the prey, in that the prey become effective in defending themselves once their population density reaches a sufficiently high level. The simplest such system, with only rational functions on the right hand side, occurs when one lets prey experience logistic growth in the absence of predation, chooses the quotient of a multiple of x with a general quadratic polynomial in x for $p(x)$, and takes, as is frequently done on the basis of biological considerations, $q(x)$ to be a multiple of $p(x)$. After a few straightforward coordinate changes and time rescaling to clear denominators, the resulting system reduces to the quartic polynomial system:

$$\begin{aligned} \dot{x} &= rx[F(x) - y] \\ \dot{y} &= -y(cx-1)(dx-1), \end{aligned} \tag{2}$$

where

$$F(x) = (1-ax)G(x) = (1-ax)\{(cx-1)(dx-1)+x\}. \tag{3}$$

The parameter $r > 0$ is the growth rate of the prey, and $a \geq 0$ is the reciprocal of the carrying capacity of the environment. In the interesting cases the remaining parameters arising from the model must satisfy $d \geq 0$ and

$$\begin{cases} 0 \leq c \leq 1 & \text{if} \quad d = 0 \\ d \leq c < c_L(d) & \text{if} \quad d > 0 \end{cases} \tag{4}$$

where $c_L(d) = 1 + 2\sqrt{d} + d$ is the larger of the two roots of $(1 - (c + d))^2 - 4cd$, the discriminant of G(x), for fixed $d \geq 0$.

The particular effect of the turning point in the predator response function $p(x)$ is the existence of a second singularity in the open first quadrant, which greatly enriches the dynamics. In this article we will discuss how, although only a quartic polynomial system, (2) exhibits a remarkable range of behavior. It contains Hopf, saddle-loop, and semi-stable bifurcation of cycles, as well as a special bifurcation of cycles from infinity. It contains the Volterra-Lotka system with its unique center, yet for every other choice of parameters has but finitely many cycles. Possible degenerate singularites are centers, saddle-nodes, and Bogdanov-Takens singularities. Moreover, it is of particular interest that all the behavior in a universal unfolding of the latter singularities in the space of smooth vector fields is actually realized by the family (2).

From the point of view of biology system (2) is interesting as well. It exhibits, for suitable choice of parameters, coexistence of a stable singularity and a stable cycle, so that in precisely the same system trajectories tend either to non-trivial (i.e., both populations non-zero) stable equilibrium or to stable oscillation, depending on initial position. An "enrichment paradox" also occurs, by which an increase in the carrying capacity of the environment leads to the extinction, rather than growth, of the predator. Two cycles (only one of them stable) can also coexist.

As a conference report, this article will be concerned mainly with illustrating the main bifurcation features of this system, and techniques used to discover them. A complete analysis of the system will appear later, together with details on finer points not discussed here. The second author wishes to express appreciation to Professors Françoise and Roussarie, and to the staff of the C.I.R.M., for their work in bringing the congress about.

2. Local Bifurcation.

We let $X_{(r,d,c,a)}(x, y)$, which we will always shorten to just $X(x, y)$, denote the parametrized quartic vector field whose horizontal and vertical components are the right hand sides of (2). The specific choice of r makes no apparent qualitative difference in the resulting family of phase portraits for X except in certain non-essential ways, so we fix r once and for all, and omit it from our considerations. We let $\mathfrak{P}$ denote the set of admissible triples (d, c, a) determined by $a \geq 0$ and (4), and let $\mathfrak{P}_0$ denote all pairs (d, c) meeting the conditions $d \geq 0$ and (4). See Figure 1, which contains a number of curves to be discussed later, and in which $\mathfrak{P}_0$ is shaded. For the most part we will discuss (2) from the point of view that d and c are fixed and that a is the one-dimensional bifurcation parameter. The biological setting of the problem we concern ourselves only with the phase portrait in the closed first quadrant $\{(x, y) \mid x, y \geq 0\}$.

As usual in predator-prey models, the coordinate axes are invariant under X, and there is a hyperbolic saddle point at the origin for all permissible parameters. The other singularities of X, when they exist, are located at $A : (1/a, 0)$, at $C : (1/c, (c - a)/c^2)$, and at $D : (1/d, (d - a)/d^2)$. The conditions (4) on the parameters insure that $G(x) > 0$ for $x \geq 0$, so from (3) $F(x)$ is positive for $0 \leq x < 1/a$ and negative for $x > 1/a$. As long as C and D are distinct and in the open first quadrant, D is a hyperbolic saddle and C is an anti-saddle. See Figures 2 and 3 for typical phase portraits.

Figure 1

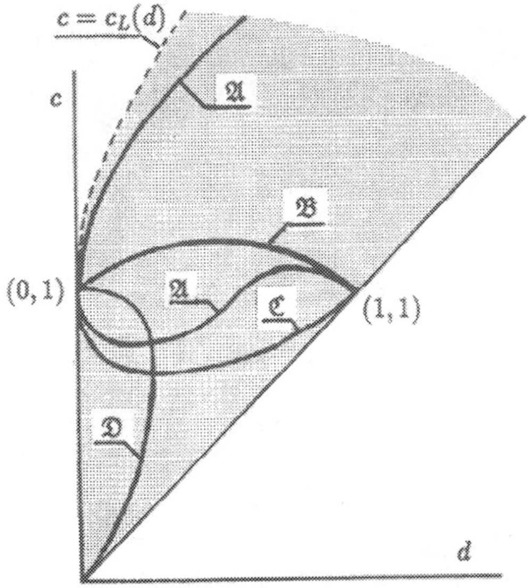

For fixed (d, c), as a increases from 0, the singularity on the positive x-axis appears from infinity and moves left. The two singularities in the first quadrant (or single singularity, if $c = d$), originally on the line $y = x$ at $a = 0$, move downward. The saddle at D passes into the fourth quadrant through the unique singularity on the x-axis in a transcritical bifurcation at $a = d$, while the anti-saddle at C does the same in a transcritical bifurcation at $a = c$.

We expect Hopf bifurcation when the trace of the singularity at C vanishes, which occurs at $a = a^* = (c - d - 1)(c - d - 2)^{-1}c$. By the Hopf bifurcation theorem, the stability of the bifurcating cycle is determined by the third focal value, which can be calculated by means of the formulas in [1], [5], or [7]. Inserting $a = a^*$ into the formulas, we obtain a cubic curve

$$\mathfrak{A}(d, c) = 3d^3 - 8d^2c + 7dc^2 - 2c^3 + 3d^2 - 5dc + 4c^2 - 2c = 0.$$

Its intersection with $\mathfrak{P}_0$ is shown in Figure 1. Label the three subsets into which $\mathfrak{A}$ divides $\mathfrak{P}_0$ by O_1 : the bounded subset; O_2 : the unbounded subset bordering part of the line $c = d$; and O_3 : the unbounded subset bordering part of the curve $c = c_L(d)$. We summarize the results of straightforward computation in the first theorem.

THEOREM 1. *When (d, c) is in the semi-infinite strip $d < c < d + 1$ [respectively, on the line $c = d + 1$; in the strip $d + 1 < c < d + 2$], then the singularity at C undergoes Hopf bifurcation as a increases across $a^* > 0$ [respectively, across $a^* = 0$; across $a^* < 0$], changing from a hyperbolic source to a hyperbolic sink. (For $c > d + 2$, the singularity has left open Quadrant I for $a \geq a^*$ and become a saddle.) In region O_1, and in the bounded* *portion of the region O_3 in which Hopf bifurcation occurs, the cycle is subcritical and stable. In the remaining region O_2 the cycle is supercritical and unstable.* □

For fixed $(d, c) \in \mathfrak{P}_0$ the trace of the linear part dX of X at D vanishes if and only if $a = a_* = (1 + c - d)(2 + c - d)^{-1}d$. In addition to $\mathfrak{A}$, two other curves of importance are

$$\mathfrak{B}(d, c) = d^2 - 2dc + c^2 + d + c - 2 = 0,$$

along which $a^* = a_*$, and

$$\mathfrak{C}(d, c) = d - (1 + d - c)^2 = 0,$$

along which $a^* = a_1$, where $a_1 = a_1(d, c) = 2dc/[c + d - 1 + (12dc - 3(d + c - 1)^2)^{\frac{1}{2}}]$ is the smallest positive value of a such that $F'(x)$ has double root. The portions of these curves lying in $\mathfrak{P}_0$ are shown in Figure 1. The correct placement and intersection of the curves (in particular, the points of intersection) become clear when the coordinate axes are rotated through 45°.

When $c = d$, there is a single singularity in open Quadrant I, which may be a cusp of Bogdanov-Takens type. The interesting point concerning its analysis is the question of which of all possible phase portraits in a universal unfolding are actually realized by the quartic family $X_{(d,c,a)}(x, y)$.

THEOREM 2. *Suppose $c = d$. When $a \neq \frac{1}{2}d$, the single singularity in open Quadrant I is a saddle-node. When $a = \frac{1}{2}d$, it is a cusp of Bogdanov-Takens type, of codimension two when $c \neq 1$, and of codimension at least three when $c = 1$. When the codimension is no more than two, the family $X_{(d,c,a)}(x, y)$ realizes all possible behavior in a universal unfolding of the singularity.*

PROOF: The statements regarding the topological type of the singularity are verified by following the scheme in §§21f of [2]. To prove the last statement we want to make a change of parameters and a singular change of coordinates to arrive at a system which is "close to Hamiltonian." In fact, one can simply follow the sequence of steps in [6]; everything done abstractly there works for our special system, and the result will follow. Here we will outline a variation on this approach, which takes into account the special form of (2). Thus we introduce parameters (d, α, γ) by writing $c = de^{\gamma}$ and $a = d(1 + \alpha)/2$, so that $\alpha = \gamma = 0$ corresponds to double eigenvalue zero. Next we choose new variables p and q and a new time scale σ by setting $x = \frac{1}{d}\exp(\gamma(p - 1))$, $y = \frac{1-\alpha}{2d}\exp(\gamma^{\frac{3}{2}}\beta q)$, and $\sigma = \frac{\gamma}{2\beta}t$, where $\beta = \sqrt{\frac{2d}{r(1-\alpha)}}$. Written in the new variables, time scale and parameters system (2) becomes

$$\frac{dp}{d\sigma} = -\frac{dK}{dq} + \frac{r}{\beta}S$$
$$\frac{dq}{d\sigma} = \frac{dK}{dp}$$

with the Hamiltonian-like function

$$K(p, q, \beta, \gamma) = \frac{\exp(\gamma^{\frac{3}{2}}\beta q) - \gamma^{\frac{3}{2}}\beta q - 1}{\gamma^3\beta^2} - \gamma^{-2}\int_0^p [\exp(\gamma\tilde{p}) - 1][\exp(\gamma(\tilde{p} - 1)) - 1]d\tilde{p}$$

and the perturbation term

$$S(p,d,\gamma,\alpha) = d\gamma^{-\frac{3}{2}}\left[F\left(\exp(\gamma(p-1))d^{-1}\right) - F\left(d^{-1}\right)\right].$$

In the new variables, the saddle D has coordinates $(p_s, q_s) = (1,0)$ and the antisaddle C has coordinates $(p_e, q_e) = (0, q_e)$. Our transformation fails in case of a saddle-node, $c = d$, $a \neq \frac{c}{2}$, corresponding to $\gamma = 0$, $\alpha \neq 0$, by producing an infinite perturbation S. Since analysis of perturbations of saddle-nodes is standard and does not lead to saddle loops or closed orbits, we exclude this case from our discussion and restrict the perturbation parameters (γ, α) to the set $\Omega = \{(\gamma,\alpha) \mid 0 \leq \gamma \leq \gamma_0, |\alpha| \leq m\gamma\}$ with an arbitrarily large m and small $\gamma_0 > 0$ to be chosen below. The main motivation for the transformation above is that it resolves the degenerate saddle-node for $c = d$, $a = \frac{d}{2}$ $(\alpha = \gamma = 0)$ into two singularities. In the limit as γ tends to 0 we get the Hamiltonian

$$H(p,q) = \lim_{\gamma \to 0} K(p,q,\beta,\gamma) = \frac{1}{2}q^2 - \int_0^p \tilde{p}(\tilde{p}-1)d\tilde{p} = \frac{1}{2}q^2 + \frac{1}{2}p^2 - \frac{1}{3}p^3.$$

An expansion of $S(p,d,\gamma,\alpha)$ in powers of α and γ is

$$S = \frac{p-1}{2d\gamma^{\frac{1}{2}}}\Big[2\alpha + \gamma\left[(1-d)p - 1\right] + \alpha\gamma\left[(2+d)p - 2\right] + \gamma^2\left[p^2 - (2+d/2)p + 1\right] + O(\gamma^2\alpha + \gamma^3)\Big].$$

Now choose a parameter set $\Omega_0 \subset \Omega$ and a $\delta > 0$ so small that for all perturbed systems with parameters $(\gamma,\alpha) \in \Omega_0$ it is true that (a) the flow is transverse to $\Sigma = \{(p,q) \mid q = 0, -\frac{1}{2} - \delta \leq p \leq -\frac{1}{2} + \delta\}$, and (b) one of the stable and one of the unstable separatices of the saddle intersects Σ.

Let K^+ and K^- be the values of the function $K(p,q,\beta,\gamma)$ at which the stable and unstable separatrices meet the section Σ, so that a homoclinic orbit corresponds to a solution of the equation

$$0 = (K^+ - K^-)(d,\gamma,\alpha) = \frac{r}{\beta}\int S dq, \tag{5}$$

where the integral is taken over the two arcs of separatrices from Σ to the saddle (the "perturbed broken loop"). Using the expansion above gives

$$K^+ - K^- = \frac{r}{2d\beta\gamma^{\frac{1}{2}}}\Big[(2\alpha - \gamma)J_1 + (1-d)\gamma J_2 + \alpha\gamma\left((2+d)J_2 - 2J_1\right) + \gamma^2\left(J_3 - (2+d/2)J_2 + J_1\right) + O\left(\gamma^2\alpha + \gamma^3\right)\Big],$$

where $J_k = \int p^{k-1}(p-1)dq$ along the broken loop, for $k = 1$, 2, and 3. We can calculate the corresponding integrals I_k along the homoclinic loop of the unperturbed system with

$S = 0$ and Hamiltonian $H(p,q) = \frac{1}{2}q^2 + \frac{1}{2}p^2 - \frac{1}{3}p^3$, obtaining $I_1 = 2/5$, $I_2 = -2/7$, and $I_3 = 2/35$. Because $J_k = I_k + O((|\alpha|+|\gamma|)/\gamma^{\frac{1}{2}}) = O(|\alpha|^{\frac{1}{2}} + |\gamma|^{\frac{1}{2}})$ for $(\gamma, \alpha) \in \Omega_0$, it follows by the implicit function theorem that for small $|c-d|$, the solutions of the equation (5) form a smooth surface in parameter space. On this surface, which we denote by $a = a^{\dagger}(d,c)$ the vector field X has closed loops. This suffices for the theorem. We note in addition that the expansions above yield

$$a^{\dagger}(d,c) = \frac{d}{2} + \frac{1}{28}(c-d)(12-5d) + O(|c-d|^{\frac{3}{2}}) \qquad \text{for all} \quad d$$
$$a^{\dagger}(1,c) = \frac{1}{2} + \frac{1}{4}(c-1) - \frac{19}{56}(c-1)^2 + O(|c-1|^{\frac{5}{2}})$$

We get an additional term when $d = 1$ because then the linear part of the expansion is independent of J_2. Indeed, in that case the expansions of $a^*(1,c)$ and $a^{\dagger}(1,c)$ for Hopf bifurcation and homoclinic bifurcation differ only in quadratic and higher order terms, which points to occurrence of multiple limit cycles. □

3. Global Bifurcation.

We now turn our attention to a global analysis of X. By the general theory of Professors Ecalle and Il'yashenko, presented at this congress, as a polynomial vector field X can have but finitely many *limit* cycles. A direct search for parameter values giving infinitely many cycles, however, indicates when there are centers, as stated in the theorem below. A fact needed in the proof of the theorem is the following, which is proved by transferring X to the sphere S^2 via the Poincaré compactification, and checking cases.

LEMMA. *For no $(d,c,a) \in \mathfrak{P}$ except $(d,c,a) = (0,1,0)$ is there a sequence of points (p_i), each on a distinct cycle of X, and tending to infinity.* □

THEOREM 3. *X has infinitely many cycles in the first quadrant if and only if the singularity at C is a center, in which case $(d,c,a) = (0,1,0)$, and X is the Volterra-Lotka system.*

PROOF: Suppose there are infinitely many cycles in the first quadrant. We must have $c > d$, else there is just one singularity in the open first quadrant, of index 0. All the cycles of course surround C. This singularity is not a center-focus, i.e., *limit* cycles do not accumulate on it, since this could happen only if tr dX vanished at C, hence only if $a = a^*$. But in that which case det dX does not vanish at C, excluding a center-focus (cf. §24 of [2]).

Since X is real analytic, the cycles cannot accumulate on a limit cycle, hence must have a point of accumulation which is either a center or a point on a separatrix cycle of the Poincaré vector field $\pi(X)$, and the first alternative implies the latter. Thus the cycles accumulate on a separatrix cycle, which either contains a point at infinity or does not. If it does not, then from the nature of the singularities of X, as described above, the cycle must be a saddle-loop at D. But the saddle is hyperbolic, hence the loop is not accumulated on by limit cycles (e.g., Theorem 4.6 of [3]). Thus every orbit near the loop and inside it is closed, hence there is a band of cycles which must have a center as inner boundary.

That the outer boundary of the annulus of cycles is the loop forces $a = a_*$, while the fact that the inner boundary a center forces $a = a^*$, hence $(d,c) \in \mathfrak{B}$. But again there is a center at C only when all focal values vanish, hence $(d,c) \in \mathfrak{A}$ as well, so that as shown by Figure 1 either $(d,c) = (0,1)$ or $(d,c) = (1,1)$, neither of which gives a saddle of X in Quadrant I. We conclude that the cycles accumulate on a separatrix cycle only if it contains a point at infinity, which by the lemma implies that $(d,c,a) = (0,1,0)$. Then (2) becomes $\dot{x} = rx(1-y)$, $\dot{y} = y(x-1)$, into which any system of Volterra-Lotka form ($\dot{x} = x(A - By)$, $\dot{y} = y(Cx - D)$, $A, B, C, D > 0$) can be transformed by coordinate change and rescaling (change of time scale). This system of course has a unique center in Quadrant I, cycles fill the quadrant, and the separatrix cycle is composed of the positive coordinate axes and an arc of the equator. □

It is also of interest, and important for further results, to locate regions in parameter space for which X has *no* cycles. Naturally, we try to use the Bendixson criterion to do this. The novelty is to first make a change of coordinates which exploits the fact that x [respectively, y] factors out of the horizontal [respectively, vertical] component of the system. The following lemma expresses a result in terms of the function $F(x)$. We will then interpret the result in solely terms of the parameters.

LEMMA. *For a, c, and d all positive, cycles and saddle loops do not exist in the following two cases:*

(I) $F'(1/c) \geq 0$ and $F'(1/d) \geq 0$.
(II) $F'(x) \leq 0$ *for all* $x \in (1/c, 1/d)$.

PROOF: To handle both cases at once, we introduce $\sigma = +1$ in case (I) and $\sigma = -1$ in case (II). We will need the third order polynomial $R(x,\alpha) = \sigma x F'(x) + \alpha(cx-1)(dx-1) = -\sigma dcax^3 + O(x^2)$, where $\alpha \in \mathbf{R}$ is to be chosen below.

Since x, $y > 0$ we may introduce new coordinates into the system (2) by setting $x = e^u$ and $y = e^v$ to obtain

$$\begin{aligned} \dot{u} &= r[F(e^u) - e^v] \\ \dot{v} &= -(ce^u - 1)(de^u - 1). \end{aligned} \tag{6}$$

If the vector field $Y(u,v)$ on the right-hand side of (6) is multiplied by the positive function $\exp(Q) = \exp(-\alpha r v)$, then the divergence of the resulting vector field satisfies

$$e^{-Q}\operatorname{div}(e^Q Y) = r[e^u F'(e^u) + \alpha(ce^u - 1)(de^u - 1)]. \tag{7}$$

Since in terms of the original coordinates all possible cycles and saddle loops must lie inside the strip $0 < x < 1/d$, by the Bendixson-Dulac criterion it is enough to demonstrate the existence of a value $\alpha_1 \geq 0$ of α for which

$$R(x,\alpha_1) \geq 0 \quad \text{for all} \quad x \in [0, 1/d). \tag{8}$$

Indeed, the existence of such an α_1 results from the different qualitative behavior of the two functions $R(x,0)$ and $R(x,\alpha)$, α sufficiently large, as we now show.

Since $R(x,\alpha)$ is nonnegative at $x = 1/c$ and $x = 1/d$, and $(cx-1)(dx-1) < 0$ for $x \in (1/c, 1/d)$, there exists $\alpha_0 > 0$ such that for all $\alpha \geq \alpha_0$ two simple real roots $x_2(\alpha)$, $x_3(\alpha)$

of the equation $R(x,\alpha) = 0$ satisfy $1/c \leq x_2 < x_3 \leq 1/d$. The third root $x_1(\alpha)$ behaves differently in cases (I) and (II). In case (I) we conclude $x_1(\alpha) > 1/d$ from $R(1/d, \alpha) \geq 0$ and the sign of the coefficient of x_3. In case (II) we conclude $x_1(\alpha) < 0$ for $\alpha > 0$ from $R(0,\alpha) = \alpha$ and the sign of the leading coefficient. Let α_1 be the supremum of all values of α for which the equation $R(x,\alpha) = 0$ has a double root. By continuity of the roots with respect to α, we have that for all $\alpha > \alpha_1$, $R(x,\alpha) < 0$ *for* $x \in (x_2, x_3)$, and

$$1/c \leq x_2(\alpha) < x_3(\alpha) \leq 1/d. \tag{9}$$

(For $F'(1/c) \cdot F'(1/d) \neq 0$ the bounds on the roots are clear; otherwise one can show that $R(1/c, \alpha)$ and $R(1/d, \alpha)$ can vanish only as double roots.) In case (I) we also have

$$x_1(\alpha) > 1/d \tag{10}$$

for all $\alpha > \alpha_1$.

We claim that $\alpha_1 \geq 0$. For $\alpha_1 < 0$ would lead to a contradiction when $\alpha = 0$ is plugged into (10) in case (I) (since $R(0,0) = 0$ and $x_1(0) = 0$) or into (9) in case (II), respectively. For the parameter $\alpha_1 \geq 0$ with the double root, one checks that (8) holds. Formula (7) together with the inequality (8) exclude existence of cycles and loops except in the case $R(x, \alpha 1) = 0$ for all x, which occurs only when $d = 0$, $c = 1$, and $a = 0$, corresponding to the Volterra-Lotka system, so the lemma follows. □

To easily express this result in terms of the parameters, let S_1, S_2, and S_3 be the open regions into which the curves $\mathfrak{B}$ and $\mathfrak{C}$ divide the strip $d \leq c \leq d+1$, as follows (cf. Figure 1): S_1 the bounded region bordered at one point by $(0,0)$; S_2 the bounded lens-shaped region; and S_3 the unbounded region.

THEOREM 4.

(1) *If* $(d,c) \in S_1$, *then* $a_* < a^* < a_1$, *and cyles or saddle loops are possible only for* $a_* < a < a^*$.
(2) *If* $(d,c) \in S_2$, *then* $a_* < a^* < a_1$, *and cycles or saddle loops are possible only for* $a_* < a < a_1$.
(3) *If* $(d,c) \in S_3$, *then* $a^* < a_*$ *and* $a^* < a_1$, *and cycles or saddle loops are possible only for* $a^* < a < a_1$.

PROOF: The equation $a^* = a_*$ has a simple solution with sign change on the curve $\mathfrak{B}$, but the equation $a^* = a_1$ has a double solution without sign change on the curve $\mathfrak{C}$. Noting this fact, the inequalities between a^*, a_* and a_1 are a simple calculation.

The theorem follows from the lemma once one verifies that for nonnegative values of a *below* the intervals mentioned in the theorem, assumption (I) holds, whereas for values of a *above* the intervals in the theorem, assumption (II) holds. Checking the second statement involves some calculations. Indeed, $a \geq a^*$ in case (1) implies $F'(x) \leq 0$ for all $x \geq \frac{1}{c}$. Furthermore, $c \geq a \geq a_1$ in case (2) or (3) even implies $F'(x) \leq 0$ for all real x. For $c > a$, no cycle or loop can occur because of the absence of an antisaddle (indeed any equilibrium) in the open first quadrant. □

We remark that by continued application of the same techniques similar results can be stated for other values of (c,d) in $\mathfrak{P}_0$.

The local study near the Bogdanov-Takens singularity shows the existence of a smooth surface $a = a^{\dagger}(d, c)$, corresponding to saddle loops of the system X at D, defined for (d, c) near $d = c$. We now wish to investigate the nature of this surface for (d, c) far from $d = c$. Thus for example let us fix (d, c) in the strip $d \leq c \leq d + 1$. Theorem 4 implies that there are no cycles when $a = 0$, and clearly there are no cycles when $a = c$, since by then both singularities have left the first quadrant. But by Theorem 1, as a increases from 0 to c, there is a unique a^* at which Hopf bifurcation occurs, changing the parity of the (finite) number of cycles. A check of the Poincaré vector field shows that cycles cannot bifurcate from infinity, and since "fold" or semi-stable bifurcation of cycles does not alter the parity, there must exist a value $a^{\dagger}$ of a at which there is a homoclinic loop bifurcation to restore the parity. Thus we have shown:

THEOREM 5. *For every (d, c) such that $d \leq c \leq d + 1$, there exists a value of a at which homoclinic loop bifurcation of a limit cycle occurs.* □

Unfortunately, $X_{(d,c,a)}$ does not form a rotated family of vector fields, so we cannot be sure that there are not multiple values of a, for a particular choice of (d, c), yielding the loop bifurcation. Nevertheless, we can make the following statement.

THEOREM 6. *In the parameter set $\mathfrak{P}$, let $\mathfrak{L}$ denote all (d, c, a) for which $X_{(d,c,a)}$ has a saddle loop. Then $\{(0, c, 0) \mid 0 \leq c \leq 1\} \subset \partial\mathfrak{L}$.*

PROOF: When $d = 0$, then the only singularities on the equator of the Poincaré sphere are a source at the "end" of the positive x-axis and a degenerate singularity at the "end" of the positive y-axis, for which the equator and the positive y-axis form separatrices bounding a hyperbolic sector. The only singularity in open Quadrant I is the anti-saddle at C. The singularity at A is a hyperbolic saddle with an unstable separatrix in open Quadrant I. For any choice $(d, c, a) = (0, c, a)$ with $0 < c < 1$ and $0 < a < a^*$, C is a source, hence there must be a stable limit cycle forming the alpha limit set of points on the separatrix in question, and thus an odd number of limit cycles (counting multiplicities) surrounding C. Increasing d slightly, the saddle moves into open Quadrant I, the behavior at infinity is unchanged, and only a semi-stable cycle could be created or destroyed. Hence the number of cycles is odd. Now decreasing a to zero, there are no cycles at $a = 0$, hence there must have occurred a loop bifurcation to change the parity. □

4. Examples.

In this final section we present three examples, the first two theoretical, and the third numerical, to exhibit some of the behavior described in the introduction.

EXAMPLE 1 (Bifurcation of a cycle from infinity). A modification of the proof of Theorem 6 shows the existence of a bifurcation of a limit cycle from infinity. For instead of increasing d, then letting a tend to 0, leave $d = 0$ and let a decrease. Since there is no saddle, there are an odd number of cycles for every value of $a \in (0, a^*)$. But at $d = a = 0$ there are no cycles, by Theorem 4. There is no saddle, and the anti-saddle is hyperbolic, hence the cycle must come from infinity. The mechanism of this bifurcation is that for $d = 0$ but $a > 0$, X is a cubic vector field, but at $d = a = 0$ it becomes quadratic. Viewing the vector field $X_{(0,c,0)}$ as a cubic with cubic coefficients zero, the scaling by the *second* power of the height function, appropriate for cubics, rather than the *first*, appropriate for quadratics, makes the equator of S^2 critical; it is from this curve of singularities that the cycle is created. □

EXAMPLE 2 (Coexistence of stable equilibrium and stable oscillation; enrichment paradox). The portion in $\mathfrak{P}_0$ of the curve $\mathfrak{D}$ curve along which $a^* = d$ is shown in Figure 1. For (d, c) in the bounded portion of $\mathfrak{P}_0$ bounded by $\mathfrak{D}$, $a^* > d$. Recall the definition of the region $O_2 \subset \mathfrak{P}_0$ just above Theorem 1, and the nature of Hopf bifurcation for $(d, c) \in O_2$ described in Theorem 1. It follows that for (d, c) in the small lens-shaped region bounded by the curves $\mathfrak{A}$ and $\mathfrak{D}$, $a^* > d$ and Hopf bifurcation creates an unstable cycle for $a > a^*$. Fix (d, c) in this region. Figure 2 shows the development of the phase portrait of X as a is varied. At $a = a^*$, shown in the lower left hand corner of the figure, since $c < a < d$ the singularity on the x-axis is a saddle with unstable manifold in Quadrant I, and since $(d, c) \in O_2$, C is a (weak) source. A key point in this example is that because of the nature of the Poincaré vector field, a discussion as in the proof of Theorem 6 gives the existence of a stable cycle as shown in the figure. (Of course there could be many cycles, but there are an odd number, with one more stable one than unstable ones.) Slight increase in a, which corresponds to a deterioration in the environment (from the point of view of the prey), yields an unstable cycle; the phase portrait is shown in the upper right hand corner of the figure. Thus as mentioned in the introduction we have coexistence of stable oscillation with stable equilibrium. Going down the right hand column in Figure 2, as a is further increased, all cycles must disappear before $a = c$ is reached, and since cycles cannot disappear at infinity, there must exist at some point a semi-stable cycle. Slight further increase in a gives a situation in which the equilibrium is globally attracting, and further increase in a leads finally to a situation in which the basin of the node on the positive x-axis is the whole open first quadrant.

On the other hand (now going up the left hand column from the bottom phase portrait), if a is decreased from the value a^*, which corresponds to an enrichment of the environment (recall that a is the *reciprocal* of the prey carrying capacity), a saddle point enters the first quadrant, creating at that moment a region of positive probability of extinction of the predator. This region of future extinction of the predator is shaded in the diagrams. Since the cycle disappears before $a = 0$ occurs, there must exist a saddle loop, after which occurs the situation in which the predator almost surely tends to extinction. Thus the system exhibits an "enrichment paradox." □

EXAMPLE 3 (Numerical study). Fix $(d, c) = (0.6, 1.1) \in O_3$, so that $a^* > d$ and the Hopf bifurcation produces an unstable cycle as a increases across $a^* \doteq 0.36667$. Numerical integration of the system as the parameter a is increased is illustrated in Figure 3. Going down the first column, the first screen ($a = 0.35000$) shows a stable separatrix of the saddle winding off the source at C; the second screen ($a = 0.36700$), just above Hopf bifurcation, shows the small unstable cycle that has just bifurcated off C; and the third screen ($a = 0.37400$) shows the cycle continuing to grow just before the separatrices come together. Going down the second column, the top screen ($a = 0.37430$) shows the coexistence of the unstable cycle and the saddle-loop; the middle screen ($a = 0.37460$), just beyond the

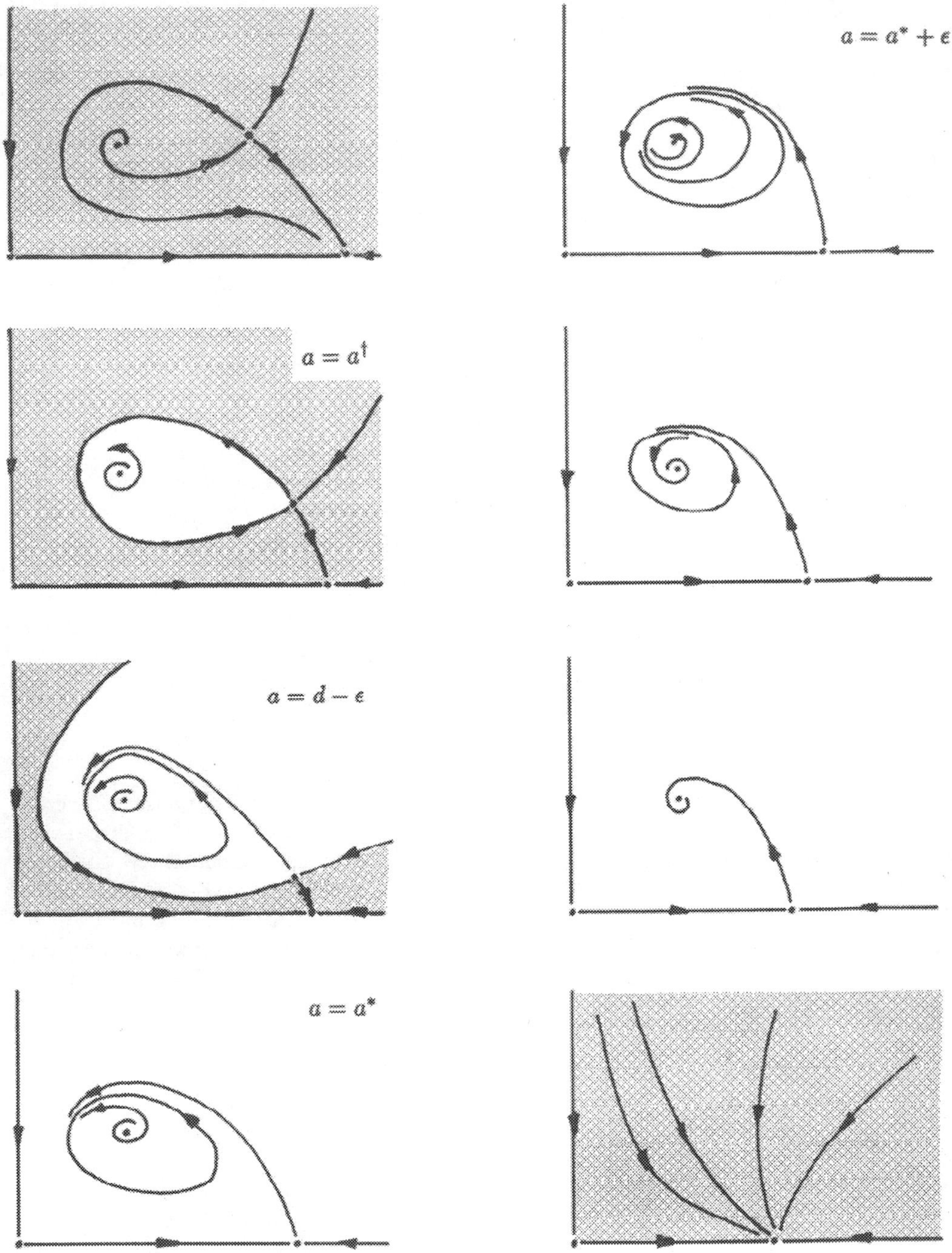

Figure 2

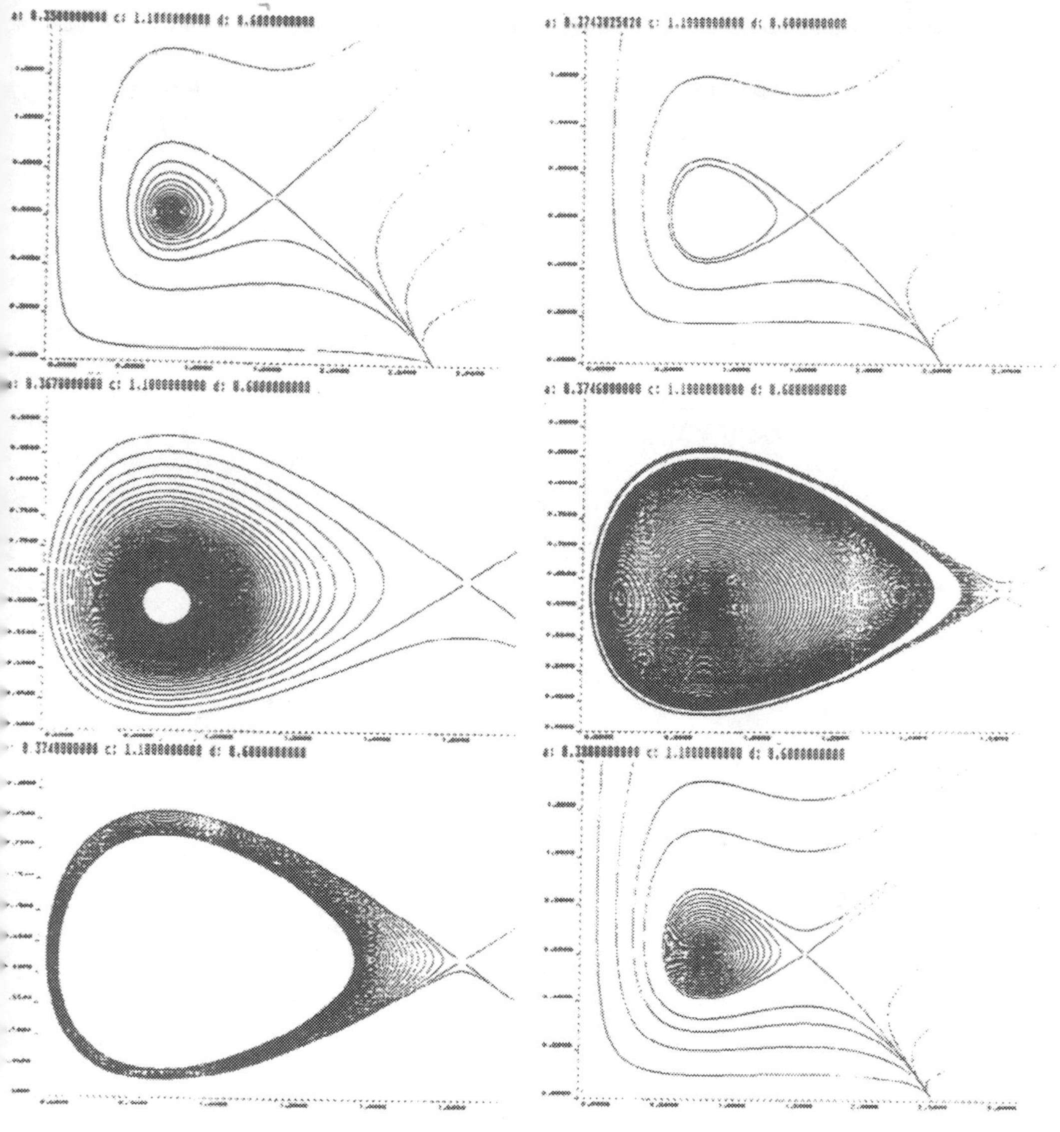
a: 0.3500000000 c: 1.1000000000 d: 0.6000000000
a: 0.3670000000 c: 1.1000000000 d: 0.6000000000
a: 0.3746000000 c: 1.1000000000 d: 0.6000000000
0.3740000000 c: 1.1000000000 d: 0.6000000000
a: 0.3800000000 c: 1.1000000000 d: 0.6000000000

Figure 3

loop, shows the coexistence of the unstable cycle growing from the Hopf bifurcation, and the stable cycle shrinking from the loop bifurcation; the final screen ($a = 0.38000$) shows the situation at or just beyond disappearance of the cycles in a fold, with the same stable separatrix now coming from infinity.

Although the system (2) is not a rotated family of vector fields along any curve in the interesting regions in parameter space, this numerical study shows it behaving as one, in that separatrices apparently rotate, and cycles apparently expand or contract, monotonically. □

References

1. A. A. Andronov, E. A. Leontovich, I. I. Gordon, and A. G. Maier, "Theory of Bifurcations of Dynamic Systems on a Plane," Israel Progam for Scientific Translations, John Wiley & Sons, New York, 1973.
2. A. A. Andronov, E. A. Leontovich, I. I. Gordon, and A. G. Maier, "Qualitative Theory of Second-Order Dynamic Systems," Israel Progam for Scientific Translations, John Wiley & Sons, New York, 1973.
3. C. Chicone and D. S. Shafer, *Separatrix and limit cycles of quadratic systems and a theorem of Dulac*, Trans. Amer. Math. Soc. **278** (1983), 585–612.
4. H. I. Freedman, "Deterministic Mathematical Models in Population Ecology," Marcel Dekker, New York, 1980.
5. J. Guckenheimer and P. Holmes, "Nonlinear Oscillations, Dynamical Systems, and Bifurcations of Vector Fields," Springer-Verlag, New York, 1983.
6. N. Kopell and I. N. Howard, *Bifurcations and trajectories joining critical points*, Adv. in Math. **18** (1975), 306–358.
7. J. E. Marsden and M. McCracken, "The Hopf Bifurcation and Its Applications," Springer-Verlag, New York, 1976.
8. H. I. Freedman and G. S. K. Wolkowocz, *Predator-Prey systems with group defence: the paradox of enrichment revisited*, Bull. Math. Biol. **48** (1986), 493–508. [*not cited*].

ON THE FINITENESS OF CERTAIN BOUNDARY CYCLES FOR NTH DEGREE POLYNOMIAL VECTOR FIELDS

Shi Songling
Centre de Recherches Mathématiques
Université de Montréal
C.P. 6128 - A
Montréal, Québec H3C 3J7 Canada

ABSTRACT

In this paper we give a precise definition of a boundary cycle for a polynomial vector field and prove the finiteness of boundary cycles which are not limit cycles for polynomial vector fields of degree n.

In his celebrated lecture before the Second International Congress of Mathematicians in Paris [1], Hilbert first used the term "boundary cycle". As far as we know, no one has ever given a definition of what a boundary cycle is. Instead of the term "boundary cycle" Lefschetz used "polycycle", Andronov et al. used "singular cycle", other people use the terms "cycle graph", "compound cycles", etc.

Inspired by L. Markus's thesis [2], A.Beck [3], M. C. Peixoto and M. M. Peixoto [4], we make more preccise definitions of the concept of parallel regions and other concepts. However, the gaps in [2] show that the precise definition of a boundary cycle which Hilbert used is very vague . Thus we feel that it is necessary to make precise the definition of a boundary cycle..

Our treatment is as follows: 1. We introduce the concept of a *parallel region*, a concept which differs little from the one introduced in [2]. 2. We introduce the concept of *noncanonical region* , partly to replace the vague concept of a limit separatrix. A boundary cycle is defined as the boundary of a canonical or a noncanonical region. A limit cycle is one of the boundary cycles. We suppose that this definition is what Hilbert had in mind when he delivered his celebrated lecture in Paris [1].

Definition 1. A parallel region is a connected invariant open set N all of whose orbits in N have the same positive and the same negative limit sets.

Definition 2. Let γ_1 and γ_2 be two orbits of a vector field. We say that γ_1 is a parallel equivalent to γ_2 if both γ_1 and γ_2 are contained in a parallel region.

Definition 3. An orbit γ of a vector field is a separatrix if there is no parallel region containing γ.

Examples of separatrices are: a singular point, a closed orbit, and the orbits which are the boundaries of a hyperbolic sector.

The parallel relation is symmetric and transitive, but it is not always reflexive. A characteristic property of a separatrix is that it is parallel to no other orbit.

Let G be the union of all separatrices of a polynomial vector field on the sphere S^2.

LEMMA 1. The set G is a closed subset of S^2.

Proof: Each non-separatrix is embedded in an open neighborhood filled in by

non-separatrices. Thus the complement of G is open in S^2. ♦

Definition 4. A canonical region is a connected component of the complement of G.

LEMMA 2. If the set G has an inner point p, then the orbit f(p,t) is either a closed orbit or a singular point.

Proof: If the orbit f(p,t) is neither a closed orbit nor a singular point, then f(p,t) is a nonclosed orbit whose limit set is a cycle or a singular point. This implies that f(p,t) is not a separatrix, which contradicts our hypothesis. ♦

For a point q of ∂G there exists an arbitrarilly small neighborhood U(q) of q such that for some q' in U(q) the orbit f(q', t) is a nonclosed orbit.

Definition 5. A noncanonical region is a connected invariant open set whose interior consists of closed orbits and whose boundary is either a singular point or a cycle formed by several separatrices, some of which are singularities.

The following system is an example with noncanonical regions:
Example 1. (Fig. 1.) $dx/dt = -y[(x-1)^2 + y^2]$
$dy/dt = x[(x-1)^2 + y^2]$.

The phase portrait consists of two noncanonical regions, as shown in Fig. 1b and 1c.

Remark 1. If we draw a phase portrait on the disk, the boundary of a noncanonical region may be the cycle at infinity. This is the case in Fig. 1c.

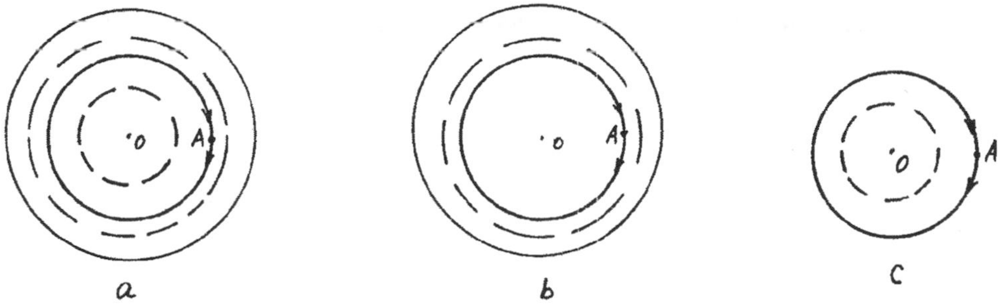

Fig. 1.

THEOREM 1. The number of noncanonical regions for a polynomial vector field with isolated singularities is finite.

Proof: The family of closed orbits surrounding a center is a disjoint union of noncanonical regions. The number of singularities for a polynomial vector field with isolated singularities is finite. In particular the number of centers is finite. This implies our conclusion. ♦

Definition 6. A boundary cycle is the boundary of either a canonical or a noncanonical region.

Clearly, a boundary cycle consists of either a singular point, or a limit cycle, or a cycle formed by several separatrices, some of which are singularities.

Definition 7. A separatrix configuration of a vector field F is a collection of curves consisting of separatrices of F plus one representative solution from each subregion.

Using the notion of topological equivalence of separatrix configurations, defined in [2], and the same arguments as in [2], one can prove the following:

THEOREM 2. Let F_1 and F_2 be two vector fields on the disk D. A necessary and sufficient condition that F_1 be topologically equivalent to F_2 is that there is an equivalence between the separatrix configurations of F_1 and those of F_2.

Theorem 2 distinguishes the separatrix configurations as complete invariants for the topological classification of vector fields.

THEOREM 3. The maximum number of boundary cycles which are not the limit cycles for polynomial vector fields of degree n with isolated singularities is finite.

Proof: For a polynomial vector field of degree n with isolated singularities there are at most n^2 singular points. For each singular point there is a finite number of hyperbolic sectors.

Therefore, the number of boundaries of these hyperbolic sectors is finite. This implies the result. ♦

The number of canonical or noncanonical regions for a polynomial vector field is a topological invariant under the group of homeomorphisms of the disk D. The maximum number of noncanonical or canonical regions for a class of polynomial vector field is an invariant. This gives only that the maximum number of noncanonical regions for quadratic systems is two, [5]. We don't know the maximum number of noncanonical regions for cubic systems.

Let n be a positive integer. Consider the set K(n) of all polynomial vector fields with n = max (degree(P), degree(Q)). Let us denote by B(n) the maximum number of boundary cycles of a system in K(n). Then, we have

PROPOSITION. B(n) is an invariant of polynomial vector fields of degree n.

So far we only know that B(1) = 4.

The following theorem is somewhat analogous to the simplification of a topological space resulting from a triangulation.

THEOREM 4. If the number of separatrices is α_1 and the number of singularities is α_0, then we have for a polynomial vector field on the disk

$$B(n) - \alpha_1 = 1 - 2\,\alpha_0 \ .$$

Remark 2: If one can prove that B(n) is finite, then the maximum number H(n) of limit cycles is finite. ♦

Acknowledgement. I am very grateful to Professor Dana Schlomiuk for helpful and stimulating discussions.

REFERENCES

1. D. HILBERT, Mathematische Probleme, Lecture, Second International Congress of Mathematicians (Paris, 1900), Nachr Ges.Wiss. Gottingen Math.-Phys. Kl. **1900**, 253-297, reprinted in [4] pp. 1-34.
2. L. Markus, Global structure of ordinary differential equations in the plane, Trans. Amer. Math. Soc. **76** (1954), 127-148.
3. A. Beck, Continuous flow in the plane, Springer Verlag.
4. M. C. PEIXOTO AND M.M.PEIXOTO, Structural stability in the plane with enlarged boundary conditions, Anais da Academia Brasileira de Ciencias, **31**(1959), 135-160.
5. D. SCHLOMIUK, The "center"-space of plane quadratic systems and its bifurcation diagram, Rapports de recherche du départment de mathématiques et de statistique, Université de Montréal, D.M.S. No 88-18.

ALGEBRAIC INTEGRALS OF QUADRATIC SYSTEMS WITH A WEAK FOCUS

Dana Schlomiuk[1]
Département de Mathématiques et de Statistique, Université de Montréal,
Montréal, P.Q., Canada

§1. Introduction. The particular integrals of the quadratic systems with a center which are algebraic curves are a key feature of these systems: these curves **control** the bifurcation diagram of such systems (cf.[1],[2]).

In this paper we first consider the computational tool serving to calculate the algebraic integrals of polynomial systems, and we discuss this tool. We next use this tool to prove here two new results saying that the existence of just one such an algebraic integral, an irreducible cubic or conic, forces a quadratic system with a degenerate weak focus F to have a center at F. (A weak focus is a singularity with purely imaginary eigenvalues). The proofs were initially obtained using MACSYMA and elimination theory. The proofs which we present here do not require computer calculations and they are entirely elementary. They provide the reader with concrete, direct evidence of the importance of the algebraic curves in the theory of the center. The geometrical content of the algebraic conditions for the center for quadratic systems can be expressed in terms of these algebraic curves. Results proving this fact (cf. [2], [3], [4]) will be published elsewhere.

The author wishes to thank J. Guckenheimer who raised her interest in the problem of the center and who asked the question of the meaning of the algebraic conditions for the center.

§2. Particular integrals which are algebraic curves, of polynomial vector fields and the specific case of the quadratic systems with a weak focus. We are interested in real polynomial vector fields on the plane

(S) $$dx/dt = P(x,y), \qquad dy/dt = Q(x,y) \qquad\qquad P,Q \in \mathbb{R}[x,y]$$

and we consider here their particular integrals i.e.real (non-empty) curves $F(x,y) = 0$ such that for all the points on this curve we have

(2.1) $$(\partial F/\partial x)(x,y)P(x,y) + (\partial F/\partial y)(x,y)Q(x,y) = 0$$

We shall be concerned with particular integrals which are algebraic curves. We shall call such particular integrals algebraic integrals. For an algebraic integral, F is of the form

(2.2) $$F(x,y) = F_n(x,y) + F_{n-1}(x,y) + \ldots + F_1(x,y) + F_0$$

where $F_i(x,y)$ is a homogeneous polynomial with real coefficients of degree i.

We discuss here a computational tool for the calculation of such algebraic integrals. We make this discussion concrete by keeping in mind the specific case which interests us, namely the case of the quadratic systems with a weak focus i.e. a singular point with purely imaginary eigenvalues. Such systems could always be brought by affine coordinate changes and positive time rescaling to the form

[1] Research partially supported by the National Sciences and Engineering of Canada and Quebec Education Ministry

$$(S_2) \qquad \begin{aligned} dx/dt &= -y - bx^2 - Cxy - dy^2 \\ dy/dt &= x + ax^2 + Axy + cy^2 \end{aligned} \qquad (x,y)\in \mathbb{R}^2$$

Let us denote by P_i, Q_i the homogeneous polynomials of degree i in P, respectively Q. If F is a divisor of $F' = (\partial F/\partial x)(x,y)P(x,y) + (\partial F/\partial y)(x,y)Q(x,y)$ over $\mathbb{R}$ i.e. if for some $H(x,y)\in \mathbb{R}[x,y]$ we have

$$\text{(CT)} \qquad (\partial F/\partial x)P + (\partial F/\partial y)Q = F(x,y)H(x,y)$$

then clearly whenever $F(x,y) = 0$, (2.1) holds and so $F(x,y) = 0$ is an algebraic integral of (S).

In the specific case (S_2) we have $P = P_1 + P_2$ and $Q = Q_1 + Q_2$ and the identity (CT) is:

$$(CT_2) \qquad (\partial F/\partial x)P + (\partial F/\partial y)Q = F(x,y)(ux + vy + w)$$

If there exist a curve $F(x,y) = 0$ satisfying (CT) for some $H(x,y)$, in such a case (CT) can be used to compute the coefficients of F. (CT) is thus a computational tool which we shall use here for the determination of the cubics and conics, algebraic integrals of (S_2).

It is known (cf.[5], [6]) that the quadratic systems with a center are "globally" integrable in the sense of the following definition:

Definition 2.1. A vector field $dx/dt = f(x,y)$, $dy/dt = g(x,y)$ defined on an open subset U of $\mathbb{R}^2$ is integrable on U if and only if there exists an analytic nonconstant function $F:U \longrightarrow \mathbb{R}$ which is a constant of motion of the field i.e. $F(x(t), y(t))$ = constant for all solution curves $(x(t), y(t))$ in U.

The constants of motion for the real quadratic systems with a center were obtained by V. A. Lunkevich and K. S. Sibirsky [5] and we have the following theorem [6]:

Theorem 2.1. Each quadratic system (S_2) with a center is integrable on $\mathbb{R}^2$ or on the complement in $\mathbb{R}^2$ of an algebraic curve.

A look at the specific constants of motion (cf. [5] or [1]) of the quadratic systems with a center makes it clear that all such systems possess algebraic integrals and in the generic case, we have (cf.[2]):

Proposition 2.1. Assume that a system (S_2) has a center at the origin. Then, generically, this system has either a straight line as a particular integral and a conic curve (not passing through the origin) satisfying an identity (CT_2) or it has no invariant straight line but it has a conic and a cubic curve, both irreducible over $\mathbb{C}$ as particular integrals or it is a hamiltonian system. Furthermore for each one of these algebraic integrals $F(x,y) = 0$, F is a divisor of $F'(x,y) = (\partial F/\partial x)P + (\partial F/\partial y)Q$, i.e.the identity (CT_2) holds.

We now discuss some points concerning the identity (CT). We note that if the curve $F(x,y) = 0$, $F\in \mathbb{R}[x,y]$, satisfies the identity (CT) for some $H(x,y)\in \mathbb{R}[x,y]$, then every complex point (a,b) on the curve $F(x,y) = 0$, considered over the complex field, also belongs to the curve $F'(x,y) = (\partial F/\partial x)P + (\partial F/\partial y)Q = 0$, so the curve is an algebraic integral for the system (S) considered over the complex field and with complex time. Viceversa, if the curve $F(x,y) = 0$ is irreducible over $\mathbb{C}$ and if it is an algebraic integral for the complex system (S) then F is a divisor of F' over $\mathbb{C}$ and consequently also over $\mathbb{R}$. Furthermore, in the case of a real curve $F(x,y) = 0$, irreducible over $\mathbb{C}$, which has an infinite number of real points and which is an algebraic integral for the real system (S), the equality (CT) holds since the curve will have a component in common with $F'(x,y) = 0$. In view of this, we have:

Remark 2.1. For irreducible curves which have an infinity of points in $\mathbb{R}^2$, (CT) (in particular (CT_2)) is equivalent to the curve being an algebraic integral for the real system (S) (in particular (S_2)).

Thus (CT) is the computational tool for the calculation of such algebraic integrals of (S). As we can see in [2] the usefulness of the polynomial identity (CT) goes beyond the calculation of the irreducible curves which have an infinite number of points in $\mathbb{R}^2$ and which are algebraic integrals of the systems. In fact (CT) also helps us understand the dynamics in the geometry of the phase portraits (cf.[2], [4]). We discuss now how this tool is used. (CT_2) reduces to n+1 equations, one for each coefficient of a non-constant term. Identifying the corresponding coefficients of the zero degree terms, we get $F_0 w = 0$. Assuming $F_0 \neq 0$, we have $w = 0$. Identifying the coefficients of the degree one terms in (CT_2) we have $u = a_{01}$, $v = -a_{10}$. Identifying the coefficients of the degree i in (CT_2), with $2 \leq i \leq n$, we have

$(CT_{2;i,F,n})$ $$(\partial F_i/\partial x)P_1 + (\partial F_i/\partial y)Q_1 + (\partial F_{i-1}/\partial x)P_2 + (\partial F_{i-1}/\partial y)Q_2 = F_{i-1}(a_{01}x - a_{10}y)$$

while identifying the coefficients of degree n+1 in (CT_2) we have

$(CT_{2;n+1,F,n})$ $$(\partial F_n/\partial x)P_2 + (\partial F_n/\partial y)Q_2 = F_n(a_{01}x - a_{10}y)$$

In the next paragraph we shall consider conics and cubics which are integrals of the system (S_2). In general if we look for a curve of degree n+1 after we discussed the curves of degree n, we note that the equations obtained by identifying the terms of degrees i with $i \leq n$ are the same for curves of degrees n and n+1. The last equation used for curves of degree n is $(CT_{2;n+1,F,n})$ while the remaining two equations obtained by identifying the coefficients of the terms of degrees n+2 and n+1 in (CT_2) for the curve $F(x,y) = 0$ of degree n+1 are new ones:

$(CT_{2;n+1,F,n+1})$ $$(\partial F_{n+1}/\partial x)P_1 + (\partial F_{n+1}/\partial y)Q_1 + (\partial F_n/\partial x)P_2 + (\partial F_n/\partial y)Q_2 = F_n(a_{01}x - a_{10}y)$$

$(CT_{2;n+2,F,n+1})$ $$(\partial F_{n+1}/\partial x)P_2 + (\partial F_{n+1}/\partial y)Q_2 = F_{n+1}(a_{01}x - a_{10}y)$$

So, all the equations used in calculating the coefficients of the algebraic curve of degree n with the exception of the one obtained by identifying the coefficients of maximum degree n+1 in (CT_2), are used in the computation of the curves of degree n+1. This last equation is replaced by the above two equations. In particular if we want to compute the conics and cubics, we first identify the coefficients of degree 2 and this will give us some of the coefficients of F_2. We then use these coeficients also when we compute the cubic curve $G(x,y) = 0$ verifying (CT_2) with G replacing F. To get the remaining coefficients for the conic, we have to use the remaining equation

$(CT_{2;3,F,2})$ $$(\partial F_2/\partial x)P_2 + (\partial F_2/\partial y)Q_2 = F_2(a_{01}x - a_{10}y)$$

while in order to compute the remaining coefficients for the cubic $G(x,y) = 0$, we need to use the remaining two equations

$(CT_{2;3,G,3})$ $$(\partial G_3/\partial x)P_1 + (\partial G_3/\partial y)Q_1 + (\partial G_2/\partial x)P_2 + (\partial G_2/\partial y)Q_2 = G_2(a_{01}x - a_{10}y)$$

$(CT_{2;4,G,3})$ $$(\partial G_3/\partial x)P_2 + (\partial G_3/\partial y)Q_2 = G_3(a_{01}x - a_{10}y)$$

Remark 2.2. The procedure for calculating the algebraic integrals of the systems by using (CT) reminds us of the one outlined by Poincaré for calculating a local constant of motion for the center or a Lyapunov function in the case of a focus, i.e. the procedure for calculating the Poincaré-Lyapunov constants. Looking for a constant of motion $F(x,y)$ with F an *analytic* function, we need to use the equation $dF/dt = 0$ i.e.:

(2.1) $$(\partial F/\partial x)P(x,y) + (\partial F/\partial y)Q(x,y) = 0.$$

Here we look for a *polynomial* F such that this equation holds for points (a,b) lying on the algebraic curve $F(x,y) = 0$. In particular this holds when F is a divisor of $(\partial F/\partial x)P + (\partial F/\partial y)Q$ over $\mathbb{R}$ in other words when the (CT) holds. The problem of solving the equations obtained from (2.1) by identifying the coefficients of terms of degrees i, depends on wether or not these equations are compatible. Saying that this is possible for each i yields the conditions for the center. Analogously, if we want to have curves satisfying the identity (CT) (in particular (CT_2)), then this will require some conditions to be satisfied by the coefficients of the systems (S) (in particular (S_2)). It will turn out that if we asked for **enough** such curves, we get in this way, as the compatibility conditions, the conditions for the center.

§3. Algebraic integrals which are conics and cubics of quadratic systems with a weak focus.

We consider a quadratic system with a weak focus. Such a system may be assumed to be of the form (S_2). Kapteyn observed [7] that after performing a rotation of axes such a system may be brought to a system of the same form (2.1) but which has $c = -a$. We shall denote by (S_2') a system (S_2) with $c = -a$. A system (S_2) has either a center or a focus at the origin. Hence there is no straight line passing through the origin, particular integral of such a system.

Bautin observed [8] that Kapteyn's work [9] gives us necessary and sufficient conditions for the center for systems (S_2') with real coefficients. We have:

Theorem 3.1 (Kaptetyn). A real quadratic system (S_2') possesses a center at the origin if and only if one of the following four conditions is satisfied.

I) $$b + d = 0$$

II) $$a = 0 = C$$

III) $$C + 2a = A + 3b + 5d = a^2 + 2d^2 + bd = 0$$

H) $$C + 2a = 0 = A - 2b$$

The system (S_2') is hamiltonian if and only if the condition H) is satisfied.

We recall the first three Poincaré-Lyapunov constants for systems (S_2') (cf.[8] or [6]):

(3.1) $$V_1 = (C + 2a)(b + d)$$

(3.2) $$V_2 = (b + d)a(A - 2b)(A + 3b + 5d)$$

(3.3) $$V_3 = (b + d)^2 a(A - 2b)(a^2 + 2d^2 + bd)$$

Definition 3.2. a) The origin is a weak focus of order one for the system (S_2') if $V_1 \neq 0$.
b) The origin is a degenerate weak focus for the system (S_2') if $V_1 = 0$.
c) The origin is a weak focus of order two for (S_2') if $V_1 = 0$ and $V_2 \neq 0$.
d) The origin is a weak focus of order three for (S_2') if $V_1 = V_2 = 0$ and $V_3 \neq 0$.

Straightforward calculations yield the following:

Proposition 3.1. A system (S_2') has a real straight line L: $ux + vy + 1 = 0$ as an algebraic integral, if and only if either $b + d = 0$ (in which case the system has at most three straight lines as particular integrals) or else $a = 0 \neq A$ in which case the straight line is unique and its equation is: $1 + Ay = 0$.

Corollarry 3.1 A system (S_2') with a straight line as integral has either a center at the origin or a weak focus of order one.

Proof: if $b + d = 0$ the origin is a center. If $a = 0 \neq A$ then $V_2 = V_3 = 0$ and hence the origin is a center or a weak focus of order one.

Since the origin is either a center or a focus for a system (S_2'), clearly such a system cannot have a real nonsingular conic curve passing through the origin as a particular integral. We have the following:

Proposition 3.2. Let P and Q be the polynomials on the right hand side of (S_2') (or (S_2)) The only conic curve $F(x,y) = 0$, passing through the origin and such that F is a divisor of $F'(x,y) = (\partial F/\partial x)P + (\partial F/\partial y)Q$ is the curve $x^2 + y^2 = 0$.

Proof: Assume the conic is $F(x,y) = 0$, with $F = F_2 + F_1$ satisfying the identity (CT_2) for some u, v, w in $\mathbb{R}$. Identifying the terms of order one we have

(3.4) $$(\partial F_1/\partial x)P_1 + (\partial F_1/\partial y)Q_1 = F_1 w$$

which yields $-a_{10} = a_{01}w$ and $a_{01} = a_{10}w$ and hence $a_{10}(1 + w^2) = 0$. So we obtain $a_{10} = 0 = a_{01}$ and $F_1 = 0$. Identifying the terms of order two we have:

(3.5) $$(\partial F_2/\partial x)P_1 + (\partial F_2/\partial y)Q_1 = F_2 w$$

Assuming that $F_2 = a_{20}x^2 + 2a_{11}xy + a_{02}y^2$, we obtain from the above equation:

(3.6) $$-y(2a_{20}x + 2a_{11}y) + (2a_{11}x + 2a_{02}y)x = (a_{20}x^2 + 2a_{11}xy + a_{02}y^2)w$$

and hence $2a_{11} = a_{20}w$, $a_{02} - a_{20} = a_{11}w$ and $-2a_{11} = a_{02}w$. Hence we have that $w(a_{20} + a_{02}) = 0$. If $w = 0$, then $a_{11} = 0$, $a_{02} = a_{20}$ and $F_2 = a_{20}(x^2 + y^2)$ so the conic is $x^2 + y^2 = 0$. If $w \neq 0$, then we must have $a_{20} + a_{02} = 0$. Adding this to $a_{02} - a_{20} = a_{11}w$, we obtain $a_{02} = a_{11}w/2$. On the other hand we have $a_{02} = -2a_{11}/w$. Hence we have $a_{11}(w^2 + 4) = 0$ and so $a_{11} = 0$ which gives us $F_2 = 0$, so there is no conic in this case. Q.E.D.

In view of the above Proposition, we consider particular algebraic integrals of systems

(S_2') (resp.(S_2)), which do not pass through the origin.

Remark 3.1. Straightforward calculations show that for a system (S_2'), if $b + d = 0$ then any conic, satisfying an identity (CT_2) is necessarily singular [2].

Theorem 3.2. Consider a system (S_2') which has a degenerated weak focus at the origin. Assume that the system has a conic curve $F(x,y) = 0$, irreducible over $\mathbb{C}$, as a particular integral of the system. Then necessarily

i) the coefficients of the system must satisfy the conditions: $b + d \neq 0$ and

$$a(2d^2 + a^2 + bd) = 0$$

and ii) the origin is a center. Furthermore:

If $a = 0 \neq d$, the conic is an ellipse or a hyperbola. (The equations of these were given in [2])

If $a = 0 = d$, then the curve is a parabola whose equation is

$$-b(A + 2b)x^2 - 2by + 1 = 0 \tag{3.7}$$

If $a \neq 0$, then the coefficients of the systems also satisfy the conditions $d \neq 0$ and

$$A + 3b + 5d = 0$$

and the conic is a parabola whose equation is

$$(d^2 + a^2)(ax - dy)^2 + 2d(d^2 + a^2)y + d^2 = 0 \tag{3.8}$$

The proof of the above theorem follows easily using the Remarks 2.1 and 3.1 from the following:

Theorem 3.3. Consider a system (S_2') which has a degenerated weak focus at the origin. Assume that the system has a conic curve $F(x,y) = 0$, satisfying the identity (CT_2) for the system. Then necessarily

i) the coefficients of the system must satisfy the equation

$$a(b + d)(2d^2 + a^2 + bd) = 0$$

ii) the origin is a center. Furthermore:

If $a = 0 \neq d$, the conic is an ellipse or a hyperbola. (The equations of these were given in [2])

If $a = 0 = d$, then the curve is a parabola whose equation is (3.7)

If $a \neq 0$, then the coefficients of the systems also satisfy the conditions $d \neq 0$ and

$$A + 3b + 5d = 0$$

and the conic is a parabola whose equation is (3.8)

Proof: In view of the Proposition 3.2, we may assume the conic to be $F(x,y) = 0$ where

$$F(x,y) = a_{20}x^2 + 2a_{11}xy + a_{02}y^2 + a_{10}x + a_{01}y + 1 \tag{3.9}$$

with $(a_{20},a_{11},a_{02}) \neq 0$, $a_{ij} \in \mathbb{R}$. Since the conic does not pass through the origin we have $w = 0$ i.e. we have

$$(\partial F/\partial x)P(x,y) + (\partial F/\partial y)Q(x,y) = F(x,y)(ux + vy) \tag{3.10}$$

Identifying the coefficients of the terms of order one we have that

(3.11) $$u = a_{01} \quad \text{and} \quad v = -a_{10}$$

Identifying the coefficients of x^2 and y^2 in (3.10) we obtain the equations:

(3.12) $$a.a_{01} + 2a_{11} - ba_{10} - a_{10}a_{01} = 0 \qquad \text{(for } x^2\text{)}$$

(3.13) $$-a.a_{01} - 2a_{11} - da_{10} + a_{10}a_{01} = 0 \qquad \text{(for } y^2\text{)}$$

From these we obtain the equation

(3.14) $$a_{10}(b+d) = 0$$

We have therefore two possibilities: $a_{10} = 0$ or else $b + d = 0$ in which case the system has a center at the origin. If $b + d \neq 0$, we have $a_{10} = 0$. Replacing this in (3.12) we obtain the equation:

(3.15) $$a_{11} = -a.a_{01}/2$$

Identifying the coefficients of xy in (3.10) we obtain:

(3.16) $$a_{02} = (2a_{20} - a_{01}A + a_{01}^2)/2$$

Replacing these in (3.10) and identifying the coefficients of y^3 in (3.10), we obtain:

(3.17) $$a(a_{01}d - 2a_{20} + a_{01}A - a_{01}^2) = 0$$

Either $a = 0$ or $a \neq 0$. In the first case, since by hypothesis we have that $C + 2a = 0$, we have $a = C = 0$. In this case, we have a center and the conics were discussed in [1] and [2]. For $d \neq 0$, we have either a parabola or a hyperbola. For $d = 0$ the conic is a parabola whose equation is (3.7). Assume now that $a \neq 0$.

From (3.17) we obtain:

(3.18) $$a_{20} = (a_{01}d + a_{01}A - a_{01}^2)/2$$

Replacing (3.15), (3.16) and (3.18) in (3.10) and identifying the coefficients of the other terms of degree three in (3.10) and then factoring, we obtain the equations:

(3.19) $$a_{01}(2d^2 - a_{01}d + 2a^2) = 0 \qquad \text{(for } y^2x\text{)}$$

(3.20) $$2a.a_{01}(3d + b + A - a_{01}) = 0 \qquad \text{(for } yx^2\text{)}$$

(3.21) $$a_{01}(2bd + a_{01}d + 2Ab - 2a_{01}b + a_{01}A - a_{01}^2 + 2a^2) = 0 \qquad \text{(for } x^2\text{)}$$

If $a_{01} = 0$, then necessarily a_{11}, a_{02}, a_{20} are all zero. So $a_{01} \neq 0$ and since $a \neq 0$, from (3.19) we must have that $d \neq 0$. Therefore we obtain

(3.22) $$a_{01} = (2d^2 + 2a^2)/d$$

Replacing the above expression in (3.20), we obtain

(3.23) $$d^2 + (b + A)d - 2a^2 = 0$$

which gives us

(3.24) $$A = -(d^2 + bd - 2a^2)/d$$

Replacing in (3.21) the values of a_{01}, A in (3.22), (3.24), factoring and since $a_{01} \neq 0$, we obtain the equation:

(3.25) $$(b + d)(2d^2 + bd + a^2) = 0$$

If we have

(3.26) $$2d^2 + bd + a^2 = 0$$

then replacing $a^2 = -2d^2 - bd$ in (3.24) we obtain

(3.27) $$A + 3b + 5d = 0$$

The conditions (3.26) and (3.27) together with the fact that the weak focus is degenerated give us that the system has a center of the type indicated by the condition III in the Theorem 3.1. In this case using the equations (3.26) and (3.27) we can eliminate the parameters A and b and the

calculations above yield a parabola for the conic, whose equation is (3.8)
Q.E.D

Proof of the Theorem 3.2: As a real (non-empty) algebraic integral for the system, which is irreducible over $\mathbb{C}$, the curve is nonsingular, hence it has an infinite number of points in $\mathbb{R}^2$. In view of the Remark 2.1 the identity (CT_2) holds and we can apply the theorem 3.3.

Corollary 3.1. A quadratic system with a weak focus of order two or three has no algebraic integral which is an irreducible conic curve.

Corollary 3.2. A quadratic system with a weak focus F and with an irreducible conic curve as an algebraic integral has either a center or a weak focus of order one at F.

Theorem 3.4. Assume that a system (S_2') has a degenerated weak focus at the origin. Let us suppose that this system has a real cubic curve $F(x,y) = 0$, irreducible over $\mathbb{C}$, not passing through the origin, as a particular integral. Then
i) the coefficients of the system necessarily satisfy the equation
(EC) $$a\,(A - 2b)(A + 3b + 5d) = 0$$
and furthermore in the case $a(A-2b) \neq 0$ they also satisfy the equation
$$2d^2 + bd + a^2 = 0$$
ii) the origin is a center.

The proof of the above theorem follows easily in view of the Remarks 2.1 and 3.1 from the following:

Theorem 3.5. Assume that a system (S_2') has a degenerate weak focus at the origin. Let us suppose that this system has a real cubic curve $F(x,y) = 0$, not passing through the origin, satisfying the identity (CT_2). Then
i) the coefficients of the system necessarily satisfy the equation
(EC) $$a(b + d)(A - 2b)(A + 3b + 5d) = 0$$
and furthermore in the case $a(b + d)(A - 2b) \neq 0$, we they also satisfy the equation
$$2d^2 + bd + a^2 = 0$$
ii) the origin is a center.

Proof: We may suppose that for the cubic curve $F(x,y) = 0$ we have $F = F_3 + F_2 + F_1 + 1$. We use the observations made in the §2 for the calculation of the coefficients of F. Clearly it will suffice to prove part i). We first show that we have (EC). We note that the terms of degrees one and two of F are exactly the same in (3.10) with a cubic F as they are in (3.10) with a quadratic F. Hence we must have $u = a_{01}$ and $v = - a_{10}$. We can use the equation ($CT_{2;2,F,2}$) and the calculations we already obtained in the proof of the Theorem 3.1 from this equation i.e. (3.14): $a_{10}(b + d) = 0$. Hence $b + d = 0$ or $a_{10} = 0$. If $b + d = 0$, clearly the origin is a center. Assuming now that $a_{10} = 0$, we obtain the coefficients a_{11} and a_{02} as given by the formulas (3.15) and (3.16). Replacing these in the identity (3.10) with $F = F_3 + F_2 + 1$, we must determine a_{01} and a_{20} from (3.10) by identifying the coefficients of degrees 3 and 4. We consider first the terms of order three in (CT_2). These give the identity:

$$(E_3) \qquad -y(\partial F_3/\partial x) + x(\partial F_3/\partial y) + (\partial F_2/\partial x)(-bx^2 + 2axy - dy^2) + (\partial F_2/\partial y)(ax^2 + Axy - ay^2) = F_2(x,y)a_{01}x$$

In this identity we use the coefficients a_{11} and a_{02} of F_2 already obtained in terms of a_{01}, a_{20} as given by the formulas (3.15), (3.16). Let $F_3 = a_{30}x^3 + a_{21}x^2y + a_{12}xy^2 + a_{03}y^3$.
Clearly the coefficients of (E_3) are polynomials of degree one in a_{ij} with $i + j = 3$. Identifying these with zero and solving these equations with respect to a_{ij}, we obtain

$$(3.28) \qquad a_{30} = a((2d + b)a_{01} + 2a_{20})/3$$

$$(3.29) \qquad a_{21} = (a_{20} + a^2)a_{01} + 2a_{20}b$$

$$(3.30) \qquad a_{12} = -a(a_{01}^2 - (d + A)a_{01} + 2a_{20})$$

$$(3.31) \qquad a_{03} = (a_{01}^3 - 3Aa_{01}^2 + (6a_{20} + 2A^2 + 6a^2)a_{01} + 4a_{20}(d + 2b - A))/6$$

Identifying the terms of degree four in (CT_2), we obtain the identity:

$$(E_4) \qquad (\partial F_3/\partial x)P_2 + (\partial F_3/\partial y)Q_2 - F_3xa_{01} = 0$$

Identifying with zero the coefficients in (E_4) we obtain the equations:

$$(E_{y^4}) \qquad a_{12}d + 3a.a_{03} = 0$$

$$(E_{xy^3}) \qquad -2a_{21}d + 3a_{03}A - a_{01}a_{03} = 0$$

$$(E_{x^2y^2}) \qquad -3a_{30}d - a_{12}b + 3a.a_{21} + 2Aa_{12} - a_{01}a_{12} + 3a.a_{03} = 0$$

$$(E_{x^3y}) \qquad -2a_{21}b + 6a.a_{30} + Aa_{21} - a_{01}a_{21} + 2a.a_{12} = 0$$

$$(E_{x^4}) \qquad -3a_{30}b - a_{01}a_{30} + a.a_{21} = 0$$

Replacing a_{ij} with $i + j = 3$ in (E_{x^4}) by their expressions found above, we obtain the equation

$$(E_{x^4}{}^*) \qquad -a.a_{01}(6bd + 2a_{01}d + 3b^2 + a_{01}b - a_{20} - 3a^2)/3 = 0$$

We have two possibilities: $a.a_{01} = 0$ or $a.a_{01} \neq 0$.

Case $a.a_{01} \neq 0$. Simplifying by $a.a_{01}$ the above equation and solving for a_{20} we obtain

$$(3.32) \qquad a_{20} = (2d + b)a_{01} + 3b^2 + 6bd - 3a^2$$

Introducing the above expression in the formulas for a_{ij} with $i + j = 3$ and then the result in ($E_{x^2y^2}$) and in (E_{y^4}) we obtain respectively the equations:

$$(E_{x^2y^2}{}^*) \qquad a(3a_{01}^3 + 3(10d + 6b - 3A)a_{01}^2 + (-4d^2 + 140bd - 20Ad + 72b^2 - 14Ab + 6A^2 - 36a^2)a_{01} + 36(4b^2d - 2Abd + 2b^3 - Ab^2 - 2a^2b + a^2A))/2 = 0$$

$$(E_{y^4}{}^*) \qquad -a(a_{01}^3 + (10d + 6b - 3A)a_{01}^2 + (2d^2 + 52bd - 6Ad + 26b^2 - 4Ab + 2A^2 - 12a^2)a_{01} + 12(4b^2d - 2Abd + 2b^3 - Ab^2 - 2a^2b + a^2A)/2 = 0$$

We observe that multiplying the second equation by three and adding to it the first equation, the terms in a_{01}^3, a_{01}^2 and the one not containing a_{01} disappear and we are left with the equation

$$(3.33) \qquad aa_{01}(-10d^2 - 16bd - 2Ad - 6b^2 - 2Ab) = 0$$

$aa_{01} \neq 0$, so we are left only with the third factor which factors:

$$(3.34) \qquad -(b + d)(A + 3b + 5d) = 0$$

Case $a.a_{01} = 0$. If $a = 0$ then we have $a = 0 = C$ and hence we have a center. Suppose now that $a \neq 0$ and $a_{01} = 0$. In this case the equations for a_{ij} with $i + j = 3$ give us

$$(3.35) \qquad a_{30} = 2aa_{20}/3, \quad a_{21} = 2a_{20}b, \quad a_{12} = -2aa_{20}, \quad a_{03} = 2a_{20}(d + 2b - A)/3$$

We replace these in the remaining equations (E_{y^4}), (E_{xy^3}), ($E_{x^2y^2}$), (E_{x^3y}) and we obtain

(3.36) $$2aa_{20}(A - 2b) = 0$$
(3.37) $$2a_{20}(A - 2b)(d - A) = 0$$
(3.38) $$- 6aa_{20}(A - 2b) = 0$$
(3.39) $$2a_{20}b(A - 2b) = 0$$

These equations hold iff $a_{20}(A - 2b) = 0$ or $a = 0 = b = d - A$. In the first case, since all coefficients are zero if $a_{20} = 0$ we have $a_{20} \neq 0$, so this yields $A - 2b = 0$ and hence the system is hamiltonian and the origin is a center. This completes the proof of (EC).

It remains to show that if $a(b + d)(A - 2b) \neq 0$, then we also have $2d^2 + a^2 + bd = 0$. Under our hypothesis, in view of (EC) we have $A + 3b + 5d = 0$. We replace $A = -3b - 5d$ in our calculation for the coefficients of F_3, F_2, F_1 which depend on A i.e. a_{03} and a_{12}. We obtain for these the expressions

$$a_{03} = (a_{01}^3 + 3(3b + 5d)a_{01}^2 + 2(25d^2 + 30bd + 9b^2 + 3a_{20} + 3a^2)a_{01} + 4a_{20}(6d + 5b))/6$$
$$a_{12} = - a(a_{01}^2 + (4d + 3b)a_{01} + 2a_{20})$$

In view of our calculation above for the case $aa_{01} = 0$, since $a(b + d)(A - 2b) \neq 0$, $a_{01} \neq 0$. We note that the equation (E_{x^4}) does not depend on A and hence because we have $aa_{01} \neq 0$ we have for a_{20} the same value as the one obtained before i.e.(3.32), which does not depend on A. Replacing these values in (E_{y^4}) and $(E_{x^2y^2})$, after dividing by a, modulo multiplication by a constant factor, we obtain the same equation which is:

$$(E_{y^4})_A: \quad a_{01}^3 + (25d + 15b)a_{01}^2 + 2(41d^2 + 75bd + 28b^2 - 6a^2)a_{01} + 60(b + d)(2bd + b^2 - a^2) = 0$$

Modulo multiplications by a constant factor, the remaining equations are respectively:

$$(E_{x^3y})_A: \quad (2d + b)a_{01}^3 + (10d^2 + 25bd + 10b^2)a_{01}^2 + (50bd^2 + 87b^2d - 6a^2d + 31b^3 - 12a^2b)a_{01} + 30b(b + d)(2bd + b^2 - a^2) = 0$$

$$(E_{xy^3})_A: \quad a_{01}^4 + (42d + 24b)a_{01}^3 + (527d^2 + 640bd + 191b^2 - 12a^2)a_{01}^2 + (1470d^3 + 3546bd^2 + 2532b^2d - 276a^2d + 564b^3 - 168a^2b)a_{01} + 540(b + d)(2d + b)(2bd + b^2 - a^2) = 0$$

We first show that $d \neq 0$. Indeed, if $d = 0$, replacing this in the equation $(E_{y^4})_A$, we get

(3.40) $$a_{01}^3 + 15ba_{01}^2 + (56b^2 - 12a^2)a_{01} + 60b(b^2 - a^2) = 0$$

We now put $d = 0$ in $(E_{x^3y})_A$ and we obtain

(3.41) $$ba_{01}^3 + 10b^2a_{01}^2 + (31b^2 - 12a^2)ba_{01} + 30b^2(b^2 - a^2) = 0$$

$b \neq 0$ since otherwise $b + d = 0$. We eliminate the cubic terms in the above two equations by multiplying the first one by b and subtracting the second. After factoring and dividing by $5b^2$ we obtain the equation

$$a_{01}^2 + 5ba_{01} + 6(b^2 - a^2) = 0$$

To obtain a second quadratic equation we subtract from the equation (3.41) multiplied by 2, the equation (3.40) multiplied by b. After dividing by ba_{01} we obtain the equation

$$a_{01}^2 + 5ba_{01} + 6(b^2 - 2a^2) = 0$$

Subtracting the above two equations we obtain $- 6a^2 = 0$ which is impossible since we assumed that $a(b + d)(A - 2b) \neq 0$. So $d \neq 0$. We now prove that $2d + b \neq 0$. Indeed, if $b = -2d$ is replaced in the equation $(E_{x^3y})_A$ then we obtain the equation $6a^2d(3a_{01} - 10d) = 0$ and since $ad \neq 0$ we obtain $a_{01} = 10d/3$. Replacing this together with $b = -2d$ in $(E_{y^4})_A$ we obtain $20d(2d^2 + 27a^2) = 0$ which is impossible.

Consider $(Exy^3)_A - ((E_{y^4})_A).a_{01}$. This gives us the equation

(E_{17d+9b}) $(17d + 9b)a_{01}^3 + (445d^2 + 490bd + 135b^2)a_{01}^2 + (1470d^3 + 3426bd^2 + + 2352b^2d - 216a^2d + 504b^3 - 108a^2b)a_{01} + 2160bd^3 + 4320b^2d^2 - 1080a^2d^2 + 2700b^3d - 1620a^2bd + 540b^4 - 540a^2b^2 = 0$

Case 17d + 9b = 0. We replace $b = -17d/9$ in the above equation and modulo a constant factor we obtain

$(E_{2,I})$ $10d^2a_{01}^2/9 - (508d^3/81 + 12a^2d)a_{01} + 2720d^4/243 + 160a^2d^2/3 = 0$

Since $d \neq 0$, we may eliminate the factor d. We also replace $b = -17d/9$ in $(E_{y^4})_A$ and obtain

$(E_{y^4})_{A,b}$ $a_{01}^3 - 10da_{01}^2/3 - 124d^2a_{01}/81 - 12a^2a_{01} + 2720d^3/243 + 160a^2d/3 = 0$

We replace $b= -17d/9$ in $(E_{x^3y})_A$ and obtain

$(E_{x^3y})_{A,b}$ $- da_{01}^3/9 + 125d^2a_{01}^2/81 - 5134d^3a_{01}/729 - 50a^2da_{01}/3 + 23120d^4/2187 + + 1360a^2d^2/27 = 0$

We form the equation $d(E_{y^4})_{A,b}/9 + (E_{x^3y})_{A,b}$ and we obtain:

$(E_{2,II})$ $95d^2a_{01}^2/81 - (5258d^3/729 + 18a^2d)a_{01} + 25840d^4/2187 + 1520a^2d^2/27 = 0$

$19(E_{2,I})/9 - 2(E_{2,II})$ gives us $32d(d^2 + 9a^2)a_{01}= 0$ which is impossible.

Case 17d+9b ≠ 0. We consider now the equations $(E_{y^4})_A$, $(E_{x^3y})_A$, $(E_{xy^3})_A$, (E_{17d+9b}) in the unknown a_{01}. We form the equation $(17d + 9b)(E_{y^4})_A - (E_{17d+9b})_A$ and we obtain a second degree equation in a_{01}:

(E) $- 2d(5(2d + b)a_{01}^2 - 2d(38d^2 + 69bd + 25b^2 - 6a^2)a_{01} - 60d(b + d)(2bd + b^2 - a^2)) = 0.$

We have $2d + b \neq 0$ and hence the equation we have obtained above is of degree two in a_{01}. To obtain another second degree equation in a_{01} we consider the equation $(2d + b)(E_{y^4})_A - (E_{x^3y})_A$. This gives us

(F) $5(4d + b)(2d + b)a_{01}^2 + (164d^3 + 332bd^2 + 175b^2d - 18a^2d + 25b^3)a_{01} + + 30(b + d)(4d + b)(2bd + b^2 - a^2) = 0$

Assume first that $4d + b \neq 0$. We divide (E) by 2d and multiply the result by $(4d + b)$ and add with the above equation. Clearly the terms in a_{01} of degrees two and zero disappear and the result factors and we have: $12a_{01}d(b + d)(2d^2 + bd + a^2) = 0$. This gives us what we need. If on the other hand $4d + b = 0$, then in the above equation (F) the terms in a_{01}^2 and the free term disappear and the equation becomes $18d(2d^2 - a^2)a_{01} = 0$. Since $da_{01} \neq 0$, we have $2d^2 = a^2$ Then $2d^2 + bd + a^2 = 2d^2 + (-4d).d + 2d^2 = 0$. Replacing b by $-4d$ and a^2 by $2d^2$ in the equation (E), this equation factors and it becomes

$$-10d(a_{01} - 9d)(a_{01} - 6d) = 0$$

Hence either $a_{01} = 9d$ or $a_{01} = 6d$, giving us the coefficients of the conic in this case. Q E. D.

Corollary 3.2. If a system (S_2') has a weak focus of order two or three, then this system has no invariant irreducible cubic curve, not passing through the origin, no invariant irreducible conic curve and no invariant straight line.

Remark 3. It is clear that as we increase the degree the possibility of having a curve which is algebraic of that degree diminishes since we have to satisfy more equations for the coefficients. In [3] the necessary conditions for the coefficients of a system (S_2') are found in order that the system have an irreducible invariant conic curve. If the origin is a weak focus of order one then this conic curve is necessarily a parabola.

Bibliography.

1. D. Schlomiuk. The "center-space" of plane quadratic systems and its bifurcation diagram. Rapport de Recherche du Département de Mathématiques et de Statisique, D.M.S. No 88-18 L'Université de Montréal, Octobre 1988, 26 p.
2. D. Schlomiuk. Invariant reducible cubics and the conditions for the center, Rapport de Recherche du Département de Mathématiques et de Statistique, D.M.S. N° 89-11, Université de Montréal, Mai 1989, 12 p.
3. D. Schlomiuk. Invariant conics of quadratic systems with a weak focus. Rapport de Recherche du Département de Mathématiques et de Statistique, D.M.S. N° 89 -19, Université de Montréal, octobre 1989, 16 p.
4. D. Schlomiuk. Sufficient geometric conditions for the center for plane quadratic vector fields. Rapport de recherche du Département de Mathématique et de Statistique, D.M.S. N° 89 - 24, Univérsité de Montréal, décembre 1989, 8 p.
5. V. A. Lunkevich and K. S. Sibirsky. Integrals of a general quadratic differential system in cases of the center. Translated from russian from Differentsial'nye Uravnenya, vol. 18, no.5, pp 786-792, May 1982.
6. D. Schlomiuk, J. Guckenheimer, R. Rand. Integrability of plane quadratic vector fields. Expositiones Mathematicae 8, p.3-25, 1990.
7. W. Kapteyn. On the midpoints of integral curves of differential equations of the first degree, Nederl. Acad. Wetensch. Verslagen, Afd.Natuurkunde Konikl. Nederland. 19, 1446-1457, 1911 (Dutch)
8. N. Bautin. "On the number of limit cycles which appear with the variation of the coefficients from an equilibrium position of focus or center type", M. Sb. 30, 1952, 181-196, 1952 (in Russian); Amer.Math.Soc. Transl. No.100, 1954.
9. W. Kapteyn. New investigations on the midpoints of integrals of differential equations of the first degree , Nederl. Acad. Wetensch. Verslagen, Afd.Natuurkunde Konikl. Nederland. 20, p.1354-1365; 21, p. 27-33, 1912. (Dutch)

ROTATED VECTOR FIELDS DECOMPOSITION METHOD AND ITS APPLICATION.

Ye Yanqian (Nanjing University, China).

The theory of rotated vector fields (RVF) founded by G.F.D. Duff in 1953 [6] was generalized and used extensively by Chinese mathematicians in the period 1956-1967 when they studied the qualitative theory of the real quadratic differential systems, see [1] §§ 3, 6, 11-15. However, up to 1981, for the system

$$\dot{x} = -y+\delta x+\ell x^2+mxy+ny^2 = P(x,y), \qquad \dot{y} = x(1+ax+by)=Q(x,y) \quad (1)$$

we had used only the whole plane RVF (when a=b=0) and the half plane RVF (when $|a|+|b|\neq 0$) with parameter δ (we will denote it by F_2). In the latter case, we have

$$\frac{\partial\theta}{\partial\delta} = -\frac{x^2(1+ax+by)}{P^2+Q^2}\begin{cases} =0, & \text{when } x(1+ax+by) = 0; \\ >0, & \text{when } 1+ax+by < 0; \\ <0, & \text{when } 1+ax+by > 0. \end{cases} \quad (2)$$

If in (1) we take one of a, b, ℓ, m, n as a variable parameter, then the whole plane will be divided by the curve P(x,y)=0, or two straight lines (at least one of which passes through O), or both them into 2, 4, 6 or 8 regions, vectors at ordinary points in the same region rotate in the same diretion, but vectors in neighbouring regions rotate in opposite directions. Limit cycles (LC) in the neighbourhood of O(0,0) is a chief objective we want to study, but O always lies on the common boundary of certain two regions. Therefore, it is impossible to use the RVF theory to study the behavior of LC's around O when a, b, ℓ, m or n varies.

When we studied the impossibility of (2,2) distribution of LC's of (1) in [2], aside from F_2, we introduced also another two RVF's F_1 and F_3 (F_1 was also used in [3] in 1982), their definitions are as follows:

F_1: to add a term $\delta_1 x(1+ax+by)$ to the right hand side of the first equation in (1). In this case:

$$\frac{\partial\theta}{\partial\delta_1} = -\frac{x^2(1+ax+by)^2}{P^2+Q^2} \leq 0,$$

so F_1 is a whole plane RVF.

Similarly, F_2 can also be defined as "to add a term $\delta_2 x$ similar to F_1".

F_3: to add a term $m_3 x(y+\frac{1}{b})$ similar to F_1 (here we assume $b\neq 0$), and we have:

$$\frac{\partial\theta}{\partial m_3} = -\frac{x^2(y+\frac{1}{b})\ (1+ax+by)}{P^2 + Q^2} \quad .$$

Vectors in different regions rotate as shown in Fig. 1 when m_3 increases.

Now let us introduce the RVF F_4:

F_4: to add a term $a_4 x^2$ to the second equation of (1), since:

$$\frac{\partial\theta}{\partial a_4} = \frac{x^2 P(x,y)}{P^2+Q^2} \quad ,$$

vectors on different sides of $P(x,y) = 0$ rotate in opposite directions as a_4 increases.

Figure 1

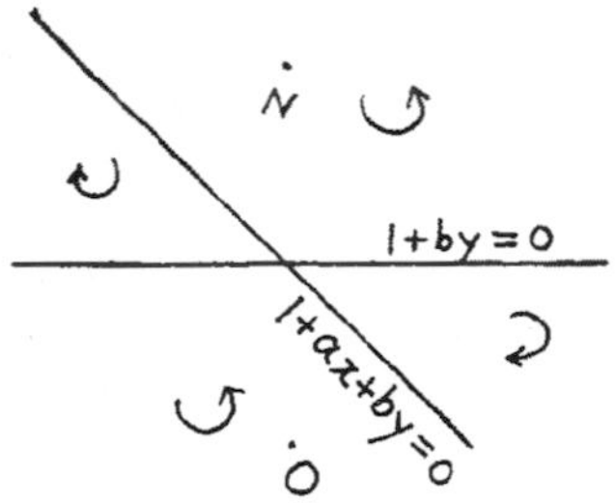

If (1) has LC's only around one critical point, we may assume the critical point to be O(0,0), in this case, the LC's are always positively oriented. We use Γ_1 (Γ_2) to denote any stable (unstable) LC around O. If (1) has LC's around two different critical points, we may assume them to be O and $N(0,\frac{1}{n})$ with $n>0$. In this case, $1+ax+by=0$ must separate O and N, so we have $b<-n<0$. In Fig. 1 LC's can only appear in the regions that O and N lie; therefore, in the study of LC problem, F_3 can be treated as a whole plane RVF. We use Γ_1' (Γ_2') to denote any unstable (stable) LC around N, they are all negatively oriented.

The behavior of LC's in the RVF's F_1, F_2, F_3 are listed in the following table.

	stable Γ_1	unstable Γ_2	stable Γ_2'	unstable Γ_1'
δ_1 increases	expands	contracts	contracts	expands
δ_1 decreases	contracts	expands	expands	contracts
δ_2 increases	expands	contracts	expands	contracts
δ_2 decreases	contracts	expands	contracts	expands
m_3 increases	contracts	expands	expands	contracts
m_3 decreases	expands	contracts	contracts	expands

Notice that O and N are fixed under any of the F_i's, but critical points on $1+ax+by=0$ and critical points at infinity will move, disappear or be generated.

Applying F_1, F_2, F_3 and F_4 to (1), we get the system:

$$\dot{x} = -y+(\delta_0+\delta_1+\delta_2+\frac{m_3}{b})x + (\ell_0+a\delta_1)x^2 +(m_3+m_0+b\delta_1)xy+ny^2,$$
$$\dot{y} = x(1+a_0x+by) + a_4x^2, \qquad (3)$$

in which we use δ_0, ℓ_0, m_0 and a_0 instead of δ, ℓ, m and a in (1).

From (3) and some of its special cases, we see at once that the following fundamental properties hold:

1) The result of the applications of F_i and F_j ($i\neq j$) to (1) is independent of the order of F_i and F_j.

2) The application of any F_i can be divided into many sub-steps. For example, if $m_3 = \sum_{i=1}^{N} m_{3i}$, and F_{3i} denote the application corresponding to m_{3i}, then $F_3 = \sum_{i=1}^{N} F_{3i}$.

3) To change the coefficient m in the first equation of (1) means to apply suitable F_2 and F_3 to (1), i. e., to take $\delta_2=-\frac{m_3}{b}$. Similarly, to change the coeffient ℓ means to take $m_3=-b\delta_1$ in F_1 and F_3.

Since in a RVF stable LC Γ_s and unstable LC Γ_v (they contain the same critical points) can move close to each other, coincide and become a semi-stable LC $\bar{\Gamma}$, and then disappear. Whether the $\bar{\Gamma}$ disappeared under F_i can reappear under F_j ($i\neq j$) , this is still an unsolved problem. Therefore, to divide F_i and F_j into substeps and apply them to (1) alternatively has the advantage such that Γ_s and Γ_v will not disappear through $\bar{\Gamma}$ in their largest possible existence regions.

Another unsolved problem in the theory of RVF is: How to determine the values of the parameter for which semi-stable LC appears suddenly (it then splits into a stable LC and an unstable one, among which one expands and another contracts)? For example, if in (1) a=b=0, $m(\ell+n)<0$, then as δ increases from zero, a stable LC Γ_1 generates from O and expands. We can use different but easy methods to prove that there will appear no semi-stable LC within Γ_1, but it is difficult to prove that outside Γ_1 there will also not appear any semi-stable LC. We must use several different transformations of coordinates to transform (1) into Lienard systems and then use the uniqueness theorem of LC's due to Zhang Zhifen. Nevertheless, we notice that the following properties still hold for system (1):

4) Corresponding to a fixed value of a parameter μ (δ_1, δ_2 or m_3) in (1) there can appear at most a finite number of semi-stable LC's (we conjecture that this number is 1). Therefore, the appearance or disappearance of semi-stable LC's does not affect the parity of the numbers of LC's around O and N.

5) If a stable (or unstable) LC disappears at a focus under the RVF with parameter μ, it will certainly reappear under the RVF with another parameter ν, only if ν varies in a suitable direction. If a stable (or unstable) LC disappears through a finite or infinite separatrix cycle under the RVF with parameter μ (e. g., δ_1), it will reappear under the RVF with parameter ν (e. g., δ_2 or a_4) through the change of relative positions of the separatrices passing through finite or infinite saddle points; or through the generation of new saddle point on $1+ax+by=0$, if on the latter there is no critical point under the former RVF. Nevertheless, during the applications of the F_{1j} or F_{3i} later on, we always demand that separatrix cycle can be formed but never be distroyed, so as to ensure the existence of the outmost LC around O.

Based upon the above properties we are now able to use the RVF decomposition method to solve some problems concerning the relative position of LC's of (1) and the maximum number of LC's around a strong or a weak focus.

The chief idea in using the RVF decomposition method to solve the relative position problem is as follows:

We assume at first that (1) has a certain type of distribution of LC's which we take to be impossible. We then apply to (1) F_1, F_2, F_3 and F_4 or some of them to transform (1) into a simpler form such that it retains certain properties concerning the existence of LC's. But on the other hand, we can prove that for this simpler system the distribution of LC's does not possess these properties. This contradiction proves therefore that the originally assumed type of distribution of LC's is impossible.

The following is a problem which we have studied carefully, i.e., to prove the impossibility of (2,2) distribution of LC's.

Assume (1) has a (2,2) distribution of LC's with strong foci O and N, such that they have different stability. In this case we may assume (1) to be:

$$\dot{x} = -y+\delta_0 x+\ell_0 x^2+m_0 xy+ny^2, \qquad \dot{y} = x(1+a_0 x-y) \tag{4}$$

in which $a_0<0$, $0<n<1$, $\delta_0>0$ and $m_0+n\delta_0<0$. Let the LC's be:

$$O \subset \Gamma_1 \subset \Gamma_2, \qquad N \subset \Gamma_1' \subset \Gamma_2',$$

where O (N) is an unstable (stable) strong focus, Γ_1 and Γ_2' are stable LC's, Γ_2 and Γ_1' are unstable LC's.

When some or all of the Γ's are semi-stable LC's, the problem can be transformed into (2,2), (2,3) or (3,3) distributions of non-semistable LC's by the application of F_i and will be mention -ed later on. We now apply F_1, F_2, F_3 and F_4 to (1) and get the system:

$$\begin{aligned}\dot{x} &= -y+(\delta_0+\delta_1+\delta_2-m_3)x + (\ell_0+a\delta_1)x^2 + (m_0-\delta_1+m_3)xy + ny^2,\\ \dot{y} &= x[\,1 + (a_0+a_4)x-y\,].\end{aligned} \tag{3'}$$

Since the F_i's are applied in sub-steps and alternatively, we assume:

$$\delta_1 = \sum_{j=1}^{M} \delta_{1j}, \quad \delta_2 = \sum_{k=1}^{G}\delta_{2k}, \quad m_3 = \sum_{i=1}^{N} m_{3i}, \quad a_4 = \sum_{h=1}^{R} a_{4h}, \tag{5}$$

where $\delta_{1j}<0$, $\delta_{2k}>0$, $m_{3i}>0$, $a_{4h}>0$ and $R+G=M+N$. We use F_{1j}, F_{2k}, F_{3i} and F_{4h} to denote the corresponding RVF's.

Moreover, we can demand that the parameters in (3)' will satisfy the relations:

$$m_0 - \delta_1 + m_3 = 0, \quad \delta = \delta_0 + \delta_2 + \delta_1 - m_3 > 0. \tag{6}$$

then (4) is transformed into:

$$\dot{x} = -y+\delta x+\ell x^2+ny^2, \qquad \dot{y} = x(1+ax-y), \qquad \delta>0 \tag{7}$$

And at the same time we can also demand that (7) has still two LC's (or in general 2n LC's, n— positive integer) around O. According to the properties listed before we know that these demands can be satisfied.

On the other hand, we can prove for (7) the following theorem [4]:

Theorem 1. If in the system (7), $\delta>0$, $a<0$ and $\ell<\frac{1}{2}$ ($>\frac{1}{2}$), then it has no LC around N (O), and can have an odd number of non-semi-stable LC's around O (N). It has no LC in the whole plane for $\ell=\frac{1}{2}$ or $a=0$ and any δ. When $\delta<0$, if "$\ell<\frac{1}{2}$ ($>\frac{1}{2}$)" is changed into "$\ell>1/2$ ($<\frac{1}{2}$)" in the first sentence, then the conclusion of the theorem is still true.

This contradiction proves that if O and N have different stability, then (2,2) distribution of LC's is impossible for (1).

Remark 1. The RVF F_4 is used only when Γ_2 becomes an infinite separatrix cycle $\bar{\Gamma}$(which is of course inner unstable) joining two non-diagonal opposite critical points at infinity after F_{1j} or F_{3i} is applied to (3)'. In this case, according to a former paper by the author and W. Y. Ye ([2], Fig.5), $\bar{\bar{\Gamma}}$ will lie completely in the region $P(x,y)<0$, and no finite critical points lie on the isoline $1+ax-y=0$, which separates the two branches of $P(x,y)=0$. So the application of F_{4h} with $a_{4h}>0$ will make $\bar{\bar{\Gamma}}$ disappear and generates a new Γ_2 (unstable). Of course, we can also use F_{2k} with $\delta_{2k}>0$.

Remark 2. If when we apply F_{1j}, F_{2k} and F_{3i} to (3)' many times, the semi-stable LC $\bar{\Gamma}_k$ generated from the coincidence of Γ_1 and Γ_2 and the separatrix cycle $\Gamma_{ji}{}^*$ generated from Γ_2 (passing through a finite saddle point $S_{ji}{}^*$) approach to each other such that when $i, j, k \to +\infty$: $S_{ji}{}^* \to S^*$ and $\bar{\Gamma}_k$, $\Gamma_{ji}{}^* \to \Gamma^*$ (a separatrix cycle passing through S^*), and the first three sums in (5) become convergent series, but still have $m_0-\delta_1+m_3<0$ in (3)', then we cannot get system (7). But this phenomenon will not happen. For otherwise, the straight line $\operatorname{div}(P,Q)=0$ passing through S^* (a weak saddle) will have another two contact points on it, one within Γ^* and one within $\Gamma_1{}'$ or $\Gamma_2{}'$, which is impossible, by a well-known property of quadratic differential systems. If at this moment $\Gamma_1{}'$ and $\Gamma_2{}'$ have already disappeared, then we may change the role of O, Γ_1, Γ_2 with that of N, $\Gamma_1{}'$, $\Gamma_2{}'$. We did not notice this fact in [4].

Remark 3. The above phenomenon can also not happen for Γ^* to be an infinite separatrix cycle passing through two critical points A and B at infinity. For we can proof [5] that div(P,Q)=0 in (1,y,z) and (x,1,z) coordinates are the quadratic curves:

$$4\overline{Q}(x,y,z)-y(\overline{P}_x+\overline{Q}_y)=0 \quad \text{and} \quad 4\overline{P}(x,y,z)-x(\overline{P}_x+\overline{Q}_y)=0$$

where $\overline{P}$, $\overline{Q}$ are P, Q in homogeneous coordinates. In the critical case mentioned above, one of A and B will lie on the terminals of both P(x,y)=0, div(P,Q)=0, and hence also on the terminals of the straight line $P_x+Q_y = 0$ ((0,1,0) is not a critical point, since $n \neq 0$), and so we can deduce contradiction as in Remark 2.

Remark 4. The case that $\overline{\Gamma}_k$ generated from the coincidence of Γ_1 and Γ_2 under the application of F_{2k} approaches 0 as $k \to \infty$, such that $\delta_1 = \sum_{j=1}^{\infty} \delta_{1j}$, $\delta_2 = \sum_{k=1}^{\infty} \delta_{2k}$ and $m_3 = \sum_{i=1}^{\infty} m_{3i}$ are all convergent series, but still $m_0 - \delta_1 + m_3 < 0$ in (3)' will not happen, too. For, on account of property 5) before, this will happen only when a finite saddle point approaches 0 or a separatrix passing through a finite or infinite saddle point approaches to 0 (they will make the convergence of the above three series) as i, j, k $\to \infty$. But the first possibility evidently can't happen, since n and b are fixed in the whole process. Now if the second possibility happens, then we will have in the limiting case an integral straight line (since system (1) is quadratic) passing through 0, which is also impossible, already shown in [4].

Since in the proof of the impossibility of (2,2) distribution of LC's we did not distinguish that around 0 there are only 2 LC's or there are 2n LC's (n—positive integer), and around N we used only the fact that $m_0<0$, hence we have proved actually:

Theorem 2. If the two foci 0 and N of (1) have different stability, then there can be no (2n,2m) distribution of non-semi-stable LC's, where n and m are positive integers.

We conjecture that this theorem is still true when 0 and N have the same stability.

Notice that if 0 and N have the same stability, there are exa-mples of (2,1) distribution of LC's. Moreover, 0 may then change its stability and a (3,1) distribution appears, see [1], § 11.

There are three other conjectures as follows:

Conjecture 1. When 0 and N have different stability (2,1) distribution of LC's is impossible for (1); when O and N have the same stability (3,1) distribution is impossible.

Conjecture 2. If there is a (1,1) distribution of non-semi-stable LC's of (1), then O and N must have different stability.

Conjecture 3. For system (1), (2,3) and (3,3) distribution of non-semi-stable LC's is impossible.

Remark 5. If we can prove that under the rotated vector fields F_1, F_2, F_3 and F_4 there can appear no multiple LC of multiplicity 3 and no separatrix loop which can generate 3 LC's, then one will be able to prove that around any focus of order 0, 1, or 2 of a quadratic differential system there can appear at most 3, 2 or 1 LC by using RVF decomposition method.

Acknoledgement. The author thanks Professor J. W. Reyn for his valuable comments and also to his kindest invitation to Delft Technical University during the period Oct. to Dec., 1989.

REFERENCES

1. Ye Yanqian and others, Theory of limit cycles, Trans. Math. Monographs, vol. 66, Amer. Math. Soc., 1986.

2. Ye Yanqian, Ye Wei Yin, On the impossibility of (2,2) distribution of limit cycles for any real quadratic differential system (II), Ann. Diff. Eqs., vol. 4, no. 1, 117-130 (1988).

3. Suo Guangjian, The algebraic critical cycles and bifurcation of limit cycles for the system $x=a+\sum_{i+j=2} a_{ij}x^i y^j$, $y=b+\sum_{i+j=2} b_{ij}x^i y^j$ Journ. of Math. Rev. and Expos., vol. 2, no. 2, 69-75 (1982).

4. Ye Yanqian, On the relative position of limit cycles of a real quadratic differential system, Chin. Ann. Math., 10, Ser.B (1989)

5. Ye Yanqian, Divergence and the uniqueness of limit cycles of a quadratic differential system. (to appear in Ann. Diff. Eqs., vol. 6, no. 1, (1990).

6. Duffy, G.F.D., Limit cycles and rotated vector fields, Ann. Math., vol. 2, no. 57, 15-31 (1953).

REMARKS ON THE DELAY OF THE LOSS OF STABILITY OF SYSTEMS WITH CHANGING PARAMETER

Henryk Żołądek
Institute of Mathematics, University of Warsaw
0090I Warsaw, PKiN, IX p., Poland

Abstract. A new kind of delay of the Andronov - Hopf bifurcation in a concrete system with slowly varying parameter is presented.

I. Introduction.

Everybody knows the typical scenario of the mild loss of stability in one-parameter families of evolutionary systems. The amplitude of selfoscilations grows like the square root of the distance to the critical value, (Andronov - Hopf bifurcation).

It turns out that when the parameter changes slowly with time a typical system's behaviour is very different from the described above. For a long period of time after the moment when the parameter passes the critical value, the system does not leave the neighbourhood of the equilibrium, which has became unstable. It is so long that the parameter becomes finite. Only then the system leaves the neighbourhood of the equilibrium point. It leaves it with a jump going to the attracting orbit created after the bifurcation. The amplitude of selfoscilations after the jump is finite. This phenomenon has been discovered by M. A. Shishkova [4] in a model example. The general theory has been developed by A. I. Neishtadt [2], [3], (see also [I]).

In the present work I want to draw attention to another kind of the delay of the loss of stability in the system with slowly varying parameter, (with a speed ε, $\mu = \varepsilon t$). Everything depends on the initial time when we begin to observe the system. If the trajectory starts at time $t_0 = -1/\varepsilon$ (the parameter then equals to $\mu_0 = -1$) then we obtain a jump in the amplitude of selfoscilations and the kind of delay of the loss of stability described by Neishtadt and Shishkova.

But if the trajectory starts at time $t_0 = -1/\sqrt{\varepsilon}$ ($\mu_0 = -\sqrt{\varepsilon}$) then the system first approaches a small neighbourhood of equilibrium and then the amplitude of selfoscilations behaves like $\varepsilon^{1/4}$ times a smooth function of $\tau = \sqrt{\varepsilon}t$, the graph of which is presented at Figure I.

The amplitude slowly starts to increase and finally the system tends to the attracting periodic orbit when the parameter becomes finite. No jumps or discontinuities are observed.

2. The results.

The results that we present here concern the following system

$$\dot{x}=x(\mu-x^2-y^2)+y$$
$$\dot{y}=y(\mu-x^2-y^2)-x\,,\qquad \mu=\varepsilon t \tag{I}$$

(slightly different from the one choosen by Shishkova).

Denote by $r=(x^2+y^2)^{1/2}$ the amplitude of selfoscilations of the system (I). Let t_0 be the initial time and let $r_0=r(t_0)$ be the amplitude at the initial time.

Theorem I. Assume that $t_0=\tau_0\cdot\varepsilon^{-1/2}<0$ and $r_0>0$ are such that τ_0 and r_0 are fixed and finite.

(a) If $u=t-t_0$ is fixed and finite then

$$r(t)\longrightarrow r_0(1+2r_0^2u)^{-1/2} \quad \text{as} \quad \varepsilon\longrightarrow 0.$$

(b) If t is of order $\varepsilon^{-1/2}$ and $\tau=\varepsilon^{1/2}t>\tau_0$ is fixed then

$$\varepsilon^{-1/4}r(t)\longrightarrow f(\tau_0,\tau) \quad \text{as} \quad \varepsilon\longrightarrow 0\,,$$

where

$$f(\tau_0,\tau)=\exp(\tau^2/2)\cdot\left\{2\int_{\tau_0}^{\tau}\exp(\varsigma^2)d\varsigma\right\}^{-1/2}.$$

(c) If t is of order ε^{-1} and $\mu=\varepsilon t$ is fixed then

$$r(t)\longrightarrow\mu^{1/2} \quad \text{as} \quad \varepsilon\longrightarrow 0\,.$$

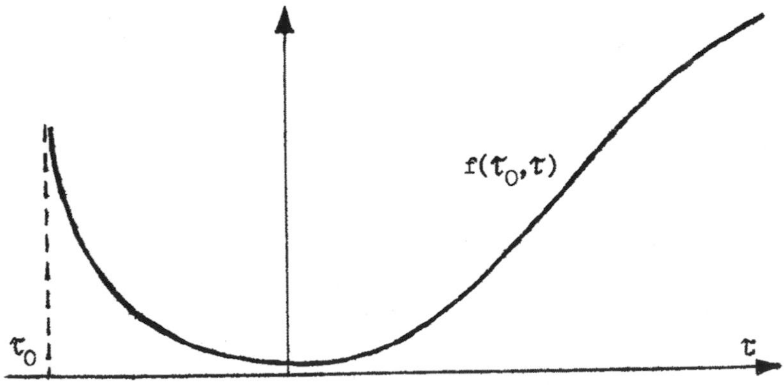

Figure I.

Remarks. 1. It is easy to check that $f(\tau_0,\tau)\sim[2(\tau-\tau_0)]^{-1/2}$ as $\tau\longrightarrow\tau_0$ and that $f(\tau_0,\tau)\sim\tau^{1/2}$ as $\tau\longrightarrow\infty$. Therefore, both asymptotics agree with those given in the points (a) and (c). Note also that the amplitude $\varepsilon^{1/4}$ is achieved relatively long period of time $\sim\varepsilon^{-1/2}$ after the initial time.

2. The theorem holds also for $\tau_0\geq 0$, i. e. if t_0 is finite for example.

3. The author hopes that a similar result holds in a general bifurcating system with slowly varying parameter.

Proof of Theorem 1. The amplitude of selfoscilations of the system (1) satisfies the following Bernoulli equation

$$\dot{r}=r(\varepsilon t-r^2).$$

The solution of this equation with the initial condition $r(t_0)=r_0$ is

$$r(t)=r_0\left\{\exp\left[\varepsilon(t_0^2-t^2)\right]+2r_0^2\int_{t_0}^{t}\exp\left[\varepsilon(s^2-t^2)\right]ds\right\}^{-1/2}. \tag{2}$$

From this formula, the asymptotics of the Theorem 1 can be obtained by straightforward calculations.

Quite analogously one obtains the result of Neishtadt for the system (1). It can be stated as follows.

Theorem 2. Assume that $t_0=\mu_0/\varepsilon<0$ and $r_0>0$ are such that $\mu_0<0$ and r_0 are fixed and finite.

(a) If $u=t-t_0>0$ is fixed and finite then

$$r(t)\longrightarrow r_0\left\{\exp(2|\mu_0|u)+r_0^2\left[\exp(2|\mu_0|u)-1\right]/|\mu_0|\right\}^{-1/2}$$

as $\varepsilon\longrightarrow 0$.

(b) If $\mu=\varepsilon t$ is fixed and $|\mu|<|\mu_0|$ then

$$r(t)<C_1\exp(-C_2/\varepsilon)\quad\text{as}\quad\varepsilon\longrightarrow 0,$$

where C_1 and C_2 are constants depending only on μ.

(c) If $v=t+t_0$ is fixed then

$$r(t)\longrightarrow r_0\left\{\exp(-2|\mu_0|v)+r_0^2\left[\exp(-2|\mu_0|v)+1\right]/|\mu_0|\right\}^{-1/2}$$

as $\varepsilon\longrightarrow 0$.

(d) If $\mu=\varepsilon t$ is fixed and $\mu\geq|\mu_0|$ then

$$r(t)\longrightarrow\mu^{1/2}\quad\text{as}\quad\varepsilon\longrightarrow 0.$$

Remark 4. From the point (c) of Theorem 2, one sees that the period of jump is relatively short. It takes about $\ln\varepsilon^{-1}$ for the amplitude to jump from ε^{a} to $\sqrt{|\mu_0|}$, as in [2].

There appears the question about behaviour of the solution (2) in the intermediary case $t_0 \sim \varepsilon^{-\alpha}$, $1/2 < \alpha < 1$. The next result shows that the amplitude undergoes a jump analogous to the one described by Neishtadt.

Theorem 3. Assume that $t_0 = \tau_0 \varepsilon^{-\alpha}$, $1/2 < \alpha < 1$ and $r_0 > 0$ are such that $\tau_0 < 0$ and r_0 are fixed and finite.

(a) If $u = t - t_0$ is fixed and finite then

$$r(t) \longrightarrow r_0(1+2r_0^2 u)^{-1/2} \quad \text{as} \quad \varepsilon \longrightarrow 0 .$$

(b) If $\tau = t\varepsilon^{\alpha}$ is fixed and $|\tau| < |\tau_0|$ then

$$r(t) < C_1 \exp(-C_2 \varepsilon^{1-2\alpha}) \quad \text{as} \quad \varepsilon \longrightarrow 0 .$$

(c) If $v = (t + t_0)\varepsilon^{1-\alpha}$ is fixed then

$$r(t) \sim [|\mu_0| / (1 + \exp(-2|\tau_0|v))]^{1/2} \quad \text{as} \quad \varepsilon \longrightarrow 0 ,$$

where $\mu_0 = \varepsilon t_0 = \varepsilon^{1-\alpha} \tau_0$.

(d) If $\tau = t\varepsilon^{\alpha}$ is fixed and $\tau \geq |\tau_0|$ then

$$r(t) \sim \varepsilon^{(1-\alpha)/2} \cdot \tau^{1/2} = \mu^{1/2} \quad \text{as} \quad \varepsilon \longrightarrow 0 .$$

The motivation for the present work comes from applications. In an experiment performed by J. Łusakowski (Warsaw), the bifurcation with appearance of selfoscilations was observed for an electrical system. The amplitude of selfoscilations grows linearly with the parameter. A possible explanation of this phenomenon might be the Theorem I (for large negative values of τ_0).

References.

[I] V.I.Arnold, V.S.Afraimovich, Yu.S.Iliashenko, L.P.Shilnikov, "Bifurcation theory". Modern Problems of Mathematics, Fundamental Directions, v.5, VINITI, Moscow I986, [Russian].

[2] A.I.Neishtadt, Study of the asymptotic loss of stability of equilibrium under slow transition of two eigenvalues through the imaginary axis, Uspiehi Mat. Nauk 40(5),300-30I,(I985),[Russian].

[3] A.I.Neishtadt, On delayed stability loss under dynamical bifurcation. I. Differential Equat., 23(I2),2060-2067,(I987); II. Differential Equat., 34(2),226-233,(I988),[Russian].

[4] M.A.Shishkova, Study of a system of differential equations with a small parameter at higher derivative, Dokl. AN USSR, 209(3), 576--579,(I973),[Russian].

Vol. 1290: G. Wüstholz (Ed.), Diophantine Approximation and Transcendence Theory. Seminar, 1985. V, 243 pages. 1987.

Vol. 1291: C. Mœglin, M.-F. Vignéras, J.-L. Waldspurger, Correspondances de Howe sur un Corps p-adique. VII, 163 pages. 1987

Vol. 1292: J.T. Baldwin (Ed.), Classification Theory. Proceedings, 1985. VI, 500 pages. 1987.

Vol. 1293: W. Ebeling, The Monodromy Groups of Isolated Singularities of Complete Intersections. XIV, 153 pages. 1987.

Vol. 1294: M. Queffélec, Substitution Dynamical Systems – Spectral Analysis. XIII, 240 pages. 1987.

Vol. 1295: P. Lelong, P. Dolbeault, H. Skoda (Réd.), Séminaire d'Analyse P. Lelong – P. Dolbeault – H. Skoda. Seminar, 1985/1986. VII, 283 pages. 1987.

Vol. 1296: M.-P. Malliavin (Ed.), Séminaire d'Algèbre Paul Dubreil et Marie-Paule Malliavin. Proceedings, 1986. IV, 324 pages. 1987.

Vol. 1297: Zhu Y.-l., Guo B.-y. (Eds.), Numerical Methods for Partial Differential Equations. Proceedings. XI, 244 pages. 1987.

Vol. 1298: J. Aguadé, R. Kane (Eds.), Algebraic Topology, Barcelona 1986. Proceedings. X, 255 pages. 1987.

Vol. 1299: S. Watanabe, Yu.V. Prokhorov (Eds.), Probability Theory and Mathematical Statistics. Proceedings, 1986. VIII, 589 pages. 1988.

Vol. 1300: G.B. Seligman, Constructions of Lie Algebras and their Modules. VI, 190 pages. 1988.

Vol. 1301: N. Schappacher, Periods of Hecke Characters. XV, 160 pages. 1988.

Vol. 1302: M. Cwikel, J. Peetre, Y. Sagher, H. Wallin (Eds.), Function Spaces and Applications. Proceedings, 1986. VI, 445 pages. 1988.

Vol. 1303: L. Accardi, W. von Waldenfels (Eds.), Quantum Probability and Applications III. Proceedings, 1987. VI, 373 pages. 1988.

Vol. 1304: F.Q. Gouvêa, Arithmetic of p-adic Modular Forms. VIII, 121 pages. 1988.

Vol. 1305: D.S. Lubinsky, E.B. Saff, Strong Asymptotics for Extremal Polynomials Associated with Weights on $\mathbb{R}$. VII, 153 pages. 1988.

Vol. 1306: S.S. Chern (Ed.), Partial Differential Equations. Proceedings, 1986. VI, 294 pages. 1988.

Vol. 1307: T. Murai, A Real Variable Method for the Cauchy Transform, and Analytic Capacity. VIII, 133 pages. 1988.

Vol. 1308: P. Imkeller, Two-Parameter Martingales and Their Quadratic Variation. IV, 177 pages. 1988.

Vol. 1309: B. Fiedler, Global Bifurcation of Periodic Solutions with Symmetry. VIII, 144 pages. 1988.

Vol. 1310: O.A. Laudal, G. Pfister, Local Moduli and Singularities. V, 117 pages. 1988.

Vol. 1311: A. Holme, R. Speiser (Eds.), Algebraic Geometry, Sundance 1986. Proceedings. VI, 320 pages. 1988.

Vol. 1312: N.A. Shirokov, Analytic Functions Smooth up to the Boundary. III, 213 pages. 1988.

Vol. 1313: F. Colonius, Optimal Periodic Control. VI, 177 pages. 1988.

Vol. 1314: A. Futaki, Kähler-Einstein Metrics and Integral Invariants. IV, 140 pages. 1988.

Vol. 1315: R.A. McCoy, I. Ntantu, Topological Properties of Spaces of Continuous Functions. IV, 124 pages. 1988.

Vol. 1316: H. Korezlioglu, A.S. Ustunel (Eds.), Stochastic Analysis and Related Topics. Proceedings, 1986. V, 371 pages. 1988.

Vol. 1317: J. Lindenstrauss, V.D. Milman (Eds.), Geometric Aspects of Functional Analysis. Seminar, 1986–87. VII, 289 pages. 1988.

Vol. 1318: Y. Felix (Ed.), Algebraic Topology – Rational Homotopy. Proceedings, 1986. VIII, 245 pages. 1988

Vol. 1319: M. Vuorinen, Conformal Geometry and Quasiregular Mappings. XIX, 209 pages. 1988.

Vol. 1320: H. Jürgensen, G. Lallement, H.J. Weinert (Eds.), Semigroups, Theory and Applications. Proceedings, 1986. X, 416 pages. 1988.

Vol. 1321: J. Azéma, P.A. Meyer, M. Yor (Eds.), Séminaire de Probabilités XXII. Proceedings. IV, 600 pages. 1988.

Vol. 1322: M. Métivier, S. Watanabe (Eds.), Stochastic Analysis. Proceedings, 1987. VII, 197 pages. 1988.

Vol. 1323: D.R. Anderson, H.J. Munkholm, Boundedly Controlled Topology. XII, 309 pages. 1988.

Vol. 1324: F. Cardoso, D.G. de Figueiredo, R. Iório, O. Lopes (Eds.), Partial Differential Equations. Proceedings, 1986. VIII, 433 pages. 1988.

Vol. 1325: A. Truman, I.M. Davies (Eds.), Stochastic Mechanics and Stochastic Processes. Proceedings, 1986. V, 220 pages. 1988.

Vol. 1326: P.S. Landweber (Ed.), Elliptic Curves and Modular Forms in Algebraic Topology. Proceedings, 1986. V, 224 pages. 1988.

Vol. 1327: W. Bruns, U. Vetter, Determinantal Rings. VII,236 pages. 1988.

Vol. 1328: J.L. Bueso, P. Jara, B. Torrecillas (Eds.), Ring Theory. Proceedings, 1986. IX, 331 pages. 1988.

Vol. 1329: M. Alfaro, J.S. Dehesa, F.J. Marcellan, J.L. Rubio de Francia, J. Vinuesa (Eds.): Orthogonal Polynomials and their Applications. Proceedings, 1986. XV, 334 pages. 1988.

Vol. 1330: A. Ambrosetti, F. Gori, R. Lucchetti (Eds.), Mathematical Economics. Montecatini Terme 1986. Seminar. VII, 137 pages. 1988.

Vol. 1331: R. Bamón, R. Labarca, J. Palis Jr. (Eds.), Dynamical Systems, Valparaiso 1986. Proceedings. VI, 250 pages. 1988.

Vol. 1332: E. Odell, H. Rosenthal (Eds.), Functional Analysis. Proceedings, 1986–87. V, 202 pages. 1988.

Vol. 1333: A.S. Kechris, D.A. Martin, J.R. Steel (Eds.), Cabal Seminar 81–85. Proceedings, 1981–85. V, 224 pages. 1988.

Vol. 1334: Yu.G. Borisovich, Yu. E. Gliklikh (Eds.), Global Analysis – Studies and Applications III. V, 331 pages. 1988.

Vol. 1335: F. Guillén, V. Navarro Aznar, P. Pascual-Gainza, F. Puerta, Hyperrésolutions cubiques et descente cohomologique. XII, 192 pages. 1988.

Vol. 1336: B. Helffer, Semi-Classical Analysis for the Schrödinger Operator and Applications. V, 107 pages. 1988.

Vol. 1337: E. Sernesi (Ed.), Theory of Moduli. Seminar, 1985. VIII, 232 pages. 1988.

Vol. 1338: A.B. Mingarelli, S.G. Halvorsen, Non-Oscillation Domains of Differential Equations with Two Parameters. XI, 109 pages. 1988.

Vol. 1339: T. Sunada (Ed.), Geometry and Analysis of Manifolds. Procedings, 1987. IX, 277 pages. 1988.

Vol. 1340: S. Hildebrandt, D.S. Kinderlehrer, M. Miranda (Eds.), Calculus of Variations and Partial Differential Equations. Proceedings, 1986. IX, 301 pages. 1988.

Vol. 1341: M. Dauge, Elliptic Boundary Value Problems on Corner Domains. VIII, 259 pages. 1988.

Vol. 1342: J.C. Alexander (Ed.), Dynamical Systems. Proceedings, 1986–87. VIII, 726 pages. 1988.

Vol. 1343: H. Ulrich, Fixed Point Theory of Parametrized Equivariant Maps. VII, 147 pages. 1988.

Vol. 1344: J. Král, J. Lukeš, J. Netuka, J. Veselý (Eds.), Potential Theory – Surveys and Problems. Proceedings, 1987. VIII, 271 pages. 1988.

Vol. 1345: X. Gomez-Mont, J. Seade, A. Verjovski (Eds.), Holomorphic Dynamics. Proceedings, 1986. VII, 321 pages. 1988.

Vol. 1346: O. Ya. Viro (Ed.), Topology and Geometry – Rohlin Seminar. XI, 581 pages. 1988.

Vol. 1347: C. Preston, Iterates of Piecewise Monotone Mappings on an Interval. V, 166 pages. 1988.

Vol. 1348: F. Borceux (Ed.), Categorical Algebra and its Applications. Proceedings, 1987. VIII, 375 pages. 1988.

Vol. 1349: E. Novak, Deterministic and Stochastic Error Bounds in Numerical Analysis. V, 113 pages. 1988.

Vol. 1350: U. Koschorke (Ed.), Differential Topology. Proceedings, 1987. VI, 269 pages. 1988.

Vol. 1351: I. Laine, S. Rickman, T. Sorvali, (Eds.), Complex Analysis, Joensuu 1987. Proceedings. XV, 378 pages. 1988.

Vol. 1352: L.L. Avramov, K.B. Tchakerian (Eds.), Algebra – Some Current Trends. Proceedings, 1986. IX, 240 Seiten. 1988.

Vol. 1353: R.S. Palais, Ch.-l. Terng, Critical Point Theory and Submanifold Geometry. X, 272 pages. 1988.

Vol. 1354: A. Gómez, F. Guerra, M.A. Jiménez, G. López (Eds.), Approximation and Optimization. Proceedings, 1987. VI, 280 pages. 1988.

Vol. 1355: J. Bokowski, B. Sturmfels, Computational Synthetic Geometry. V, 168 pages. 1989.

Vol. 1356: H. Volkmer, Multiparameter Eigenvalue Problems and Expansion Theorems. VI, 157 pages. 1988.

Vol. 1357: S. Hildebrandt, R. Leis (Eds.), Partial Differential Equations and Calculus of Variations. VI, 423 pages. 1988.

Vol. 1358: D. Mumford, The Red Book of Varieties and Schemes. V, 309 pages. 1988.

Vol. 1359: P. Eymard, J.-P. Pier (Eds.), Harmonic Analysis. Proceedings, 1987. VIII, 287 pages. 1988.

Vol. 1360: G. Anderson, C. Greengard (Eds.), Vortex Methods. Proceedings, 1987. V, 141 pages. 1988.

Vol. 1361: T. tom Dieck (Ed.), Algebraic Topology and Transformation Groups. Proceedings, 1987. VI, 298 pages. 1988.

Vol. 1362: P. Diaconis, D. Elworthy, H. Föllmer, E. Nelson, G.C. Papanicolaou, S.R.S. Varadhan. École d'Été de Probabilités de Saint-Flour XV–XVII, 1985–87. Editor: P.L. Hennequin. V, 459 pages. 1988.

Vol. 1363: P.G. Casazza, T.J. Shura, Tsirelson's Space. VIII, 204 pages. 1988.

Vol. 1364: R.R. Phelps, Convex Functions, Monotone Operators and Differentiability. IX, 115 pages. 1989.

Vol. 1365: M. Giaquinta (Ed.), Topics in Calculus of Variations. Seminar, 1987. X, 196 pages. 1989.

Vol. 1366: N. Levitt, Grassmannians and Gauss Maps in PL-Topology. V, 203 pages. 1989.

Vol. 1367: M. Knebusch, Weakly Semialgebraic Spaces. XX, 376 pages. 1989.

Vol. 1368: R. Hübl, Traces of Differential Forms and Hochschild Homology. III, 111 pages. 1989.

Vol. 1369: B. Jiang, Ch.-K. Peng, Z. Hou (Eds.), Differential Geometry and Topology. Proceedings, 1986–87. VI, 366 pages. 1989.

Vol. 1370: G. Carlsson, R.L. Cohen, H.R. Miller, D.C. Ravenel (Eds.), Algebraic Topology. Proceedings, 1986. IX, 456 pages. 1989.

Vol. 1371: S. Glaz, Commutative Coherent Rings. XI, 347 pages. 1989.

Vol. 1372: J. Azéma, P.A. Meyer, M. Yor (Eds.), Séminaire de Probabilités XXIII. Proceedings. IV, 583 pages. 1989.

Vol. 1373: G. Benkart, J.M. Osborn (Eds.), Lie Algebras, Madison 1987. Proceedings. V, 145 pages. 1989.

Vol. 1374: R.C. Kirby, The Topology of 4-Manifolds. VI, 108 pages. 1989.

Vol. 1375: K. Kawakubo (Ed.), Transformation Groups. Proceedings, 1987. VIII, 394 pages, 1989.

Vol. 1376: J. Lindenstrauss, V.D. Milman (Eds.), Geometric Aspects of Functional Analysis. Seminar (GAFA) 1987–88. VII, 288 pages. 1989.

Vol. 1377: J.F. Pierce, Singularity Theory, Rod Theory, and Symmetry-Breaking Loads. IV, 177 pages. 1989.

Vol. 1378: R.S. Rumely, Capacity Theory on Algebraic Curves. III, 437 pages. 1989.

Vol. 1379: H. Heyer (Ed.), Probability Measures on Groups IX. Proceedings, 1988. VIII, 437 pages. 1989

Vol. 1380: H.P. Schlickewei, E. Wirsing (Eds.), Number Theory, Ulm 1987. Proceedings. V, 266 pages. 1989.

Vol. 1381: J.-O. Strömberg, A. Torchinsky. Weighted Hardy Spaces. V, 193 pages. 1989.

Vol. 1382: H. Reiter, Metaplectic Groups and Segal Algebras. XI, 128 pages. 1989.

Vol. 1383: D.V. Chudnovsky, G.V. Chudnovsky, H. Cohn, M.B. Nathanson (Eds.), Number Theory, New York 1985–88. Seminar. V, 256 pages. 1989.

Vol. 1384: J. Garcia-Cuerva (Ed.), Harmonic Analysis and Partial Differential Equations. Proceedings, 1987. VII, 213 pages. 1989.

Vol. 1385: A.M. Anile, Y. Choquet-Bruhat (Eds.), Relativistic Fluid Dynamics. Seminar, 1987. V, 308 pages. 1989.

Vol. 1386: A. Bellen, C.W. Gear, E. Russo (Eds.), Numerical Methods for Ordinary Differential Equations. Proceedings, 1987. VII, 136 pages. 1989.

Vol. 1387: M. Petković, Iterative Methods for Simultaneous Inclusion of Polynomial Zeros. X, 263 pages. 1989.

Vol. 1388: J. Shinoda, T.A. Slaman, T. Tugué (Eds.), Mathematical Logic and Applications. Proceedings, 1987. V, 223 pages. 1989.

Vol. 1000: Second Edition. H. Hopf, Differential Geometry in the Large. VII, 184 pages. 1989.

Vol. 1389: E. Ballico, C. Ciliberto (Eds.), Algebraic Curves and Projective Geometry. Proceedings, 1988. V, 288 pages. 1989.

Vol. 1390: G. Da Prato, L. Tubaro (Eds.), Stochastic Partial Differential Equations and Applications II. Proceedings, 1988. VI, 258 pages. 1989.

Vol. 1391: S. Cambanis, A. Weron (Eds.), Probability Theory on Vector Spaces IV. Proceedings, 1987. VIII, 424 pages. 1989.

Vol. 1392: R. Silhol, Real Algebraic Surfaces. X, 215 pages. 1989.

Vol. 1393: N. Bouleau, D. Feyel, F. Hirsch, G. Mokobodzki (Eds.), Séminaire de Théorie du Potentiel Paris, No. 9. Proceedings. VI, 265 pages. 1989.

Vol. 1394: T.L. Gill, W.W. Zachary (Eds.), Nonlinear Semigroups, Partial Differential Equations and Attractors. Proceedings, 1987. IX, 233 pages. 1989.

Vol. 1395: K. Alladi (Ed.), Number Theory, Madras 1987. Proceedings. VII, 234 pages. 1989.

Vol. 1396: L. Accardi, W. von Waldenfels (Eds.), Quantum Probability and Applications IV. Proceedings, 1987. VI, 355 pages. 1989.

Vol. 1397: P.R. Turner (Ed.), Numerical Analysis and Parallel Processing. Seminar, 1987. VI, 264 pages. 1989.

Vol. 1398: A.C. Kim, B.H. Neumann (Eds.), Groups – Korea 1988. Proceedings. V, 189 pages. 1989.

Vol. 1399: W.-P. Barth, H. Lange (Eds.), Arithmetic of Complex Manifolds. Proceedings, 1988. V, 171 pages. 1989.

Vol. 1400: U. Jannsen. Mixed Motives and Algebraic K-Theory. XIII, 246 pages. 1990.

Vol. 1401: J. Steprāns, S. Watson (Eds.), Set Theory and its Applications. Proceedings, 1987. V, 227 pages. 1989.

Vol. 1402: C. Carasso, P. Charrier, B. Hanouzet, J.-L. Joly (Eds.), Nonlinear Hyperbolic Problems. Proceedings, 1988. V, 249 pages. 1989.

Vol. 1403: B. Simeone (Ed.), Combinatorial Optimization. Seminar, 1986. V, 314 pages. 1989.

Vol. 1404: M.-P. Malliavin (Ed.), Séminaire d'Algèbre Paul Dubreil et Marie-Paul Malliavin. Proceedings, 1987 – 1988. IV, 410 pages. 1989.

Vol. 1405: S. Dolecki (Ed.), Optimization. Proceedings, 1988. V, 223 pages. 1989.

Vol. 1406: L. Jacobsen (Ed.), Analytic Theory of Continued Fractions III. Proceedings, 1988. VI, 142 pages. 1989.

Vol. 1407: W. Pohlers, Proof Theory. VI, 213 pages. 1989.

Vol. 1408: W. Lück, Transformation Groups and Algebraic K-Theory. XII, 443 pages. 1989.

Vol. 1409: E. Hairer, Ch. Lubich, M. Roche. The Numerical Solution of Differential-Algebraic Systems by Runge-Kutta Methods. VII, 139 pages. 1989.

Vol. 1410: F.J. Carreras, O. Gil-Medrano, A.M. Naveira (Eds.), Differential Geometry. Proceedings, 1988. V, 308 pages. 1989.